Woody Plants and Woody Plant Management

BOOKS IN SOILS, PLANTS, AND THE ENVIRONMENT

Soil Biochemistry, Volume 1, edited by A. D. McLaren and G. H. Peterson
Soil Biochemistry, Volume 2, edited by A. D. McLaren and J. Skuji š
Soil Biochemistry, Volume 3, edited by E. A. Paul and A. D. McLaren
Soil Biochemistry, Volume 4, edited by E. A. Paul and A. D. McLaren
Soil Biochemistry, Volume 5, edited by E. A. Paul and J. N. Ladd
Soil Biochemistry, Volume 6, edited by Jean-Marc Bollag and G. Stotzky
Soil Biochemistry, Volume 7, edited by G. Stotzky and Jean-Marc Bollag
Soil Biochemistry, Volume 8, edited by Jean-Marc Bollag and G. Stotzky
Soil Biochemistry, Volume 9, edited by G. Stotzky and Jean-Marc Bollag
Soil Biochemistry, Volume 10, edited by Jean-Marc Bollag and G. Stotzky

Organic Chemicals in the Soil Environment, Volumes 1 and 2, edited by C. A. I. Goring and J. W. Hamaker
Humic Substances in the Environment, M. Schnitzer and S. U. Khan
Microbial Life in the Soil: An Introduction, T. Hattori
Principles of Soil Chemistry, Kim H. Tan
Soil Analysis: Instrumental Techniques and Related Procedures, edited by Keith A. Smith
Soil Reclamation Processes: Microbiological Analyses and Applications, edited by Robert L. Tate III and Donald A. Klein
Symbiotic Nitrogen Fixation Technology, edited by Gerald H. Elkan
Soil–Water Interactions: Mechanisms and Applications, Shingo Iwata and Toshio Tabuchi with Benno P. Warkentin

Soil Analysis: Modern Instrumental Techniques, Second Edition, edited by Keith A. Smith
Soil Analysis: Physical Methods, edited by Keith A. Smith and Chris E. Mullins
Growth and Mineral Nutrition of Field Crops, N. K. Fageria, V. C. Baligar, and Charles Allan Jones
Semiarid Lands and Deserts: Soil Resource and Reclamation, edited by J. Skuji š
Plant Roots: The Hidden Half, edited by Yoav Waisel, Amram Eshel, and Uzi Kafkafi
Plant Biochemical Regulators, edited by Harold W. Gausman
Maximizing Crop Yields, N. K. Fageria
Transgenic Plants: Fundamentals and Applications, edited by Andrew Hiatt
Soil Microbial Ecology: Applications in Agricultural and Environmental Management, edited by F. Blaine Metting, Jr.
Principles of Soil Chemistry: Second Edition, Kim H. Tan
Water Flow in Soils, edited by Tsuyoshi Miyazaki
Handbook of Plant and Crop Stress, edited by Mohammad Pessarakli
Genetic Improvement of Field Crops, edited by Gustavo A. Slafer
Agricultural Field Experiments: Design and Analysis, Roger G. Petersen
Environmental Soil Science, Kim H. Tan
Mechanisms of Plant Growth and Improved Productivity: Modern Approaches, edited by Amarjit S. Basra
Selenium in the Environment, edited by W. T. Frankenberger, Jr., and Sally Benson
Plant–Environment Interactions, edited by Robert E. Wilkinson
Handbook of Plant and Crop Physiology, edited by Mohammad Pessarakli
Handbook of Phytoalexin Metabolism and Action, edited by M. Daniel and R. P. Purkayastha
Soil–Water Interactions: Mechanisms and Applications, Second Edition, Revised and Expanded, Shingo Iwata, Toshio Tabuchi, and Benno P. Warkentin
Stored-Grain Ecosystems, edited by Digvir S. Jayas, Noel D. G. White, and William E. Muir
Agrochemicals from Natural Products, edited by C. R. A. Godfrey
Seed Development and Germination, edited by Jaime Kigel and Gad Galili
Nitrogen Fertilization in the Environment, edited by Peter Edward Bacon
Phytohormones in Soils: Microbial Production and Function, William T. Frankenberger, Jr., and Muhammad Arshad
Handbook of Weed Management Systems, edited by Albert E. Smith
Soil Sampling, Preparation, and Analysis, Kim H. Tan
Soil Erosion, Conservation, and Rehabilitation, edited by Menachem Agassi
Plant Roots: The Hidden Half, Second Edition, Revised and Expanded, edited by Yoav Waisel, Amram Eshel, and Uzi Kafkafi
Photoassimilate Distribution in Plants and Crops: Source–Sink Relationships, edited by Eli Zamski and Arthur A. Schaffer
Mass Spectrometry of Soils, edited by Thomas W. Boutton and Shinichi Yamasaki

Handbook of Photosynthesis, edited by Mohammad Pessarakli
Chemical and Isotopic Groundwater Hydrology: The Applied Approach, Second Edition, Revised and Expanded, Emanuel Mazor
Fauna in Soil Ecosystems: Recycling Processes, Nutrient Fluxes, and Agricultural Production, edited by Gero Benckiser
Soil and Plant Analysis in Sustainable Agriculture and Environment, edited by Teresa Hood and J. Benton Jones, Jr.
Seeds Handbook: Biology, Production, Processing, and Storage, B. B. Desai, P. M. Kotecha, and D. K. Salunkhe
Modern Soil Microbiology, edited by J. D. van Elsas, J. T. Trevors, and E. M. H. Wellington
Growth and Mineral Nutrition of Field Crops: Second Edition, N. K. Fageria, V. C. Baligar, and Charles Allan Jones
Fungal Pathogenesis in Plants and Crops: Molecular Biology and Host Defense Mechanisms, P. Vidhyasekaran
Plant Pathogen Detection and Disease Diagnosis, P. Narayanasamy
Agricultural Systems Modeling and Simulation, edited by Robert M. Peart and R. Bruce Curry
Agricultural Biotechnology, edited by Arie Altman
Plant–Microbe Interactions and Biological Control, edited by Greg J. Boland and L. David Kuykendall
Handbook of Soil Conditioners: Substances That Enhance the Physical Properties of Soil, edited by Arthur Wallace and Richard E. Terry
Environmental Chemistry of Selenium, edited by William T. Frankenberger, Jr., and Richard A. Engberg
Principles of Soil Chemistry: Third Edition, Revised and Expanded, Kim H. Tan
Sulfur in the Environment, edited by Douglas G. Maynard
Soil–Machine Interactions: A Finite Element Perspective, edited by Jie Shen and Radhey Lal Kushwaha
Mycotoxins in Agriculture and Food Safety, edited by Kaushal K. Sinha and Deepak Bhatnagar
Plant Amino Acids: Biochemistry and Biotechnology, edited by Bijay K. Singh
Handbook of Functional Plant Ecology, edited by Francisco I. Pugnaire and Fernando Valladares
Handbook of Plant and Crop Stress: Second Edition, Revised and Expanded, edited by Mohammad Pessarakli
Plant Responses to Environmental Stresses: From Phytohormones to Genome Reorganization, edited by H. R. Lerner
Handbook of Pest Management, edited by John R. Ruberson
Environmental Soil Science: Second Edition, Revised and Expanded, Kim H. Tan
Microbial Endophytes, edited by Charles W. Bacon and James F. White, Jr.
Plant–Environment Interactions: Second Edition, edited by Robert E. Wilkinson
Microbial Pest Control, Sushil K. Khetan
Soil and Environmental Analysis: Physical Methods, Second Edition, Revised and Expanded, edited by Keith A. Smith and Chris E. Mullins

The Rhizosphere: Biochemistry and Organic Substances at the Soil–Plant Interface, Roberto Pinton, Zeno Varanini, and Paolo Nannipieri
Woody Plants and Woody Plant Management: Ecology, Safety, and Environmental Impact, Rodney W. Bovey

Additional Volumes in Preparation

Handbook of Postharvest Technology, A. Chakraverty, Arun S. Mujumdar, and G. S. V. Raghavan
Metals in the Environment, M. N. V. Prasad
Plant Pathogen Detection and Disease Diagnosis, Second Edition, Revised and Expanded, P. Narayanasamy

Woody Plants and Woody Plant Management

Ecology, Safety, and Environmental Impact

Rodney W. Bovey

Texas A&M University
College Station, Texas

MARCEL DEKKER, INC. NEW YORK • BASEL

ISBN: 0-8247-0438-X

This book is printed on acid-free paper.

Headquarters
Marcel Dekker, Inc.
270 Madison Avenue, New York, NY 10016
tel: 212-696-9000; fax: 212-685-4540

Eastern Hemisphere Distribution
Marcel Dekker AG
Hutgasse 4, Postfach 812, CH-4001 Basel, Switzerland
tel: 44-61-261-8482; fax: 44-61-261-8896

World Wide Web
http://www.dekker.com

The publisher offers discounts on this book when ordered in bulk quantities. For more information, write to Special Sales/Professional Marketing at the headquarters address above.

Current printing (last digit):
10 9 8 7 6 5 4 3 2 1

PRINTED IN THE UNITED STATES OF AMERICA

To my four sons, Seth Ivan, Todd Evin, Shawn Erin, and Cary Lane, who, along with their families, are all talented professionals in widely divergent careers. They have greatly enhanced my life and career. It's been a fun journey and I am most grateful for their love and support.

Preface

The book is about managing woody plants that may be undesirable for wildlife, livestock, or wood production. Managing woody plants is sometimes very challenging and requires judicious use of the best management practices in balancing recreation, wildlife needs, watershed yield, livestock use, economics, conservation of resources, and human needs and goals.

Some woody plants that have no apparent redeeming value are difficult to manage because they resist all forms of suppression and dominate the landscape. They may also be costly to treat. Many, however, are desirable for food and shelter for wildlife, human recreation, aesthetics, and soil and water conservation. But when a woody plant species becomes so dominant that it markedly reduces animal and plant diversity, interferes with recreation, livestock, and wildlife production, or becomes a fire hazard, some control method may be needed. On forested land, undesirable brush may interfere with wood production, wildlife habitat, and herbaceous plant and animal diversity. On many noncrop areas, such as utility rights-of-way, roadways, industrial sites, airports, and areas around buildings and structures, woody plant control may be necessary to improve visibility, safety, and aesthetics and to protect expensive facilities.

Although the literature on woody plant management is extensive, it is scattered throughout many diverse sources that span more than 50 years. The purpose of this book is to bring together the most significant literature and data into one reference.

The book covers the significance and botanical nature of woody plants, the history and use of fire, biological, mechanical, and chemical control methods, and combinations of these methods where appropriate. Also examined are herbicide chemistry and properties, toxicology and safety, residues and environmental impact, and how herbicides are applied. The fate and activity of herbicides in plants and their effects on plants and animals are described. The response of over 370 woody plants in North America to commercially available herbicides is presented in Chapter 12. Finally, Chapter 8 provides an update on the phenoxy herbicide controversy (Agent Orange).

To complete the book, the economics of woody plant control and growing woody plants for experimental purposes is discussed, as well as future research needs and recommendations.

Special thanks are extended to Mary Alice Peel and Julie Preiss for typing and correcting the manuscript and to Beth Ellison for her encouragement and administrative assistance.

Rodney W. Bovey

Contents

Woody Plants and Woody Plant Management

1

Significance and Botanical Nature of Woody Plants

I. INTRODUCTION

Williams et al. (1968) indicated that about one-third of the Earth is land area, or 14 billion hectares (ha). Approximately 10% is farmed, 28% is in forest (which is grazed at least part-time), and 15% is covered with icecaps or fresh water, leaving 47% or nearly half of the globe for grazing by livestock or game animals. Such land is usually only suitable for grazing because it is too steep, shallow, sandy, and wet, cold, or saline for crops. Williams et al. (1968) indicated that 75% of the domestic animals and most wildlife depend upon grazing lands for survival. Such land is also extremely important as watersheds, for conservation, for wood, medicinal, and industrial compounds, for mining resources, and for recreational purposes. Forests and grasslands are also apparently more efficient in carbon dioxide removal from the atmosphere than cropland, which according to some investigators may be increasing to the point of causing unacceptable climatic change unless reversed (Woodwell, 1978; Mayeux et al., 1991).

Of the 0.4 billion ha of range and pastureland in the United States (Thomas and Ronningen, 1965), about one-third is estimated to be infested with undesirable woody plants (Allred and Mitchell, 1955; Williams et al., 1968). Klingman (1962) indicated that woody plant infestations of rangeland included 30 million ha of juniper (*Juniperus* spp.), 28 million ha of mesquite (*Prosopis* spp.), and 38 million ha of sagebrush (*Artemisia* spp.). The ha infested with junipers, sagebrush, and mesquite have probably changed little or have increased in the last

35 years, since during this time little change occurred in the 23 million ha of honey mesquite growing on Texas rangelands in spite of control efforts (Bovey, 1998). Platt (1959) reported ha infested from a survey of 36 range authorities in the western United States and Canada. A total of nearly 240 million ha of land covered with problem woody plants were reported, with an additional 105 million ha of herbaceous weeds (many poisonous to livestock) indicated, for a grand total of over 345 million ha infested. Platt (1959) indicated some acreage was counted more than once because of interspersed stands of two or more undesirable species. Platt (1959) further stated, however, that it was not total number of ha that were important, but the ha requiring treatment for undesirable plants. Platt (1959) reported that sagebrush (*Artemisia* spp.), snakeweed (*Gutierrizia* spp.), juniper (*Juniperus* spp.), creosote brush (*Larrea* spp.), cactus (*Opuntia* spp.), mesquite (*Prosopis* spp.), and scrub oak (*Quercus* spp.) occurred on 16 million ha or more each of U.S. western rangelands, with about 13 major herbaceous weeds included.

Le Clerg et al. (1962) estimated that annual losses in forage production due to weeds on pastureland and rangeland in the 31 eastern states was 20% and 13% for rangeland in the 17 western states. When the cost of control measures was included with forage loss, the total loss for the United States was $1 billion. The loss did not include weed problems in establishing forage plants or losses due to poisonous plants or plants that cause mechanical injury to animals from needles, thorns, or other means.

Additional benefits of woody plant control in fostering sound range and pasture management and improved forage production include herbaceous weed control, increased quality and quantity of animal products, increased ranch and domestic water, better recreational opportunities, such as hunting, fishing, hiking, and picnicking, decreased number of pests, and decreased pollution (poisonous plants), resulting in both a better dollar return to the rancher and consumer and improved conservation practices.

Walker (1973) indicated timberlands supporting important amounts of undesirable vegetation in the United States total 120 million ha, or a conservative estimated annual loss of 0.4 billion cubic meters of lumber due to weeds alone. Walstad (1973) stated that about 36 million ha of forestland in the South needs timber stand improvement, including suppression of low-quality hardwood species. On an operational level foresters indicate that weed control can increase timber volume production in southern pine plantations by 14% and by 25% in natural stands.

Brush or undesirable woody plants are not confined to forest and rangeland but may become majors problems around industrial buildings and structures, railroads, roadways, overhead power lines, fences, vacant lots, airports, military installations, canals, ditch banks, cropland, and similar locations where they may interfere with safety, visibility, transportation, irrigation, recreation, or other activities of animals and man.

Woody plant problems are not confined to the United States; they are a worldwide problem. Little and Ivens (1965) cited the work of the IBEC Research Institute in New York, indicating that leiteiro (*Tabernae/montana fuchsiifolia*), amendoim de campo (*Pterogyne nitens*), a cyperaceae, and two shrubs (*Acacia polyphylla* and *A. paniculata*) are serious pasture weeds in Brazil. In Cuba, marabu (*Dichrostachys nutans*) is an important brush weed with growth characteristics reported similar to leiteiro. Herbicide 2,4,5-T or the substituted ureas were effective as basal sprays and soil treatments, respectively. Tschirley (1968) reported on herbicides to control a number of woody plants growing in Texas and Puerto Rico. The Puerto Rican treatments included guara (*Cupania americana*), mango (*Mangifera indica*), pomarrorsa (*Eugenia jambos*), camasey (*Miconia prasina*), common bamboo (*Bambusa vulgaris*), palma de sierra (*Prestoea montana*), and mixed semievergreen forest and evergreen rain forest. Although some of the species may not be considered brush problems the research program greatly expanded our knowledge of the response of tropical and subtropical woody plants to herbicides. Picloram was one of the more effective herbicides of the many investigated.

Willard (1973) studied the effectiveness of a backfire and rapidly moving headfire on dense shrubland in Argentina. Main woody trees included calden (*Prosopis caldenio*), algarobo (*P. flexuosa*), and sombra de toro (*Jodinia rhombifolia*). Shrubs included piquillin (*Condalia microphylla*), molle (*Schinus fasaiculatus*), chanar (*Geoffroea decorticans*), jarilla (*Larrea divaricata*), and alpataco (*Prosopis alpataco*). The Monte region of central and western Argentina contains nearly 6.0 million ha of thorny shrubland. Wildfires are common in the region during the dry summers and are a hazard to livestock, wildlife, and man. Willard (1973) suggests prescribed burning for brush control in the Monte region of Argentina.

While land managers and scientists were optimistic about herbaceous and woody plant control in the 1950s to the 1980s when costs were reasonable and new herbicides and nonchemical methods were coming on stream, time has proven that wholesale eradication of weeds and brush was not possible or desirable.

Today land managers can use brush management to attain multiple benefits for society, including wildlife habitat management, watershed enhancement, aesthetics, and improved livestock-carrying capacity (Hanselka, 1997). The brush sculptor concept is the need to assess an integrated pest management (IPM) approach then integrate various brush management technologies (mechanical, herbicide, fire, and biological), monitor results, and adjust management strategies as needed. Sculpting brush allows the landowner or manager to optimize the value of his resource for livestock, wildlife, aesthetics, recreation, water, and real estate, while providing the desired products and services.

II. WHAT IS A WOODY PLANT?

Woody plants are plants that produce secondary growth in the form of wood. The mechanical support provided by the wood allows them to grow taller and effectively compete for available sunlight. They are always perennial, and the wood is produced over the lifetime of the plant (Rubin, 1997).

Because of the perennial and vigorous nature of woody plants they are sometimes difficult to manage when they become overabundant and dominate landscapes with purposes counter to land management.

Woody plants provide valuable wood products, browse for wildlife and livestock, cover and shade for wildlife, livestock, and man, water and soil conservation, medicinal extracts, hunting, recreation, human food, beauty, and many other benefits. When some woody plants become overly aggressive and dominant, possess no apparent redeeming value, and cause economic loss and management problems some control practice may be desired, however.

III. CHARACTERISTICS OF WOODY PLANTS

A. Perennial Nature

Control of undesirable woody plants is sometimes difficult because they may be vigorous perennials that live many years. They may reproduce by seed and vegetatively by basal stem buds, root sprouts, or rhizomes such as mesquite (*Prosopis glanlulosa* Torr.), huisache [*Acacia faresiana* (L) Wild.], and Macartney rose (*Rose bracteata*) Wendle.) Once these plants are injured by cutting, fire, animals, or chemicals they have the ability to regenerate from buds, root sprouts, or rhizomes. By removing top growth, apical dominance may be removed and dormant buds may be activated and produce new shoots. The new shoots grow uninhibited to produce mature plants unless they are further disturbed by top removal. Woody plants are adapted to survive injury from frost, fire, cutting, and other disturbances.

B. Competition

Woody plants may be strong competitors with other woody plants and herbaceous vegetation. They may grow tall and shade and deprive lower-stature plants of light. Some have deep and extensive root systems that allow them to grow in hostile environments to extract soil water and nutrients more efficiently than other plants. Some have the ability to shed leaves, produce smaller, thicker, cutinized leaves that resist heat, harvest light more efficiently, and resist desiccation in times of drought. By shedding excess leaves or slowing down metabolically, the plant may become dormant until a more favorable environment exists, while neighboring plants may suffer or die of desiccation. In areas devoid of vegetation from drought or other reasons the seeds of woody plants may establish seedlings during periods of ample rainfall and further infest the area.

C. Unpalatable

Many woody plants on rangeland have thorns and appendages or chemical composition unattractive to grazing animals. Because of causing possible injury, dense impenetrable stands, or offensive taste, these plants are essentially unmolested and grow and reproduce without interference. If grazing animals selectively utilize other woody plants and herbaceous vegetation near them, they are allowed to grow inhibited and spread, and may become serious weeds.

IV. CAUSES FOR ENCROACHMENT

Vallentine (1989) lists eight primary factors causing or contributing to the increase, spread, and invasions of noxious plants on grazing lands in the western United States, including: 1) grazing of domestic livestock (where the more desirable forage is reduced because of selective grazing, overgrazing, improper grazing season, or rigid livestock members), 2) reduction of fire, 3) seed transport by grazing animals, 4) weed seed dissemination by small animals, 5) climate fluctuations, 6) cultivation and subsequent abandonment, 7) local denudation (includes roads, railroads, stock trails, industrial areas, mining, farmsteads, and other locally denuded areas), and 8) increase in commerce (transport of weed seed into new areas).

Archer (1994) argues that selective grazing by large numbers and high concentrations of livestock has been the primary force in altering plant life-form interactions to favor unpalatable woody species over graminoids. Mayeux et al. (1991), however, indicated that increasing atomospheric CO_2 levels favor the C3 broadleaf herbaceous and woody species over warm-season perennial grasses (C4 plants) on rangelands. They hypothesize that increased atmospheric CO_2 causes vegetation change on rangelands rather than overgrazing, suppression of fire, and climate changes.

Archer (1994), however, cites many exceptions that limit the utility of the atmospheric CO_2 enrichment hypothesis as a robust explanation of the cause of woody plant encroachment into grasslands.

Archer (1994) cites numerous historical accounts and photographic records indicating that in the last 50 to 100 years shrublands, woodlands, and forests of North America have expanded and replaced what were grasslands and savannas at the time of European settlement. Early settlers in north-central and southern

Texas indicated little woody vegetation in the mid-1800s, with only a few scattered honey mesquite trees. Today honey mesquite occurs on at least half of Texas rangeland along with many other species of woody plants, despite some control practices (Bovey, 1998). Archer (1994) suggests that replacement of grasslands and savannas with shrub and woodlands dominated by unpalatable species appears to have been rapid (50 to 100 years), nonlinear, accentuated by climatic fluctuation, locally influenced by topoedaphic factors, and irreversible. Archer (1994) states that past industrial atomospheric CO_2 enrichment and climate change may have facilitated shifts from grass to woody plants but that there is a strong link between livestock grazing and woody plant encroachment. This has occurred because of livestock preference for grasses versus woody plants, alteration of soil structure and chemistry, seed dispersal, and fire reduction.

V. LONG-TERM WOODY PLANT ENCROACHMENT

A. Mesquite and Associated Plants

Brown (1950) studied woody plant encroachment for an 18-year period (from 1931 to 1949) on a desert-grassland in the Santa Rita Experimental Range in southern Arizona. Changes in velvet mesquite and burroweed (*Haplopappus tenuisectus*) were directly correlated with grazing pressure. Total protection from grazing did not retard mesquite encroachment or decrease burroweed. Velvet-pod mimosa (*Mimosa dysocarpa*) and fern acacia (*Acacia angustissima*) were reduced by drought-aggravated grazing injury. The desert grassland is indicated by this study to be subclimax to a desert shrub climax in southern Arizona.

Branscomb (1958) studied shrub invasion over a 30-year period in southern New Mexico semidesert grassland on the Jornada Experimental Range. Twelve percent of the total area formerly classed as grassland was dominated by shrubs. Honey mesquite was the principal invader, having increased its original acreage by 107%. Tarbush-creosote-type vegetation occupies 8% less area than before; snakeweed (*Gutierrezia sarothrae*)-dominated acreage was reduced by one-half. Grazing pressure has disseminated noxious plant seed and weakened the grass plants and removed fuel for fire. Wildfires prior to white settlement was indicated as the factor keeping grasslands free of shrubby invaders.

From 1935 to 1980 honey mesquite attained complete dominance, and many new mesquite dunes formed on the study area on the Jornada Experimental Range (Hennessy et al., 1983). Black grama (*Bouteloua eriopoda*) had relatively high frequency in 1935 but completely disappeared by 1980 on both grazed and nongrazed areas. Mesa dropseed (*Sporobolus flexuosus*), fluffgrass (*Erioneuron pulchellum*), and broom snakeweed increased in abundance, even during the drought period between 1950 and 1955. Only 25% of the perennial forbs present in the period from 1935 to 1955 were found in 1980.

Bogusch (1952) indicated that much of the region lying between the Nueces River and the Rio Grande was originally grassland. Bogusch (1952) indicated that honey mesquite was the primary invader, and wild game and buffalo had no importance in the spread of mesquite. He further stated fencing restricted cattle movement and increased grass damage, providing opportunity for shrub invasion.

B. Sagebrush Domination

Lommasson (1948) indicated that the big sagebrush of the high grasslands of the Gravelly Range of the Beaverhead National Forest in southwestern Montana will maintain itself indefinitely under natural conditions. This conclusion was the result of a 31-year-old study by the U.S. Forest Service. From 1882 until about 1914 little grazing use (buffalos killed off) was made of the range, and sagebrush became dominant. Since 1914 the area has been grazed by sheep. By 1945 sagebrush plants averaged 61 years in age by growth ring count. They became established in 1885, and in 1915 when the study began they were 31 years old.

In contrast, Robertson (1971) stated that a 30-year rest enabled a 20-acre tract of eroded sagebrush-grass range in northern Nevada to increase its vegetal cover in all life forms. The cover of perennial forbs increased the most, 85%. Thurber needle grass increased 7-fold. Only annual forbs and locoweed declined. Bluebunch wheatgrass reestablished naturally in favored spots. Newly cleared and seeded range outside the exclosure produced three times as much grass forage as was produced after long rest without clearing. Robertson (1971) further indicated that while the plot data showed improved forage cover by long rest, restoration would be quicker by brush control and seeding.

Data from permanent vegetation transects, established on the Idaho National Engineering Laboratory Site in 1950, were analyzed to determine what changes had taken place in the vegetation complex over the past 25 years in the absence of grazing by domestic livestock (Anderson and Holte, 1981). Cover of shrubs and perennial grasses nearly doubled. Shrub cover in 1975 was 154% greater than in 1950; this change was almost entirely due to increases in cover of big sagebrush between 1957 and 1965. Cover of perennial grasses increased exponentially over the 25-year period, from 0.28% in 1950 to 5.8% in 1975. This was paralleled by significant increases in density and distribution of the four most important grasses on the study area. The 20-fold increase in perennial grass cover has not been at the expense of the shrub overstory.

Hull and Hull (1974) indicated that explorers and early settlers found abundant grass and little sagebrush in Cache Valley in northeastern Utah and southeastern Idaho. Excessive grazing by livestock after settlement caused the grass to decrease and the sagebrush to increase. Most grassland areas were eventually plowed for dry-land or irrigated farming. In the dry-farm belt, however, there are many steep or rocky slopes, inaccessible corners, and similar areas that have not been plowed, irrigated, heavily grazed, or burned in recent years. Many of these areas support vegetation that, except for increased sagebrush, is undoubtedly similar to that described by explorers, early settlers, and historians.

In contrast, Vale (1975) reviewed 29 journals and diaries for their vegetation descriptions of the sagebrush-grass area in an attempt to assess the relative importance of herbaceous plants and woody brush in the northern intermountain west. The early writings suggest a pristine vegetation visually dominated by shrubs. Stands of grass apparently were largely confined to wet valley bottoms, moist canyons, and mountain slopes, with more extensive areas in eastern Oregon near the Cascade range. The major area was apparently covered by thick stands of brush.

C. Juniper Invasion

As a means of studying inter- and intrazonal invasion in black sagebrush (*Artemisia nova*) communities, six maturity classes were established for pinyon (*Pinus monophylla*) and juniper (*Juniperus osteosperma*) in east-central Nevada (Blackburn and Tueller, 1970). Pinyon and juniper invade and increase in black sagebrush communities until the understory is eliminated. Juniper invades first and tends to be replaced by pinyon. Accelerated invasion by both species started in about 1921 and is closely related to overgrazing, fire suppression, and climatic change.

Johnsen and Elson (1979) studied vegetation changes over 60 years in central Arizona grassland-juniper woodland ecotone sites by matched photograph pairs. Grazing use markedly affected understory species. Juniper numbers and sizes increased markedly on hillside and rocky ridges, but did poorly on bottom land sites. Utah juniper rapidly reestablished on areas cleared in the 1950s and 1960s. Stands of shrub live oak, cliforse, mountain mahogany, and Apache-plume did not spread, but shrub crown cover increased greatly.

D. Kansas Bluestem Prairie

Postsettlement invasion of trees and shrubs in Geary County, Kansas Flint Hills, was assessed by aerial photos (Bragg and Hulbert, 1970). Tree cover increased 8% from 1856 to 1969 throughout the county except on regularly burned sites, where trees and shrubs were maintained at presettlement amounts. On unburned sites, woody plants cover increased 34% from 1937 to 1969. Herbicide spraying only slowed the invasion rate. Lowland soils rapidly increased in brush from 1856 to 1937. The authors concluded that on the Flint Hills, bluestem prairie rangeland burning was effective in restricting woody plants to presettlement amounts and soil type, and topography affected woody plant invasion.

VI. INDIVIDUAL WOODY PLANT PROBLEMS

There are many references to identify woody plant species, therefore taxonomic descriptions will be very brief and are provided only as an aid to diagnosing a brush problem. Woody plant problems of great importance will be given first, followed by other important brush species. (See Chap. 12 for more information on woody plants and their control.)

For positive identification (if needed) of the brush and proper management of the brush problem one can consult the local county agricultural extension agent or extension weed specialists or a private consultant, chemical company representatives, federal or state university weed scientists, or personnel of the Natural Resources Conservation Service of the USDA. Species given in this text will be numbered for reference.

1. Honey mesquite (*Prosopis glaudulosa* Torr.).
   Description: a woody, thorny, legume shrub or tree of varying height and longevity; has natural resistance to fire, drought, and livestock grazing; competes aggressively with other woody and herbaceous plants for water and plant nutrients because it is a deep-rooted phreatophyte. Leaves on honey mesquite are an alternate, deciduous, long-petioled, bipennately compound of two (occasionally three or four) pairs of pinnae; with 12 to 20 leaflets. Leaflets are glabrous, linear, acute, or obtuse at the apex, 3 to 4 cm long, and 0.5 to 1 cm wide. Legumes (pods) with seed are 10 to 22 cm long. Foliage is low in palatability, and excessive consumption of beans can cause livestock health problems (Jacoby and Ansley, 1991; Meyer et al., 1971; Vallentine, 1989; Vines, 1960; Welch and Hyden, 1996).
   Distribution: Honey mesquite is distributed from Kansas and New Mexico east into Oklahoma and Arkansas and across Texas (except for Piney Woods) into Louisiana (Vines, 1960).
   Reproduction: Produces seed pods abundantly; seed are brown oval 5 mm wide, 7 mm long, and 2 mm thick in the center (Meyer et al., 1971). Seed can readily germinate under favorable environment especially if they have passed through the digestive tracts of foraging animals (cattle), and will sometimes germinate and become established in dung (Jacoby and Ansley, 1991; Meyer et al., 1971). Seed is usually destroyed by insects, fungi, or rodents, but experiments have shown small numbers of seeds many lie dormant in the soil for several years. Established plants will usually sprout and regrow from the extensive basal buds and buds along the stem if injured. Cutoff plants may resprout at one to several places just below the point severed (Meyer et al., 1971).

Control: Destroy underground buds by mechanically uprooting, killing with diesel fuel oil, or using translocating herbicides such as triclopyr in diesel fuel as basal sprays or foliar sprays of triclopyr, clopyralid, or a 1:1 mixture of triclopyr and clopyralid. Successful control depends upon follow-up treatments to control new and missed plants from previous treatments (Welch and Hyden, 1996; Bovey, 1991).

2. Velvet mesquite (*Prosopis velutina* Woot.).
   Description: Velvet mesquite leaves are bipinnately compound with mostly four pairs of pinnae. Leaflets are pubescent and smaller than those of honey mesquite. Velvet mesquite is very deeprooted, sprouts from the stump, and is not easily damaged by disease or insects. Velvet mesquite foliage and pods are eaten by livestock and the seeds are an important wildlife food. Mesquite beans were important in the diet of the southwestern Indian (Vines 1960; Jacoby and Ansley, 1991).
   Distribution: Velvet mesquite occurs in mainly in southern Arizona, but is found in California and northern Mexico.
   Reproduction: Seeds readily pass through the digestive tracts of grazing animals and may germinate and grow where they fall. Established plants readily sprout from the underground basal bud and stem bud system when the top is injured.
   Control: Similar to honey mesquite.
3. Creosotebush [*Larrea tridentata* (Sesse & Moc. ex DC.) Coville].
   Description: Creosotebush is an evergreen shrub with an extensive lateral root system that dominates the landscape. It can attain a height of 3.3 m. The leaves are bifoliate; leaflets, small, opposite, divaricate, strongly falcate, united at the base, and pointed at the apex. They are oblong to obovate and 0.5 to 1 cm long, thick, dark green to yellowish green. They have a sticky resin that is strong scented. The flowers are also small, 0.8 to 1 cm long, the resulting carpels are one-seeded, indehiscent. The twigs are brown and the bark is dark gray to black. It has medicinal uses as an antiseptic and is employed as a treatment for rheumatism, venereal disease, tuberculosis, intestinal disorders, and emetic (Vines, 1960).
   Distribution: Crosotebush is a widespread desert shrub in most arid regions of the southwest, ranging from West Texas to California and south to Mexico and north to Nevada and Utah. It is poisonous to sheep but is not consumed by cattle. It is consumed by small mammals and antelope. It is often found with tarbush. Creosotebush produces germination and growth inhibitors that inhibit associated desert grasses (Vallentine, 1989).

Reproduction: Reproduction is by seed.
Control: Root plowing or soil-applied herbicides such as tebuthiuron.

4. Cholla [*Opuntia imbricata* (Haw.) DC.].
Description: Also referred to as walking stick cholla, it is an arborescent cactus with a short woody trunk and many erect candelabrumlike branches. It can attain a height of 2.7 meters and has a 25-cm-diameter trunk. It sometimes forms dense thickets. It has terminal purple flowers 4 to 6 cm long and 5 to 8 cm broad. The fruit is yellow, 2.5 to 4 cm long, and near hemispheric, and sometime falls from the plant. The seeds are small, 0.2 to 0.4 cm in diameter. It has spines 2 to 3 cm in length that are barbed (Vines, 1960).
Distribution: Texas north to Oklahoma and Kansas, south and west through New Mexico (Vines, 1960).
Reproduction: Spread by seed and vegetatively by the cylindric joints removed from the parent plant (Welch and Hyden, 1996).
Control: Individual plant treatment with foliar-applied herbicides such as picloram (Welch and Hyden, 1996).
5. Tasajillo (*Opuntia leptocaulis* DC.).
Description: Tasajillo cactus, also known as tesajo or pencil cactus, has a bushy appearance and is usually less than 1.5 meters tall. The stems are cylindric, with small, inconspicuous greenish flowers. The fruit is globular, small (less than 2.5 cm), red, and fleshy. The leaves (pads) are 1.3 cm or less, acute, early deciduous. The stems are various shades of green, branches slender, ascending, cylindric 0.6 to 1.2 cm in diameter or larger with joints varying in length from 2.5 to 25 cm. Red fruit is attractive to wildlife (game birds) (Vines, 1960).
Distribution: Grows at elevations to 915 meters in Texas, usually west of the Brazos river, west through New Mexico to California; south in Mexico to Puebla.
Reproduction: Spreads by seed and vegetatively by joints removed from the parent plant (Welch and Hyden, 1996).
Control: Foliar-applied herbicide, such as picloram.
6. Prickly pear cactus (*Opuntia* spp.) There are many species of prickly pear in the United States, and they respond in a similar manner to herbicides.
Lindheimer pricklypear (*Opuntia lindheimeri* Engelm.).
Description: A thicket-forming cactus with heavy thick large pads growing in clumps 1.5 to 3 meters tall. It has a definite trunk, often prostrate; flowers bright and showy, yellow to orange to red; fruit large (to 8 cm long), red to purple; joints green to bluish green, obovate; spines variable and absent on some joints (Vines, 1960).

Ranchers may burn off spines and use as supplemental cattle feed. It provides protection and some feed for wildlife from fruit and from the cladophylles. It is also troublesome to ranchers as a weed (Scifres, 1980).
Distribution: Southern and western Texas.
Reproduction: Spread by seed and vegetatively by pads (cladophylles).
Control: Foliar-applied picloram sprays. Prescribed burning used in combination with low rates of foliar-applied picloram is very effective (Welch and Hyden, 1996).
Engelman pricklypear (*Optunia engelmannii* Salm-Dyck) is a common pricklypear of the Southwest (Vines, 1960). It is a bushy cactus and grows to nearly 2 meters tall without a definite trunk. It is found in Trans-Pecos Texas, west through New Mexico and Arizona, north to Nevada and Utah, and south to Mexico. Control is similar to Lendheimer pricklypear cactus.

7. Broom snakeweed (*Gutierrezia sarothroe* Britt. & Rusby).
   Description: Plant bushy, herbaceous above and woody toward the base, attaining a height of 10 to 40 cm. The branches are numerous, erect, and redivided into slender branchlets. It has numerous small yellow flowers. Fruit are oblong achenes.
   Distribution: West Texas, New Mexico, Arizona, California, Utah, Montana, Idaho, Nevada, Kansas, and Saskatchewan (Vines, 1960).
   Reproduction: By seed.
   Control: Soil-applied herbicides such as tebuthiuron (Welch, 1997) or foliar-applied herbicides such as picloram, picloram plus 2, 4-D, picloram plus dicamba or metsulfuron.
8. Sand sagebrush (*Artemisia filifolia* Torr.).
   Description: Rounded aromatic shrub, freely branched, usually less than 1 meter tall: flowers in dense, leafy panicles; fruit an achene, glabrous, without pappus; alternate leaves sessile, often fascicled, usually entire; lower leaves often divided into threadlike divisions; twigs slender, pubescent dark gray to black (Vines, 1960).
   Distribution: An indicator of sandy soils to an altitude of 1800 meters in the panhandle of Texas, New Mexico, Arizona, Nevada, Colorado, Utah, and Chihuahua, Mexico (Vines, 1960).
   Reproduction: Spreads by seed, persists by sprouts from a shallow basal bud zone (Welch and Hyden, 1996).
   Control: Deep plow with a disk plow or apply foliar herbicides, such as the low volatile ester of 2, 4-D. (Welch, 1997).
9. Big sagebrush (*Artemisia tridentata* Nutt.).

Description: Big sagebrush is the most widely distributed shrub of the western United States. It varies in height from 0.5 to 3 meters. It is often dwarfed or prostrate from cattle grazing. Stems are mostly erect with ascending branches. Young parts are silvery, canescent, aromatic, and bitter to the taste. Flowers are borne in panicles 10 to 25 cm long or 2 to 10 cm wide, sometimes spikelike, often leafy-bracted; heads numerous, bracts 8 to 18. Fruit are cylindric, turbinate, achenes, border-raised, 4- to 5-ribbed, resinous granuliferous. Leaves are very leafy, sessile or slightly petioled, cuneate at or flabelliform, narrowed at base, rounded apex or truncate, and 3–7 toothed (usually 3) silvery-canescent. Twigs are slender and gray or white at first and later gray to black. It can cause hay fever and has some medical uses.
Distribution: Dry and stoney soils on the arid plain of the Great Basin up to the timberline; also in British Columbia and Northern Mexico. It grows in Texas, New Mexico, Arizona, Colorado, California, North Dakota, Montana, Wyoming, and Washington.
Reproduction: Spreads and reestablishes by seed.
Control: Foliar-applied herbicides, such as 2, 4-D or soil-applied tebuthiuron.

10. Tarbush (*Flourensia cernua* DC.).
    Description: Shrub 0.5 to 3 meters tall, highly branched, and leafy. Flowers are small, in groups of 12 to 20, yellow; fruit an achene, 0.6 cm or less long; leaves simple, alternate, persistent, elliptic to oblong or ovate to oval; margin entire; upper surface green, resinous; lower surface paler and glabrous; aromatic and hoplike odor associated with leaves. May be grazed during drought by sheep and deer, but branch tops, flowers, and fruit can be toxic (Welch and Hyden, 1996). It develops thick, persistent stands when in poorly managed rangeland. Twigs are light brown to gray (Vines, 1960).
    Distribution: Tarbush is found in West Texas, New Mexico, Arizona, and Mexico.
    Reproduction: It spreads by seeds and resists top removal by resprouting from a persistent crown. (Welch and Hyden, 1996).
    Control: Soil-applied herbicides such as tebuthiuron and plowing will control tarbush.
11. Blackjack oak (*Quercus marilandica* Muenchh.).
    Description: Blackjack oak is a shrub or round-topped symmetrical tree attaining a height of 18 meters and a trunk diameter of 0.6 meters. Flowers as catkins; fruit light brown acorns, closed one-third to two-thirds in a cup about 2 cm long; leaves obovate, usually with 3 lobes at apex; bark usually black, rough with grayish brown stiff twigs (Vines, 1960).

Distribution: Dry, sandy, sterile soils of central Texas, Oklahoma, and Arkansas; eastward through Louisiana to Florida, north to New York and west to Minnesota, Michigan, Illinois, and Kansas. Blackjack oak is usually found in close association with post oak, especially in the Post Oak Savannah of Texas. Although a serious management problem in some areas it serves as shade and cover for livestock and wildlife (Scifres, 1980; Vines, 1960).
Reproduction: By acorns and it readily sprouts from the trunk after top injury or removal.
Control: Soil-applied hexazinone and tebuthiuron are effective.

12. Black oak (*Quercus velutina* Lam.).
Description: Stout tree attaining a height of 30 meters with a spreading open crown. Bark is dark brownish black. Leaves are cut into usually 7 oblique lobes with sinuses of different depths. The species name *velutina* refers to the velvety pubescence of the lower leaf surface. The wood of black oak is used for rough lumber, crossties, and fuel. Flowers occur in April to May in staminate and pistillate catkins. Fruit matures in September and October as solitary or paired ovoid-oblong acorns 1.3 to 2.5 cm long with one-half to three-fourths of length enclosed in cup. Twigs are reddish brown (Vines, 1960).
Distribution: Often on poor, dry, sandy, heavy clay or gravelly soils in East Texas, Louisiana, Oklahoma, and Arkansas eastbound to Florida, north to Maine, and west to Ontario, Canada, and Wisconsin and Iowa (Vines, 1960).
Reproduction: The minimum seed-bearing age is 20 years, optimum is 40 to 75 years, and the maximum is 100 years. Good acorn crops are borne every 2 to 3 years with intervening light crops. Reproduction is by acorns.
Control: Soil-applied herbicides such as hexazinone and tebuthiuron.

13. Gambel oak (*Quercus gambelii* Nutt.).
Description: Very variable and widespread western species. Sometimes only a thicket-forming shrub or in favorable locations; can be a 15-meters-tall tree 0.6 meters in trunk diameter with a rounded crown. Flowers in May with young leaves in separate staminate and pistillate catkins. Fruit is solitary or several together, sessile or on tomentose peduncles. Acorns rounded, light brown, and glabrous, 1.3 to 1.9 cm long, enclosed about one-third to one-half in cup. Leaves are deciduous, oblong, obovate, oval, or elliptic, 6 to 15 cm long, and 3.8 to 8 cm wide with 5 to 9 lobes on margin. Twigs are reddish brown, pubescent to glabrous; older twigs are grayish brown. Bark is gray with deep fissures with small and appressed scales (Vines, 1960).
Distribution: At altitudes of 1200 to 2400 meters in western Texas;

northward to New Mexico, Arizona, Colorado, Utah, Nevada, and Wyoming; southward to Mexico in Coahuila and Chihuahua.
Reproduction: Reproduction is by acorns.
Control: The foliage is sometimes browsed by livestock, deer, and porcupine, and is sometimes controlled biologically by sheep and goats. Soil-active herbicides can be used.

14. Live oak (*Quercus virginiana* Mill.).
Description: Live oak is an evergreen tree or shrub. Flowers are staminate and pistillate borne in separate catkins on same tree and are 5 to 8 cm long, calyx yellow with 4 to 7 ovate lobes. Fruit are acorns on peduncles 0.6 to 10 cm long in clusters of 3 to 5. Acorns are brownish black, shiney, 0.8 to 1.3 cm long enclosed about one-half their length in the cup. Leaves are simple, alternate, dark green and lustrous above, paler and glabrous to pubesent beneath. The leaves are 5 to 13 cm long and 1.3 to 6.3 cm wide. Twigs are grayish brown, glabrous, slender, and rigid. Bark is dark brown to black (Vines, 1960).
Distribution: Live oak is usually found on sandy-loam soils, but occurs in heavier clays in Texas, Oklahoma, and Louisiana; east to Florida and north to Virginia (Vines, 1960).
Reproduction: Spreads by acorns and sometimes underground stems (rhizomes) and sometimes forms mottes. On the Gulf Coast of Texas it can assume a low, running-type growth form referred to as ''running'' live oak and highly restricts forage production in grazing lands.
Control: Soil-applied tebuthiuron is very effective.

15. Post oak (*Quercus stellata* Wangenh.).
Description: Shrub to tree to 23 meters tall. Flowers appear with leaves in March to May borne on the same tree in separate catkins 5 to 10 cm long, calyx yellow, hairy, 5-lobed; lobes acute, laciniately segmented. Fruit ripens in September to November, acorns oval or avoid to oblong, in pairs 1.3 to 1.9 cm long set in cup one-third to one-half their length. Leaves are simple, alternate, deciduous, oblong to obovate, blade 10 to 18 cm long and 8 to 10 cm wide, 5-lobed with deep rounded sinuses, lobes short and wide, obtuse or truncate at the apex, dark green, rough and glabrous above; paler and tomentose beneath. Twigs are brown and stout. Bark is gray to reddish brown (Vines, 1960).
Distribution: The Edward Plateau of Texas, adjacent Oklahoma and Arkansas; east to Florida, north to New England, and west to Iowa and Kansas.
Reproduction: Reproduces by acorns and can resprout from buds on the tree trunk.

Control: Soil-applied herbicides and some foliar growth regulator herbicides.

16. Sand shinnery oak (*Quercus havardii* Rydb.).
Description: Low shrub, hardly over 1 meter tall, forming thickets by underground rhizomes in deep, sandy soils. Rarely a small tree. Flowers in separate catkins. Fruit is a large acorn 1.3 to 2.5 cm long and 1.3 to 1.9 cm wide in cup. Leaves are alternate, deciduous, leathery, 1.9 to 10 cm long and 1.9 to 3.8 cm wide, oblong, coarsely toothed or lobed. Twigs are rounded or sulcate, gray to reddish brown. Bark is gray, smooth, or scaly (Vines, 1960). Acorns are sought by wildlife but are poisonous to livestock in the bud stage. The plant severely limits forage production (Welch and Hyden, 1996).
Distribution: Sandy plains of the Texas panhandle into eastern New Mexico.
Reproduction: By acorns and sprouts from the rhizomes.
Control: Deep plowing or goats are more effective than foliar-applied herbicides. Soil-applied herbicides such as tebuthiuron are effective (Welch and Hyden, 1996).

17. White oak (*Quercus alba* L.).
Description: Large tree to 45 meters tall with a broad, open head. Flowers appear in April to May, stamimate catkins about 7 cm long, calyx yellow, pubescent. Fruit ripens in September and October and is 1.9 to 2.5 cm long in cup. Leaves are alternate, simple, deciduous, oblong to obovate, 12 to 23 cm long, 7- to 11-lobed. Twigs are reddish brown to gray and bark is light gray to reddish brown (Vines, 1960).
Distribution: On bottom lands, rich uplands, and gravelly ridges in East Texas, Oklahoma, Arkansas, and Louisiana east to Florida, north to Maine, Ontario, Canada, and Minnesota, and west to Nebraska.
Reproduction: By acorns.
Control: Foliar and soil-applied herbicides.

18. One-seed Juniper [*Juniperus monosperma* (Engelm.) Sarg.].
Description: Evergreen tree sometimes 15 meters tall with a trunk to 1 meter in diameter, often several trunks. Flowers occur in March to April, dioecious, terminal, and axillary borne on branches of previous year. Fruit develops in September, fleshy, dark blue to brownish, 0.3 to 0.6 cm long. Seeds usually 1 to 2 sometimes extruded from fruit apex. Leaves minute grayish green, 0.15 to 0.3 cm long, opposite or in threes. Twigs are slender reddish brown. Bark is gray on trunks; ridges flattened and irregular. The fibrous bark is used for mats, saddles, and breechcloths. The wood is used for fuel and fence posts. The fruit is consumed by wildlife.
Distribution: At altitudes of 900 to 2000 meters in New Mexico, West

Texas, and Oklahoma, northward to northern Arizona, and south to Mexico (Vines, 1960).
Reproduction: Fruit persists 1 to 2 years. It is propagated by seed (Vines, 1960).
Control: Soil-applied herbicides.

19. Redberry juniper (*Juniperus pinchotii* Sudw.).
Description: Scraggly shrub or evergreen tree rarely over 8 meters tall. The numerous branches that spread are often close to the ground. Flowers are small cones dioecious, terminal or axillary on short branchlets. Fruit is a berrylike cone 0.6 to 0.8 cm long. Seed is avoid, obtuse at one end, and rounded at the other, solitary and 0.3 to 0.5 cm long. Leaves are aromatic, scalelike, yellowish green, appressed in ranks of two or three 0.15 to 0.3 cm long. Twigs are greenish; older twigs are red. Bark is reddish brown peeling off in shaggy longitudinal strips. Redberry juniper has low browse value but may furnish wildlife cover (Vines, 1960; Welch and Hyden, 1996).
Distribution: Dry hillsides and canyons of West Texas and the panhandle of Texas.
Reproduction: Spreads by seed but can sprout from the crown after top removal (Welch and Hyden, 1996).
Control: Chaining and cabling give acceptable control. Individual plant treatment with soil- and foliar-applied herbicides control redberry juniper.

20. Ashe or blue-berry juniper (*Juniperus ashei* Buchholz).
Description: Shrub or evergreen tree rarely over 8 meters tall; low-branched and trunk twisted. Flowers are minute, dioecious about 0.4 cm long; fruit is fleshy, berrylike cone about 0.6 cm long, blueish green, formed by compression of enlarged fleshy scales; leaves are long, scalelike, opposite. Twigs are gray to reddish and bark is gray to reddish brown, often in shaggy strips. The wood is used for fuel, posts, crossties, and woodenware. The foliage is occasionally consumed by sheep, goats, and deer. The fruit is eaten by wildlife, and wildlife receive cover from this tree (Vines, 1960; Welch and Hyden, 1996).
Distribution: Found in Texas, Arkansas, Oklahoma, and Missouri. It is common in central Texas but found southward and westward into Mexico and Guatemala (Vines, 1960).
Reproduction: Spreads only by seed; does not sprout from the crown after top removal.
Control: Prescribed fire and mechanical top removal are effective. Individual plant treatment with soil- and foliar-applied herbicides is effective (Welch and Hyden, 1996).

21. Eastern red cedar (*Juniperus virginiana* L.).
Description: Evergreen tree varying in shape, occasionally to 14 meters tall. Flower are dioecious; catkins of golden brown, female cones are fleshy purplish. Fruit is a berrylike cone, pale blue; leaves are scalelike or awl-shaped, sharp-pointed, glandless, and dark green. Twigs are reddish brown and the bark is light reddish brown, occasionally separating into long, fibrous strips. The wood is used for paneling, chests, pencils, poles, posts, and other uses. The aromatic nature of the wood makes it a good insect repellent. Cedar oil has various commercial uses (Vines, 1960).
Distribution: Eastern red cedar is found throughout the eastern United States, and west into Texas, Oklahoma, Arkansas, Nebraska, and North and South Dakota (Vines, 1960).
Reproduction: Spreads only by seed; does not sprout from the crown after top removal (Welch and Hyden, 1996).
Control: Prescribed fire and mechanical top removal are effective. Individual plant treatment with soil- and foliar-applied herbicides is effective.
22. Chamise (*Adenostoma fasciculatum* Hook. & Arn.).
Description: Evergreen shrubs with small needle-shaped and heathlike leaves about 0.6 cm long. White flowers in terminal panicles; fruit is an acheme. The plant can attain a height of 3 meters. The herbage is somewhat resinous and sweet-smelling (Bailey and Bailey, 1959).
Distribution: Southern California.
Reproduction: By seed, and can sprout from the stem and crown if top removed.
Control: Foliar- or soil-applied herbicides (Bovey, 1987).
23. Manzanita (*Arctostaphylos* spp.).
Description: The manzanitas are evergreen woody plants that vary from prostrate ground cover to small trees. Usually they have very crooked branches with smooth pinkish or reddish brown bark that may become shreddy on old branches. The leaves are simple and alternate. The small white or pinkish flowers are urn-shaped and borne in simple or compound clusters. The berrylike or drupelike fruit consist of several hard nutlets surrounded by a soft pulp. The manzanitas are highly ornamental plants but are not popular as such. They do not provide high-quality browse, but fruit is eaten by wildlife (Sampson and Jespersen, 1963).
Distribution: Of nearly 50 species of manzanita, 36 are native to the United States; the others are largely Mexican. California recognizes

43 species and 24 varieties of manzanita. Manzanita are constituents of the California chaparrel and are also found in Arizona, New Mexico. Texas, Colorado, Utah, and Oregon, as well as Mexico (Sampson and Jespersen, 1963).
Reproduction: Reproduction is by seed. Some manzanitas produce sprouts if the top is removed; others are nonsprouters (Sampson and Jespersen, 1963).
Control: Foliar spray during rapid spring growth with 2, 4-D amine (Bovey et al., 1984).

24. Rabbitbrush (*Chryothamnus* spp.).
Description: Rabbitbrushes are evergreen shrubs or subshrubs with smooth or tomentose foliage that is often resinous or aromatic. The leaves are simple, alternate, and entire. Heads have 4 to 20 yellow disk flowers that are bisexual and fertile. The heads are usually borne in panicles or cymes, rarely solitary. The achenes are rounded or somewhat angled, smooth to densely hairy. The pappus consists of dull white or brownish soft bristles. The genus *Chrysothamnus* contains 16 species and 41 subspecies (Call, 1991). The terminal portion of the flower stalks are browsed mostly in fall and winter (Sampson and Jespersen, 1963).
Distribution: Rabbitbrush is confined to North America, mainly in the western United States. Approximately 9 species and 18 varieties occur in California (Sampson and Jespersen, 1963).
Reproduction: By seed.
Control: Foliar sprays of 2,4-D ester, picloram, or picloram plus 2, 4-D when new growth appears (Bovey et al., 1984). Results with foliar- and soil-applied herbicides are variable (Call, 1991).

25. Snowberry (*Symphoicarpos* spp.).
Description: This genus includes low- or medium-height deciduous shrubs that often spread by suckers. The leaves are simple and opposite. The bisexual, bell-shaped, or tubular flowers are pink or white and borne in small axillary or terminal clusters. The fruits are roundish white berries. There are 10 to 15 variable species, all nature to North America. They are grazed by livestock and deer and provide wildlife cover (Sampson and Jespersen, 1963).
Distribution: Western United States and North America.
Reproduction: By seed.
Control: Foliar application with 2, 4-D or dichlorprop (Bovey, 1977).

26. Willow (*Salix* spp.).
Description: Willows vary in size from low creeping plants to large shrubs and small trees. Winter buds are covered by a single scale.

Leaves are mostly narrow. Willows have some value for browse for livestock and big game animals. Some are used in landscaping. There are many species of willow.
Distribution: Widespread in North America.
Reproduction: By seed, and produces sprouts if the top is removed.
Control: Foliar applications of 2, 4-D and other growth-regulator herbicides and soil-applied herbicides (Bovey, 1987).

27. Aspen poplar (*Populus tremuloides* Michx).
Description: Aspen poplar is a tree 6 to 12 meters tall with a trunk diameter of 38 to 50 cm. Rarely 30 meters tall and 1 meters in diameter. It usually has a long, slender trunk and narrow, rounded top. Bark is conspicuously whitened. Flowers occur in April and May on drooping catkins. Fruit is a capsule on short stalks, seed is obovoid, light brown, less than 10 mm long. Leaves are simple, alternate, deciduous, ovate and broad-ovate or reniform, 2.5 to 10 cm long. Twigs are reddish brown to gray and slender. Bark is thin, smooth, greenish to gray or white marked with rows of leaf scars. Foliage is consumed by livestock, deer, and elk (Simpson and Jespersen, 1963; Vines, 1960).
Distribution: Occurs from sea level up to 3000 meters, widely distributed in North America.
Reproduction: The tree spreads by root sprouts and seed.
Control: Foliar- and soil-applied herbicides.

28. Yucca (*Yucca* spp.).
Description: About 30 bayonet-leaved, showy flowered species. They are stemless or rising to stature of small trees. The leaves are stiff and long-pointed, often toothed or fibrallose on margins; mostly in rosettes at the surface of the ground or ends of trunk or branches. Flowers are cup-shaped or saucer-shaped with waxy texture, white cream or violet, opening and fragrant at night (Vines, 1960).
Distribution: Tablelands of Mexico and northward and in the West Indies and eastern United States.
Reproduction: By seed and root stock.
Control: Individual plant treatment with foliar sprays of triclopyr (Welch, 1997).

REFERENCES

Allred BW, Mitchell HC. Major plant types of Arkansas, Oklahoma, Louisiana and Texas and their relations to climate and soil. Tex J Sci 7:7–19, 1955.

Anderson JE, Holte KE. Vegetation development over 25 years without grazing on sagebrush-dominated rangeland in Southeastern Idaho. J Range Mgt 34:25–29, 1981.

Archer S. Woody plant encroachment into southwestern grasslands and savannas: Rates, patterns and proximate causes. In: Vara M, Laycock WA, Pieper RD, eds. Ecological Implications of Livestock Herbivory in the West. Denver: Society of Range Management, 1994. pp. 13–68.

Bailey LH, Bailey EZ. Hortus Second. New York: Macmillan, 1959.

Blackburn WH, Tueller PT. Pinyon and juniper invasion in black sagebrush communities in East-Central Nevada. Ecology 5:841–848, 1970.

Bogusch ER. Brush invasion in the Rio Grande Plain of Texas. Tex J Sci 4:85–91, 1952.

Bovey RW. Herbicide absorption and transport in honey mesquite and associated woody plants in Texas. Texas Agric. Exp. Stn. B-1728, 1998.

Bovey RW. Principles of chemical control. In: James LF, Evans JO, Ralphs MH, Child RD, eds. Noxions Range Weeds. Boulder, CO: Westview Press, 1991, pp. 103–114.

Bovey RW. Weed control problems, approaches, and opportunities in rangelands. In: Foy CL, ed. Rev. Weed Sci., vol. 3 Champaign, IL: Weed Sci. Soc. Am., 1987, pp. 57–91.

Bovey RW. Response of selected woody plants in the United States to herbicides. Agric. handbook no. 493. USDA-ARS. 1977.

Bovey RW, Wiese AF, Evans RA, Morton HL, Alley HP. Control of Weeds and Woody Plants on Rangelands. AO-BU. 2344. USDA and University of Minnesota, 1984.

Bragg TB, Hulbert LC. Woody plant invasion of unburned Kansas bluestem prairie. J Range Mgt 29:19–24, 1976.

Branscomb BL. Shrub invasion of a southern New Mexico desert grassland range. J Range Mgt 11:129–132, 1958.

Brown AI. Shrub invasion of southern Arizona desert grasslands. J Range Mgt 3:172–177, 1950.

Call CA. Rabbitbrush classification, distribution, ecology and control. In: James LF, Evans JO, Ralphs MH, Child RD, eds. Noxious Range Weeds. Boulder, CO: Westview Press, 1991, pp. 342–351.

Hanselka WC. Brush sculpting: Applied landscape management for multiple objectives, 9th Ann. Texas Plant Protect. Assoc. Conf., Texas Agric. Ext. Serv. and Texas Agric. Exp. Stn. and Agribusiness Industries, College Station, TX: 1997, p. 16.

Hull AC Jr, Hull MK. Presettlement vegetation of Cache Valley Utah and Idaho. J Range Mgt 27:27–29, 1974.

Hennessy JT, Gibbens RP, Tromble JM, Cardenas M. Vegetation changes from 1935 to 1980 in mesquite dunelands and former grasslands of southern New Mexico. J Range Mgt 36:370–374, 1983.

Jacoby PW, Ansley RJ. Mesquite: Classification, distribution, ecology and control. In: James LF, Evans JO, Ralphs MH, Child RD, eds. Noxious Range Weeds. Boulder, CO: Westview Press, 1991, pp. 364–376.

Johnsen TN, Elson JW. Sixty Years Change on Central Arizona Grassland-Juniper Woodland Ecotone. Agric. Reviews and Manuals-ARM-W-7, USDA-SEA. 1979.

Klingman DL. Problems and progress in woody plant control on rangelands. Proc South Weed Conf 15:35–43, 1962.

LeClerg EL. Losses in Agriculture. Agric. handbook no. 291, Washington, DC: USDA-ARS, 1965.

Little ECS, Ivens GW. The control of brush by herbicides in tropical and subtropical grassland; Part I. The Americas, Australia, the Pacific, New Zealand. Afr Herbage Abstr 35:1–12, 1965.

Lommasson T. Succession in sagebrush. J Range Mgt 1:19–21, 1948.

Mayeux HS, Johnson HB, Polley HW. Global change and vegetation dynamics. In: James LF, Evans JO, Ralphs MH, Child RD, eds. Noxious Range Weeds. Boulder, CO: Westview Press, 1991, pp. 62–74.

Meyer RE, Morton HL, Haas RH, Robison ED, Riley TE. Morphology and anatomy of honey mesquite. tech bull. no 1423, USDA-ARS in coop. with Texas Agric Exp Stn, 1971.

Platt KB. Plant control-some possibilities and limitations I: The challenge to management. J Range Mgt 12:64–68, 1959.

Robertson JH. Changes in a sagebrush-grass range in Nevada ungrazed for 30 years. J Range Mgt 24:397–408, 1971.

Rubin N. Dictionary of Modern Biology. Hauppauge, NY: Barrons Educational Series, 1997.

Sampson AW, Jespersen BS. California range brushland and browse plants. manual 33 Davis, CA: University of California Div. Agric. Sci., 1963.

Scifres CJ. Brush Management: Principles and Practices for Texas and the Southwest. College Station, TX: Texas A&M University Press, 1980, pp. 41–123.

Thomas GW, Ronningen TS. Rangelands—Our billion acre resource. Agric Sci Rev 3: 11–17, 1965.

Tschirley FH. Response of Tropical and Subtropical Woody Plants to Chemical Treatments. ARS-USDA, ARPA order no. 424, CR-13-67, U.S. Dept. Defense, 1968.

Vale TR. Presettlement vegetation in the sagebrush-grass area of the intermountain west. J Range Mgt 28:32–36, 1975.

Vallentine JF. Range Development and Improvements. 3rd ed. New York: Academic, 1989.

Vines RA. Trees, Shrubs and Woody Vines of the Southwest. Austin: University of Texas Press, 1960.

Walker CM. Rehabilitation of forestlands. J For 71:136–137, 1973.

Walstad JD. Weed Control for Better Southern Pine Management. Weyerhaeuser Forestry Paper no. 15, 1976.

Welch TG. Chemical Weed and Brush Control Suggestions for Rangeland. Texas Agric. Ext. Serv. College Station, TX: Texas A&M University, B-1466. 1997.

Welch TG, Hyden SH. Weed and brush control for pastures and rangeland. Texas Agric. Ext. Serv., College Station, TX: Texas A&M Univ., B-6034. 1996. pp 11–19.

Willard EE. Effect of wildfires on woody species in the Monte region of Argentina. J Range Mgt 26:97–100, 1973.

Williams RE, Allred BW, Denio RM, Paulsen HA Jr. Conservation, development and use of the world's rangelands. J Range Mgt 21:355–360, 1968.

Woodwell GM. The carbon dioxide question. Sci Am. 283:34–43, 1978.

2

History and Development of Woody Plant Management

I. INTRODUCTION

A comprehensive review of the *Journal of Range Management* indicated that there were far more scientific papers published on chemical woody plant control than any other method in the last 50 years (Table 1). This is true regardless of the time period investigated. Even in the last 10-year period (1990 to 2000) more papers were published about using herbicides than about using fire or biological or mechanical means. Economic constraints and governmental restrictions have been particularly operative against chemicals, beginning in the 1970s. More than 54 papers were published on chemical brush control in each of the two decades from 1970 to 1990. Even during the 1980s, when prescribed burning was reemphasized, there were still more than twice as many papers published about using herbicides (55) versus fire (26). These numbers do not include chemicals used in integrated brush management systems (IBMS) or in papers in which several methods were compared (Table 1). It clearly shows the importance of herbicides or the perceived importance of herbicides in woody plant control, whether used alone or with other methods.

It is interesting that all methods were researched and published in each 10-year period from 1949 to 2000. Herbicide use was the most researched method, then fire, mechanical, and biological control, in descending order. The number of papers on fire and biological control was fairly evenly distributed between each 10-year period except for the small increase for fire in the 1980s.

More than twice as many papers were published about fire than about biological control over the 50-year period, although in the 1990s paper numbers

TABLE 1 The Number of Scientific Papers Dealing with Different Woody Plant Control Methods Published in the *Journal of Range Management* (1949–2000)

	Control method						
Years	Chem.	Fire	Bioc.	Mech.	IBM	Several	Reveg.
1949–1959	31	18	5	8	4	7	7
1960–1969	37	13	7	8	1	1	3
1970–1979	59	15	6	18	5	1	4
1980–1989	55	26	7	19	4	8	3
1990–1999	23	10	9	15	3	0	4
Total	205	82	34	68	17	17	21

Note: Chem. = chemical control; Fire = prescribed burning or wildfire; Bioc. = biological control; Mech. = mechanical control; IBM = integrated brush management; Several = several control methods used but not IBM; Reveg. = reseeding and/or revegetation of range and/or forest land. A few papers listed under fire do not specifically include woody vegetation, but emphasize the rangeland environment.
Source: *Journal of Range Management*, 50 years of research on woody plant control.

were similar (10 versus 9). In the 1970s to the 1990s papers on mechanical brush control doubled, compared to the 1950s and 1960s.

We think of IBMS as a concept developed in the 1980s but such concepts had their beginning much earlier (Table 1), even though they were not called by the same name. The attempts at comparing several single brush control methods on the same site or on the woody species (or combining some methods) were merely a search for the most economical and effective method(s).

Reseeding and revegetation is not a control method, but should be part of the brush control effort to restore desirable forage and grazing opportunity, as well as to stabilize the soil. A vigorous forb and grass stand should discourage weed and brush invasion or reinvasion, as indicated in Chapter 11.

As stated, one woody plant control method may be combined with another for the most efficient and economical method of woody plant control and maintenance. Two control methods are usually not applied in the same year, but may be administered a year or two later or used to maintain control every few years (e.g., the use of prescribed burning following herbicides).

II. FIRE

A. Early History

Fire has been a tool for woody plant control and has influenced woody and herbaceous species composition on a given site for centuries. Stokes (1980) indicated that it is possible to use tree rings—dendrochronology—to study the fire history of a given area. Ahlstrand (1980) studied fire-scarred southwestern white pine (*Pinus strobiformis*) cross sections from a 1700-hectare (ha) area in the Guadalupe mountains in New Mexico. At least 71 fires have occurred on the site since 1554. From 1696 to 1922 the mean interval between major fires was 17.6 years. No samples were scarred after 1922, coinciding with occupancy and use patterns in the mountains during the past century. In contrast, Arno and Davis (1980) gathered data on the fire history for the western red cedar/hemlock forests. On upland habitat types fires of variable intensities generally occurred at 50- to 150-year intervals, often having a major effect on forest succession. On wet and subalpen habitat types, fires were infrequent and generally small. The average interval between successive fires in Yellowstone National Park is estimated to be greater than or equal to 300 years (Romme, 1980).

Prior to 1860 the maximum individual fire intervals usually did not exceed 35 years in the forests of western Montana (Barrett, 1980). Indians set fires for a number of reasons, including weed and brush control, improved berry production, forage production, hunting, camping, and travel. During the period from 1861 to 1910 Indian fires still occurred in western Montana, but to a much lesser degree. Prospectors and others caused many fires in the late 1800s, but after 1910 fire suppression attempted to eliminate all fires in the region. In the McCalla Creek area fires occurred on an average of every 8.6 years, but sample trees have not recorded a fire for the last 91 years (Barrett, 1980).

Madany and West (1980) developed a fire chronology for the last 480 years from 119 partial cross sections of fire-scarred ponderosa pine. Large fires (>400 ha) occurred nearly every 3 years prior to 1881 on the Horse Pasture Plateau in Utah. A sharp decline in fire frequency began thereafter, some 40 years before the area was obtained by the National Park Service. Changes in land use triggered fire decline.

Lorimer (1980) used records from early government land surveys to estimate the proportion of stands killed by fire in a 15- to 25-year period preceding the survey for vast areas of presettlement forest in eastern North America. Identification of postfire stands was possible in some regions. In South Florida historic fires were identified by charcoal deposits and endemic plants (Taylor, 1980). Lightning caused fires, probably occurring during drought cycles of about 8 years. The fires were several thousand ha in size and burned most fire-adapted vegetation every two cycles.

Wright and Bailey (1982) indicated that there are no reliable historical records of fire frequency in the Great Plains grassland because there are no trees to carry fire scars from which to estimate fire frequency. Fire frequency was high because explorers and settlers were concerned about the danger of prairie fires. We can also extrapolate fire frequency data for grasslands from forests having grassland understories, such as pondersosa pine (*Pinus ponderosa*) in the western

United States and longleaf pine (*Pinus palustris*) in the southeastern United States. Fire frequency of grasslands has been estimated at 5 to 10 years and up to 30 years where topography is dissected with breaks and rivers such as the rolling plains and Edwards Plateau of Texas (Wright and Bailey, 1982). Wright and Bailey (1982) suggested that natural fire every 15 to 30 years in the southern mixed prairie has significantly reduced shrubs and trees, although drought and biotic factors were also dominant factors in maintaining North American grasslands.

These examples and many others given at the Fire History Workshop in Tucson, Arizona, in 1980 indicated that fire has played a very significant role for centuries in the control of woody vegetation (e.g., control of underbrush in conifers) and influenced the type and frequency of vegetation that occurs in a given area. There was great variation in fire frequency and the resulting development of vegetation within the same area, depending upon the type of original vegetation, topography, soil type, climate, and many other factors. In many areas during the last 100 years or more, after settlement land use has changed fire frequency, and in many areas fire prevention is practiced.

B. Recent History

Fire continued to have a major impact on North American plant communities after the arrival of Europeans because of lightning, deliberate use of fires, and carelessness (Wright and Bailey, 1982). European-trained foresters indicated that fire was bad because it killed trees, and fire suppression was strictly practiced starting about the turn of the century. Wright and Bailey (1982) cite a report by Leopold et al. (1963a) as a turning point in informing the public of the benefits of using fire in the national parks. Without fire bad effects included excessive fuel buildups, stagnant young pine trees, dense understories of shrubs and trees, catastrophic stand-replacing fires, less diversity in wildlife, and devastating fires that cannot be readily controlled. The benefits of fire included fuel reduction, seedbed preparation, disease control, thinning, suppression of shrubs, removal of litter, increased herbage yield, increased availability of forage, and increased wildlife (Wright and Bailey, 1982).

Since fire was a natural component of plant community evolution other individuals also supported natural fire where possible. For example, Stewart (1951) indicated that although the influence of burning on vegetation is well known, it is not always fully understood. What European settlers observed in vegetation in North America was significantly influenced by fire since its use was widespread and probably frequent.

Aside from fires set by lightning, Indian burning in the United States was almost universal (Stewart, 1951). More than 200 references to Indians setting fire to vegetation in aboriginal times indicated all major geographic and cultured areas were burned. Once started, fires were allowed to burn unchecked until they burned out or were extinguished by natural causes. Reasons for burning were many, including brush control and fire for personal safety or protection of habitations. Fire was sometimes directed toward enemies and away from retreating Indians (Stewart, 1951). Fires were started to aid in hunting, to improve pasturage, to allow for the growth of berry bushes, tobacco plants, and other crop space, to facilitate travel, and to facilitate offense and defense in war.

Stewart (1951) cites southeastern and south Texas as an example of an extensive region now covered with mesquite and other woody plants that was once open prairie, probably because of extensive burning by Indians. When livestock grazing increased, however, woody plants began to increase. Extensive grazing left insufficient fuel to produce fires hot enough to kill trees. Many areas in California now covered with chaparral have a similar history.

Stewart (1951) indicated that the origin of the tall-grass prairies is difficult to explain, but fire may have played a primary role. The absence of charred roots and stumps has been regarded by some to indicate that forests never existed on prairie land in aboriginal times. Others have suggested that fires may have been so intense and rapid—frequently aided by high winds—that trees and shrubs were never allowed to establish.

In contrast to the benefits indicated by Stewart (1951), the U.S. Forest Service indicated that in the decade that ended in 1950 nearly 2 million forest fires occurred in the United States (Forest Service, 1954). They occurred at a rate of about 500 per day, and burned an average of 8.5 million ha each year, an area larger than Maine. The cost to timber and property in 1950 was estimated at nearly $400 million, with scores of human lives lost. At the same time this bulletin describes beneficial fire uses, such as prescribed burning to aid the regeneration of longleaf pine for weed and brush control in the southern pine region (Forest Service, 1954). It was also used to remove heavy ground cover to reduce destructive wildfire. In the Northwest, burning was used at a safe season to eliminate logging slash and debris.

It therefore became apparent that some areas needed to be protected from fire, while other areas benefited from fire. Sampson (1957) therefore proposed a change in burning terminology as follows:

Management burning. A general term covering the deliberate use of fire on land for the purpose of removing unwanted plant material. Management burning includes convenience burning, control burning, and prescribed burning.

Convenience burning. The simplest form of management burning, in which the only elements planned are the time and place of firing.

Control burning. The application of fire to a preselected land area according to a definite plan, utilizing control forces adequate to confine the fire to the area selected.

Prescribed burning. The ultimate in careful use of fire as a tool for land clearing, involving: the use of fire as a silvicultural tool—burning under rigid restrictions with respect to the humidity and the temperature of the air and fuel; burning within rigidly specified ground limits; burning with the fire under control at all times; and not burning when weather or other conditions are unfavorable at the time planned for firing.

Apparently these definitions were adopted, since similar definitions are indicated by Vallentine (1989).

Wright and Bailey (1982) indicated that in the 1960s biologists of all disciplines began taking a constructive view of fire in North America. Fire reintroduction as a natural force began in the Southeast and Northeast and spread on a limited scale to all North America. Interest in fire accelerated because alternative methods such as herbicides and mechanical and biological means were environmentally unacceptable or ineffective. In the case of insect and disease outbreaks, alternatives to fire have sometimes been unsatisfactory. Further, with 80 years of fire protection in our western forests, understory growth and the resulting wildfire are difficult to control. These massive fuel buildups must be removed in forests by highly trained personnel for maximum safety and reduced cost. Prescribed burning on private and public lands could restore forest, rangeland, and wildlife habitat instead of wasting vast funds on fighting fire.

McCord and McMurphy (1967) indicated that prairie fires burn thousands of grass and woodland ha in Oklahoma every year. In the early 1960s they indicated that most fires were accidental, but some were purposely burned, with some beneficial and some detrimental effects. Most accidental fires occur under dry conditions, damaging rangeland forbs and grasses. Annual late spring burning reduced forage about 10%, but winter burning left soil bare and subject to erosion, and reduced forage yield about 20%. Burned areas needed protection from overgrazing. Fire can increase seed production of desirable native grasses, and can control (spring burn) many cool-season weeds. It can also cause some undesirable woody species to sprout profusely, such as smooth sumac, blackjack, and post oak. Spring burns prior to the initiation of new growth may reduce forage yield, but the increase in steer gain has been about 6 to 9 kg/ha of beef annually.

In the mid-1970s Wright (1974) indicated that few land managers had the training or courage to conduct a burn. Most were exposed to catastrophic fires, which are untimely, have undesirable effects, and are frightening. Prescribed burning has many uses in the management of forests, chaparral, grasslands, watersheds, and wildlife. To minimize harmful effects, fire should never be used during extended dry periods; burns should always take place when the soil is damp or wet (Wright, 1974). Moreover, the user should be an experienced professional with a thorough knowledge of ecosystems, weather, and fire behavior.

Kilgore (1976) stated that many of the present wildlife problems began when attempting to ban all fires from forests, yet control of wildfire was essential in the late nineteenth century since forest resources were being destroyed by careless logging and catastrophic fires (Forest Service, 1954). Two such wildfires were Peshtigo in Wisconsin in 1871, which killed 150 people and burned 1.2 million acres (0.5 million ha), and Hinkley in Minnesota in 1894, which killed 400 people and burned an undetermined area of land (Kilgore, 1976). Such events set the stage for rigid fire-control policies. Efforts to more effectively control forest fires in America began with the founding of such organizations as the American Forestry Association in 1875. The policy to suppress fires in national parks began in Yellowstone in 1886 and was incorporated in the National Park Act of 1916 (Agee, 1974). The establishment of the Forest Reserves in 1891 and the Forest Service soon after had public support (Clepper, 1975). Fire suppression was based on claims that fire of any kind damaged trees and killed seedlings, destroyed forage plants, depleted soil fertility, promoted floods, droughts, and erosion, and destroyed bird and animal habitat (Komarek, 1973).

Early research in the South and West changed government policy and public opinion about prescribed burning. The southern fire scientists showed that controlled burning can be beneficial to longleaf pine, cattle, and quail, while total fire exclusion in the South led to considerable problems, including a tremendous increase in fuel and fire hazards (Kilgore, 1976). Not until 1943 did the weight of this evidence bring about adoption of a prescribed burning policy for southern forests (Shiff, 1962). A similar challenge was raised in the West by the combined research and experimental management efforts of two foresters at the Bureau of Indian Affairs (BIA) and a forestry professor at the University of California (Kilgore, 1976). Conclusions from their studies indicated that vegetation in the ponderosa pine forests of the western United States developed in nature with frequent light fires, that fire exclusion has resulted in extreme fire hazards today, and that prescribed burning by means of light fires can reduce fuels while simulating other ecological impacts of natural burning (Kilgore, 1976).

As indicated by Wright and Bailey (1982) earlier in this chapter, the Leopold report was the document of greatest significance to present National Park Service fire policy (Kilgore, 1976). The report was presented at the North American Wildlife and Natural Resource Conference in 1963 and suggested that "a reasonable illusion of primitive America would be recreated using the utmost in skill, judgment, and ecologic sensitivity" (Leopold et al., 1963b). Data in this report were largely adopted as National Park Service policy in 1968, bringing about a major reorientation in attitudes toward fire suppression (Kilgore, 1976).

As early as 1976 the U.S. Forest Service completed a large-scale computer model for evaluating alternative fire-management plans (Davis and Irwin, 1976).

The model—called FOCUS (fire operational characteristics using simulation)—has been widely tested and became operational in 1976. FOCUS can help fire management organizations evaluate the performance of alternative plans and their impact on the fire protection job. The model can also be applied to fire-related environments, economic and political problems, and overall land-use planning.

In several western national forests, the Forest Service was heavily criticized by the public after the 1979 fire season for allowing certain forest fires to burn rather than extinguishing them (Fischer 1980). Fischer (1980) stated that current fire-management policy requires an appropriate suppression action to be taken in each fire. The potential resource damage by fire is weighed against potential benefits and cost. If the analysis indicates high escape potential for serious resource damage an all-out suppression effort is launched; however, if potential damage is low, the manager may elect to limit suppression. The fire is allowed to burn under preselected conditions, and the use of prescribed burning (planned) is encouraged. Fischer (1980) indicated that early results in the northern region illustrated by events of Forest Services fire-management policy are encouraging but by no means conclusive. The ultimate success or failure of fire management will most likely depend upon how well fire managers master improved fire-management techniques and how accurately they can predict long-term fire effects on forest and rangeland resources.

Controversy still exists for chaparral management in southern California, even with prescribed burning (Laisz and Wilson, 1980). Other means are also needed in addition to prescribed burning since frequent burns on steep slopes can degrade the soils and flora and cause serious off-site damage.

C. Fire History by Location and Vegetation Type

1. Great Plains

Wright and Bailey (1980) indicated that fires were prevalent in grasslands and climate was the major factor in their maintenance. Fire frequency probably varied from 5 to 10 years on level to rolling topography and from 15 to 30 years on rougher terrain. In the short-grass prairie fires do not benefit grasses but can be used to control small juniper and cactus. In the mixed and tallgrass prairies prescribed burning can control undesirable trees and shrubs, burn debris, increase herbage yields, can increase coarse grass utilization and available forage, can improve wildlife habitat, and can control exotic, cool-season grasses.

2. Semidesert Grass-Shrub Type

Historical evidence indicates that fires were present in the semidesert grass-shrub type in southeastern Arizona, but there is less supportive evidence for southern New Mexico and southwestern Texas (Wright, 1980). The change from grass to brush in the last 100 years was due to a combination of factors related to the intensification of grazing. During dry seasons that follow 1 or 2 years of above-average summer precipitation, fire can be used to control burroweed, cactus, broom snakeweed, creosote bush, and young mesquite plants. False mesquite, velvet-pod minosa, wright baccharis, and fourwing saltbush recover quickly after burning. Natural fire frequency was about 10 years for southeastern Arizona, but probably less than every 10 years in southern New Mexico and southwestern Texas.

3. Sagebrush-Grass and Pinyon-Juniper

In sagebrush-grass communities fire was less frequent than in the grasslands of the Great Plains or in the semidesert grass-shrub type (Wright et al., 1979). In Yellowstone National Park estimates were 20 to 25 years but was probably about every 50 years based on the vigorous response of horseweed (*Tetradymia canescens*) and rabbitbrush to fire. Although it is generally known how most shrubs and herbaceous plants respond to fire in the sagebrush-grass communities more data are needed on bluebunch wheatgrass, Idaho fescue, big sagebrush, and bitterbrush (Wright et al., 1979).

The pinyon-juniper association covers 17 to 31 million ha in western North America. The historic role of fire in controlling the distribution of pinyon-juniper cannot be separated from the effects of drought and competition from grass (Wright et al., 1979). Fires every 10 to 30 years probably kept junipers restricted to shallow, rocky soils and rough topography, but for the last 90 years heavy grazing has reduced grass competition as well as fuel for fire that checked pinyon and juniper invasion.

4. Chaparral and Oakbrush

a. California Chaparral California chaparral consists of about 3.5 million ha in California and south-central Oregon (Wright and Bailey, 1982). When woody vegetation reaches 20 years of age dead fuels become great enough to support big fires under adverse conditions. As a consequence, the recurrence interval of large fires (>2000 ha) is 20 to 40 years, but most fires larger than 12,000 ha occur in brush 30 years old or older. At high elevations (1200 meters) on northern aspects, fire frequency in Eastwood manzanita (*Arctostaphylos glandulosa*) chaparral may be every 50 to 100 years.

b. Arizona Chaparral Arizona chaparral burns periodically, but has a lower fire frequency than California chaparral (Wright and Bailey, 1982).

c. Oak Brush The oak-brush areas are just above the pinyon-juniper zone in the central Rocky Mountains and have a fire frequency of 50 to 100 years (Wright and Bailey, 1982). Most fires in Gambel oak (*Quercus gambelii*) occur after a buildup of litter and mulch under the shrub mottes during dry periods (Vallentine, 1989). Most fires are spotty and irregular.

5. Ponderosa Pine

Fire frequency varies considerably, depending upon region and site (Wright and Bailey, 1982). In Arizona and New Mexico frequency for climax and seral communities was between 5 and 12 years. The frequency was 10 years in California and the Blue Mountains of eastern Oregon, but ranged from 2 to 23 years in the Sierra Nevada. In the Bitterroot National Forest of eastern Idaho and western Montana fire frequency averaged from 6 to 11 years for climax stands and 7 to 19 years for ponderosa pine that was seral to Douglas fir (*Pseudotsuga menziesii*).

6. Douglas Fir and Associated Communities

Low-frequency, high-intensity crown fires were the norm before settlement in West Coast coniferous forests (Wright and Bailey, 1982). Fires of 50,000 to 1 million ha were common (Martin et al., 1976). Lower-intensity surface fires were frequent, but generally burned over relatively small areas. Fire frequency in West Coast forests of Douglas fir has been estimated from 150 to over 500 years (Franklin et al., 1981). The free-fire interval in western hemlock forests has been estimated at <150 years (Martin et al., 1976). Pacific silver fir requires a fire-return interval of 700 to 800 years to maintain stand dominance (Schmidt, 1957).

The fire history of the northern Rocky Mountains and adjacent plateaus east of the Cascade range is very varied. Fire-return intervals varied from greater than 500 years in moist subalpine forests to about 6 years in dry forest on grassland of valleys and lower exposed slopes (Wright and Bailey, 1982). Both crown surface fires occurred. Fire severity varied with vegetation type, fuel, weather, and topography.

In the central and southern Rocky Mountains most forests established after fire disturbance (Wright and Bailey, 1982). Before settlement fire maintained open stands of Douglas fir every 25 to 100 years.

7. Southeastern Forest

Historical evidence suggests that fire was a common and widespread occurrence across the South (Harper, 1962). Fire was generally not accepted as a tool in the southeast United States until 50 years after it was first recommended as a management tool by Chapman in 1909, however (Riebold, 1971).

8. Other Forested Regions

Other forested regions mentioned by Wright and Bailey (1982) included the spruce-fire community, in which fires are rare. Fire, however, is a natural force that can influence regeneration of all spruce species in the northwestern and northeastern United States. Sitka spruce occurs in a narrow zone along the Pacific coast of North America and includes trees of western hemlock and western red cedar with some fir species. Prescribed fire is seldom used in the spruce-hemlock type because of the wet climate.

The Engelman spruce (*Picea engelmannii*) is associated with subalpine fir (*Abies lasiocarpa*) and occurs throughout the Rockies. These species are susceptible to severe damage by fire during drought years, but do recover (Wright and Bailey, 1982). Fire can be used as a silvicultural tool if done wisely.

Red spruce (*Picea rubens*) grows from sea level to 1370 meters in the northeast United States. Many other tree species grow with it. Fires occur in red spruce, but are rare, distructive, and of little silvicultrual valve.

Fire was used to maintain stands of red pine (*Pinus resinosa*) and eastern white pine (*Pinus strobuis*) in southern Canada, the lake states, New England, and the southern Appalachians (Wright and Bailey, 1982). Fire frequency has been documented for red pine between 29 to 37 years. Fires have been documented for both red and white pines through historical records, and since they are natural ''fire types,'' controlled burning can be used in their management.

Coastal redwood (*Sequoia sempervirens*) is a climax forest in which redwood is sustained by low rates of reproduction in tree replacement (Wright and Bailey, 1982). It does not depend upon recurrent fires for its status, but is tolerant of low-intensity fire. On mesic sites fire frequency was from 200- to 500-year intervals, and inland and at higher elevations was from 50- to 100-year intervals. Where fires are more frequent, Douglas fir, a redwood associate, appears more frequently. Inland fires occurred about every 25 years.

Giant sequoias (*Sequoiadendron giganteum*) do not sprout; they depend upon seed for regeneration. Fire frequency averaged every 10 to 18 years before 1875, depending upon the site. Fire provided a mineral seedbed for regeneration, recycling of nutrients, and removal of thickets of climax species, such as incense cedar (*Calocedrus decurrens*) and white fir.

D. Prescribed Burning on Rangelands

Since this report is concerned with woody plant management we will discuss the effect of fire on our major woody weed species and associated plants in some of the vegetation types mentioned earlier in this chapter.

1. Sagebrush-Grass

Pechanec and Stewart (1944) provided research in southeastern Idaho that indicated increased grazing capacity averaged 69% on experimental burns compared to unburned areas, with perennial grass and weed increases of 60%. Big sagebrush was completely killed, and only 6% of three-tip sagebrush sprouted. Soil losses were minimal, and range, with little understory of perennial grasses and weeds, should not be burned. Valuable plants, however, such as Idaho fescue, bitterbrush, and some shrubby weeds, were badly damaged. In contrast to the benefits from planned burning, accidental or haphazard burning nearly always produced great damage and loss of soil by erosion and forage. Pechanec and Stewart (1944) recommended burning only where sagebrush is dense, where there are firm and gentle slopes, where fire-resistant perennial grasses and weeds are abundant, and where principal use of the range is for livestock grazing. Severely damaged, slightly damaged, and undamaged species are shown in Table 2.

TABLE 2 Plant Damage from Experimental Burns on Southeastern Idaho Rangeland

Severely damaged species	
Idaho fescue	*Hoary phlox*
Threadleaf sedge	Saskatoon serviceberry
Low pussytoes	Big sagebrush
Littleleaf pussytoes	Threetip sagebrush
Uinta sandwort	Granite gilia
Englemann fleabane	Broom snakeweed
Wyeth eriogonum	*Antelope bitterbrush*
Mat eriogonum	
Slightly damaged species	
Bluebunch wheatgrass	Timber poisonvetch
Prairie Junegrass	*Milkvetch*
Indian ricegrass	*Northwestern painted-cup*
Sandberg bluegrass	*Tapertip hawksbeard*
Nevada bluegrass	Sticky geranium
Cusick bluegrass	*Tailcup lupine*
Subalpine needlegrass	*Royal penstemon*
Needle-and-thread	Munro globemallow
Thurber needlegrass	
Undamaged species	
Crested wheatgrass	Velvet lupine
**Thickspike wheatgrass*	**Stansbury phlox*
*Bluestem wheatgrass	*Flaxleaf plainsmustard
Cheatgrass brome	*Lambstongue groundsel
*Purple pinegrass	Foothill deathcamas
**Douglas sedge*	*Downy rabbitbrush*
*Western yarrow	Spineless gray horsebrush
Wild onion	*Orange arnica
Arrowleaf balsamroot	*Common comandra
**Purpledaisy fleabane*	

Note: Names in italics are important because of their abundance and moderate to high palatability. Those undamaged species marked with an asterisk are spread by rootstocks or root shoots.

Nonresistant plants may require years of careful management for recovery. Pechanec and Stewart (1944) suggested when, where, and how to burn.

Blaisdell (1953) considered big sagebrush a serious management problem in extensive areas of the west. Blaisdell (1953) indicated that all grasses were injured by burning in the upper Snake River region of Idaho, but thickspike wheatgrass, plains reedgrass, and bluebunch wheatgrass recovered rapidly. After 12 to 15 years burned areas produced as much or more herbage as unburned areas even though other grasses were shown to recover. Forbs were injured, but by 3 years more herbage was produced on burned versus unburned range. Rabbitbrush and horsebrush quickly regained or surpassed their original size. The estimated grazing capacity of burned range was 40% greater than unburned in Fremont County and 100% in Clark County, Idaho.

Blaisdell and Mueggler (1956) found that bitterbrush [*Purshia tridentata* (Pursh) DC.], a widely distributed forage shrub in the western United States, commonly resprouted and recovered after the burning of a big sagebrush range or mechanical top removal. Some mortality occurred from burning or top removal. One big problem in more recent times in the Snake River plains is that cheatgrass (*Bromus tectorum L.*), an introduced annual, increases with fire frequencies (to less than every 5 years) by creating a more continuous fuelbed (Whisenant, 1990). More frequent fire and reduced patchiness greatly retard normal vegetation replacement. Presettlement fire frequency was probably between 60 and 110 years. Reducing fire frequency and fire size on these areas should be a primary management objective.

In Utah, Ralphs et al. (1975) indicated that millions of ha of western rangeland are dominated by stands of sagebrush and juniper. Prescribed burning is an efficient and economical range-management tool for these low-value brush species. Late summer, fall, or spring burns are made, but adequate fuel breaks and fire-suppression equipment must be available to prevent escape of the fire. Ralphs et al. (1975) list the effects of fire on foothill plants based on usability for livestock forage (Table 3).

In Wyoming, Smith et al. (1985) found that big sagebrush was killed where uniform fire spread occurred. Green rabbitbrush (*Chrysothamnus viscidiflorus*) and horsebrush (*Tetradymia canesecens*) resprouted and increased in production but did not increase in density. Greasewood (*Sareobatus vermiculatus*) had high mortality. On a more mesic site, aspen, snowberry, and serviceberry resprouted, while bitterbrush did not. Dominant grasses tended to increase. Rhizomatous wheatgrass outperformed other species. Good follow-up management is necessary for the best vegetation responses.

2. Juniper

Burkhardt and Tisdale (1976) indicated that the invasion of western juniper (*Juniperus occidentalis* subsp. *occidentalis*) into vegetation dominated by mountain big sagebrush and perennial bunchgrass on the Owyhee Plateau of southeast

TABLE 3 Foothill Plant Damage from Fire in Sagebrush- and Juniper-Dominated Stands in Utah

Severely damaged	Moderately damaged	Slightly damaged
Desirable	*Desirable*	*Desirable*
Bitterbrush	Bluebunch wheatgrass	Arrowleaf balsamroot
Cliffrose	Indian paintbrush	Crested wheatgrass
Curlleaf mountain mahogany	Indian ricegrass	Douglas sedge
Eriogonum	Needle-and-thread	Sandberg bluegrass
Idaho fescue	Nevada bluegrass	Serviceberry
Threadleaf sedge	Penstemon	Snowberry
	Prairie Junegrass	True mountain mahogany
	Squirreltail	Western wheatgrass
Undesirable	Thurber needlegrass	Yarrow
Sagebrush		
Juniper	*Undesirable*	*Undesirable*
Pinyon pine	Tailcup lupine	Broom snakeweed
Pussytoes		Cheatgrass
		Deathcamas
		Horsebrush
		Rabbitbrush
		Velvet lupine

Idaho appears to be directly related to the cessation of periodic fire. Juniper seedlings became established most readily on areas supporting well-developed herbaceous and shrubby vegetation. Evidence from adjacent climax juniper stands indicated that fires were frequent for at least several hundred years preceding settlement.

Prescribed burning benefits became evident in the pinyon (*Pinus monophylla*) juniper (*Juniperus osteosperma*) stands in Utah and Nevada (Ralphs et al., 1975; Blackburn and Bruner, 1975; Wright et al., 1979) in the late 1960s and early 1970s. Blackburn and Bruner (1975) indicated in Nevada the various prescribed pinyon-juniper burnings were categorized into 1) burning slash and debris, 2) burning individual trees, 3) burning grassland or sagebrush/grassland to kill invading trees, and 4) using broadcast burning when the fire hazard is low and impact is minimal on vegetation, and using natural firebreaks to control fire.

Wright et al. (1979) indicated that the pinyon-juniper association covers from 17 to 31 million ha in western North America, from the east slope of the Sierra Nevada, eastward throughout the mountains of the Great Basin in Nevada and Utah, and on both flanks of the Rocky Mountains in Colorado, as well as on the mesas of the Colorado Plateau and the interior valley. It also occurs southward into Arizona, New Mexico, and northern Mexico. Dominant tree species are Utah juniper (*Juniperus osterosperma*), one-seeded juniper (*J. monosperma*), Rocky Mountain juniper (*J. seopulorum*), Alligator juniper (*J. deppeana*), doubleleaf pinyon (*Pinus edulis*), and singleleaf pinyon (*P. monophylla*). Dense stands of juniper alone that join the pinyon-juniper woodlands extend further north, into eastern Oregon, southern Idaho, and Wyoming.

Wright et al. (1979) indicated that in addition to fire, drought and competition from other vegetation have played a role in controlling the pinyon-juniper distribution. For the last 90 years, however, heavy livestock grazing has reduced grass competition as well as fuel for fires. Reduced competition from grasses has permitted pinyon and juniper to invade adjacent communities rapidly.

Experimental burns by Jameson (1962) in 1956 in the Coconino National Forest in Arizona of galleta [*Hilaria Jamesii* (Torr.) Benth.] and black grama [*Bouteloua eriopoda* (Torr.) Torr.] grasslands and a wildfire in June of 1956 in the Wupalki National Monument indicated that fires caused considerable damage to small juniper (*J. monosperma*) trees (70 to 100% kill), but less damage to larger trees (30 to 40% kill). Kill on larger trees (>1.2 m tall) depended upon ample fuel beneath the tree and correct wind direction (Jameson, 1962; Wright et al., 1979). Trees in closed stands of pinyon juniper with no grass or sagebrush in the understory are difficult to kill because fires do not carry easily (Blackburn and Bruner, 1975).

Bunting (1996) indicated that fire history studies for juniper-dominated areas show that a fire-free interval of 50 years or less would probably have checked juniper invasion during the pristine period. The number of fire ignitions currently received do not adequately check juniper, given the dissected nature of topography and the discontinuous fuels of these areas, however. Kittams (1972) studied 10 burn areas (wildfire) in the Carlsbad Caverns National Park, New Mexico, in the Chihuahuan Desert region. The response of different woody and herbaceous species to fire was recorded as shown in Table 4.

Although redberry juniper (*J. pinchotii*), alligator juniper (*J. deppeana*), earyleaf oak (*Quercus undulata*), and low Mohr's oak (*Q. mohriana*) were not controlled by hot burns, the investigator concluded that fires recurring as often as every 10 years would probably maintain the grassland aspect of the burned sites. Deer-forage quality would be increased by rejuvenation of hairy mountain mahogany and oak and natural reseeding of ceanothus.

In other studies Ahlstrand (1982) found that the coverage and frequency of redberry juniper and whiteball acacia (*Acacia texensis*) were lower, while frequencies of catclaw mimosa (*Mimosa biuncifera*) and skeleton goldeneye (*Viguiera stenoloba*) were higher on burned sites compared to unburned paired plants on the Chihuahuan Desert. Other data on plants species agree with Kittams's (1972) data. Sleuter and Wright (1983) indicated that burning intervals of 7 to 20 years should prevent new redberry juniper from becoming established in north and west Texas. These burning intervals should top-kill established plants and

TABLE 4 Effect of Fire on Vegetation, Carlsbad Caverns National Park

Plant	Burns hot	Usually killed by fire	Types of sprouts after fire	Speed of recovery after fire	Improved as deer forage by fire	Remarks
Lechuguilla	Yes	Yes	None	Nil	No	Often main carrier of fire
Smooth sotol–old plants	Yes	Yes	None	Nil	No	Often "catches" ightning strikes
Young plants	No	No	None	Rapid	No	
Redberry juniper	Yes	No	Crown, stem, seldom	Slow	Yes	
Alligator juniper	No	No	Stem	Slow	Yes	Foliage often too high to burn
Oaks	No	No	Root	Moderate	Yes	Oak clumps do not carry fire
Hairy mountain mahogany		No	Crown	Moderate	Yes	Often rejuvenated by fire
Desert ceanothus	Yes?	No	Crown	Nil	No	Roots burn out
Catclaw mimosa	No?	No	Crown	Rapid	Yes	
Sacahuista	Yes	No	Crown	Variable	Yes	Sometimes killed
Skeleton goldeneye		No	Crown, root, stem	Rapid	Yes	
Silver dalea		No	Root and crown	Rapid	Yes	
Skunkbush	Yes?	No	Crown	Rapid	Yes	
Datil yucca	Yes	Yes	Root	Moderate	Yes	Deer seek new leaves
Grasses	Seldom	No	Crown	Rapid	Yes	Often main carrier of fire

maintain juniper regrowth below the height at which livestock handling becomes difficult. Total forage for livestock and wildlife increases where redberry juniper stands are burned.

Adams et al. (1982) found that in the tallgrass prairie of Oklahoma late winter burning (March) was more effective in reducing density of woody species than summer (July) burning and that herbaceous vegetation was less affected than woody plants. *Schizachyrium scoparium* (little bluestem) was dominant. Most woody species' density decreased with summer or late-winter burn. Decreases were found in poison ivy, roughleaf dogwood, black willow, green ash, winged elm, eastern cottonwood, eastern red cedar, black hickory, and post oak. Increases were found in smooth sumac and common persimmon. Mixed response was found in chickasaw plum and dwarf sumac.

Phillips (1987) indicated that wildfires once controlled eastern red cedar in Oklahoma, but by 1985 it had become a problem in 33 of 77 total counties. Prescribed burning controls eastern red cedar under 1.5 meters tall plus provides other benefits by removing all growth to improve grazing distribution and wildlife habitat. To kill trees that survived prescribed burns, Engle and Stritzke (1992) developed a propane torch technique. Igniting scorched trees after a prescribed burn in several positions killed 90% of the crown and two-thirds of the trees, regardless of tree size. Reburning was more effective on trees highly damaged after prescribed burning. Effectiveness of a single-point ignition declined with increasing tree size.

Engle and Stritzke (1995) concluded that understory eastern red cedar can be controlled successfully by burning leaf-litter firebeds in either late fall or winter after natural leaf fall from hardwood trees or in late summer, fall, or winter following a spring application of tebuthiuron for the control of overstory hardwoods.

3. Aspen

Gruell and Loope (1974) postulated that suckering of aspen and growth of palatable grasses, herbs, and shrubs following extensive fires, particularly on winter range, produced a forage supply sufficiently large to overcome biotic effects of ungulates, thereby allowing successful regeneration of aspen stands. Because of the advanced stage of plant succession, current production of aspen suckers and associated palatable forage is drastically reduced from former levels on elk winter range. Gruell and Loope (1974) concluded that the current decline of aspen stands is primarily due to virtual elimination of fire as an ecological agent in the twentieth century. Adopting fire management policies would reintroduce fire into the aspen–sagebrush habitat of Jackson Hole, Wyoming, and help maintain aspen communities.

Bartos and Mueggler (1979) also found that aspen suckers on high-intensity burns decreased the first postburn year and doubled the second. On moderate-intensity burns aspen suckers tripled the first year, increased sevenfold the second year, and were still three times as numerous as before burning by the third year.

The effect of short-duration heavy grazing by cattle was evaluated 3 and 6 years after the burning and seeding of an aspen grove in Canada (Bailey et al., 1990). Replicated paddocks of June-grazed (early), August-grazed (late), and ungrazed treatments were established. Regardless of treatment, the density of all woody species was lower 6 years after burning than after 3 years. Early- or late-season grazing reduced the density of aspen and wild raspberry (*Rubus strigosus* Michx.) Late-season grazing promoted a greater density of unpalatable western snowberry (*Symphoricarpos occidentalis* Hook.). Grazing reduced the height of aspen, preventing the development of a forest canopy. Herbage production averaged 1,700 kg/ha, not differing between years 3 and 6, but the proportion of smooth brome (*Bromus inermis* Leyss.) increased, while orchard grass (*Dactylis glomerata* L.) declined. Burning of aspen forest in central Alberta followed by forage seeding and short-duration heavy grazing is an effective, economical range-improvement tool.

4. Chaparral—California

Early improvement of brushland for livestock production was expanded in the 1940s and 1950s, most commonly using fire (Fenner et al., 1955). Area ignition applied during sage-burning periods is a safe and effective method. Area ignition is the distribution of many individual fires over an area simultaneously or in quick succession. These individual fires are spaced so that they influence and support each other. This system is particularly suited to land clearing on many California brush ranges. It can also be combined with brush smashing in advance of burning to create concentrations of fuel accessibility. Area ignition spreads fire at a controlled rate and can be accomplished quickly (within a few hours). It is not well suited to areas in which dry grass is the principal surface fuel, but rather to brush and brush litter (chamise and mixed chaparral).

Conversely, Countryman et al. (1969) described a wildfire in which seven Los Angeles County firefighters and their foremen were overrun by a fire flareup and fatally burned. The Canyon Fire burned over 20,000 acres (8,900 ha) before it was finally controlled. The Canyon Fire started on August 23, 1968, near Canyon Inn. An analysis of the fire load index (FLI) was made for the period from August 21 through August 24. On August 23, a sharp rise in air temperature and a drop in relative humidity caused a large rise in the FLI. The chief chaparral fuels at the disaster site were sumac, scrub oak, chamise, and sagebrush, with a few sycamore and introduced shrubs and trees. The last fire in this area was in 1919. No fire had occurred for 70 years or more, and a heavy accumulation of litter and standing dead material was present. Because of the steep slope and strong convective wind currents, the flames were held close to the ground so firefighting crews were subjected to maximum temperatures with fire whirls. The increase in the speed of the local airflow may have resulted from the sea-breeze front reaching the fire area or caused by turbulence by a low-level air tanker flying through the area (or both) just before fire flareup.

Countryman (1974) indicated that the best prospect for alleviation of conflagration fires in southern California is modification of the vegetation to reduce fuel energy output. Creation of the fuel-type mosaic would require coordinated area-by-area planning and a variety of techniques.

Countryman (1983) described the physical characteristics of five northern California brush species—greenleaf, manzanita, snowbrush, chinkapen, mountain whitethorn, and bitter cherry. Ash content ranged from 3% for chinkapen to 8.5% for bitter cherry in foliage. Average ash content in woody material was 1.6%. The ash content of dead material was 1.4%. Leached plants with a low percentage of ash and phosphorus burn more readily than unleached plants of high ash and phosphorus content. Fuel density (weight per unit volume) affects ignition. Fuels with low density can be ignited in a shorter time or with less heat than high-density fuels. Density of foliage varied widely among species, from 0.05 g/cm^3 for bitter cherry to 0.88 g/cm^3 for greenleaf manzanita. Living wood was similar for all species (0.5 to 0.7 g/cm^3). Deadwood density was highly variable, but averaged about 90% of that of living fuel. Solvent extractives, surface-to-volume ratios, heating values, fuel loading by size classes of materials, relative amounts of dead and living fuels, vertical distribution of fuel elements, and the amount of litter fuel were provided.

Vallentine (1989) indicated that controlled burning was the most widely used range-improvement tool for California chaparral. Complete conversion of chaparral to herbaceous range has generally required burning, grass seeding, and herbicidal control of brush sprouts and seedlings. Late fall burning results in less sprouting of brush than spring burning. Rainfall of 43 cm or more is recommended on land being converted to grass.

According to Wright and Bailey (1982), chaparral is a major plant association in California and a small part of south-central Oregon composed almost entirely of shrubs 0.6 to 3.0 meters tall covering about 4 million ha. Chaparral communities are bounded by forests above and grasslands below. Chamise (*Adenostema fasciculatum*) is the most abundant and widespread of all chaparral shrubs in California, and the genus manzanita (*Arclostaphylos* spp.) is the second most abundant group of shrubs, with both sprouting and nonsprouting species. Other common shrubs include Christmas berry, wedgeleaf ceanothus, desert ceanothus, scrub oak, and western mountain mahogany. Only the ceanothus species are nonsprouters. In addition to chamise, on south-facing slopes plants such as California sagebrush, black sage, white sage, California buckwheat, and deerweed occur.

5. Arizona Chaparral

Low-value shrubs have been invading southern Arizona grassland ranges for many years (Humphrey and Everson, 1951). These woody plants generally produce much less forage than the grasses they replace. In a study by Humphrey and Everson (1951), burning before summer rainfall killed most of the burroweed (*Haplopapus tenuisectus*) and snakeweed (*Gutierrezia lucida*). Jumping cholla (*Opuntia fulgida*), and cane cholla (*O. spinosior* and pricklypear catus *O. engelmannii*) had 61, 32, and 44% mortality. Lehmann lovegrass (*Eragrostis lehmanniana*) stand was reduced about 33%, but after 1 year the grass had recovered.

Reynolds and Bohning (1956) burned two grass-shrub areas in the semidesert grassland type in southern Arizona on the Santa Rita Experimental Range near Tucson. Burroweed was reduced 90% from a single burn, while cholla was reduced 50% and prickly pear by about 25%. Velvet mesquite, the most undesirable shrub, was reduced only 9%. Of the mesquite trees killed, all had basal stem diameters <15 cm. All trees <5 cm were affected, but 60% of them resprouted. Black grama was seriously damaged by burning and did not recover during the study period, but the perennial three awns and other grasses recovered by the second to fourth growing season.

Pond and Cable (1960) evaluated the effects of fire on several chaparral species in central Arizona. Shrub live oak (*Q. turbinella*) was difficult to kill by burning; skunkbush sumac (*Rhus trilobata*) could be killed by burning but it took about one burn each year for 4 or 5 years to completely kill the plants. Wrights's silktassel (*Garrya wrightii*) was killed by four annual burns or two burns spaced 2 years apart; burning less frequently did not kill sprouts. Two annual burns or two burns spread 2 years apart completely killed hollyleaf buckthorn (*Rhamnus crocea*). One burn killed old desert ceanothus plants, pointleaf manzanita (*Arctostaphylos pungen*), and larchleaf goldenweed (*Aplopappus laricifolius*). Pond and Cable (1960) concluded that the possibilities of reducing less desirable shrubs by broadcast burning were remote. The species more easily killed are of the most value to livestock and deer.

Chaparrel crown copy and total shrub weights were still increasing six growing seasons after a wildfire on Mingus Mountain (Pase and Pond, 1964). Pointleaf and pringle manzanita and desert ceanothus plants were greatly reduced by fire, but their seedlings were numerous after 5 years. Herbaceous species production was small except where shrub canopy was controlled by herbicide.

Approximately 25% of each of three small watersheds was treated in strips of 15, 30, and 60 meters wide in each of 4 years (Pase and Lindenmuth, 1971). A fourth watershed was not treated. Treatment consisted of late-summer spraying with a commercial mixture of 2,4-D and 2,4,5-T, and prescribed fire in late September or October. Shrub crown cover was reduced an average of 94%. Most shrubs resprouted and quickly reestablished control over the site. Seedlings of desert ceanothus and manzanita were abundant. Herbaceous cover, low before the treatment, increased greatly in the early postfire years. Grasses were uncommon both before and after treatment. Litter mass averaged 14 tons per ha before treatment; 66% remained after the prescribed fire. Results show good but very temporary control of oak-mountain mahogany chaparral with carefully prescribed fire. The technique appears less damaging to the site than wildfires or those broadcast fires with less carefully controlled prescription and execution.

Lindenmuth and Davis (1973) developed usable guidelines for predicting fire spread in Arizona oak chaparral. In a bulletin Pase and Granfelt (1977) describe the methods and proper use of prescribed burning in Arizona. They indicated that fire has long been an important force shaping plant communities on Arizona ranges. The introduction of domestic livestock to Arizona's ranges greatly reduced the frequency of wildfires by reducing the quantity of herbaceous fuel. Woody plants such as mesquite and juniper then increased. Their report is about planned burning, and includes recommendations for using fire to accomplish desired benefits on pinyon-juniper woodland, interior chaparral, sonoran desert shrub, desert-grassland, short-grass, sagebrush, and ponderosa pine ranges.

Pase and Ingebo (1965) indicated that in addition to chaparral control and conversion of land to grasses, runoff was highest on all watersheds the first few years following fire or herbicide application. Sediment yield increased sharply on burned watersheds but rapidly declined to near prefire levels. Zwolinski and Ehrenreich (1967) further restated the importance of watershed management with fire and described its use in the spruce-fir, ponderosa pine, pinyon-juniper, desert grassland, and chaparral. Arnold (1963) indicated that fire has a place in achieving the desired goals of vegetation management, but only in combination with mechanical and chemical methods of treatment, by tailoring the right combinations of treatment to each specific site.

6. Mesquite and Associated Woody Plants

Humphrey (1949) found that broadcast burning controlled burroweed. Effective control of 1.5- to 3-meter velvet mesquite trees has been obtained. These general conclusions were reached as early as 1910 by workers of the Bureau of Plant Industry and the University of Arizona. Groundfires in velvet mesquite, however, are rarely hot enough to kill large shrubs or trees, but may burn through all tissues outside the cambium and kill the cambium, effectively girdling the plant. Resprouting and regeneration of the plant, however, can occur from buds on the stem or from the crown of the mesquite tree. Observations 15 years after the burn indicated complete reinvasion of burroweed and cholla cactus, but mesquite was not as abundant on burned areas (Humphrey, 1949). On a second site there was little invasion by any of the shrubs. The total number of grass plants increased on both burned areas.

Blydenstein (1957) concluded from winter and summer burns in southern Arizona that single fires do not cause high mortality rates in velvet mesquite, but many trees survive by basal sprouting. Mortality was restricted to trees under 2.5 cm in diameter. Blydenstein (1957) stated that as a tool in eradicating established mesquite stands, fire appeared uneconomical due to the need for recurrent burns and adequate fuel but may have promise for young invading mesquite in grasslands.

Cable (1961) showed that burning killed about two-thirds of the mesquite in southern Arizona, some of which were 10 to 15 cm tall and up to 1 year old. The other one-third was top-killed, but sprouted from the base. Data from Cable (1961) agrees with Blydenstein (1957) in that control of velvet mesquite by burning is difficult in southern Arizona.

Box (1967) indicated that the major brush species in west Texas rangelands today is honey mesquite. In describing the extensive grasslands, early explorers (1850) indicated no trees or shrubs occurred on the southern high plains. The Texas high plains were apparently still free of honey mesquite at the turn of the century, but by 1927 Richard (1927) reported the same southern high plains with mesquite trees but much smaller than further east. By 1964, nearly 2 million ha of the high plains grasslands were infested with brush (Smith and Rechenthin, 1964).

Smith and Rechenthin (1964) reported that in the 1960s, honey mesquite grew on more than half of all Texas rangelands (about 22.7 million ha). Surveys in the mid-1980s suggest that the ha infested had changed little, at about 50% of all Texas rangeland (Texas State Soil and Water Conservation Board, 1991). Box (1967) suggested the decreased use of fire is responsible for increased honey mesquite infestations on the Texas high plains. Data of Britton and Wright (1971) and Wright et al. (1976) support the conclusions of Box (1967) that fire kills only a small percentage of mesquite trees. On upland sites in the rolling plains, 27% of the trees were killed following single fires (Wright et al., 1976). When repeated on upland sites on the high plains at 5- to 10-year intervals the potential exists to kill 50% of the old mesquite trees, but honey mesquite is very difficult to kill along river bottoms. Mesquite seedlings are easy to kill with moderate fires until they are 1.5 years old. At 3.5 years old they are tolerant to intense fires (Wright et al., 1976).

In a tobosagrass (*Hilaria mutica*)–mesquite community, fire reduced the canopy cover, height, and biomass of the conspicuous shrubs (Neuenschwander et al., 1978). Tobosagrass dominates on young burns, but the effect of spring burning is temporary since composition of the tobosagrass–mesquite community reaches equilibrium within 5 years after the burn. About 27% of the mesquite was killed by fire, and several years are required for mesquite to recover to equal unburned areas. None of the lotebush was killed by fire, but it was slower to recover than honey mesquite. About 70% of the unburned height and canopy recovered by the end of the fifth growing season after the burn. The mortality of prickly pear ranged from 98% during a wet year (1973) to 13% during a dry year (1974).

Potter et al. (1983) found that maximum temperatures generated within pricklypear cladophylls that sprouted after fire averaged 81°C, compared to 117°C for those that did not sprout. Cladophylls that sprouted after fire remained above 60°C an average of only 2 min, compared to 9 min for those that were killed. Fire temperatures at the same elevation as the cladophylls averaged 478°C for cladophylls that sprouted compared to 595°C for those that were killed. Average fine fuel loads, air and preburn internal cladophyll temperature, and wind speeds were higher for fires that killed cladophylls compared to those that did not.

Box et al. (1967) indicated that fall burning of south Texas chaparral communities significantly reduced the canopy of all brush species in untreated areas and in areas previously treated by roller chopping, shredding, and scapling. Brush species included lotebush [*Condalia obtusifolia* (Hook.) Weberbi.], lycium (*Lycium berlandieri* Dunal.), creeping mesquite [*Prosopis repens* Benth. var *cinerascens* (Gray) Burkart], brasil (*Condalia obovata* Hook.), and Mexican persimmon (*Diospyros texana* Scheele). Total grass production on burned and unburned areas was not different, while forb production decreased on burned versus unburned areas.

Elwell et al. (1970) reported that burning alone without herbicide on large post and blackjack oak trees contributed little to woody plant control. Where 2,4,5-T was used, herbaceous vegetation increased and provided fuel for fire, which controlled some tree sprouts. Sumac and dogweed sprouting increased from burning. Decreased grass species increased only in plots treated with 2,4,5-T. Burning alone or with 2,4,5-T resulted in an immediate increase in annual weeds. The weeds were controlled by the perennial grasses in succeeding years. Soil moisture was significantly increased where herbicide was applied.

In east-central Texas Scifres et al. (1987) burned areas supporting post and blackjack post and associated woody plants previously treated with tebuthiuron two growing seasons earlier. Prescribed burning as headfire in late winter suppressed secondary woody plant stand development. Follow-up burns were done as needed. Tebuthiuron can be applied in strips or other patterns to prevent negative alterations in wildlife habitat.

Tester (1996) burned nine oak forest sites from 1964 to 1984 two to 19 times. The percentage cover of 13 of 14 true prairie grasses was positively correlated with burn frequency. Of these, eight have C4 and six have C3 photosynthetic pathways. Cover of 34 of 39 true prairie forbs increased with the frequency of burning. Cover of six of seven native species, all C3, decreased with increasing burn frequency. These data suggest that the adaptation of tree-prairie species to repeated burning outweighs the effects of their photosynthetic pathways.

The response of live oak-dominated vegetation on the Gulf Coast of Texas was evaluated to spring or fall burning (Scifres and Kelley, 1979). Burning uplands in the fall generally increased grass standing crop compared to spring burns, but spring burns were more productive than unburned areas by the second growing season. Herbaceous species diversity was greatest on fall burns. Fire initially top-killed live oak, but new sprouts the second growing season exceeded preburn stem densities as much as five times. Even though live oak stem numbers usually returned to preburn densities by 2 years following the fires, the release of herbaceous species occurred the first 2 years after the burn. Burning thickitized live oak savannahs every 2 to 3 years depending upon rainfall may be necessary to maintain vegetation at a successional stage suitable for quality wildlife habitat and livestock use.

Other examples of fire use alone and/or in combination with herbicides includes control of common goldenweed (*Iscoma coronopifolia*), a subshrub in south Texas, where fire plus 1 kg/ha of either tebuthiuron or picloram controls the weed and enhances forage production. The fire plus herbicides is a synergistic combination (Mayeux and Hamilton, 1983).

Fire can be used to remove Macartney rose (*Rosa bracteata* Wendl.) canopy, but kills few plants (Gordon and Scifres, 1977). The almost complete control of Macartney rose resulted from herbicide use followed by prescribed burn 18 months later. The landowner must respray or burn the areas periodically to maintain satisfactory control to prevent reinvasion (Scifres, 1975).

Prescribed burning in the winter about 31 months after herbicide application in the spring and burning at about 57 months after spraying herbicide maintained canopy reduction of running mesquite more effectively than herbicide or prescribed burning alone (Scifres et al., 1983). Grass standing crop was greater and a higher proportion of good to excellent grass species occurred in treated versus untreated areas. Aerial application of pelleted tebuthiuron in the fall effectively controlled whitebrush (*Aloysia gratissima*) (Scifres et al., 1983). Subsequent burning did not improve brush control, but botanical composition of forage and forage production was improved by fire compared to areas treated with herbicide alone.

In their book on prescribed burning for south Texas, Scifres and Hamilton (1993) define prescribed burning, the historical role of fire in south Texas, fire behavior, the effects on soils, plants, and wildlife, techniques and fire plans, and other subjects related to prescribed burning. The book provides excellent information for south Texas and other similar areas.

7. Southern Wax-Myrtle

Southern wax-myrtle (*Myrica cerifera* L.) is an undesirable shrub that is invading thousands of ha of rangeland in south Florida (Terry and White, 1979). Prescribed burning has been considered a potential management tool for maintaining pastures free of wax-myrtle. The results of this study show wax-myrtle to be easily crown-killed by a single winter fire. Most plants survive through basal sprouts, however. Users prescribed winter fire to reduce wax-myrtle competition will require repeated periodic burns coordinated with cattle-grazing programs.

E. Prescribed Burning in Forests

1. Southern United States

Walstad (1976) indicated the principal objectives of prescribed burning are slash disposal, flammable fuel reduction, seedbed preparation, browse and forage production, and disease control in southern pines. Prescribed burning is also used to suppress hardwood competition. It can also augment mechanical or chemical brush control in site preparation, but the principal limitation is the shortage of days suitable for prescribed burning. Hot spring and summer burns are more effective than winter burns, although none eradicates the brush. Stems of hardwoods less than 5 cm dbh are readily killed by spring and summer burns, but rootstock may remain viable and regain its position within 5 to 7 years. Hardwood stems larger than 10 cm diameter at breast height (dbh) are rarely killed by fire. If prescribed burning is used to release established pines from understory hardwood competition, pines should be 3.7 to 4.7 meters tall, with bark thick enough to insulate the cambium (Walstad, 1976).

Chen et al. (1975) concluded in Alabama studies that prescribed winter burning top-killed understory hardwoods up to 7.6 cm dbh and increased browse, forage, and edible leguminous fruit supply. Summer burning offered equivalent hardwood control, but less wildlife food supply with greater risk of pine (loblolly and short leaf) damage and soil erosion. Winter burning to enhance wildlife food is highly important for deer, turkey, and quail. Good burning of the understory is dependent on well-distributed pine needle litter or grass litter for fuel. Hardwoods included sweetgum, red oak, winged elm, laurel-water oaks, and hickories.

Studies were conducted in southeast Texas to determine prescribed burning effects on control of undesirable hardwoods invading pine-hardwood stands and in slash pine plantations (Silker, 1955). Sweetgum, blackgum, and dogwood were common, with southern wax-myrtle, yaupon, and holly also present. Burning should be done early, when pine stands are between 10 and 20 years of age and when hardwoods are small. Frequency of burn would be determined by the character of the stand. Where fuel was adequate most understory hardwoods were top-killed, with stems under 10 cm. Research indicated that two burns in 25-year-old slash pine stands provided 77% top kill when hardwoods were 2.5 to 12.7 cm in diameter, 15 cm above the ground line. For pine-hardwood stands, it was 40%. Prescribed burning can be made without risk using qualified personnel (Silker, 1955). Ferguson (1957) indicated from studies in east Texas that hardwoods under 3.8 cm in diameter were controlled much more readily than larger trees and tops were killed more effectively by headfires than by backfires during the growing season.

The response to prescribed burning of plant communities ranging from dry to wet habitats was monitored using permanent plots sampled from 1989 to 1993 (Liu et al., 1997). Temporal controls for fire effects were provided by matched sets of plots protected from fire by newly constructed fire breaks. Changes in species composition were studied by ordination of strata of trees (>5 cm dbh), small tree (2–5 cm dbh), large saplings (1–2 cm dbh), and small saplings and seedlings (50–140 cm tall). Results show that changes occurred largely in the small tree stratum, in which xeric species increased in importance. Although there were changes in sapling and seedling strata, no clear direction of change was recognized. Fire had little effect on the tree stratum.

Of the seven community types under study, three types—sandhill, upland pine, and upperslope pine-oak—were most strongly affected, as indicated by the postfire change in both the positions of samples representing these communities in ordination space and the reduction in understory species abundance. Samples representing the other four mesic and wet communities showed little or no change in their positions.

These short-term results indicate that changes in vegetation resulting from fire were small and mostly restricted to the dry types, in which possible compositional change is expected to occur. This differential effect of fire suggests that the influence of fire is secondary to that of topographical and soil gradients in determining vegetation pattern under current fire regimes. Fire seems to reinforce an overall vegetation gradient controlled by soil in southeastern Texas.

2. Southwest United States

Weaver (1952) concluded that large-scale burning operations in virgin ponderosa pine stands in the Fort Apache Reservation produced good results in reducing the fire hazard and preventing damaging wildfires that accelerated soil erosion. Thinning dense reproduction stands has provided benefits based on sound ecological concepts. Weaver (1955) further stated that fire has played a significant role in the development of ponderosa pine and that prescribed burning can correct some of the adverse conditions that have developed since the coming of the white man. Weaver (1955) indicated that in the southern pine region, for example, longleaf pine (*Pinus palustris* Mill) survives and thrives under conditions of frequent periodic burning. Weaver (1955) also stated that prescribed burning should be used under proper control to correct adverse conditions in the ponderosa pine forests. Such tests were conducted by the School of Forestry, University of California, in the 1950s.

Kallander (1969) simply states that the objective of controlled burning in the ponderosa pine region is to reduce the flammability of the forest. Controlled fire has reduced the rate of spread of wildfires. As indicated by the records, however, 3 to 4 years after the burn new litter accumulation presents the same fire hazard as before. In spite of new improved methods and equipment for fire control, fuel reduction is still the most effective method to reduce wildfire.

Biswell et al. (1973) state that controlled burning, especially after high fuel accumulations have been eliminated, has been shown to be a low-cost solution to destrictive wildfire. Research has shown that controlled burning at intervals of 6 to 7 years meets the ecological requirements of ponderosa pine. Education programs are needed to acquaint the public with the purposes and need for controlled burning and maintenance of low-level fire hazards.

Fiollott et al. (1977) indicated that benefits other than reducing wildfire hazard include overstory thinning from below, site preparation for seedling establishment, and increased forage production. Burning on more production sites may benefit wildlife and/or domestic grazing.

Campbell et al. (1977) evaluated the effects of a wildfire in north-central Arizona. Where fire was intense it killed 90% of the small trees and 50% of the sawtimber, burned 6.5 cm deep in the forest floor to mineral soil, and induced a water-repellent layer in the sandier soils. The reduced infiltration rates increased water yield from severely burned areas, but heavy rainfall eroded soils and removed nutrients mineralized by the fire. Water yields declined back to prefire levels after several years, but some soluble nutrients were further removed by snowmelt. Herbage production was about three times higher on burned versus unburned areas.

Dieterich (1976) indicated that ponderosa pine occurs in nearly every western state, and a great deal of research has been published. Dieterich (1976), however, stated that urgent research needs were still present, including 1) improved prescriptions for fuel reductions, 2) better use of seasons for prescribed fire, 3) more detailed studies, 4) expanded training opportunities for land managers, 5) improved financing for fuels, 6) improved guidance for areas to be treated, 7) better utilization of waste material after logging, and 8) better evaluation of fuel inventory.

3. Northwest United States

Lotan (1979) indicated that fire management in the United States is practiced at several national parks, such as Grand Teton, Glacier, Yellowstone, Yosemite, Sequoia, King Canyon, and the Everglades. The U.S. Department of Agriculture (USDA) Forest Service has programs in several wilderness areas, such as the Selway-Butterroot Wilderness in Idaho, Gila Wilderness in New Mexico, and the Teton Wilderness in Wyoming. In addition, two national forests in the South burn to meet land-management objectives. This policy has been expanded to all national forests based on management strategies.

Fahnestock (1973) stated that on the West Coast the most productive fire-related type of all is the Pacific Douglas fir and redwood. Prescribed burning in the South is to arrest progress toward a less valuable hardwood climax. In the inland Northwest the aim is to maintain the understory species (big-game browse) and its size by prescribed burn to kill overly tall stems and produce sprouts.

Lyon (1966) described a prescribed burn in a Douglas fir stand in Neal Canyon near Ketchum, Idaho. Prefire vegetation consisted of 54 species, including six types of trees and 12 types of shrubs. After the burn natural vegetation rapidly recovered. Three species uncommon or not recorded prefire were the most important in resurgence, including *Moldavica parviflora, Ceanothus velutinus*, and *Iliamna rivularies*. Within 2 years the total density of shrubs had nearly doubled the prefire density. The important result of the fire was the rehabilitation of big-game forage plants. Although shrubs in the second year had reached only 63% of the total crown volume recorded before the fire, forage values had doubled. Maple and willow accounted for 95% of the shrub volume prefire. Shrub dominance shifted from maple to the more palatable willow postfire. The effects of the prescribed burn were exceptionally favorable. Timber-management objectives were achieved, wildlife habitat was markedly improved, and watershed protection was not compromised.

Fahnestock (1976) studied the fire/fuel relationship in the Pasayten Wilderness (north-central Washington). Fifty three percent of the commercial forest area was typed as Engelmann spruce (*Picea engelmannii* Parry), 35% lodgepole pine, and 10% Douglas fir. Fire management as opposed to fire control appeared scientifically sound, environmentally safe, and economically attractive. Forest succession is long term, but Engelmann spruce and subalpine fir gradually invade, if not present initially, from surviving trees in intermingled moist areas that did not burn, and Douglas fir from dry sites where fire intensely was low. Lodgepole pine may initually take over the site, but Englemann spruce and subalpine fir gradually dominate. Shrubs and herbs occur in great numbers and species. Both types of vegetation are usually short (<0.6m); alder and/or willow clumps are sometimes taller.

4. Other Forested Areas

Bock and Bock (1983) conducted a 3-year study (1979–1981) on the effects of prescribed burning of ponderosa pine forests at Wind Cave National Park in the southern Black Hills of South Dakota. The fires were largely restricted to surface fuels and forest understory vegetation. Effects upon understory shrubs and deciduous trees were modest. In eight study plots, two cool-season (autumn and spring) fires consistently reduced densities of *Ribes* spp. and stimulated *Amorpha canescens* Pursh., while other shrubs were unaffected. These burns significantly reduced the density of immature and smaller mature ponderosa pines. The reductions were consistent across all eight study plots. By contrast, an autumn (1974) crown fire in ponderosa pine killed pines of all sizes, and most shrub species increased dramatically. We attribute these differences in postfire vegetation response to variable fire intensities.

III. BIOLOGICAL CONTROL OF WOODY PLANTS

Biological control of woody and herbaceous plants has been practiced since the beginning of time, since most plants have natural enemies that feed on them or injure them in some way. This natural feeding or predation on a given species may or may not hold some plants in check, especially plants removed from their native habitat and placed in an environment in which few or no natural enimies exist. On grazing and forested lands some plants are so unpalatable that livestock and sometimes wildlife selectively graze other more palatable species, leaving certain plants to dominate or become weed problems. Some exotic plants are especially troublesome if unpalatable and do not have natural pathogens or insects that attack them. Other plants, such as creosotebush, may contain natural inhibitors that limit the growth of other organisms, sometimes including their own species. One plant species producing natural allelopathic compounds may favor its existence against neighboring plants.

By understanding the mechanisms of competitiveness and organisms useful in biocontrol one might use this information to manipulate difficult-to-control problem weeds on vast and inaccessible range and forested areas. (See Chapter 14 for more detail on definitions and procedures of biocontrol.)

A. Control of Woody Plants by Insects and Plant Pathogens

1. Lantana

The first published report on the deliberate use of insects to control an unwanted plant species was made by Perkins and Swezey (1924). This work was initiated in 1902 in Hawaii, where *Lantana camera* L., an introduced woody ornamental plant, had escaped cultivation and dominated large areas of rangeland, causing great concern (Holloway, 1964).

Holloway (1964) indicated that at the turn of the century an accidentally introduced scale insect, *Orthezia insignis* Dough., caused considerable damage to lantana, so this insect and others from central America (native home) were evaluated. Lantana has become a pest of rangeland and coconut plantations in Hawaii, Fiji, India, and Australia. The lace bug was the most important control agent in Australia (Harley, 1973); however, a complex of organisms that attack lantana may be required for control (Schroeder, 1983; Julien, 1997).

2. Pricklypear Cactus

Pricklypear cactus (*Opuntia vulgaris*) was introduced from southern South America to India, where it eventually naturalized (Deloach, 1997). In 1795 a scale insect (*Dactylopius ceylonicus*) was introduced to India from Brazil in the mistaken belief that it was the cochineal insect (*D. coccus*), from which the British hoped to produce red dye. The insect increased rapidly and soon controlled the cactus. In 1836 *D. ceylonicus* was introduced to southern India and in 1865 to Ceylon (now Sri Lanka) as the first deliberate attempt at biological control of a weed. It provided complete control of the cactus over vast areas (Julien, 1992).

When up to 11 species of cacti (from North and South America) became serious pests in Australia and several other countries, biological control efforts were organized (Deloach, 1997). Released in South Africa in 1913 (Moran and Zimmerman, 1991), *Dactylopius ceylonicus* gave complete control of prickly pear, as it did the next year in Australia. Several other insects from the United States released in Australia from 1921 to 1933 controlled *Opuntia inermis*. Released in Australia in 1926, the Argentine phycitid moth (*Cactoblastis cactorum*) completely controlled *O. inermis* and three other species of *Opuntia*. Most of the infested land returned to production within 5 years (Dodd, 1940; Mann, 1969). Releases have continued to the present; the 22 species of insects introduced to Australia have provided complete or substantial control of 11 cacti. Some of these insects have now been released in 15 countries and islands, with excellent control in most areas (Julien, 1992). Attempts to control prickly pear cactus in Texas by biological means have been discouraged because some ranchers use prickly pear for livestock feed during periods of severe drought (Bovey, 1995).

3. Gorse

Gorse (*Ulex europaeus* L.) is a thorny leguminous shrub that probably originated in Western Europe and North Africa. It is used extensively as a hedge and ornamental (Holloway, 1964). It can now be found in California, Oregon, Washington, and Hawaii and potentially is a very serious weed on rangeland. *Apion ulicis*, the gorse weevil, was introduced to control gorse into the United States in 1953 and 1954 (Holloway, 1964). It was also introduced in New Zealand in 1927 from England and in Hawaii from France. The gorse weevil reduces seed production but does not affect established plants. Several other insects have been tested, but none is as useful as *A. ulicis*, the gorse weevil (Hollway, 1964).

A European spider mite (*Tetranychus lintearius*) was recently introduced to California from New Zealand to control gorse (Deloach, 1997). Two oecophorid moths also have been introduced: *Agonopterix nervosa* accidentally into Canada and the western United States and *A. ulicitella* into Hawaii. In 1991, a thrips (*Sericothrips staphlinus*) was released in Hawaii (Rees et al., 1996).

4. Koster's Curse

A native plant of the West Indies and central and northern South America, *Clidemia hirta* has proved to be a serious pest in Fiji (Crafts, 1975). It is a shrubby plant with hard, tough stems and produces quantities of berries with small seeds. The berries are relished by many birds, particularly the Indian myna that proved so active in spreading *Lantana*. Leaves of the plant are pubescent and are not eaten by cattle. In Trinidad, a thrips, *Liothrips urichi*, was found that proved specific on *C. hirta*, and this insect has been liberated, free of parasistes, in Fiji. The first importation was in 1930. The thrips have spread over the islands and are controlling the weed. Although they do not kill it outright, they inhibit its growth so that other plants successfully compete with it. Under the influence of the introduced thrips, *C. hirta* is no longer considered a noxious weed pest. Under severe grazing, competition is reduced and the plant may become dominant again. In Hawaii, *C. hirta* has been parasitized with *L. urichi*, and the weed is reduced in seriousness, particularly in sunny, exposed sites, although not in forests.

A pyralid moth (*Ategumian atulinalis*) was established in Hawaii in 1966, a buprestid beatle (*Lius poseidon*) in 1988, and a leaf-spot fungus (*Colletotrichum gloeosporioides clidemiae*) in 1988 (Deloach, 1997). All damage but do not completely control *C. hirta*. Several other natural enemies from Trinidad are possible candiates for release on Koster's curse.

5. Pamakani

This shrub, termed pamakani or Crofton weed (*Eupatorium adenophorum*), has become widespread in Hawaii and Australia (Crafts, 1975). In Hawaii a gall fly, *Procecidochares utilis*, was introduced from Mexico in 1945. It effected control and reclaimed some 25,000 acres of grazing land. The same insect was introduced into Australia in 1952 and showed promise for satisfactory suppression of the weed.

In 1975, scientists from the University of Hawaii released a plant pathogen (*Cercosporella* sp.) from Jamaica on Oahu to control Crofton weed (Deloach, 1997). Control was spectacular in zones of high rainfall and optimal temperatures for disease development. More than 50,000 ha of pastureland has been restored.

6. Banana Poka

Banana poka (*Passiflora mollissima*), a climbing vine native to the Andes of northern South America, was introduced on Kauai before 1920 and today occupies over 40,000 ha of Hawaiian forests, where it spreads through the canopy, killing the trees. For potential control, two moths and a fly from Colombia have been released. Only the pyralid moth is established, but it is not yet causing much damage to the weed (Markin and Pemberton, 1995).

From 1945 to 1965, research was expanded by the Hawaiian Department of Agriculture (HAD), and natural enemies were introduced to control 18 weed species (Deloach, 1997). After the late 1960s, little work was done until 1986 when research on gorse (Markin and Yoshioka, 1990) and other weeds was resumed by HAD and the U.S. Forest Service. Of the 21 weed species for which biological control was attempted in Hawaii, 12 have been completely or substantially controlled, a success rate of 57%. Seventy-one species of insects and one fungus were released; of these, 43 insects and the fungus became established (Funasaki et al., 1988; Goeden 1978; Julien 1992). The weeds that were completely or substantially controlled were blackberry (*Rubus argutus*), Crofton weed (*Eupatorium adenophorum*), hamakua pa-makani (*Ageratina riparia*), three-cornered jacks (*Emex australis*), lesser jacks (*E. spinosa*), Koster's curse (*Clidemia *hirta*), lantana, two species of prickly pear (*Opuntia ficus-indica, O. cordobensis*), puncturevines (*Tribulus terrestris, T. cistoides*), and St.-John's-wort.

7. Scotch Broom

Scotch broom (*Cytisus scoparius*) is a European leguminous shrub introduced as an ornamental and for soil stabilization on road banks. It now displaces indigenous species and is a weed in replanted forest from British Columbia to California (Deloach, 1997). Two species of moths were accidentally introduced and started giving some control: *Agonopeterix nervosa*, found in the 1920s, and *Leucoptera spartifoliella*, found in the 1960s. A seed-feeding weevil (*Apion fuscirostre*) from Italy was tested and released in California in 1964 and is now established in Oregon and Washington. None of these insects gives effective control, although the weevil greatly reduces seed production (Andres and Coombs, 1995; Rees et al., 1996).

8. Blackberry

Crafts (1975) indicated that the common blackberry (*Rubus fruticosus*) was considered the worst weed pest in New Zealand and a serious pest in Australia. It also occurs in orchards and pastures in Oregon and California. In New Zealand the introduced blackbird was important in its spread. Blackberry is closely related to raspberry, boysenberry, loganberry, and other commercial varieties and hybrids. The buprestid beetle *Coraebus rubi* was promising, but may cause damage to strawberries and roses.

Julien (1992) lists several insect releases in Hawaii, Washington, Oregon, and California with limited establishment and effect on Rubus spp. The rust fungus (*Phragmidium violaceaum*) was reported to persist in a small isolated pocket east of the township of Stanthrope in Queensland, Australia, and was spreading very slowly, causing minimal damage to *Rubus fruticocus* (Wilson, 1988). (Other examples of biocontrol with plant pathogens are indicated in Chapter 14.)

9. Woody Plants of Southwestern Rangelands

Except for the work of DeLoach (1997), little attention is given to the control of woody plants on southwest rangelands with insects. Several of the most damaging weeds have related species native to southern South America, where they are attacked by indigenous insects and pathogens. Several of these natural enemies are sufficiently host-specific and could be introduced into North America. Targets included mesquite (*Prosopis*), snakeweeds and broomweeds, willow baccharis (*Baccharis neglecta*) (Boldt and Robbins, 1987), and creosote bush (*Larrea tridentata*). Two stem borers were tested for control of snakeweed: a weevil (*Heilipodus ventralis*) and clearwing moth (*Carmenta haematica*). The weevil was released in New Mexico and Texas, but establishment is not confirmed (DeLoach 1995).

DeLoach (1997) indicated that salt cedars—native to Asia and introduced into the United States in the early 1800s as ornamentals—pose many ecological problems in riparian areas. An international team of researchers is testing several natural enemies to control salt cedar. A mealybug (*Trabutina mannipara*) from Israel and a leaf beetle (*Diorhabda elongata*) from China have been recommended for field release. Two other species are being tested in quarantine in Temple, Texas. A gall midge has been approved for quarantine testing, and overseas testing has been completed on two foliage-feeding weevils and a pterophorid moth. Other host-specific species that attack salt cedar in the Old World could also be tested.

Russian olive—introduced from Israel as an ornamental and windbreak—causes damage very similar to that of salt cedar, but it does not increase soil salinity (DeLoach, 1997). It is little used by white-winged doves or by bees for honey, and is planted more as an ornamental. It is rapidly invading and displacing indiginous species in riparian areas of the West. Opportunistic explorations in China in connection with the salt-cedar project have revealed an exceptionally promising psyllid (*Trioza magnisetosa*) damaging to Russian olive over wide areas that is apparently host-specific.

The rationale and explanation of procedures in biological control of weeds with insects are given by DeLoach (1997) and in Chapter 14. There are many serious nonwoody plants that are problems on grazing lands, forests, cropland, and noncropland. Many are indicated by DeLoach (1997) and in Chapter 14. Plants are selected if they are viable candidates for biocontrol and if promising insects or pathogens exit that are host-specific and may have potential whether native or exotic to control them.

B. Selective Grazing

Selective grazing by livestock and wildlife has probably been practiced for centuries in an attempt to control unwanted woody vegetation. Today selective grazing is a wide-scale practice to control brush with cattle, sheep, goats, horses, poultry, and wildlife species.

1. Texas

Goats are especially effective for brush suppression. Magee (1957) used Angora goats to control regrowth on recently cleaned brushland by bulldozing or cabling or chaining on 15 farms on the grand prairie in Texas. Major brush species were oaks and cedar (juniper). During 1950 to 1955 the goat enterprise paid for the goats, the fencing, the shelter, and all year-to-year costs, including brush clearing. The number of goats was maintained during this period without reducing cattle numbers. Askins and Turner (1972) found that Angora goats were remarkably similar in their vegetation preferences. Weeds (herbaceous) and grass were used the most (28%), followed by woody plants of lotewood (*Condalia obtusifolia*; 27%), catclaw (17%), shin oak (10%), and persimmon (6%), with a small percentage of a variety of other woody plants.

Merrill and Taylor (1976) indicated that goats were one of first animals domesticated by man. It is suggested that goats were brought to Europe and Asia from the Mediterranean countries but had an earlier origin in the East. They are favored for meat and specialized use as milk animals, and their value as a tool for biological brush control has been well established in some parts of the world. Merrill and Taylor (1976) indicated that Spanish goats were probably introduced into the New World by the early Spaniards and can survive on vegetation too harsh for other domestic animals. The Spanish goat is more frequently used for biological brush control than other goats. The Angora goat has been developed for its mohair production and was introduced into Europe during the seventeenth century from Turkey. It originated in the mountains of Tibet. The Angora goat, however, is a good browser and is used for its meat in addition to mohair. The Angora goat is more fragile than the Spanish goat and is subject to cold weather loss.

Merrill and Taylor (1976) stated that goats are far superior to other animals for brush control and can survive on either forbs or grasses. Taylor (1986) discusses results from a 3-year study of Angora and Spanish goats diets measured from a heavily stocked exclosure that indicate that the animals' diets averaged 50, 10, and 40% for grass, forbs, and browse. In other studies in which Angora and Spanish goats diets were compared they were similar to each other in grass, forbs, and browse consumed.

When brush cover was chained and grazed by goats to control brush the percentage of reduction of the canopy cover for live oak, shin oak, juniper, and mesquite was 92, 99, 97, and 77%, respectively. Taylor (1986) indicated that Spanish goats are larger than Angoras, can browse at great heights, and have less hair, enabling them to browse in dense brush. The nutritional requirements of Spanish goats are lower than the Angora. These differences give the Spanish goat significant advantage in foraging ability, especially under heavy grazing pressures.

Cumulative use by Angora goats was 79% in the middle canopy strata of guajillo (*Acacia berlandieri*) compared to 63% in the low strata and 28% in the high strata (Owens, 1991). Blackbrush (*A. rigidula*) also had the highest use in the middle canopy strata, with 39% used compared to 27 and 9% for the low and high canopies, respectively. On heavily used sites, the average grazed twig diameter increased in the two highest canopy layers as the season progressed. Twig sizes grazed in the middle zone on the most heavily grazed sites was significantly higher than in any other canopy strata.

Villena and Pfister (1990) found that Angora and Spanish goats consumed similar amounts of shinnery oak, grass, and forbs. Shin oak consumption increased from 31% in June to 55% in August. Spanish goats had higher forage intakes during July and August than Angora goats. Digestible energy (DE) intake (Mcal/dog) did not differ between breeds, but increased during the summer. Other dietary measurements were also made. The investigators conclude that shinnery oak substantially contributed to the nutrition of foraging goats on these ranges, but that some supplemental feeding may be necessary.

Riddle et al. (1999) found that both ashe juniper and live oak foliage can provide nutrients for Angora mutton goats, but only as portions of their diet.

2. Arkansas

In Arkansas, on Kentucky 61 tall fescue pasture and native pasture, Angora goats suppressed black oak, blackjack oak, post oak, red oak, dogwood, hickory, and maple, but grazed very little on red cedar (Stoin, 1970). If the stocking rate of goats was high they grazed on the grass more heavily.

3. California

Firebreaks offer a promising way to control wildfires (Green and Newell, 1982). After native vegetation has been removed on fuelbreaks, herbicides have been the primary tool for brush control. Using goats to control regrowth appears to be a promising alternative. When averaged over a year, goat diets usually contain at least half browse and the rest grasses and forbs. Goats ate first-year regrowth of chamise, desert ceanothus, California bush, buckwheat, and Eastwood manzanita, but scarcely touched 5-year-old plants of these species unless confined, then mountain mahogany and scrub oak were preferred. As indicated earlier, Spanish goats are larger and better able to fend off predators, and marketable kids are larger than Angoras. Supplemental feeding is desirable in winter.

Murphy (1986) indicated that California rangelands were in excess of 10 million ha and support about 10 million head of livestock. Cattle, sheep, and goats can economically use weeds as feed, when attractive, and later the area can be seeded with palatable plants to replace the weeds.

Scifres (1981) indicated that most quantitative information exists on the use of goats to improve rangeland-supported brush. Scifres (1981) stated that this method has been particularly successful in the oak woodland-oak savannah and mixed brush areas of Texas and scrub bush areas of Africa. Goats are highly selective and use browse in preference to herbaceous plants, whereas sheep prefer forbs in their diet. Sheep are therefore less useful in brush management. Cattle have the least utility for selective plant management since they prefer grasses.

4. Colorado and Utah

Davis et al. (1975) found that a high degree of Gambel oak control can be attained by mechanical treatment followed by goating. Mechanical treatment is necessary to pull oak down for goat use. High stocking rates and proper timing of browsing are important in the management of the oak. Gambel oak is an important component of several million ha of foothill rangelands in Arizona, Colorado, New Mexico, and Utah.

In northern Utah the cumulative effects of Spanish goat browsing caused declines in serviceberry and Gambel oak productivity and an increase in big sagebrush (Riggs et al., 1990). Frischknecht (1979) suggested that sheep offer the best possibility for controlling sagebrush and should be considered in management programs. Hamid (1981) showed that inexperienced sheep and goats refused to eat sagebrush but experienced sheep ate substantial quantities in June and moderate amounts in August and November. Monoterpene concentration in sagebrush was highest later in the growing season (August), and the data suggest that the monoterpene content affects palatability. Sheep and goats discriminate against monoterpene odor and taste, respectively, but intraumal infusion of as much as 3 g per day of a monoterpene mixture did not affect fermentation capacity of the rumen fluid and was rapidly lost from the rumen. Hamid (1981) suggested that the monoterpene content of big sagebrush reduces its palatability to grazing animals, and the control of big sagebrush by sheep and goats is remote.

When domestic goats used oak-brush communities, reduction of deciduous browse by the goats the prior summer resulted in the increased use of Wyoming big sagebrush by male deer in winter when snow cover precluded the use of understory species (Riggs et al., 1990). As a result, goatbrowsed pasture contained less fiber and tannins and were more digestible than those of deer on control pastures. Diet quality under snow-free conditions was not substantially different from snowcovered conditions. The authors concluded that goats could be used to periodically manipulate composition of oakbrush winter range to enhance deer diet quality under snowcovered conditions. Enhancement of deer diets under snow-free conditions probably requires annual manipulation of the understory, however.

5. Arizona

Severson and Debano (1991) indicated goating could reduce total shrub cover (scrub live oak, desert ceanothus, Pringle manzanita, skunkbush, sumac, wait-a-bit brush), but problems may result, because shrubs preferred by goats were also preferred by native deer. Reduced forage diversity and nutritional stress could result if these species were eliminated from the stand. Also, trampling by goats affected nitrogen accumulation in the litter and soil, and heavy browsing may eliminate nitrogen-fixing shrubs.

6. West Virginia

Knowing that sheep or goats can be used to control brush, Dabaan et al. (1997) quantified pasture changes during and after brush control. Over an 8-year period the effects of control were measured using no soil amendment, medium soil amendment (4,500 kg lime and 40 kg phosphorus/ha), and high soil amendment (9,000 kg lime and 117 kg P/ha) on soil fertility, pasture botanical composition, and production of brushy pasture grazed by sheep or goats. Botanical composition was estimated from clipped samples. More animal grazing days were obtained on paddocks treated with lime and P. Grazing with sheep or goats, treatment of lime, and application of P resulted in pastures with a grass and broadleaf plant composition similar to that of brush-free natural pasture after 4 years.

7. Canada

Aspen forest occupies potentially useful grazing land in the aspen parkland of western Canada, and is expanding. The replacement of forest with grassland involves the removal of trees and the control of suckers, which invariably emerge following overstory removal (Fitzgerald and Bailey, 1984).

In order to evaluate the effectiveness of browsing by cattle, aspen forest was burned and seeded to forages, after which the regrowth was heavily grazed by cattle either after the emergence of suckers (early) or just prior to leaf fall (late). Grazing treatments were conducted over two growing season.

A single heavy late grazing practically eliminated aspen regeneration, and two quite different plant communities resulted from two grazing regimes. After the first year, the plant biomass in early-grazed plots consisted of 29% aspen and 28% grass (mainly sown species), while late-grazed plots had only 2.5% aspen and 18% grass, with a higher proportion of shrubs, especially snowberry. Trends established after the first year were still evident after the second year. The results indicated that heavy browsing by cattle in August may be an effective technique for control of aspen suckers following initial top kill. (See Chapter 14 for further details on selective grazing.)

C. Woody Plant Control by Plant Competition

This subject will be discussed in detail in Chapters 11 and 14.

IV. MECHANICAL CONTROL OF WOODY PLANTS

Like fire and biological control methods, hand methods of mechanical control have probably been practised since antiquity. Although such methods are sometimes effective, they are slow, costly, and laborious. Hand methods include grubbing, cutting, and girdling. Grubbing is done with, a grubbing hoe, shovel, or similar tool to partially or totally expose the root system to injure or kill the plant. Axes or saws effectively remove the top of woody plants but may not kill the plant if the species can resprout.

Girdling, cutting a ring through the bark and cambuim layer to prevent movement of nutrients to the roots, is practical in scattered stands. Portable chain saws and girdlers reduce labor, time, and cost. Herbicides can be applied to the cut ring on girdled plants or to cut surfaces if the top is removed to improve effectiveness.

Large or power equipment has been developed primarily in the twentieth century and includes bulldozing, grubbing, chaining, railing, chopping, mowing and shredding, root plowing, and disking. The history of each method will be discussed. (See Chapter 14 for the various methods of mechanical control and their use in woody plant control.)

A. Texas

Hamilton (1993) indicated that the use of combustion engine-driven tractors to push or pull brush from the soil dates back to the 1920s and 1930s, and practices such as chaining and root plowing in south Texas increased dramatically in the mid-1940s. Crawler tractors and anchor chains became available after World War II. Root plowing of honey mesquite and mixed brush stands was common in the 1940s and is still practiced today. In those early days heavy competition for brush control work allowed for extremely inexpensive treatment. By the mid- to late 1950s over 100,000 ha were root plowed in the Rio Grande plains alone (Carter, 1958). Root plowing is very effective, and chaining (dragging a large anchor chain or cable between two tractors over the brush) at best kills 85% of the brush, even under ideal conditions. These practices, however, resulted in dramatic increases in prickly pear if present, by transplanting the plants by mechanical disturbance (Dodd, 1968). Other mechanical innovations were therefore developed, such as stacking to remove prickly pear where there were associated woody plants, or by railing where prickly pear occurred mainly as the dominant species (Hamilton, 1993).

Although mechanical treatments were considered and used in experimental brush control research by institutions of higher learning in Texas, it was not until 1969 that the Texas agricultural experiment station hired a full-time agricultural engineer to develop aerial and mechanical control equipment and revegetation practices (Bovey, 1998). As a result of hiring an engineer, a tractor-mounted low-energy grubber was developed to control redberry juniper, Ashe juniper, and other small weed trees and brush (Wiedeman and Cross, 1981). Wiedemann et al. (1977) discussed the historical development of grubbing in North Texas leading to the tractor-mounted low-energy grubber for controlling brush. A new brush grubber for small brush was also reported by Churchill and Schuster (1971). Wiedeman and Cross (1984) also developed a disk chain with a triangular pulling configuration that reduced draft requirements by 36% and increased the operating width by 23%, compared with the two-tractor diagonal-pulling method. The device is cost-effective for preparing seedbeds on rough, log-littered, and root-plowed rangeland.

In 1964, Rechenthin et al. (1964) described the primary mechanical methods to treat brush problems in Texas. Hand methods included grubbing, cutting or axing, girdling, and burning. Power methods included power saws and girdlers, bulldozing, chaining, railing, chopping, disking, mowing and shredding, and root plowing. (All these methods are described in detail in Chapter 14, including their advantages and disadvantages.) All these methods are still widely used today in Texas and neighboring states.

Scifres (1980) treated raking and stacking as a separate category, but it is discussed under bulldozing in Chapter 14. Raking and stacking is used primarily as a follow-up to collect and pile debris following other mechanical operations; rakes are used effectively following chaining of honey mesquite. A brush rake 3 to 4.5 meters wide mounted on a crawler tractor can be used. Stacker rakes are usually 4.5 to 8 meters long and have closely space tines or prongs and a 15-cm-thick steel plate welded near the base of the teeth to shear off or pull up small plants. Root rakes are 6 to 8 meters long and penetrate soil up to 25 cm deep to pull up woody plants and roots following root plowing.

As early as 1943, Bell and Dyksterhuis (1943) described a bulldozer with a triangular bumper about 1.5 meters above the soil and in front of a dozer blade to break and partially uproot stems of mesquite and cedar trees. The dozer blade completed the uprooting and pushed the debris to one side. They also described horizontal Jacques saws on a steel-wheeled tractor that had an upper bar that pushed the canopy forward before the circular saw engaged the tree trunk near ground level.

Smith (1956) indicated that in southwest Texas in the 1950s much brush control was accompanied by overseeding with bufflegrass, blue panicum, or Rhodesgrass. The seed was usually broadcast by airplane ahead of chaining or broadcast immediately behind the root plow or chopper. With the latter method, the seeder was mounted on the machine.

Brock et al. (1970) found that in north Texas most successful grass stands occurred where drilled. Disking plus roller chopping compared to no treatment, tandem disking, roller chopping, and direct seeding without seedbed preparation was best for both introduced and native grasses.

Data by Stuth and Dahl (1974) from 62 ranches in Texas support that of Brock et al. (1970) except that in drier areas as soil depth decreased the amount of rainfall received at planting aided seedling establishment more than seedbed preparation.

Box and Powell (1965) suggested that livestock and game animal forage in south Texas could be improved by mowing, roller chopping, bulldozing with the K-G blade, root plowing, and root plowing and raking. All methods reduced cover significantly, but root-plowed and root-plowed and raked plots reduced cover to where it had no forage value. Mowing of brush increased forage values eightfold over the check, roller chopping sixfold, and K-G blade by fourfold on clay soils. Browse preference was increased by all treatments, and removal of brush tops by these methods increased forage values and preference ratings of brush but did not permanently destroy wildlife cover.

Drawe (1977) found that herbage production was increased most on root-plowed plus raked areas. Major reinvasion of brush was most apparent on less severe treatments, such as stack and K-G blade and K-G blade plus disking. Major reinvading plants were mesquite, granjeno, aster, and false willow.

Trew (1952) indicated that mowing is one of the most important weed- and brush-management practices. Mowing or shredding removed the top growth of weeds, allowing more palatable forbs and grasses to grow on pastures and rangeland. A ''hydromower,'' a heavy-duty self-propelled brush cutter, effectively shredded all mixed brush in south Texas (Texas Agric. Exp. Stn., 1980). In 1 hr the hydromower shredded three times the area as the conventional shredder. Following brush shredding, the first herbage yields doubled, with better forage species as well.

Carpenter (1971) evaluated a commercial flail forage harvester for harvesting brush for potential utilization of mesquite and other brush as a protein supply or help in converting brushland to cropland. Herndon (1971) tested the Kershaw brush cutter (self-propelled) by Kershaw Mfg. Co. and satisfactorily cut mesquite trees and stumps up to 0.5 meters in diameter as a method to remove the trees.

Ulich (1982) developed a mobile harvester to cut trees (several woody species) near the soil surface, to retain the whole tree or cut it into small chips, to convey it, to elevate it into a basket, and to dump it into transport vehicles. This harvester would be highly useful for harvesting the biomass of woody plants for any commercial production as well as for removing brush from the land.

B. Oklahoma and New Mexico

Porterfield and Roth (1957) described a heavy-duty stalk shredder to clear land of brush and small trees, a portable wood chipper to dispose of woody growth for land clearing and tilling, and seeding operations following mechanical land clearing. Mechanical treatments consisted of tree pulling, bulldozing, tree shearing, and broaching with portable circular and chain saws. The time and cost required to control brush on a given area was obtained.

Greer (1967) described grubbing for mesquite control in Oklahoma as removing the sprout buds (crown) on seedlings by severing the plant 10 to 15 cm below the soil surface any season of the year. Similar results and practices worked for mesquite in New Mexico (Herbel et al., 1958; U.S. Department of Agriculture-Agriculture Research Service, 1958).

C. Arizona

As early as 1938 Streets and Stanley (1938) discovered that grubbing was effective on velvet mesquite in southern Arizona but impractical where plants or seedlings were abundant. They reported the results of experiments using 29 different methods. The application of sodium arsenite to the sapwood by frilling the base of the stump or tree with downward strokes of an ax killed all trees. Sodium arsenite is no longer used because of its toxicity.

Experiments were conducted in 1957 to determine if a double-loop combination of chaining would control all age classes of junipers without dragging the second time in the opposite direction (Bureau of Indian Affairs, 1958). Tests were conducted on the Fort Apache Reservation, White River, Arizona. Double-looped chaining had several disadvantages, including problems in operation, excessive drag, destruction of the sod and soil surface, and piled-up debris, and did not remove young whip-size trees. The technicians concluded that double-loop drag was not commercially practical and that use of a single-loop drag passing over the stem in the opposite direction would be more effective than use of a double-loop drag in one direction.

In other studies Schmutz et al. (1959) found that chaining opened the shrub stand by knocking down the woody plants. Due to a large number of young cholla plants establishing after chaining, however they became much more dense than in the original stand. Native perennial grasses did not benefit, and seeded lovegrass failed to establish. Some shrubby species such as young jumping chollas and burroweed increased in abundance, while others, such as old jumping cholla and pencil cholla, decreased. Martin and Tschirley (1969) found that cholla cactus numbers 3 to 4 years after chaining were lower than before treatment but had little impact on prickly pear. Martin et al. (1974) found that staghorn and jumping cholla and barrel cactus decreased dramatically on chained areas, but prickly pear numbers remained about the same as before chaining. Herbage production of grasses was increased on the chained area. Perennial grass production doubled where mesquite was controlled.

The primary species in the San Simon Valley in southeastern Arizona are sand-dune mesquite, creosote bush, tarbush and subdomiates yucca, snakeweed, ephedra, fourwing saltbush, and cactus (Jordan and Maynard, 1970). Fourwing saltbush resisted mechanical treatment, which is fortunate because it is a palatable browse plant and desirable vegetative component. Root plowing was very effective on mesquite, creoste bush, and tarbush. Root plowing was highly successful in other studies on shrub live oak (Pond et al., 1965). Brush chopping or chaining was ineffective on mesquite and creoste bush, as was the pitting disk. A combination of chaining and pitting decreased density of mesquite about 40% and creosote bush about 50%, but control was not adequate. Disk plowing and root plowing were successful because they severed the crown of the plant from the roots. Mechanical shrub control was best when soil moisture was low. Deer spent one-fourth to one-half as much time on root-plowed chaparral pastures seeded to lovegrass (32 ha) as in adjacent brush fields (Urness, 1974). Although high-quality forbs–of low availability in intact brush—greatly increased on treated southerly exposures, deer showed no apparent preference for these slopes. Northern and southeast exposures were used most heavily during the spring and summer in intact brush, while south-facing slopes were used more heavily in the fall and winter. Pellet-group counts showed no marked relationship between deer use and distance from cover up to 275 meters.

Cotner (1963) indicated that about 30 million ha of pinyon-juniper woodland in Arizona, New Mexico, Nevada, and California are woody weed problems. The leading control methods in 1963 were cabling, bulldozing, individual tree burning, and hand chopping. The technique selected depends upon cost, tree density, size, terrain features, and many other factors.

O'Rourke and Ogden (1979) suggested that pinyon-juniper control should take place on land on which the greatest gain is forage cover and production will be derived. Upper, more moist sites were preferred.

D. California–Nevada

In 1963 the U.S. Forest Service evaluated a machine that masticated woody vegetation (Burbank et al., 1970). The investigators concluded that the machine would contribute significantly to resource management progress involving release, thinning, slash disposal, fire-hazard reduction, right-of-way clearing, and range and watershed management cleaning. The machine, which was available from tree Eater Corp. of Gurdon, Arkansas, had a front and flail cutter powered by a 325-hp GMC diesel engine mounted on a Case 750 tractor. The machine had some limitations on some terrain and in terms of operating maintenance costs and safety.

Equipment for land clearing of California chaparral by crushing and compacting brush included the straight-blade bulldozer, the bulldozer with a Tomahawk crusher, the brush rake, and the chain (Roby and Green 1976). The tractor with straight bulldozer blades is valuable for crushing brush in preparation for burning. The Tomahawk compactor increases the amount of brush crushed. The tractor with the brush rake is useful for uprooting and piling brush for burning. Brush can also be prepared for burning on favorable terrain by one or more passes of an anchor chain drawn between two large tractors or between a tractor and a ball. The best crushing, chopping, and uprooting of brush occurred when the chain was modified by welding crossbars across every link or every third link. The ball and chain are used on steep side slopes below a ridgetop. A steel ball 15 meters in diameter is filled with water or water and gravel. A large crawler tractor moving along the ridge and towing the ball crushes brush on the slope in a swath about half the length of the chain. Mature chamise chaparral was most susceptible to crushing. Oaks did not crush well. Late summer or fall is best, and slopes of 50% or steeper are preferred.

Equipment for chopping or shredding brush included the brushland disk, roller chopper, and Tritter shredder. The Tritter shredder (developed in Australia) worked well on 1.5-meter-tall dense manzanita. A Homelite brush cutter (handheld) features a motor carried in a pack on the operator's back and has a circular saw at the end of a flexible drive shaft that speeds up hand cutting. Cyclone seeders were used by hand and all-terrain vehicles. Aerial seeding is used on large areas.

Young et al. (1983) indicated that the brushland plow (developed by the Range Seeding Equipment Committee) was used to make the most of wheatgrass seeding in the 1930s in the intermountain area. Mechanical brush control was necessary if sagebrush ranges were to be successfully reseeded. The Wheatland-type disk plow was most effective for controlling big sagebrush, but was subject to breakage. Other equipment developed in the 1940s to the 1950s included brush beaters, both front-end mounted and a trailing cutter. During the same time root plowing was developed, as discussed in Section IV.A.

The first implement of the brush roller chopper to be tested on rangeland was patterned after the palmetto chopper developed in Florida and manufactured by E. L. Caldwell and Sons of Corpus Christi, Texas (Pechanec, 1950).

In 1952 a subcommittee was established under the Range Seeding Committee for control of "heavy brush" (Young et al., 1983). In 1953, region 3 of the U.S. Forest Service reported on trials using a heavy cable to knock down and uproot juniper trees (*Juniperus* sp.) (Hull, 1953). Cabling had previously been reported as a technique for controlling mesquite in Texas and Oklahoma (Allred, 1949). Experiments with cabling were expanded to include sagebrush control in the 1955 range seeding report (Dresskell, 1955).

Cabling evolved into chaining with the advent of surplus anchor chains from World War II. Allan Johnston, manager of the Kapapla Ranch on the island of Hawaii, developed a chain drag for clearing brush-invaded ranges (Lyma and Sykes, 1954). Anchor chains were introduced to the mainland in the late 1950s and used at several southwest and intermountain locations (McKenzie et al., 1984).

Generally it was found that smooth anchor chains produced better brush control than cables but often failed to give adequate brush control and seedbed preparation. Modified chains were developed at Ely, Nevada, by the Bureau of Land Management (called the Ely chain) (Cain, 1971), and St. George, Utah, by the Forest Service (called the Dixie-Sager chain) (Jensen, 1969). Both of these chains involved welding railroad rails either along the links (Dixie-Sager) or across the links (Ely). Both modified chains included swivels to allow the chains to rotate.

The current level of technology for chains involves the development of disk chains. Large disks from disk harrows are welded to the chain (Wiedeman and Cross, 1984). Private industry has always played an important role in the development of brush-control implements for rangeland (Young et al., 1983). Usually the Range Seeding Equipment Committee evaluated commercially available products and assisted in developing modifications to meet the special needs of rangelands. This is especially true in the case of brush and tree control implements for attachment to large tractors. Early in this century the Holt Tractor Company built prototype sagebrush grubbers for land-reclamation projects. A host of rakes, stump splitters, grubbers, rooters, choppers, and cutting blades has been designed for use with large tractors. Personnel of the Texas Agricultural Experiment Station have reversed the heavy industry trend and developed hydraulic-assisted grubbing implements suitable for use on smaller farm tractors (Wiedeman et al., 1977). These implements were designed for control of scattered stands of brush reinvading areas previously treated by mechanical brush control.

E. Utah

In 1946 Robertson and Plummer (1946) produced a publication giving specific instructions on getting the most from the use of the Wheatland plow for brush control. The suggestions were based on both the experience of the authors and the manufacturer's recommendations. It was considered the best implement for sagebrush control and seedbed preparation at that time. By 1979, however, there was a wide spectrum of mechanical equipment and methods suitable for maximizing sagebrush control and selecting less severe environment treatments (Parker, 1979). The choices were the moldboard plow, heavy-duty offset discs, the brushland plow, the Wheatland plow, the dixie harrow, beaters (mowers), heavy anchor chains, rails, cables, and light rapid anchor chains. The rancher and technician must know the site's needs and choose the most suitable methods.

In the pinyon-juniper woodlands in southern Utah chaining became an accepted practice (Christensen et al., 1966). Many of the browse shrubs recovered well. Deer grazing did not damage seeded or native perennial grasses. Lanner (1977) questioned the benefits of chaining for watershed, wildlife, and livestock forage improvement in these treated areas and felt that the benefits were costly and of marginal value.

F. Florida

Lewis (1972) indicated both cross-chopping and webbing (root plowing) controlled saw palmetto in south Florida. Webbing was less effective on a moist site. Many other shrubs were also effectively reduced by these treatments.

G. Southern Pine Area

Koch and McKenzie (1976) proposed that biomass from noncommercial thinnings and from logging slash residual after harvest be hogged and recovered for fuel and fiber. Such a procedure might yield two dividends of biomass totaling as much as 90,000 kg per ha (40 tons per acre, green-weight basis) from each rotation of southern pine. For sites deficient in organic matter, it is alternatively proposed that residual biomass be hogged and returned to the forest floor—either uniformly distributed or concentrated in bands. Mobile machines to hog and retrieve logging slash or standing trees to be removed by row thinning were proposed to various manufacturers, and various equipment was field-tested (McKenzie and Miller, 1978).

H. U.S. Forest Service

The U.S. Forest Service has been a major player in development and use of mechanical brush-control techniques and reseeding operations on forest and rangelands as exemplified by the many publications cited in this discussion. Other examples of the early involvement of the USDA in this work have been cited (USDA, 1961; Costello, 1941; Pechanec, 1952; Pond et al., 1965; Reed, 1958). A more modern treatise of equipment available for land clearing and range seeding can be found in these publications (USDA, 1974; USDA, 1982; Larson, 1980).

I. Special Equipment

1. *Rotary plowing*. The rotary plow works on the principle of a rotary tiller. Such a plow has been developed commercially but is not widely used in the United States (Rotary Plow Co., 1971). Rotoclear Manufacturing, Ltd., of Alberta, Canada, developed a large plow capable of preparing a good seedbed in shrub-covered areas with one pass of the machine. Large brush must be removed first, and its feasibility on U.S. rangelands is unknown at this time.
2. *Range restorer*. Dr. Carlton Herbel and George Abernathy of the USDA-ARS and the New Mexico Agricultural Experiment Station developed equipment that uproots shrubs, prepares a seedbed, plants seed, and distributes the brush over the seeded strip for protection in a single pass over the land (USDA-ARS, 1972).
3. *Cactus harvester*. A cactus harvester is basically a farm tractor with a front-mounted side-delivery rake and a side-mounted collector (USDA-ARS, 1982). As the machine moves across the land, specially designed rake teeth uproot the cactus and deposit it in a windrow that the collector picks up. The harvester reduced prickly pear biomass 86% in moist soil and 91% in dry soil. The machine can remove cactus for livestock feed.

J. General Comments

Young et al. (1983) indicated that techniques for brush control on range and forest lands have been exported from North America to other countries and that we have borrowed ideas from other countries. Heavy machines developed for brush crushing and control in the United States have been exported to tropical areas where site conversion is often highly controversial.

Ideas on mechanical control have often been developed or modified by land managers, machinery manufacturers, engineers, scientists, and technicians interested in brush control. Some equipment will never become commercialized because of limited use and/or cost.

The cost of mechanical treatment usually is closely correlated with the degree of soil disturbance and the size and density of brush to be removed. Bulldozing, disking, grubbing, and root plowing are among the most effective mechanical brush-control treatments, but they are the most costly to perform; chaining, roller chopping, and mowing are less expensive. Hand methods, such as sawing, axing, or grubbing sometimes are effective, but they are slow, costly, and laborious. (See Chapter 14 for details on the use of each method.)

V. WOODY PLANT MANAGEMENT WITH CHEMICALS

The history and development of herbicides used in woody plant management are discussed in Chapter 3 and summarized in Table 1, Chapter 3. Also, the response of over 370 woody plant species to the herbicides used in woody plant control are listed in Table 2 of Chapter 12. The history of phenoxy herbicides and Agent Orange has also been given in Chapter 8, with a smattering of history on herbicides in other chapters, namely Chapters 4 and 11.

As observed in Table 1 of this chapter the number of papers on chemical brush control far exceeds papers presented on any other control method regardless of time since the inception of publication of the *Journal of Range Management* in 1949. Numerous research data have also been published in many other publications, such as *Weed Science, Weed Technology, Weed Research*, agricultural extension and experiment station leaflets and bulletins of most state universities, USDA publications, books, and industry technical brochures and labels.

VI. INTEGRATED BRUSH MANAGEMENT SYSTEMS

The use of two or more brush-management methods are referred to as integrated brush management systems (IBMS). Methods are selected on the basis of maximum brush-control effectiveness with the lowest possible cost and environmental impact. Reference to Table 1 indicated that IBMS was practiced in the 1950s and earlier but was not called IBMS. Also, several brush-management practices were tested but were not paired or used with each other.

A. California and Nevada

Hedrick (1951) suggested the use of heavy browsing by livestock and big game in combination with fire for brush control in California. He also pointed out the advantages of bulldozing and windrowing live oak followed by goat grazing for controlling sprouts in the Sierra foothills. Biswill (1954) indicated that combinations of brush-control methods appeared most effective for California brush control. Bulldozing was often used to break down or remove brush to facilitate burning for complete removal, and chemicals were often used to kill sprouting species in combination with fire and mechanical means. Chemicals were often used to kill certain undesirable shrubs to favor those that are more readily browsed.

By 1955 IBMS had become widely used in California in foothill rangeland (Burcham, 1955). On the better sites the use of fire may be limited to a single treatment to remove the mature brush stand with subsequent regrowth controlled by chemicals or other means. In some cases one or two returns were made to control sprouts and seedlings at intervals of 2 to 4 years after the first treatment. Mechanical methods were also widely used to prepare brush for controlled burning, such as bulldozing, railing, or chaining down brush 6 months or more before burning. From 1945 to 1953 over 300,000 ha were treated by these methods in California.

Manley and Walker (1956) described brush-control work on the Keith Ranches in central California. Brush crushing was done with a bulldozer prior to burning in oak woodland. Other ranches helped during the day of burning to contain the fire. Some areas with oak trees were killed by herbicide using the cut-surface treatment. Grazing was delayed until late summer the first year after seeding burned areas with perennial grasses.

Roberston and Cords (1957) indicated that at Elko, Nevada, repeat treatment using either burning followed by the use of 2,4-D spray the following year or spraying the same area in successive years appeared promising for rabbitbrush control.

Bently (1967) indicated that several methods were available in California to remove chaparral and scrub oak woodland. For chaparral, compaction by crushing, looping, or falling is recommended. If present, trees can be felled onto the crushed brush. Compacted chaparral fuel of sufficient volume will burn well even if stems are not broken off. Crushing is best done in old stands of chamise, manzanita, and ceanothus. Young stands of chaparral must be severed with a roller or chopper. Chemical sprays make standing chaparral burn more readily. Chemicals can also be used to control brush regrowth after mechanical treatment, chemical treatment, or as a follow-up treatment to prescribed burning. Reseeding is done on the better sites to convert the area from brush to grass after brush control.

B. Arizona

Tiedemann and Schmutz (1966) found that burning-reseeding and burning-reseeding-herbicide treatments near Dewey, Arizona, significantly reduced oak chaparral (pointleaf manzanita, wait-a-minute bush, skunkbush, sumac, desert ceanothus, Wright's mountain mahogany, and hollyleaf buckthorn) for over 7 years and resulted in significantly more grass for 5 to 7 years. Turbinella oak was the dominant shrubby species.

C. Great Basin

Pechanic et al. (1965) summarized information on the state of the art of sagebrush control in USDA agricultural handbook no. 277, and no mention was made of IBMS. Lancaster et al. (1987) indicated that most treatments to remove sagebrush are single treatments; however, in northwestern California and northwestern Nevada a popular combination is to mechanically remove sagebrush by rotobeating, seed the area with an improved grass species, and then follow up with a herbicide 2 or 3 years later to control undesirable sprouting brush. It is not practical to use this method on large areas because of the cost.

Davis et al. (1975) showed that a high degree of Gambel oak control can be attained by a combination of mechanical treatment followed by goat grazing. Gambel oak is an important brush species on several million ha of foothill range in Arizona, Colorado, New Mexico, and Utah. The brush is first treated by roller chopping, bulldozing, or undercutting to allow goats full access to the foliage. This treatment resulted in increased forage availability and livestock production.

D. Oklahoma

Stritzke et al. (1991) applied tebuthiuron and triclopyr alone and in combination with burning at 2.2 kg/ha in March and June of 1983. Pastures were burned in 1985, 1986, and 1987. Both herbicides were effective on the dominant blackjack oak and post oak overstory; however, triclopyr effects were short-lived on American elm, gum bumelia, hackberry, roughleaf dogwood, and buckbrush. Better brush control associated with tebuthiuron resulted in better fine fuel release, and by 1988 burning had a significant effect on woody plant control in tebuthiuron-treated plots.

Engle et al. (1988) found that paraquat pretreatments increased postfire damage to small and medium-size trees of eastern red cedar and partially compensated for light fine fuel loading in the tallgrass prairie of Oklahoma. Paraquat alone at 0.6 g/L killed about 90% of the crown of small trees but as little as 30% of the crown of large trees.

E. Texas

Some early work in Texas by Hosley and Stoddart (1947) using IBMS involved cutting down brush during the hot summer months (roller-chopper) and using fire in two successive annual treatments to kill sprouts of regenerating brush on the Aransas National Wildlife Refuge. Scientists indicated that the benefits derived for wildlife were very favorable in food resources cover and animal behavior.

Dodd and Holtz (1972) roller chopped and shredded honey mesquite, spiny hackberry, agarito, lotebush, blackbrush, Mexican persimmon, post oak, and live oak in Goliad County. Mechanical treatments reduced overall stature, canopy cover, and density, but stem density increased due to basal sprouting. Burning 2 consecutive years in late summer following mechanical treatment did not lower woody plant or stem densities. Mechanical clearing with burning did result in the highest total herbage production, grass production, and herbaceous basal cover over other treatments. Using goats after mechanical clearing of brush has also been mentioned earlier in this chapter (Darrow and McCully, 1959; Rechenthin et al., 1964).

Sosebee (1974) caused high mortality in shredded honey mesquite 1 and 2 years after shredding by spraying 2,4,5-T with low rates of B vitamins. Beck et al. (1975) indicated that shredding and spraying 2,4,5-T + picloram at the same time in May was also effective. Further work by Boyd et al. (1978) indicated shredding and spraying honey mesquite simultaneously could also be effective during other months of the year, depending upon water content and temperature in the upper 15 cm of soil. The highest root mortality for all months was the greatest from picloram: 2,4,5-T mixtures followed by dicamba: 2,4,5-T mixtures. Root mortality was the lowest from 2,4,5-T amine or ester treatments.

Scifres (1975) indicated that undisturbed dense stands of Macartney rose can be controlled in south Texas by mechanical treatment followed by herbicide treatment 2 years later. Prescribed burning or herbicides can be used for maintenance. In other studies Gordon et al. (1982) applied picloram pellets in June or September after burning Macartney rose. Picloram pellets at 1 kg/ha also effectively controlled Macartney rose when applied directly into the ash immediately after burning in winter or early spring. Picloram pellet application extended the beneficial effects of prescribed burning. Prescribed burning eliminated the debris and allowed uniform grazing distribution and better livestock handling.

Scifres et al. (1983) concluded that the herbicide–fire combination improved rangeland supporting dense stands of running mesquite or white brush. The herbicide reduced the brush cover and released herbaceous vegetation either for livestock or for fuel for burning. The burn expedited forage production, improved botanical composition of herbage, and suppressed brush. A burn at 3- to 5-year intervals was suggested to maintain desirable species.

Wink and Wright (1973) used prescribed burning to reduce woody debris, suppress missed and newly established Ashe juniper plants, and enhance forage production after tree bulldozing.

Mayeux and Hamilton (1983) found that burning prior to the application of tebuthiuron synergistically enhanced the control of common goldenweed compared to each treatment applied alone. Buffelgrass yield was increased as much as threefold with the burn–herbicide combination.

Concern for pricklypear control was initiated in the early 1930s on the Edwards plateau of Texas (Dameron and Smith, 1939). Pulling and grubbing was recommended, and in some situations burning and grazing gave some control. Grubbing that included piling and poisoning (arsenic pentoxide plus sulphuric acid) were considered the most practical and effective prickly pear controls at that time. The best months for chemical control were June, July, August, and September with this mixture. Nearly one half million ha were cleared of prickly pear in 1938, mostly by grubbing.

Hoffman and Dodd (1967) indicated excellent control resulted when prickly pear plants were mechanically bruised then sprayed with 2,4,5-T. Where plants were over six pads tall, plants were bruised by dragging a bridge eyebeam or a railroad rail over them. Stacking was also used as a mechanical treatment before spraying.

Research by Ueckert et al. (1988) found that controlling prickly pear cactus after fires in lighter fine fuel vegetation such as sideoats grama and buffalo grass was often not satisfactory. Fire followed by aerial sprays of picloram at 0.14 kg/ha reduced prickly pear cover by an average of 98% and represents an economically significant level of control.

F. Florida and Georgia

The best control of both gallberry [*Ilex glabra* (L.) Grey] and saw palmetto [*Serenoa repens* (Bartr.) Small] was obtained by burning in March and spraying 2,4,5-T in August (Burton and Hughes, 1961). By 2 years after treatment grass yield on treated areas increased threefold on the Georgia site. In Florida burning and 2,4,5-T applied in June decreased saw palmetto cover and increased grass cover from 29% at the beginning of the study to 68% at the end (Kalmbacher et al., 1983).

G. Tennessee

Significantly more poison ivy growing points were killed by glyphosate applications on burned plots than on unburned plots (Faulkner et al., 1989). Honeysuckle, however, showed less damage on burned than unburned plots, whereas prevet showed little difference in control on burned and unburned plots.

H. North Carolina

Romancier (1971) showed the rosebay rhododendron (*Rhododendron maximum* L.) controlled by fire-chemical treatment. Two years after prescribed burning, vigorous sprouts were very susceptible to basal spraying with 2,4,5-T esters in oil or to mist blower applications of either 2,4,5-T esters in water with ammonium thiocyanate or a mixture of 2,4-D and 2,4,5-T and 2,3,6-TBA in oil.

I. Pacific Northwest

Gratkowaski (1977) indicated that after burning, and mechanical, or chemical control of weeds and brush for site preparation in forests, herbicide application is necessary to control resprouting shrubs and brush seedlings in order to maintain planted conifer dominance. To release conifers already established, aerial spray is needed to defoliate and decrease brush competition.

J. Louisiana

Haywood (1993) prepared two upland sites for planting of loblolly pine and applied several treatments to suppress weeds and brush. After several growing seasons, chopping plus burning versus picloram resulted in comparable average dbh volume per tree, and number and yields per ha for loblolly pine. Hexazinone can be effectively applied either before or after planting as a site-preparation method, but this formulation was less effective than a combination of chopping-burning treatments.

K. Canada

Balsam poplar, aspen, and willows that invaded subirrigated sandy rangeland were treated with 1) prescribed spring burning, 2) the herbicide 2,4-D ester, and 3) prescribed burning followed by 2,4-D ester (Bailey and Anderson, 1979). After 5 years, burning and spraying had reduced the most brush. Brush reinvasion occurred rapidly on all treated areas. Stand openings of about one-quarter ha in an 8-meter-high poplar forest resulting from these treatments persisted for at least 5 years. Treatments reduced forest cover and in some cases increased forage production. Repeated burning and spraying substantially reduced the density of reinvading woody suckers.

L. Australia

Eremophila gilesii, a woody pioneer species, is very responsive to disturbance in the mulga communities of southwestern Queensland (Burrows, 1973). Under present management it is increasing in density. The ability of the plant to invade areas cleared of mulga trees is demonstrated.

Ploughing out stands of *E. gilesii*, slashing at ground level, and applying any one of many common herbicides as a high-volume foliar spray were effective in killing the plant. A 1% active ingredient (a.i.) 2,4,5-T ester-diesel distillate combination was particularly effective. The plant is periodically attacked by a wingless grasshopper (*Monistria pustulifera*), and when insect populations are high, large areas of *E. gilesii* are killed.

The feasibility of applying a previously hypothesized grazing management technique to prevent future regeneration of *E. gilesii* from seed is demonstrated.

Spraying brigalow regrowth following a pasture burn resulted in significantly better control than spraying the unburnt suckers (Johnson and Back, 1977). Using 1.12 kg/ha acid equivalent (a.e.) 2,4,5-T ester, reductions in density of 70% and 88% were achieved following aerial spraying and tractor-mounted misting respectively of young regrowth after burning. This compared with reductions of 20% and 16% following spraying of unburnt suckers.

On burnt suckers 0.56 kg/ha 2,4,5-T in diesel distillate gave similar results to 1.12 kg/ha 2,4,5-T in water.

In a field trial in southern Victoria, boom spraying with 2,4,5-T gave a higher level of control of blackberry than mowing, and mowing plus spraying did not improve ($P = 0.05$) the level of control (Amor and Harris, 1981). In a second trial, spraying was superior to mowing, although the difference was smaller than in the first trial, and the combination of mowing and spraying resulted in a greater reduction in the percentage at ground cover of blackberry.

Sprayed thickets were colonized by grasses and weedy species but not by *Trifolium repens*, which is a basic component of pastures in areas in which blackberry is often a weed.

Repeated mowing—or boom spraying with 2,4,5-T—and topdressing with fertilizer are only moderately successful in attempting to replace blackberry with productive pasture. This is attributed to the difficulty in killing the extensive root system of blackberry and the effect of 2,4,5-t on *T. repens*.

M. Venezuela

A study of ecosystem recovery following forest cutting, forest cutting and burning, and forest clearing by bulldozing was conducted in Amazon caatinga forest in the upper Rio Negro region of southern Venezuela (Uhl et al., 1982). Ecosystem recovery was evaluated by measuring vegetation composition, biomass, nutrient accumulation, soil characteristics, and nutrient leaching. As disturbance increased in intensity, the early successional vegetation changed from primary forest trees (cut treatment) to successional woody species (cut and burned treatment) to forbs and grasses (bulldozed treatment). Soil nutrient levels were greater in both the cut and the cut and burned treatment plots than in the control forest 3 years after the disturbances, presumably because of steady transfers of nutrients from the forest slash to the soil. Soil nutrient levels in the bulldozed plot were much lower than the control forest because of topsoil removal. Above-ground biomass at 3 years was 1291 g m^{-2} in the cut site, 870 g m^{-2} in the cut and burned site, and 77 g m^{-2} in the bulldozed site. Given these rates of biomass accumulation, approximately 100 years will be required for both the cut and the cut and burned sites to reach biomass levels characteristic of mature caatinga forest, while more than 1000 years will be necessary in the case of the bulldozed site.

VII. SUMMARY

A comprehensive review of the *Journal of Range Management* indicated that there were far more scientific papers published on chemical control of woody plants than on any other method in the last 50 years. This was true regardless of the time period investigated, including the last 10 years (from 1990 through 1999). These numbers do not include chemicals used in IBMS or in scientific papers in which several methods were compared that were not IBMS. The number of papers on fire and biological control were fairly evenly distributed between each 10 year period since 1949. In the 1970s to the present, papers on mechanical brush control doubled compared to the 1950s and 1960s.

Fire has been a tool for woody plant control and has influenced woody and herbaceous species composition on a given site for many centuries. North American fires started by natural means or by native people have been frequent (every 10 years or more) and of great benefit to the indigenous vegetation. Upon the arrival of European man fire suppression was practiced on rangeland and forests at the turn of the last century, but disaster fires, experience, historic records, and research have shown that fire has a place in vegetation management. In the 1960s the national parks started using fire in their management programs and more recently in California chaparral, Texas brush, and other sites for rangeland and woodland maintenance and weed and brush management. Southern fire scientists showed that controlled burning could benefit longleaf pine, cattle, and quail, while total fire exclusion in the South led to a tremendous increase in fuel accumulations and fire hazard. In the early 1940s this evidence brought about the adoption of a prescribed burning policy in both southern forests and the western United States.

Biological control of woody and herbaceous plants has been practiced since the beginning of time, since most plants have natural enemies of some kind that feed on them or injure them in some way. This natural feeding or predation may or may not hold them in check, especially if the plant is introduced into a new environment free of these natural enemies. On grazing lands some plants are so impalatable that wildlife or livestock do not graze them and they may become problems.

Insects have been found that control or partially control lantana, prickly pear cactus, gorse, Koster's curse, pamakani or Crofton weed, banana poka, Scotch broom, blackberry, salt cedar, Russian olive, and many others. There are many insects that also control herbaceous weeds. The search continues for other insects and plant pathogens for woody plants that infest vast areas and cannot be economically controlled by other means. Searches are made in the native environment (country) from which these woody plant problems originated and the organisms are released on the introduced plant once it is determined by extensive testing that the introduced organisms will feed only on the target weed. Not all woody or herbaceous weeds are good candidates for biological control.

Selective grazing by livestock and wildlife has probably been practiced for centuries in an attempt to control unwanted woody vegetation. Cattle, sheep, goats, horses, poultry, and wild animals have been used. Of all animals mentioned goats are the most effective species for woody plant destruction. Goats are also valuable for their meat, milk, and mohair production, and they can utilize browse and weeds unpalatable to other grazing animals. Spanish goats are more aggressive grazers than Angora goats, but both are very useful for this purpose.

Such mechanical methods of woody plant control as fire and biological control have been practiced since antiquity. Hand methods include grubbing, cutting, girdling, and burning. Grubbing is done with a grubbing hoe, shovel, or similar tool to dig out the root system. Girdling is cutting a ring through the bark and cambium layer around the tree to prevent nutrient flow to the roots. It can be done mechanically. Axes or saws effectively remove the top of the woody plant but may not kill the plant because of resprouts.

Large equipment for woody plant control includes bulldozing, grubbing, chaining, railing, chopping, mowing and shredding, root plowing, and disking. Bulldozing, disking, grubbing, and root plowing are most effective but most costly; chaining, roller chopping, and mowing are less expensive but also less effective. Hand methods are sometimes very effective but they are slow, costly, and laborious.

There are several chemicals effective for woody plant control. Their method of application depends upon their chemistry and mode of action. The growth regulators–hormone type-herbicides are usually effective as foliar sprays, or they can be wiped on smaller woody plants with a carpeted roller or wick-type applicator. Other herbicides are more effective through root uptake, such as bromacil and tebuthiuron. Others are effective through both foliage and roots, such as clopyralid, dicamba, hexazinone, imazapyr, metsulfuron, and triclopyr. The herbicide is selected based on cost, selectivity, effectiveness, and ease of application. The biggest problem with herbicides is cost and sometimes lack of effectiveness.

In some situations the use of two or more control methods are more effective and economical to use than single treatments. Such use can be referred to as an IBMS. An example of IBMS is chaparral control in California, where heavy browsing by livestock and big game can be followed by fire. Another method is crushing the brush with a bulldozer or roller chopper before burning, followed by herbicides to kill sprouting species in the next year or two after the fire. Other examples include application of soil-applied herbicides in oak brush in combination with burning. Mechanical clearing has been used to pull brush down so goats can better utilize it for feed and control it at the same time. The methods selected depend upon the brush species, soil type, water availability, equipment availability, cost, and goals of the land manager. As discussed in this chapter, many combinations may be available to fit the land situation and woody plants to be controlled, with consideration for desirable forage plants and wildlife food and habitat.

REFERENCES

Adams DE, Anderson RC, Collins SL. Differential response of woody and herbaceous species to summer and winter burning in an Oklahoma grassland. Southwest Natural 27:55–61, 1982.

Agee JK. Fire management in the national parks. West Wildlands 1:27–33, 1974.

Ahlstrand GM. Fir history of a mixed conifer forest in Guadalupe Mountains National Park. In: Stokes MA, Dieterich JH, tech. coords. Proceedings Fire History Workshop. Tucson, AZ: gen. tech. rept. RM-81. USDA, Rocky Mt. For. Range Exp. Stn., 1980, pp. 4–7.

Ahlstrand GM. Response of Chihuahuan desert mountain shrub vegetation to burning. J Range Mgt 35:62–65, 1982.

Allred BW. Distribution and control of several woody plants in Oklahoma and Texas. J Range Mgt 2:17–29, 1949.

Amor RL, Harris RV. Effects of mowing and spraying of 2,4,5-T on blackberry (*Rubus fruticosus* agg.) and pasture species. Austr J Exp Agric Anim Husb 21:614–617, 1981.

Andres LA, Coombs EM. Scotch broom. In: Nechols JR, Andres LA, Beardsley JW, Goeden RD, Jackson CG, eds. Biological Control in the Western United States: Accomplishments and Benefits of Regional Research Project W-84, 1964–1989. Div. Agric. Natural Resources publ. 3361. University of California, 1995, pp. 303–305.

Arno SF, Davis DH. Fire history of western redcedar/hemlock forests in Northern Idaho. In: Stokes MA, Dieterich JH, tech. coords. Proceedings Fire History Workshop. gen. tech. rept. RM-81. Tucson, AZ: USDA, Rocky Mt. For. Range Exp. Stn., 1980, p. 21–26.

Arnold JF. Use of fire in the management of Arizona watersheds. Tall Timbers Fire Ecol Conf 2:99–111, 1963.

Askins GD, Turner EE. A behavioral study of Angora goats on West Texas range. J Range Mgt 25:82–87, 1972.

Bailey AW, Anderson HG. Brush control on sandy rangelands in central Alberta. J Range Mgt 32:29–32, 1979.

Bailey AW, Irving BD, Fitzgerald RD. Regeneration of woody species following burning and grazing in aspen parkland. J Range Mgt 43:212–215, 1990.

Barrett SW. Indian fires in the pre-settlement forests of western Montana. In: Stokes MA, Dieterich JH, tech. coords. Proceedings Fire History Workshop. gen. tech. rept. RM-81. Tucson, AZ: USDA, Rocky Mt. For. Range Exp. Stn., 1980, pp. 35–44.

Bartos DL, Mueggler WF. Influence of fire on vegetation production in the aspen ecosystem in western Wyoming. In: Boyce MS, Hayden-Wing LD, eds. North American Elk, Ecology, Behavior and Management. Laramie, WY: Univ of Wyoming, 1979, pp. 75–78.

Beck DL, Sosebee RE, Herndon EB. Control honey mesquite by shredding and spraying. J Range Mgt 28:487–490, 1975.

Bell HM, Dyksterhuis EJ. Fighting the mesquite and cedar invasion on Texas ranges. Soil Conserv 9:111–114, 1943.

Bently JR. Conversion of chaparral areas to grassland: Techniques used in California. Agric. handbook no. 328. Berkeley, CA: USDA, For. Serv., 1967.

Biswell HH. The brush control problem in California. J Range Mgt. 7:57–62, 1954.

Biswell HH, Kallander HR, Komarek R, Vogl RJ, Weaver H. Ponderosa fire management: A task force evaluation of controlled burning in ponderosa pine forests of central Arizona. misc. pub. no. 2. Tallahassee, FL: Tall Timbers Res. Stn., 1973.

Blackburn WH, and Bruner AD. Use of fire in manipulation of the pinyon-juniper ecosystem. In: The Pinyon-Juniper Ecosystem: A Symposium. Logan, UT: 1975, pp. 91–96.

Blaisdell JP. Ecological effects of planned burn of sagebrush-grass range on the upper Snake River plains. tech bull no. 1075. USDA, 1953.

Blaisdell JP, Mueggler WF. Sprouting of bitterbrush (*Purshia tridentata*) following burning and top removal. Ecol 3:365–370, 1956.

Blydenstein J. The survival of velvet mesquite (*Prosopis juliflora* var. *velutina*) after fire. J Range Mgt 10:221–223, 1957.

Bock JH, Bock CE. Effect of fires on woody vegetation in the pine-grassland ecotone of the southern Black Hills. Amer Midland Naturalist 112:35–42, 1983.

Boldt PE, Robbins TO. Phytophagous and pollinating insects fauna to *Baccharis neglecta* (Compositae) in Texas. Environ Entomol 16:887–895, 1987.

Bovey RW. Weed management systems for rangelands. In: Smith AC, ed. Handbook of Weed Management Systems. New York: Marcel Dekker, 1995, pp. 519–552.

Bovey RW. A fifty year history of the weed brush program in Texas and suggested future direction. B-1729. Texas Agric. Exp. Stn., College Station. 1998.

Box TW. Brush, fire and West Texas rangeland. Tall Timbers Fire Ecol Conf 6:7–19, 1967.

Box TW, Powell J. Brush management techniques for improved forage values in South Texas. Trans N Amer Wildlife Conf 30:285–296, 1965.

Box TW, Powell J, Drawe DL. Influence of fire on South Texas chaparral communities. Ecol 48:955–961, 1967.

Boyd WE, Sosebee RE, Herdon EB. Shredding and spraying honey mesquite. J Range Mgt 31:230–233, 1978.

Britton CM, Wright HA. Correlation of weather and fuel variables to mesquite damage by fire. J Range Mgt 24:136–141, 1971.

Brock JH, Robinson ED, Fisher CE, Cross BT, Meadors CH. Reestablishment of grasses following mechanical removal of brush. prog rept. 2810. Texas Agric. Exp. Stn. College Station: 1970, 39–46.

Bunting SC. (1996). Effects of fire on juniper woodland ecosystems in the Great Basin. Symposium on Ecology, Management and Restoration of Intermountain Annual

Rangelands. Boise, ID, May 18–22, 1992, For. Serv. gen. tech. rept. INT-313. USDA, Ogden, UT: Intermountain For. Range. Exp. Stn., 1996, pp. 53–55.

Burbank FM, Johanson EE, Sirois DL. Results of field trials of the tree eater, a tree and brush masticator. Equip. develp. & test rept. 7120-1. San Dimas, CA: USDA, For. Serv., Equip. Develp. Center. 1970.

Burcham LT. Recent trends in range improvement on California foothill ranges. J Range Mgt 8:121–125, 1955.

Bureau of Indian Affairs. Brush Removal by Chain Drag. Project rept. on TEB-602. Range Impr. Notes. vol. 3. Bureau of Indian Affairs. U.S. For. Serv. Intermtn. Reg., 1958, pp. 1–5.

Burkhardt JW, Tisdale EW. Causes of juniper invasion in southwestern Idaho. Ecol 57: 472–484, 1976.

Burrows WH. Studies in the dynamics and control of woody weeds in semi-arid Queensland. Queensland J Agric Anim Sci 30:57–64, 1973.

Burton GW, Hughes RH. Effects of burning and 2,4,5-T on gallberry and saw palmetto. J For 59:497–500, 1961.

Cable DR. Small velvet mesquite seedlings survive burning. J Range Mgt 14:160–161, 1961.

Cain D. The Ely Chain. Ely, NV: Bureau of Land Management, USDI, 1971.

Campbell RE, Baker MB Jr, Fiolliott PF, Larson FR, Avery CC. Wildfire effects on a ponderosa pine ecosystem: An Arizona case study. For. Serv. res. paper RM-191. Fort Collins, CO: USDA, For. Serv., Rocky Mountain For. Range Exp. Stn., 1977.

Carpenter TG. Brush harvesting techniques. In: Noxious Brush and Weed Control Research Highlights 1971. sp. rept. no. 31. Lubbock: Texas Tech University and Texas Dep. Agric., 1971.

Carter MG. Reclaiming Texas brushland range. J Range Mgt 11:1–5, 1958.

Chen MY, Hodgkins EJ, Wilson WJ. Prescribed burning for improved pine production and wildlife habitat in the hilly Coastal Plain of Alabama. bull. 473. Auburn: Alabama Agric. Exp. Stn., 1975.

Christensen DR, Monsen SB, Plummer AP. Response of seeded and native plants six and seven years after eradication of Utah juniper by cabling and hula dozing followed by pipe harrowing as an after treatment on portions of the major treatments. Proceedings West Assoc Game Fish Comm 41:169, 1966.

Churchill FM, Schuster JL. A new brush grubber for the Southwest. J Soil Water Conserv 25:200, 1971.

Clepper H. Crusade for conservation: The centennial history of the American Forestry Association. Amer For 81:17–113, 1975.

Costello DF. Pricklypear control on short-grass range in the Central Great Plains. leaflet no. 210. Washington, DC: USDA, 1941.

Cotner ML. (1963). Controlling pinyon-juniper on southwestern rangelands. rept. 210. Tucson, AZ: Arizona Agric. Exp. Stn. and USDA, ERS, Farm Econ. Div., 1963.

Countryman CM. Can southern California wildland conflagrations be stopped? For. Serv. gen. tech. rept. PSW-7. Berkeley, CA: USDA, For. Serv. Pacific Southwest For. Range. Exp. Stn., 1974.

Countryman CM. Physical characteristics of some northern California brush fuels. gen. tech. rept. PSW-61. Berkeley, CA: USDA For. Serv. Pacific Southwest For. Range Exp. Stn. 1983.

Countryman CM, McCutchan MH, Ryan BC. Fire weather and fire behavior at the 1968 Canyon Fire. For. Serv. res. paper PSW-55.; Berkeley, CA: USDA, For. Serv., Pacific Southwest For. Range. Exp. Stn., 1969.

Crafts AS. Modern Weed Control. Berkeley, CA: University of California Press, 1975.

Dabaan ME, Magedlela AM, Bryan WB, Arbogast BL, Prigge EC, Flores G, Skousen JG. Pasture development during brush clearing with sheep and goats. J Range Mgt 50:217–221, 1997.

Dameron WH, Smith HP. Prickly pear eradication and control. College Station: Texas A&M University, Texas Agric. Exp. Stn. bull. 575, 1939.

Darrow RA, McCully WC. Brush control and range improvement in the post oak–blackjack oak area of Texas. College Station: Texas A&M University, Texas Agric. Exp. Stn. bull. 942, 1959.

Davis GG, Bartel LE, Cook CW. Control of Gambel oak sprouts by goats. J Range Mgt 28:216–218, 1975.

Davis JB, Irwin RL. Focus: A computerized approach to fire management planning. J For 74:615–618, 1976.

Deloach CJ. Progress and problems in introductory biological control of native weeds in the U.S. Proceedings of 8th International Symposium Biological Control of Weeds, Lincoln University, Caterbury, New Zealand, Feb. 2–7, 1992. Delforse ES, Scott RR, eds. Melbourne, Australia: CSIRO, 1995, pp. 111–112.

Deloach JC. Biological control of weeds in the United States and Canada. In: Luken JO, Thieret JW, eds. Assessment and Management of Plant Invasions. New York: Springer-Verlag, 1997, pp. 172–193.

Dieterich JH. Prescribed burning in ponderosa pine–State of the art. Prescribed Fire Workshop, Bend, OR, 1976.

Dodd AP. The biological campaign against prickly-pear. Brisbane, Australia: Commonwealth Pricklyl Pear Board, 1940.

Dodd JD. Mechanical control of pricklypear and other woody species in the Rio Grand Plains. J Range Mgt 21:366–370, 1968.

Dodd JD, Holtz ST. Integration of burning with mechanical manipulation in South Texas grassland. J Range Mgt 25:130–135, 1972.

Drawe DL. A study of five methods of mechanical brush control in South Texas. Rangeman's J 4:37–39, 1977.

Dresskell WW. Report Range Seeding Equipment Committee. Washington, DC: U.S. For. Serv., 1955.

Elwell H, McMurphy WE, Santelmann PW. Burning and 2,4,5-T on post and blackjack oak rangeland in Oklahoma. bull. B-675. Oklahoma Agric. Exp. Stn. and USDA, 1970.

Engle DE, Stritzke JF, Claypool PL. Effect of paraquat plus prescribed burning on eastern redcedar (*Juniperus virginiana*). Weed Tech 2:172–174, 1988.

Engle DM, Stritzke JF. Enhancing control of eastern redcedar through individual plant ignition following prescribed burning. J Range Mgt. 45:493–495, 1992.

Engle DM, Stritzke JF. Fire behavior and fire effects on eastern redcedar in hardwood leaf-litter fires. Internat J Wildland Fire 5:135–141, 1995.

Fahnestock GR. Use of fire in managing forest vegetation. Trans ASAE 16:440–413, 419, 1973.

Fahnestock GR. Fires, fuels and flora as factors in wilderness management: The Pasayten case. Tall Timbers Fire Ecol Conf 15:33–69, 1976.

Faulkner JL, Clebsach EEC, Sanders WL. Use of prescribed burning for managing natural and historic resources in Chickamauga and Chattanooga National Military Park, USA. Environ Mgt 13:603–612, 1989.

Fenner RL, Arold RK, Buck CC. Area ignition for brush burning. tech. paper no. 10. Berkeley: USDA, For. Serv. California For. Range Exp. Stn., 1955.

Ferguson ER. Stem-kill and sprouting following prescribed fire in a pine–hardwood stand in Texas. J For 55:426–429, 1957.

Fiollott PF, Clary WP, Larson FR. Effects of a prescribed fire in an Arizona ponderosa pine forest. Res. note RM-336. Fort Collins, CO: USDA, For. Serv., Rocky Mountain For. Range Exp. Stn., 1977.

Fischer WC. Fire management in the 1980's. AMES Forester 23–28, 1980.

Fitzgerald RD, Bailey AW. Control of aspen regrowth by grazing with cattle. J Range Mgt. 37:156–158, 1984.

Forest Service. Protecting the Forests from Fire. Agric. Info. bull. no. 130, For. Serv., USDA, 1954.

Franklin JF, Cromack K Jr, Denison W, McKee FC, Maser C, Sedell J, Swanson F, Juday H. Ecological characteristics of old-growth forest ecosystems in the Douglas-fir region. For. Serv. gen. tech. rept. PNW. Portland, OR: USDA, Pac. Northwest For. Range Exp. Stn., 1981.

Frischknecht NC. Biological methods: A tool for sagebrush management. Sagebrush Ecosystem Symp. 300, Utah State University, Logan, 1979, pp. 121–128.

Funasaki GY, Lai PY, Nakahara LM, Beardsley JW, Ota AK. A review of biological control introduction in Hawaii: 1890 to 1985. Proceedings Hawaiian Entomol Soc 28:105–160, 1988.

Goeden RD. Part II. Biological control of weeds. In: Clausen CP, ed. Introduced Parasites and Predators of Arthropod Pests and Weeds. USDA-ARS handbook no. 480. Washington, DC: U.S. Government Printing Office, 1978, pp. 357–414.

Gordon RA, Scifres CJ. Burning for improvements of Macartney rose infested Coastal Prairie. College Station: Texas A&M University, Texas Agric. Exp. Stn., B-1183. 1977.

Gordon RA, Scifres CJ, Mutz JL. Integration of burning and picloram pellets for Macartney rose control. J Range Mgt 35:427–431, 1982.

Gratkowski H. Site preparation and conifer release in Pacific Northwest forests. Proceedings 27th Annual Weed Conference, Yakima, WA, 1977, pp. 29–32.

Green LR, Newell LA. Using goats to control brush regrowth on fuelbreaks. gen. tech. rept. PSW-59. Berkeley, CA: USDA, For. Serv., Pacific Southwest For. Range. Exp. Stn., 1982.

Greer HA. Mesquite control in Oklahoma. OSU Ext. Facts, no. 2760. Stillwater, Ok: Oklahoma State University, 1967.

Gruell GE, Loope LL. Relationship among aspen, fire and ungulate browsing in Jackson

Hole, Wyoming. Jackson and Moore, WY: USDA, For. Serv., Intermount. Region and USDI, Nat. Park. Serv., Rocky Mountain Region, 1974.

Hamid N. Acceptability of big sagebrush to sheep and goats: Role of monoterpenes. PhD dissertation, Utah State University, Logan, 1981.

Hamilton WC. Past history of brush control in Texas: A personal perspective of efforts in South Texas. Workshop on brush management in Texas. Texas Agric. Exp. Stn., Dallas, 1993.

Harley KLS. Biological control of lantana in Australia. Proceedings of the 3rd International Symposium on the Biological Control of Weeds. Montpelier, Canada, 1973, pp. 23–29.

Harper RW. Historical notes on the relation of fire to forests. Proceedings Tall Timber Fire Ecol Conf 1:11–19, 1962.

Haywood JD. Preparing plant sites for loblolly pine with hexazinone, picloram or by chopping and burning. New Orleans: USDA, For. Serv. Southern Forest Exp. Stn., 1993.

Hedrick DW. Brushland management in California. J Range Mgt 4:181–183, 1951.

Herbel C, Ares F, Bridges J. Hand-grubbing mesquite in the semidesert grassland. J Range Mgt 11:267–270, 1958.

Herndon EB. Kershaw brush cutter. In: Noxious Brush and Weed Control Research Highlights 1971. Sp. rept. no. 31. Lubbock: Texas Tech University and Texas Dep. Agric., 1971.

Hoffman GD, Dodd JD. How to whip pricklypear: Texas Agric. Progress. TAP 475 13: 16–18, 1967.

Holloway JK. Projects in biological control of weeds. In: DeBach P, ed., Biological Control of Insects, Pests and Weeds. New York: Reinhold, 1964, pp. 650–570.

Hosley NW, Stoddart LA. The role of brush control in habitat improvement on the Aransas National Wildlife Refuge. Trans N Amer Wildlife Conf 12:179–185, 1947.

Hull AC Jr. Report of Range Seeding Committee. Washington, DC: U.S. Forest Service, 1953.

Humphrey RR. Fire as a means of controlling velvet mesquite, burroweed, and cholla on southern Arizona ranges. J Range Mgt 2:175–182, 1949.

Humphey RR, Everson AC. Effect of fire of mixed grass-shrub range in southern Arizona. J Range Mgt 4:264–266, 1951.

Jameson DA. Effects of burning on a galleta-blackgrama range invaded by juniper. Ecol 43:760–763, 1962.

Jensen FR. Development of the Dixie-Sager Chain. USDA For Serv Intermtn Reg Range Impr Notes 14:1–10, 1969.

Johnson RW, Back PV. Relationship between burning and spraying in the control of brigalow (*Acacia harpophylla*) regrowth. Queensland J Agric Anim Sci 34:179–188, 1977.

Jordan GL, Maynard ML. The San Simon Watershed: Shrub control. Prog Agric AZ 22: 6–9, 1970.

Julien MH. Biological Control of Weeds: A World Catalogue of Agents and Their Target Weeds. 3rd ed. Wallington, Oxon, UK: CAB International, 1992.

Kallander H. Controlled burning on the Fort Apache Indian Reservation, Arizona. Tall Timbers For Ecol Conf 9:241–249, 1969.

Kalmbacher RS, Boote KJ, Martin FG. Burning and 2,4,5-T application on mortality and carbohydrate reserves in saw-palmetto. J Range Mgt 36:9–12, 1983.

Kilgore BM. From fire control to fire management: An ecological basis for polices. Trans N Amer Wildlife Natl Res Conf 41:407–420, 1976.

Kittams WH. Effect on fire on vegetation of the Chihuahuan desert region. Tall Timbers Fire Ecol Conf 12:427–444, 1972.

Koch P, McKenzie DW. Machine to harvest slash, brush, and thinnings for fuel and fiber a concept. J For 74:809–812, 1976.

Komarek EV Sr. Comments on the history of controlled burning in the southern United States. Proceedings 17th Ann Watershed Symp 17:11–17, 1973.

Laisz DR, Wilson CC. To burn or not to burn: Fire and chaparral management in southern California. J For 78:94–95, 1980.

Lancaster DL, Young JA, Evans RA. Weed and brush control tactics in the sagebrush ecosystem. In: Onsager JA, ed. Integrated Pest Management of Rangeland: State of the Art in the Sagebrush Ecosystem. USDA, ARS-50, 1987.

Lanner RM. The eradication of pinyon-juniper woodland: Has the program a legitimate purpose? West Wildlands 3:13–17, 1977.

Larson JE. Revegetation equipment catalog. Vegetation Rehabilitation and Equipment Workshop, USDA, For. Serv. Equip. Develop. Center, Missoula MT, 1980.

Leopold AJ, Cain SA, Cottam CM, Gabrielson IN, Kimball TL. Wildlife management in the national parks. Amer For 69:32–35, 61–63, 1963a.

Leopold AS, Cain SA, Cottam CM, Gabrielson IN, Kimball TL. Study of wildlife problems in national parks: Wildlife management in the national parks. Trans N Amer Wildlife Natl Res Conf 28:28–45, 1963b.

Lewis CE. Chopping and webbing control saw-palmetto in South Florida. For. Serv. res. note SE-77. Asheville, NC: USDA, For. Serv. Southeastern For. Exp. Stn., 1972.

Lindenmuth AW Jr, Davis JR. Predicting fire spread in Arizona's oak chaparral. res. paper RM-101. Fort Collins, CO: USDA, For. Serv. Rocky Mountain For. Range Exp. Stn., 1973.

Liu C, Harcombe PA, Knox RG. Effects of prescribed fire on the compostion of woody plant communities in southeastern Texas. J Veg Sci 8:495–504, 1997.

Lorimer GG. The use of land survey records in estimating presettlement fire frequency. In: Stokes MA, Dieterich JH, tech. coords. Proceedings Fire History Workshop. gen tech. rept. RM-81. Tucson: USDA, Rocky Mt. For. Range Exp. Stn., 1980, pp. 57–62.

Lotan JE. Integrating fire management into land-use planning: A multiple-use management research, development, and applications program. Environ Mgt. 3:7–14, 1979.

Lyma RA, Sykes WE. The Johnston chain drag for clearing brush from rangeland. J Range Mgt 7:31–32, 1954.

Lyon LJ. Initial vegetal development following prescribed burning of Douglas-Fir in South-Central Idaho. res. paper INT-29. Ogden, UT: USDA, For. Serv., Intermountain For. Range Exp. Stn., 1966.

Madany MH, West NE. Fire history of two Montana forest areas of Zion National Park. In: Stokes CA, Dieterich JH, tech. coords. Proceedings Fire History Workshop. Gen. tech. rept. RM-81. Tucson: USDA, Rocky Mt. For. Range Exp. Stn., 1980, pp. 50–56.

Magee AC. Costs pay for clearing grand prairie rangeland. misc. pub. 206. College Station, TX: Texas Agric. Exp. Stn., 1957.

Manley K, Walker CF. Brush control and reseeding for range improvement in central California. J Range Mgt 9:278–280, 1956.

Mann J. Cactus-feeding insects and mites. United States National Museum bulletin 256. Washington, DC: Smithsonian Institution Press, 1969.

Markin GP, Pemberton RW. Banana poka. In: Nechols JR, Andres LA, Beardsley JW, Goeden RD, Jackson CG, eds. Biological Control in the Western United States: Accomplishments and Benefits of Regional Research Project W-84. 1964–1989. publ. 3361. University of California, Division of Agriculture and Natural Resources, 1995, pp. 309–311.

Markin GP, Yoshioka ER. Present status of biological control of the weed gorse (*Ulex europaeus*) L. in Hawaii. In: Delfosse ES, ed. Proceedings of the 7th International Symposium Biological Control of Weeds, Rome. Melbourne, Australia: CSIRO, 1990, pp. 357–362.

Martin RE, Robinson DD, Schaeffer WH. Fire in the Pacific Northwest–Perspectives and problems. Proc Tall Timbers Fire Ecol Conf 15:1–23, 1976.

Martin SC, Techirley FH. Changes in cactus numbers after cabling. Prog Agric AZ 21: 16–17, 1969.

Martin SC, Thames JL, Fish EB. Changes in cactus numbers and herbage production after chaining and mesquite control. Prog Agric AZ 26:3–6, 1974.

Mayeux HS Jr, Hamilton WT. Response of common goldenweed (*Iscoma coronopifolia*) and buffelgrass (*Cenchrus ciliaris*) to fire and soil-applied herbicides. Weed Sci 31:355–360, 1983.

McCord EW, McMurphy WE. Grassland management after fire. OCU Ext. Facts. no. 2853. Stillwater, IK: Oklahoma State University, 1967.

McKenzie DW, Miller M. Field equipment for precommercial thinning and slash treatment. San Dimas, CA: USDA, For. Serv. Equip. Develop Center, 1978.

McKenzie DW, Jensen FR, Young JA. Chains for mechanical brush control. Rangelands 6:122, 1984.

Merrill LB, Taylor CA, Jr. Take note of the versatile goat. Rangeman's J 3:74–76, 1976.

Moran VC, Zimmerman HG. Biological control of cactus weeds of minor importance in South Africa. Agric Eco-syst Environ. 37:37–55, 1991.

Murphy AH. Significance of rangeland weeds for livestock management strategies. Proc Calif Weed Conf Sacramento 38:114–116, 1986.

Neuenschwander LF, Wright HA, Bunting SC. The effect of fire on a tobosa grass–mesquite community in the rolling plains of Texas. Southwest Natural 23:315–338, 1978.

O'Rourke JT, Ogden PR. Pinyon-juniper control: Where? Why? Prog Agric AZ 22:12–15, 1979.

Owens KM. Utilization patterns by Angora goats within the plant canopies of two *Acacia* shrubs. J Range Mgt. 44:456–461, 1991.

Parker KG. Use of mechanical methods as a management tool. Sagebrush Ecosystem Symposium, Utah State University, Logan, 1979, pp. 117–120.

Pase CP, Granfelt CE. The use of fire on Arizona rangelands. Pub. no. 4. Fort Collins, CO: USDA, For. Serv. Rocky Mt. For. Range. Exp. Stn., Arizona Interagency Range Committee, 1977.

Pase CP, Ingebo PA. Burned chaparral to grass: Early effects on water and sediment yields from two granitic soil watersheds in Arizona. Proceedings of the Arizona Watershed Symposium, Tempe, 1965, pp. 8–11.

Pase CP, Lindenmuth AW Jr. Effects of prescribed fire on vegetation and sediment in oak–mountain mahogany chaparral. J For 69:800–805, 1971.

Pase CP, Pond FW. Vegetation change following the Mingus Mountain Burn. res vegetation note RM-18. Tempe, AZ: USDA, For. Serv, 1964.

Pechanec JF. Report of the Range Reseeding Committee. Portland, OR: Pacific Northwest Forest and Range Exp. Stn., 1950.

Pechanec JF. Machinery for clearing brush lands. Internat Grassland Cong Proceedings 6:1279–1284, 1952.

Pechanec JF, Stewart G. Sagebrush burning—good and bad. Farmer bull. no. 1948. USDA, 1944.

Pechanec JF, Plummer AP, Robertson JH, Hull AC. Sagebrush Control on Rangelands. Agric. Handbook no. 227. Washington, DC: USDA, 1965.

Perkins RCL, Swezey OH. The introduction into Hawaii of insects that attack lantana. Ent. ser. bull. 16. Hawaiian Sugar Planter's Assoc. Exp. Stn., 1924.

Phillips FD. Burning improves Oklahoma rangelands. Soil and Water Conserv News 8: 5, 1987.

Pond FW, Cable DR. Effect of heat treatment on sprout production of some shrubs of the chaparral in central Arizona. J Range Mgt. 13:313–317, 1960.

Pond FW, Lillie DT, Holbo HR. Shrub Live Oak Control by Root Plowing. res. note RM-38. Tucson, AZ: USDA, For. Serv. Rocky Mountain For. Range Exp. Stn., 1965.

Porterfield JG, Roth LD. Some Machines and Methods for Removal and Control of Brush. bull no. B-496. Stillwater, OK: Oklahoma Agric. Exp. Stn., Oklahoma State University, 1957.

Potter RL, Ueckert DN, Peterson JL. Internal temperatures of pricklypear cladophylls during prescribed fire in West Texas. PR 4132. Texas Agric. Exp. Stn., 1983.

Ralphs M, Schen D, Busby F. Prescribed burning–effective control of sagebrush and open juniper. Utah Sci 36:94–98, 1975.

Rechanthin CA, Bell HM, Pederson RJ, Polk DB. Grassland restoration Part II. Brush control. Temple, TX: USDA-Soil Conservation Service, 1964.

Reed IF. Disk plows. USDA Farmers bull. no. 2121. Washington, DC: 1958.

Rees NE, Quinby PC, Piper GL, Turner CE, Coombs EM, Spencer NR, Knutson LV. Biological Control of Weeds in the West. West. Soc. Weed Sci., Bozeman: Montana Dep. Agric., USDA-ARS and Montana State University, 1996.

Reynolds HG, Bohning JW. Effects of burning on a desert grass-shrub range in southern Arizona. Ecol 37:769–777, 1956.

Richard JA. The ranch industry of the Texas South Plains. MA thesis, Austin: University of Texas, 1927.

Riddle RR, Taylor CA Jr, Huston JE, Kothmann MM. Intake of Ashe juniper and live oak by Angora goats. J Range Mgt 52:161–165, 1999.

Riebold RJ. The early burning of wildfires and prescribed burning. Prescribed Burning Symposium Proceedings. Asheville, NC: USDA, For. Serv. Southeastern For. Exp. Stn., 1971, pp. 11–20.

Riggs RA, Urness PJ. Effect of goat browsing on Gambel oak communities in northern Utah. J Range Mgt 42:354–360, 1989.

Riggs RA, Urness PJ, Gonzalez KA. Effects of domestic goats on deer wintering in Utah oakbrush. J Range Mgt 43:239–234, 1990.
Robertson JH, Cords HP. Survival of rabbitbrush, *Chrysothammus* spp., following chemical, burning and mechanical treatments. J Range Mgt 10:83–85, 1957.
Robertson JH, Plummer AP. Hints for use of the wheatland-type plows for brush eradication in connection with range reseeding. res. paper no. 13. Ogden, UT: USDA, For. Serv. Intermountain For. Range Exp. Stn., 1946.
Roby GA, Green LR. Mechanical methods of chaparral modification. Agric. handbook no. 487. USDA, For. Serv., 1976.
Romancier RM. Combining fire and chemicals for control of rhododendron thickets. USDA For. Serv. res. note BE-149. Asheville, NC: USDA, For. Serv., Southeastern For. Exp. Stn., 1971.
Romme WH. Fire frequency in subalpine forest of Yellowstone National Park. In: Proc. Stokes MA, Dieterich JH, tech. coords. Proceedings Fire History Workshop. Gen. tech. rept. RM-81. Tucson, AZ: USDA, Rocky Mt. For. Range Exp. Stn., 1980, pp. 27–38.
Rotary Plow Co. Rotary plow. Rotary Plow Co., Alberta, Canada, 1971.
Sampson AW. Change in burning technology. Letters to the editor. J Range Mgt 10:104, 1957.
Schmidt RL. The silvics and plant geography of the genus *Abies* in the coast forests of British Columbia. B.C. For. Serv. tech. pub. T. 46, 1957.
Schmutz EM, Cable DR, Warwick JJ. Effects of shrub removal on vegetation of semidesert grass-shrub range. J Range Mgt. 12:34–37, 1959.
Schroeder D. Biological control of weeds. In: Fletcher WW, ed. Recent Advances in Weed Research. Slough, England: Common Wealth Agric. Bureaux, 1983, pp. 41–78.
Scifres CJ. Systems for improving Maccartney rose infested coastal prairie rangeland. MP-1225. Texas Agric. Exp. Stn., 1975.
Scifres CJ. Mechanical methods of brush management. In: Brush Management Principles and Practices for Texas and the Southwest. College Station, TX: Texas A&M University Press, 1980, pp. 124–135.
Scifres CJ. Selective grazing as a weed control method. In: Pimentel D, ed. CRC Handbook of Pest Management in Agriculture, vol. II. CRC Series in Agric. Boca Raton, FL: CRC Press, 1981, pp. 369–376.
Scifres CJ, Hamilton WT. Prescribed burning for Brushland Management: The South Texas Example. College Station, TX: Texas A&M University Press, 1993.
Scifres CJ, Kelley DM. Range vegetation response to burning thicketized live oak savannah. B-1246. College Station: Texas Agric. Exp. Stn., 1979.
Scifres CJ, Mutz JL, Rasmussen GA, Smith RP. Integrated brush management systems (IBMS): Concepts and potential technologies for running mesquite and whitebrush. B-1450. College Station: Texas Agric. Exp. Stn., 1983.
Scifres CJ, Stuth JW, Koerth BH. Improvement of oak-dominated rangeland with tebuthiuron and prescribed burning. B-1567. College Station: Texas Agric. Exp. Stn., 1987.
Severson KE, Debano LF. Influence of Spanish goats on vegetation and soils in Arizona chaparral. J Range Mgt. 44:111–117, 1991.
Shiff AL. Fire and Water: Scientific Heresy in the Forest Service. Cambridge, MA: Harvard University Press, 1962.

Silker TH. Prescribed burning for the control of undesirable hardwoods in pine-hardwood stands and slash pine plantations. bull no. 46. Texas For. Serv., 1955.

Sleuter AA, Wright HA. Spring burning to manage redberry juniper rangelands—Texas Rolling Plains. Rangelands 5:249–251, 1983.

Smith HN, Rechenthin CA. Grassland restoration: The Texas brush problem. Temple, TX: USDA Soil Con. Serv., 1964.

Smith JE, Jr. New rangeland in Texas with brush control and seeding. Crops Soil 9:7–8, 1956.

Smith MA, Dodd JL, Rodgers JD. Prescribed burning on Wyoming rangeland. bull. 810. Laramie, WY: Agric. Ext. Serv., University of Wyoming, 1985.

Sosebee RE. Herbicide plus various additives for follow-up control of shredded mesquite. J Range Mgt 27:53–55, 1974.

Stewart OC. Burning and natural vegetation in the United States. Geo Rev 41:317–320, 1951.

Stoin HR. Controlling brush with goats. AR Farm Res 19:12, 1970.

Streets RB, Stanley EB. Control of mesquite and noxious shrubs on southern Arizona grassland ranges. tech. bull. 74. Arizona Agric. Exp. Stn., 1938.

Stritzke JF, Engle DM, McCollum FT. Vegetation management in the Cross-Timbers: Response of woody species to herbicides and burning. Weed Tech 5:400–405, 1991.

Strokes MA. The dendrochronology of fire history. In Stokes MA, Dieterick JH, tech. coords. Proceedings Fire History Workshop. USDA gen. tech. rept. RM-81. Tucson, AZ: Rocky Mt. For. Range Exp. Stn., 1980, pp. 1–3.

Stuth JW, Dahl BE. Evaluation of rangeland seedings following mechanical brush control in Texas. J Range Mgt 27:146–149, 1974.

Taylor CA Jr. Multispecies grazing-vegetation manipulation. PR-4426. College Station: Texas Agric. Exp. Stn., 1986, pp. 17–18.

Taylor DL. Fire history and man-induced fire problems in subtropical south Florida. In: Stokes MA, Dieterich JH, tech. coords. Proceedings Fire History Workshop. USDA gen. tech. rept. RM-81. Tucson, AZ: Rocky Mt. For. Range Exp. Stn., 1980, pp. 63–68.

Terry SW, White LD. Southern wax-myrtle response following winter prescribed burning in south Florida. J Range Mgt 32:326–327, 1979.

Tester JR. Effect of fire frequency on plant species in oak savanna in east-central Minnesota. Bull Torrey Bot Club 123:304–308, 1996.

Texas Agric. Exp. Stn. Brush Clearance Machines Compared, vol. 6. Brief repts. on agric. research in Texas, College Station: Texas Agric. Exp. Stn. 1980, p. 2.

Texas State Soil and Water Construction Board. A Comprehensive Study of Texas Watersheds and Their Impacts on Water Quality and Quantity. Temple, TX: Texas State Soil & Water Conservation Board, 1991.

Tiedemann AR, Schmutz EM. Shrub control and reseeding effects on the oak chaparral of Arizona. J Range Mgt 19:191–195, 1966.

Trew EM. Mowing pastures. cir. 216. College Station: Texas Agric. Ext. Serv., 1952.

Ueckert DN, Petersen JL, Potter RL, Whipple JD, Wagner MW. Managing pricklypear with herbicides and fire. Sheep & Goat, Wool & Mohair 1988. prog. rept. 4570. College Station: Texas Agric. Exp. Stn., 1988, pp. 10–20.

Uhl C, Jordan C, Clark K, Clark H, Herrera R. Ecosystem recovery in Amazon caatinga forest after cutting, cutting and burning, bulldozer clearing treatments. Oikos 38: 313–320, 1982.

Ulich WL. Economics and technology of harvesting whole-tree mesquite. Research Highlights 1982. Noxious Brush and Weed Control; Range and Wildlife Management, vol. 13. Lubbock, TX: Texas Tech University, 1982, pp. 37–38.

Urness PJ. Deer use changes after root plowing in Arizona chaparral. USDA For. Serv. res. note RM–25. Tucson, AZ: USDA, For. Serv. Rocky Mountain For. Range Exp. Stn., 1974.

USDA. Equipment for Clearing Brush from Land. Farmers Bull. no. 2180. Washington, DC: USDA, 1961.

USDA. History-range seeding equipment committee: 1946–1973. Washington, DC: USDA-USDI, 1974.

USDA. History of the vegetative rehabilitation and equipment workshop (VREA): 1946–1981. Missoula, MT: USDA For. Serv. Equip. Develop Center, 1982.

USDA-ARS. It may pay to grub mesquite, Agric. Res. Beltsville, MD: USDA-ARS, 1958, pp. 12–13.

USDA-ARS. Machine that restores rangeland. Agric. Res. Beltsville, MD: USDA-ARS, 1972, pp. 4–5.

USDA-ARS. Cactus Harvester. Agric. Res. Beltsville, MD: USDA-ARS, 1982, p. 5.

Vallentine JF. Range Development and Improvements. 3rd ed. Provo, UT: Brigham Young University Press, 1989, pp. 168–214.

Villena F, Pfister JA. Sand shinnery oak as forage for Angora and Spanish goats. J Range Mgt. 43:116–122, 1990.

Walstad JD. Weed control for better southern pine management. Weyerhauser For. Paper No. 15, Hot Springs, AR: South For. Res. Center, Weyerhauser Co., 1976.

Weaver H. A preliminary report on prescribed burning in virgin ponderosa pine. J For 50:662–668, 1952.

Weaver H. Fires as an enemy, friend, tool in forest management. J For 53:499–504, 1955.

Whisenant SG. Changing fire frequencies on Idaho's Snake River plains: Ecological and management implications. Proceedings Symposium on Cheatgrass Invasion, Shrub Dieoff and Other Aspects of Shrub Biology and Management. Las Vegas, April 5–7, 1989, USDA, For Serv. gen. tech. rept. INT-276. Ogden, UT: Intermount. Res. Stn., 1990, pp. 1–10.

Wiedemann HT, Cross BT. Low-energy grubbing for control of juniper. J Range Mgt. 34:235–237, 1981.

Wiedemann HT, Cross BT. Influence of pulling configuration on draft of disk-chains. Trans ASAE 28:79–82, 1984.

Wiedemann HT, Cross BT, Fisher CE. Low energy grubber for controlling brush. Trans ASAE 20:210–213, 1977.

Wilson BJ. A Report on Weed Research. Annual Report 1987–88. Queensland, Australia: Biol. Branch, Queensland Dept. Lands, Alan Fletcher Res. Stn. & Trop Weed Res. Center, 1988, p. 19.

Wink RL, Wright HA. Effects of fire on an Ashe juniper community. J Range Mgt 26: 326–329, 1973.

Wright HA. Range burning. J Range Mgt 27:5–11, 1974.

Wright HA. The role and use of fire in the semidesert grass-shrub type. USDA gen. tech rept. INT-85. Ogden, UT: USDA Intermount. For. Range Exp. Stn., 1980.

Wright HA, Bailey AW. Fire ecology and prescribed burning in the the Great Plains—A research review. USDA, gen. tech. rept. INT-70. Ogden, UT: USDA, Intermount. For. Range Exp. Stn., 1980.

Wright HA, Bailey AW. Fire Ecology, United States and Southern Canada. New York: Wiley, 1982.

Wright HA, Bunting SC, Neuenschwander LF. Effect of fire on honey mesquite. J Range Mgt 29:467–471, 1976.

Wright HA, Neuenschwander LF, Britton CW. The role and use of fire in sagebrush-grass and pinyon-juniper plant communities: A state-of-the-art review. USDA, For. Serv. gen. tech. rept. INT-58. Ogden, UT: USDA, Intermount. For. Range Exp. Stn., 1979.

Young JA, Evans RA, McKenzie DW. History of brush control on western U.S. rangelands. In: McDaniel KC, ed. Proceedings—Brush Management Symposium Albuquerque, NM: Soc. Range Management, 1983, pp. 17–25.

Zwolinski MJ, Ehrenreich JH. Prescribed burning on Arizona's watersheds. Tall Timbers Ecol Fire Conf 7:195–205, 1967.

3

Herbicide Development

I. INTRODUCTION

The fact that small amounts of a chemical (herbicide) can be applied to unwanted vegetation and make ecological swifts favorable to man's purposes is remarkable and a modern-day wonder, especially for woody plants. For centuries man has attempted to control weeds with his hands and crude devices to favor food production, many times without success. In his book *Are Pesticides Really Necessary*?, Barrons (1981) quotes an old Byzantine proverb that states "He who has bread may have many troubles; he who lacks it has only one."

In spite of our great success in modern agriculture and farm chemicals we have become our own worst critics in condemning sometimes innocent chemicals for want of allegedly 100% no-risk products. When used as directed by the herbicide label a public risk does not exist and allegations by the uninformed may trigger mistrust and eventual loss of an otherwise worthy product.

When the public health, water supplies, food, or the environment are truly at risk then corrective measures are necessary. Sometimes herbicides are used to preserve the environment by removing allergenic plants or producing favorable conditions for endangered plant species and other desirable plants. Their extensive use in cropland attests to this.

The critics also state that based on canopy cover of brush in Texas and other areas that little progress has been made using chemical and nonchemical control methods since World War II. During the 1970s, however, attitudes changed from brush eradication to retention of acceptable brush levels for wildlife habitat, temporary knockdown of brush for wildlife and livestock browse, and selective control of some woody plants in mixtures without significantly reducing

TABLE 1 History and Development of Herbicides Used in Woody Plant Management

Herbicide	Discovered or reported (year)	Country, researcher, or company	First U.S. use	Basic producer (1998)
Amtirole	1953	Union Carbide	1954	CFPI Agro
Bromacil	1961	DuPont	1963	DuPont
Clopyralid	1961	—	1978	Chimac-Agriphar, S.A.
2,4-D	1942	Zimmerman and Hitchcock	1944	Several
Dicamba	1958	S. B. Richter, Velsicol Chem. Corp.		Novartis
Dichlorprop	1944	Zimmerman and Hitchcock	1961	BASF AG
Diesel fuel			1930s	Several
Diquat	1955	Dyestuffs Div. ICI, G.B.	1955	Zeneca
Fosamine	1974	DuPont	—	DuPont
Glyphosate	1971	Monsanto Co.	1974	Several
Hexazinone	1975	DuPont	—	DuPont
Imazapyr	Late 1970s	American Cyanamid	1981	American Cyanamid
Metsulfuron	1983	DuPont	1986	DuPont
Paraquat	1958	Dyestuffs div. ICI, G.B.	1959	Several
Picloram	1960	Dow AgroSciences	1963	Dow AgroSciences
Tebuthiuron	1970	Air Supply, Inc.	1974	Dow AgroSciences
Triclopyr	1975	Dow AgroSciences	—	Chimac-Agriphar, S.A.

Sources: Bovey, 1996, 1980; Brehrens, 1953; Burnside et al., 1996; Carter and Amitrole, 1975; Geissbühler et al., 1975; Hamilton, 1993; Hammer, 1944; Klingman, 1961, 1975; Kraus, 1947; Martin and Worthing, 1974; Pokorny, 1941; Weed Science Society of America, 1994, 1970; Young and Fisher, 1949; Zimmerman, 1943; Zimmerman and Hitchcock, 1942, 1944.

overall canopy. Land valves are also sometimes enhanced by the presence of certain woody plants (Hamilton, 1993).

Some excellent chemicals have been developed for woody plant control and are listed in Table 1. Limitations on their use are usually economical, and one must fit the proper herbicide to the target woody plant or plants while considering the herbaceous weed population present. Agricultural chemicals (herbicides) undergo extensive toxicological, environmental, and use benefit tests before ever being released for widespread use.

This chapter will explore the history of herbicides used in woody plant control, the requirements for widespread use, the herbicide developmental process and safety requirements, and the meaning of chemical residues in the environment.

II. EARLY HISTORY

The use of herbicides for woody plant control has sometimes been discovered after use for herbaceous annual and perennial weed control. The history of 2,4,5-T is an exception, in that Dow AgroSciences researchers found it effective against certain brambles (*Rubus* spp.) and other weeds resistant to 2,4-D as early as 1947 (Barrons and Coulter, 1947). Similarly, it was reported in 1949 that the ester of 2,4,5-T applied to the foliage of honey mesquite was more effective than several formulations of 2,4-D or other chemicals (Young and Fisher, 1949).

As early as 1948, Fisher and Young (1948) reported that sodium arsenite, sodium arsenate, ammonium sulfumate, sulfamic acid, ammonium thiocyanate, 2,4-D, and 2,4,5-T were the only chemicals out of several hundred tested that were absorbed by mesquite foliage and translocated in sufficient amounts to kill all dormant buds on the underground stem. Fisher and Young (1948) indicated that kerosene and diesel fuel had been used extensively for mesquite control by application around the base of the plant to completely wet the underground stem. They also mentioned that oils (including diesel fuel) had been studied as diluents alone and with water with the butyl and isopropyl esters of 2,4-D and 2,4,5-T for mesquite control.

At the same time, control methods were being developed by the Texas Agriculture Experiment Station for cedar, pricklypear cactus, oaks, and mixed brush in east-central and south Texas (Darrow and McCully, 1959).

In 1961, Klingman (1961) stated that the use of herbicides to control brush and undesirable trees in the United States had expanded rapidly since about 1950. He further stated that the use of chemicals had generally proven to be more effective and less costly than most other methods. Although chemicals are still highly effective today and are used in pastures, rangeland, industrial sites, rights-of-way, and forestry, they are no longer inexpensive. The cost of 2,4-D is still reasonable (2,4-D ester at 0.56 kg/ha is about $1.75). The cost of more effective herbicides for brush management such as triclopyr and dicamba is five times as much as 2,4-D for the same application rate, however (Bovey, 1996). Is the cost associated with inflation, high production costs, or requirements associated with compliance to government laws and regulations or all of the above? The reasons for the high cost maybe difficult to sort out, but herbicide use is greatly hampered in the marketplace and use may be limited to the more productive land sites or special use situations.

If the data in Table 1 are examined, it is clear that there have been no major herbicides developed for woody plant management since the 1970s. The list contains essential chemicals for woody plant control and it is important that they be reregistered by EPA and retained for use. The prospects of new herbicides for woody plant control appears remote and will probably result from spinoffs from other herbicide uses. The search for more effective, low-cost materials should continue, however.

III. INDIVIDUAL HERBICIDES

A. 2,4-D

The herbicide 2,4-D will be discussed first since it is the oldest one listed in Table 1. The potent effects on plant growth of these derivatives were first described by Zimmerman and Hitchcock (1942) in 1942. This herbicide was first synthesized by Pokorny (1941) and is produced by the interaction of 2,4-dichlorophenol and sodium monochloracetate to produce the sodium salt of 2,4-D (Martin and Worthing, 1974). Common formulations of 2,4-D are diethylamine salts or isooctyl ester, although other formulations are available. This herbicide has been the subject of more physiological and field research than any other herbicide and is extremely important worldwide. It was used commercially in 1944 (Bovey and Young, 1980; Burnside at al., 1996). It is still used today and is extremely important for weed and brush control when applied alone and in mixtures with dicaniba, picloram, or triclopyr.

When 2,4,5-T was available, 2,4-D was a common component of it. The commercial preparations were commonly in 1:1 2,4-D:2,4,5-T mixtures. (See Weed Science Soc. Am., 1994; Bovey and Young, 1980; Burnside et al., 1996 for further historical details.)

B. Dichlorprop

Dichlorprop is the proprionic acid form of 2,4-D and is also called 2,4-DP. Although the growth-regulating properties of dichlorprop were described by Zimmerman and Hitchcock (1944), it was not commercially introduced until 1961 by the Boots Co., Ltd. (Martin and Worthing, 1974).

Dichlorprop is synthesized by the condensation of 2-choropropionic acid with 2,4-dichlorophenol or by the chlorination of 2-phenoxypropionic acid. Having an asymmetric carbon atom, it exists as two optically active forms, of which only the (+) form is biologically active. The commercial product contains equal amounts of both forms (Martin and Worthing, 1974). Dichlorprop is formulated as a dimethylamine salt, butoxyethyl ester, or isooctyl ester. As a brush killer it is commonly mixed with 2,4-D, or mecoprop (MCPP) or both. Dichlorprop can be applied as wetting sprays for brush in fence rows, highway or railroad rights-of-way, and utility rights-of-way. It controls some oaks, sand sagebrush, and many hardwood and coniferous brush species (Weed Science Soc. Am., 1994).

C. 2,4,5-T and Silvex

Silvex and 2,4,5-T are no longer manufactured or used in the United States. Both materials provided inexpensive brush control of many woody species. The herbicide 2,4,5-T was also synthesized by Pokorny (1941). Zimmerman (1943) indicated that 2,4,5-T was active for cell elongation but did not cause formative effects in leaves. In 1944, however, Hammer and Tukey (1944) reported that 2,4-D and 2,4,5-T applied with carbowax were effective herbicides on bindiweed (*Convolvulus arvensis* L.), with several additional weeds being affected by 2,4-D. Kraus and Mitchell (1947) reported in 1947 that 2,4,5-T and 2,4-D were extremely phytotoxic to several broadleaf crop plants, whereas cereal crops were insensitive to the herbicides, indicating selectivity. As indicated earlier, 2,4-D and 2,4,5-T were introduced in 1944 by Amchem Products as weed killers (Martin and Worthing, 1974). There appeared to be little interest in 2,4,5-T until 1947, when Dow AgroSciences researchers found it effective on *Rubus* spp. and weeds resistant to 2,4-D (Barrons and Coulter, 1947). Soon after, the ester of 2,4,5-T was reported more effective than several formulations of 2,4-D or other chemicals when applied to the foliage of honey mesquite (Young and Fisher, 1949).

By 1951 the *Farm Chemicals Handbook* (Meister Publishing Co., Willoughly, OH) listed four U.S. companies as sources of 2,4,5,T; sources increased to over 50 companies by 1970. The number of basic producers in the United States declined after cessation of use in Vietnam in 1970. In addition, adverse publicity against this compound and pesticides in general forced U.S. companies to drop certain products only to be recouped by foreign companies. From 1951 to 1971 domestic disappearance (use) of 2,4,5-T ranged from 1.3 to 7.8×10^6 kg annually. The high use in 1962 through 1968 reflects usage in Vietnam (3.7 to 7.8×10^6) kg as a component of Agent Orange (Bovey and Young, 1980).

Production chemistry of 2,4,5-T involves the reaction of 2,4,5-trichlorophenol with chloroacetic acid, NaOH, and high temperature (~105°C) to yield the sodium salt. The sodium salt of 2,4,5-T is then acidified to produce 2,4,5-T. Production of 2-(2,4,5-trichlorophenoxy) propironic acid (silvex)—a closely related herbicide to 2,4,5-T—is made by the interaction of 2,4,5-trichlorophenol with sodium 2-chloroproprionate (Martin and Worthing, 1974). During the manufacturing of 2,4,5-T a highly toxic impurity TCDD (2,3,7,8-tetrachlorodibenzo-*p*-dioxin) is formed in production of 2,4,5-trichlorphenol, the precusor for 2,4,5-T (Bovey and Young, 1980). TCDD caused adverse public opinion against Agent Orange, although TCDD was later found to be a ubiquitous contaminant in the environment from burning fossil fuels and from other combustion situations in ultratrace amounts. Nevertheless, despite the benefits of 2,4,5-T and silvex, EPA eventually canceled all uses of 2,4,5-T and silvex in the United States because they were considered a potential health hazard. On February 28, 1979, the EPA announced emergency suspension of 2,4,5-T use on forests, rights-of-way, and pastures, and suspensed registered uses of silvex on forests, rights-of-way, pasture, and home, aquatic, and recreation areas. In 1985 all 2,4,5-T and silvex products were canceled for use in the United States by the U.S. District Court in Washington, D.C.

The United States lost some excellent products in 2,4,5-T and silvex, and after examining several hundred documents the author has concluded that no health hazard exists with these products when used as recommended. (See Chapters 7 and 8 for more details.)

D. Ammonium Sulfamate

Another herbicide no longer in use is ammonium sulfamate. It is effective for controlling most woody plants and is nonselective. It has a low acute oral LD_{50} for rats of 3900 mg/kg. It was used for basal bark, cut-stump, or frill-girdle treatments with 40% w/v water solution or by direct application of crystals. When applied directly, a 7.5% w/v solution was used. Ammonium sulfamate was introduced in 1945 by DuPont under the trade name Ammate. It has not been available for several years. The reason for its demise is not clear, but it has probably been replaced by more effective and selective materials.

E. Amitrole

Amitrole was first reported in 1953 by (Brehrens, 1953). It was suggested that certain crops were tolerant to preemergence applications of amitrole, while poison ivy was controlled by amitrole. In 1953 it was suggested as a potent new herbicide and that in combination with 2,4-D or 2,4,5-T it might be useful on certain weeds and brush. Since its introduction, amitrole has been widely used for a variety of agricultural and industrial applications. Residues of this pesticide led to withdrawal of certain lots of cranberries from the market in 1959 when it was proposed that the pesticide could induce thyroid tumors in rats. All registered uses of amitrole on food crops were canceled in 1971 (Carter, 1975). Amitrole has limited use in woody plant control, but controls poison ivy and poison oak.

F. Diquat and Paraquat

Diquat and paraquat were introduced in 1955 and 1958, respectively. Paraquat has achieved greater use because of its wider spectrum of activity against grasses, as well as most broadleaved weeds. Diquat controls cattails and other aquatic weeds and algae. Paraquat is also used as a postdirected herbicide in many crops, and as weed control in the conservation reserve and federal set-aside programs. Diquat and paraquat have limited use in woody plant control (Weed Science Soc. Am., 1994).

G. Dicamba

Dicamba was discovered in 1958 and introduced about 1965 by the Velsicol Chemical Corporation under the code number Velsicol 58-CS-11 and trade names Barvel and Mediben (Martin and Worthing, 1974). Dicamba is made from 1,2,4-trichorobenzene by a three-stage process: the trichorobenzene is converted to 2,5-dichlorophenol by methanol and NaOH; the phenol is treated with CO_2 under pressure to give 3,6-dichloro-2-hydroxybenzoic acid, which is converted to dicamba by treatment with dimethyl sulphate. Dicamba is widely used for weed control in cereals and grazing lands. It is sometimes formulated with 2,4-D. Although it is not as effective on woody plants as some other brush herbicides it is a very potent and useful product.

H. Picloram

Picloram was discovered in 1960 and made available for evaluation in about 1963. It was recognized early that picloram was a powerful hormonelike herbicide highly effective on broadleaved plants, including woody plants. It is made by the chlorination of $\propto$picoline, hydrolysis, and reaction with ammonia (Martin and Worthing, 1974). The low toxicity and efficacy on weeds make it a highly desirable herbicide for deep-root perennials and brush, but its relatively long persistence and mobility in water sources make it a restricted-use herbicide.

I. Clopyralid

Clopyralid was discovered in 1961. The original patent has expired. Clopyralid was first marketed in 1978 in Europe (Weed Science Soc. Am., 1994). We evaluated it in the 1970s for honey mesquite control. It controls many important weeds, but composite, polygonacea, and legume weeds, including honey mesquite and acacias, are particularly suseptable. It is highly effective alone or with triclopyr for honey mesquite control. It is also used with 2,4-D for weed control. It is related to picloram, but is different with respect to the weeds it controls and its shorter residue in the environment.

J. Bromacil

Bromacil was introduced in 1961, and its use started about 1963. It is prepared by the bromination of 3-*sec*-butyl-6-methyl-uracil. Bromacil is a nonselective inhibitor of phytosynthesis, absorbed mainly through the roots, and is recommended for general weed control on noncrop land at 5 to 15 kg/ha. It is useful for woody plant control in situations in which total vegetation control is desired or in which injury to herbaceous vegetation is not a concern. Bromacil can be selective at lower rates to maintain grass cover, however. Care should be taken when using it around desirable shrubs and trees, since roots may absorb the chemical if they extend into treated areas.

K. Tebuthiuron

Tebuthiuron was discovered in 1970 by Air Supply, Inc., and was introduced by Eli Lilly in 1974. Synthesis is accomplished when pivalic acid and MTSC (4-methyl-3-thiosemicarbazide) are combined and a mix of sulfuric and polyphosphoric acids are added over time. Toward the end of this reaction, ammonium hydroxide is added to neutralize pH. The lower aqueous layer is removed, and isopropylbenzene is added to form an intermediate amine. Then 1,3-dimethylurea is added with HCI. All reactions are conducted under nitrogen. Finally, solvents are removed under vacuum and the resulting crystals are washed in water and dried (Weed Science Soc. Am., 1994).

For woody plant control, tebuthiuron is most useful in pellet form and can be applied by hand, ground, or aerial equipment. Pellets are more effective on brush than the wettable powder in water sprays and is less damaging to the turf. Precautions are necessary in retaining large trees and shrubs near treated areas since roots may absorb the chemical from adjacent treated areas. Tebuthiuron is most effective on woody plants by soil application.

Herbaceous vegetation, especially broadleaved plants, may show injury the first season of application but usually recover vigorously by the second growing season. Tebuthiuron can be soil-or foliar-applied in pasture and rangeland and noncrop situations, such as industrial sites. Tebuthiuron controls certain broadleaved weeds and woody plants at low rates, but controls most vegetation at higher rates. Tebuthiuron is a potent woody plant control herbicide.

L. Substiluted Ureas and TCA

Substiluted ureas such as fenuron (Dybar) (*N N*-dimethyl-*N′*-phenylurea), monuron (Telvar) [*N′*-(4-chlorophenyl)-*N,N*-dimethyl urea], and diuron (Karmex)

[*N*′-(3,4-dichlorophenyl-*N*′,*N*′-dimethylurea] were developed by DuPont shortly after the end of World War II. These compounds were effective herbicides, since they inhibited the Hill reaction (Geissbühler et al., 1975).

These compounds are no longer available, but fenuron and fenuron-TCA (trichloroacetic acid) (Urab) were used for temporary sterilization of the soil in noncrop areas and in brush control (Weed Science Soc. Am., 1970). Monuron selectively controlled germinating broadleaved weeds and grass in cotton, sugarcane, pineapple, asparagus, and citrus. High monuron rates were used as a general weed killer. Monuron-TCA (Urox) was a nonselective herbicide for soil sterilization in noncrop areas. Diuron use was similar to Monuron. The higher water solubility of fenuron (3550 ppm) made it more effective as a soil treatment for deep-rooted woody perennials.

M. Glyphosate

Glyphosate is a nonselective, foliar-applied, versatile herbicide useful in many crop and noncrop situations. It can be applied preplant or preemergence to control emerging weeds, or postemergence- (POST) directed for selectivity in crops. It can be applied with conventional, recirculating sprayers or with shielded or wiper applicators. It can be used in industrial situations but has limitations on pastures and rangeland because it is phylotoxic to most annual- and perennials, including grasses. It is sometimes used in renovation programs on grazing lands. Glyphosate was experimentally introduced in 1971 by Monsanto Co. under code number MON-0573. The principal formulation tested was the isopropylamine salt, with the trade name Roundup.

N. Fosamine

Fosamine was first described in 1974 by DuPont, and has the trade name Krenite. Fosamine is foliar-applied at 9 to 54 kg/ha for brush control on noncrop lands or on sites in preparation for conifers. It can also be applied in a spray-to-wet treatment using 1.8 to 3.6 kg/100 L water. An oil adjuvant may increase efficacy. Fosamine has limited use on grazing lands (Weed Science Soc. Am., 1994).

O. Hexazinone

Another DuPont material, first described in 1975, is hexazinone. Hexazinone has both foliar and soil activity. It is used for weed control in dormant alfalfa, pineapple, sugarcane, Christmas tree planting, site preparation for reforestation, and noncrop lands. It is very useful as a soil treatment around individual woody plants, such as honey mesquite and blueberry or redberry cedars. We first evaluated hexazinone in 1975 for woody plant control in Texas.

P. Triclopyr

Triclopyr was first reported in 1975 by the Dow Chemical Co. We also evaluated triclopyr for honey mesquite control and other brush beginning in 1975. Triclopyr is applied foliar at 1.1 to 10.1 kg/ha or in spray-to-wet applications at 0.4 to 0.9 kg/ha/100 to 400 L of spray solution for noncrop areas on utility and pipeline rights-of-way, roadsides, railroads, industrial sites, forestry sites, rangeland, and permanent pastures. It can be applied POST-directed for conifer release, injected into stem cuts for controlling large trees, or applied to freshly cut stumps or mixed with oils for bark treatment on young trees. It also controls many annual broadleaved weeds. It controls honey mesquite alone or with clopyralid and as a foliar spray on honey mesquite and other brush and as a basal spray. (See Chapter 12.)

Q. Imazapyr

Imazapyr was introduced in the late 1970s by American Cyanamid. It is produced by 2,3 pyridine carboxylic acid undergoing dehydration (to form the anhydride). A condensation reaction, hydrolysis, is cyclized (to form the imidazoline ring), and is precipitated as the manufactured product. Testing began in 1981 for total vegetation control (Weed Science Society, 1994). Imazapyr controls a wide variety of annual and perennial weeds and brush in noncrop situations. It can also be used for sugarcane and plantation crops such as rubber and oil palm. Imazapyr is available as an oil-soluble formulation called Chopper for cut-stump, basal bark, and frill application.

R. Metsulfuron

Metsulfuron was first reported in 1983 and was introduced by DuPont. Metsulfuron is of interest in woody plant control because it is labeled for some woody brush and vine species.

IV. HOW HERBICIDES ARE PRODUCED

Laws and regulations concerning toxicology and safety are covered in Chapter 7. Laws and safety regulations have become more complex with greater pesticide use as the need dictated. A great deal of information must be known about a compound before it can be registered with EPA (Klingman, 1975). EPA can request more data from the company if questions are unanswered about its toxicology or environmental fate.

The development of effective and safe pesticides is a complex and costly undertaking because the data needed for registration have increased considerably since the 1940s (American Crop Protection Association, 1994). Such an undertaking is very time-consuming, and costs may exceed $50 million. For the company to recoup its expenses and remain profitable it depends upon beneficial products and a sound financial and diverse base to weather the failures and successes of agricultural chemicals. As indicated before, the development of herbicides for woody plant control have usually resulted from herbicides developed for other uses and through secondary testing, brush control activity has been discovered.

Pesticides are strictly regulated under the Federal Insecticide, Fungicide and Rodenticide Act (FIFRA) and the Federal Food, Drug and Cosmetic Act (FFDCA). Since 1970 EPA is responsible for both acts and major revisions were made to FIFRA in 1972, 1975, 1978, and 1988 (American Crop Protection Association, 1994). Congress has mandated the reevaluation and reregistration of all chemicals registered prior to November 1984, which includes all herbicides listed in this publication. Most have been reregistered or will be reregistered. (See Chapter 7 on toxicology and safety for details.)

Because the development of new pesticide chemistry is slow and costly, the use of older, effective products is essential. In the United States, food production and low consumer food costs depends directly upon agricultural chemicals. Such chemicals control weeds, insects, fungi, bacteria, plant diseases, rodents, and other organisms that threaten our food supply. On the average, only one in 20,000 chemicals tested becomes a commercially successful pesticide (American Crop Protection Association, 1994). Each compound is subject to rigorous health, safety, and environmental testing before being approved by EPA for widespread use. Development, testing, and registration typically require 8 to 10 years and can cost the manufacturer up to $50 million for each pesticide. A booklet describing the pesticide registration process written by the American Crop Protection Association, in Washington, D.C., is partially reproduced in the next few pages (American Crop Protection Association, 1994).

A. Primary Screening

Potential pesticides are first synthesized in small quantities, then tested in the laboratory and greenhouse for biological activity (insecticidal, fungicidal, herbicidal, or growth-regulatory activity, etc). Promising chemicals are tested for toxic effects on laboratory animals. If the chemical causes genetic damage, research is terminated. Hazards to human and animals are determined if accidental exposure occurs. Pesticide research is conducted according to good laboratory practices (GLPs) by federally prescribed standards (American Crop Protection Association, 1994).

B. Secondary Screening

Once past the primary screening, promising pesticides are tested in small field plots under actual crop conditions. Studies are conducted by company personnel, research consultants, and state and federal scientists. Data such as application methods, pesticide rates, spectrum of pest control, crop tolerance, compatibility with other pesticides, half-life, biological longevity, dissipation, and loss by run-off and leaching are determined (Klingman, 1975; American Crop Protection Association, 1994).

The massive research conducted over several years provides scientific and economic data to decide if commercialization of the product is warranted. Further data are obtained from this research concerning human and environmental safety and market feasibility.

C. Safety Considerations

Safety is the primary consideration in pesticide research, and the EPA requires that the label list precise application methods, rates, and specific instructions for protective clothing, training of handlers, warning statements, time limits on reentry of farm workers into treated fields, environmental warnings, and proper container disposal.

D. Preparing for Registration

To register a new product the manufacturer proceeds with long-term research required by EPA that supports product label registration. The tests include data on the chemistry, toxicology, residues in food and feed, environmental fate, ecological effects, and efficacy. The pesticide may be subject to more than 140 tests in order to receive and maintain EPA product label registration.

E. Tests Required by EPA for Pesticide Registration

Pesticide products are subject to the tests listed here in order to receive and maintain EPA product label registration. The exact tests that must be performed for each pesticide depend on how it will be used. For example, a pesticide that would not be used on food or feed crops would not require extensive residue and metabolism tests. The list of tests required for biochemical pesticides is a somewhat modified version of the list below (American Crop Protection Association, 1994).

1. Product Chemistry Data Requirements

Product identification and disclosure of ingredients
Description of beginning materials and manufacturing process
Discussion of formation of impurities
Preliminary analysis

Certification of limits
Analytical methods to verify certified limits

2. Physical and Chemical Characteristics

Color
Physical state
Odor
Melting point
Boiling point
Density, bulk density, or specific gravity
Solubility
Vapor pressure
Dissociation constant
Octanol/water partition coefficient
Ph
Stability
Oxidizing or reducing action
Flammability
Explodability
Storage stability
Viscosity
Miscibility
Corrosion characteristics
Dielectric breakdown voltage
Submittal of samples

3. Wildlife and Aquatic Organisms Data Requirements

Acute avian oral toxicity (LD_{50}) in bobwhite quail or mallard duck
Acute avian oral toxicity (LD_{50}) in bobwhite quail or mallard duck (using typical end-use product)
Acute avian dietary toxicity (LC_{50}) in bobwhite quail
Acute avian dietary toxicity (LC_{50}) in mallard duck
Wild mammal toxicity test
Avian reproductive toxicity in bobwhite quail
Avian reproductive toxicity in mallard duck
Simulated terrestrial field study
Actual terrestrial field study
Fish toxicity in bluegill sunfish
Fish toxicity in bluegill sunfish (using typical end-use product)
Fish toxicity in rainbow trout
Fish toxicity in rainbow trout (using typical end-use product)

Invertebrate toxicity freshwater LC_{50} (daphnia preferred)
Invertebrate toxicity freshwater LC_{50} (daphnia preferred—using typical end-use product)
Toxicity to estuarine and marine organisms in fish, mollusks, and shrimp using active ingredient and typical end-use product (6 tests)
Early life stage in fish
Life cycle in aquatic invertebrates (daphnia/mysid)
Fish life cycle study
Aquatic organism accumulation study
Simulated field tests for aquatic organisms
Actual field tests for aquatic organisms

4. Toxicology Data Requirements

Acute oral toxicity in the rat
Acute dermal toxicity
Acute inhalation toxicity in the rat
Primary eye irritation in the rabbit
Primary dermal irritation
Dermal sensitization
Acute delayed neurotoxicity in the hen
90-day feeding study in the rodent
90-day feeding study in the nonrodent
21-day dermal toxicity
90-day subchronic dermal toxicity
90-day inhalation in the rat
90-day neurotoxicity in the hen
90-day neurotoxicity in the mammal (rat preferred)
Chronic feeding study in the rodent
Chronic feeding study in the nonrodent
Oncogenicity study in the rat
Oncogenicity study in the mouse
Teratogenicity in the rat
Teratogenicity in the rabbit
2-generation reproduction study in the rat
Chronic feeding/oncogenicity in the rat
Gene mutation
Structural chromosome aberration
Other genotoxic effects
General metabolism
Dermal penetration
Domestic animal safety

5. Plant Protection Data Requirements

Seed germination and seeding emergence
Vegetative vigor
Aquatic plant growth
Seed germination and seedling emergence
Vegetative vigor
Aquatic plant growth
Terrestrial field
Aquatic field

6. Reentry Protection Data Requirements

Foliar residue dissipation
Soil residue dissipation
Dermal passive dosimetry exposure
Inhalation passive dosimetry exposure

7. Nontarget Insect Data Requirements

Honey bee acute contact (LD_{50})
Honey bee toxicity of residues on foliage
Field testing for pollinators

8. Environmental Fate Data Requirements

Chemical identity
Hydrolysis
Photodegradation in water
Photodegradation on soil
Photodegradation in air
Aerobic soil metabolism study
Anaerobic soil metabolism study
Anaerobic aquatic metabolism study
Aerobic aquatic metabolism study
Leaching and adsorption/desorption
Laboratory volatility study
Field volatility study
Soil field dissipation study
Aquatic sediment field dissipation study
Forestry field dissipation study
Combinations and tank mixes
Long-term soil dissipation study
Confined rotational crop study

Field rotational crop study
Accumulation in irrigated crops
Accumulation in fish
Accumulation in aquatic nontarget organisms
Groundwater studies requirements
Small-scale prospective groundwater monitoring study
Small-scale retrospective groundwater monitoring study
Large-scale retrospective groundwater monitoring study

9. Residual Chemistry Data Requirements

Chemical identity
Directions for use
Nature of residue in plants
Nature of residue in livestock
Residue analytical method (plants)
Residue analytical method (animals)
Storage stability
Magnitude of the residue in potable water
Magnitude of the residue in fish
Magnitude of the residue in irrigated crops
Magnitude of the residue in food handling
Magnitude of the residue in meat/milk/poultry/eggs (feeding/dermal treatment)
Crop field trials
Magnitude of the residue in processed food/feed
Reduction of residues
Proposed tolerance
Reasonable grounds in support of petition
Analytical reference standard

10. Spray Drift Data Requirements

Droplet size spectrum
Drift field evaluation

F. Other Needs for Registration

During the time tests are being conducted for label registration large-scale field testing requires that the manufacturer acquire an experimental use permit (EUP) from EPA, expand research on toxicological, environmental, and residue tolerance, and determine the risk assessment of the product. Once EPA approves the product label, commercial production and sale of the product can proceed. The decision to proceed with manufacturing the product depends upon patent protection (17 years) and if the market is large enough to justify development costs.

G. Additional Uses

The search by manufacturers for additional beneficial uses of the newly approved pesticide does not end with EPA registration, so additional testing continues. New uses are added to the label only after all required research has been conducted by the manufacturer and reviewed and approved by EPA (American Crop Protection Association, 1994).

1. Special Local Needs Registration

Normally pesticide products are registered for use on widely grown crops, such as corn and soybeans. When potential uses involve small-acreage crops, highly specialized commodities, or pests found only in limited areas, manufacturers must closely evaluate the expense and time required to develop and register the pesticide for these uses. The discovery and registration for such uses, however essential to a number of growers, may be limited. Fortunately, FIFRA section 24(c) provides special local needs (SLN) registrations to assist in such situations (American Crop Protection Association, 1994).

Special local needs registrations are issued by individual states under EPA supervision only for pesticides that have federal registrations and crops with established residue tolerances. They cannot be registered by EPA or for any uses previously denied, disapproved, or canceled by the EPA. Finally, an SLN registration only allows use of a pesticide within the state granting that registration.

2. Emergency Exemptions

Crop emergencies due to weeds, insects, or diseases can arise from time to time. A good example is the unexpected infestation in 1991 of the poinsettia whitefly in California that threatened crops throughout the state. Such threats to crops can spring up with no effective registered pesticide control available. EPA may grant an emergency exemption under section 18 of FIFRA for use of pesticides not otherwise registered for the specific need (American Crop Protection Association, 1994).

These exemptions may be granted only at the request of state governments, the USDA, or certain other federal agencies. After evaluating all the health and environmental aspects of the proposed emergency use, exemptions are given only 1) for the duration of the emergency; 2) for precisely defined geographic areas; and 3) with rigid use restrictions. In addition, reports must be submitted to EPA regarding progress in the emergency, including the use of the chemical and the results achieved.

3. Pesticides Used on Minor Crops

''Minor use'' pesticides are used on specialty crops, crops grown only on limited acreage (most fruits and vegetables in the United States, and geographically limited pest problems of large-acreage crops. Although the combined economic value of minor crops is substantial—estimated at $35 billion per year in harvested produce—they are grown on only about 5 million acres. By comparison, the U.S. corn crop is grown on some 70 million acres (American Crop Protection Association, 1994).

The economic realities of supporting registration of pesticides for minor uses present special problems for growers, regulators, and manufacturers. The pesticide industry, grower organizations, EPA, and USDA are cooperating to develop cost-effective means of supporting registration of crop protection products for minor uses.

4. Exporting U.S.-Produced Pesticides

Some pesticides that are manufactured in the United States and exported to other countries are not registered with EPA. Typically, they lack U.S. registration for reasons unrelated to human health or environmental concerns. For example, the product may control pests that are not a problem in the United States may be used on crops that are not grown in the United States, and/or may be currently under consideration by EPA for registration (American Crop Protection Association, 1994).

Although not registered in the United States, these products are fully tested and subject to the health and environmental standards of the importing nations. Registration in countries with regulations similar to those in the United States already has been granted. Additionally, U.S. law requires that all U.S.-exported pesticides be labeled in the language of the countries to which they are exported. They also must bear the words ''not registered in the United States.'' Manufacturers comply with international standards that require informing the governments of importing nations about the product's U.S. registration status.

5. Reregistration: Reevaluating Pesticide Safety

Acknowledging the advances in analytical science and technology that had taken place in the 25 years since FIFRA enactment in 1947, in 1972 Congress directed EPA to reregister all pesticides to assure they met current scientific and regulatory standards (American Corp Protection Association, 1994).

By 1988, it was evident that more resources were needed for the complex job of reregistration, so Congress amended FIFRA again to accelerate reregistration and to provide additional funds and personnel specifically for this program. A large part of these funds come from industry-supported user fees. Reregistration now is being implemented over five phases, with completion by the late 1990s.

H. Reregistration of Pesticides

Phase I—EPA
Publish lists of active ingredients subject to reregistration
Phase II—Registrant
Notify EPA of intent to seek reregistration
Identify missing and inadequate data (''data gaps'')
Agree to fill data gaps under prescribed guidelines
Begin paying annual reregistration product maintenance fee
Pay one-time reregistration active ingredient fee
Phase III—EPA
Issue guidelines for submitting reregistration data
Monitor data acquisition and submission
Registrant
Summarize and reformat studies to facilitate EPA review requirements
Certify possession of raw data generated by studies
''Flag'' data showing potential adverse effects
Commit to fill data gaps
Phase IV—EPA
Review registrants' phase II and III submissions
Identify each outstanding data gap revealed in phases II and III
Publish outstanding data gaps for each active ingredient
Issue data call-in (DCI) requiring registrants to fill data gaps
Registrant
Conduct required studies
Submit additional data
Phase V—EPA
Thorough review of all data submitted for the various active ingredients
Determine eligibility of each active ingredient for reregistration
Final Result
Reregistration of commercial product

I. Meaning of Pesticide Residues

Pesticide residues in crops, soils, food, feed, or water may be expressed in several ways, but parts per million (ppm) or parts per billion (ppb) are common.

ppm = mg/kg, mg/L, or μg/g

ppb = μg/kg or μg/L

To show how small ppm, ppb, or parts per trillon (ppt) are, the following examples are presented (source unknown).

One part per million is the same as

1 minute in 2 years
1 inch in 16 miles
1 ounce in 32 tons
1 bad apple in 2,000 barrels

One part per billion is the same as

1 second in 32 years
1 inch in 16,000 miles
1 drop in 10,000 gallons
1 bad apple in 2,000,000 barrels

One part per trillion is the same as

1 second in 32,000 years
1 inch in 16,000,000 miles (642.56 times around the Earth)
1 grain of sugar in an Olympic-sized pool
1 bad apple in 2,000,000,000 barrels

Analytical methods are available from the manufacturer or other sources to determine chemical residues in plants, animals, water, air, feed, or food sources at these very small amounts.

V. SUMMARY

The phenoxy herbicides were developed in the 1940s and became commercially available in the 1950s. The herbicide 2,4,5-T was available for weed and brush control until negative publicity from its use in Vietnam resulted in EPA suspension of certain uses in 1974 and all uses in 1985. Silvex and 2,4,5-T were inexpensive and effective herbicides. Dichlorprop and 2,4-D are still widely used (Burnside et al., 1996).

Many excellent herbicides still exist for weed and brush control (Table 1), but no new herbicides have been developed for woody plant management since the late 1970s. It is essential that these materials be reregistered and made available to land managers and growers.

Increasingly complex laws, regulations, and costs to develop new herbicide chemistry for forestry, grazing, and noncrop land have no doubt caused this situation. The older herbicides are now more expensive to use because of reregistration requirements. Developing new herbicides for commercial use and securing federal label registration is complex and costly, but this system provides scientific and regulatory checks and balances that safeguard pesticide users and consumers.

Herbicides, however, will continue to be needed for the benefit of food, fiber, and animal feed production for many years to come.

REFERENCES

American Crop Protection Association. From Lab to Label. The Research, Testing and Registration of Agricultural Chemicals. Washington, DC: American Crop Protection Association, 1994.

Barrons KC. Are Pesticides Really Necessary? Chicago: Regnery Gateway, 1981.

Barrons KC, Coulter LL. The specific effect of 2,4,5-trichlorophenoxyacetic acid on members of the genus Rubus and certain other 2,4-D resistant species. North Central Weed Control Conf Res Rept 4:255, 1947.

Bovey RW. Use of 2,4-D and other phenoxy herbicides on pastureland, rangeland, alfalfa forage, and noxious weeds in the United States. In: Burnside OC, ed. Biological and Economic Assessment of Benefits from Use of Phenoxy Herbicides in the United States. USDA-NAIAP rept. no. 1-PA-96, 1996, pp. 76–86.

Bovey RW, Young AL. The Science of 2,4,5-T and Associated Phenoxy Herbicides. New York: Wiley, 1980.

Brehrens R. Amino triazole. Proc North Central Weed Control Conf 10:61, 1953.

Burnside OC, Bovey RW, Elmore CL, Johnson RA, Knake EL, Lembi CA, Nalewaja JD, Newton M, Szmedra P, Wattenberg EV. Biological and Economic Assessment of Benefits from Use of Phenoxy Herbicides in the United States. USDA-NAPIAP rept. no. 1-PA-96, 1996.

Carter MC. Amitrole. In: Kearney PC, Kaufman DD, eds. Herbicides: Chemistry Degradation and Mode of Action. 2nd ed., vol. 1. New York: Marcel Dekker, 1975, pp. 377–398.

Darrow RA, McCully WG. Brush control and range improvement in the post oak–blackjack oak area of Texas: bulletin 942. Texas Agric. Exp. Stn. and Texas Agric. Ext. Serv., College Station: Texas A&M University, 1959.

Fisher CE, Young DW. Some factors influencing the penetration and mobility of chemicals in the mesquite plant. Proc North Central Weed Control Conf 6:197–202, 1948.

Geissbühler H, Martin H, Voss G. The substituted ureas. In: Kearney PC, Kaufman DD, eds. Herbicides: Chemistry, Degradation and Mode of Action. 2nd ed., vol. 1. New York: Marcel Dekker, 1975, pp. 209–291.

Hamilton WT. Past history of brush control in Texas: A personal perspective of efforts in South Texas. Texas Agric. Exp. Stn. Workshop, Dallas, 1993.

Hammer CL, Tukey HB. The herbicidal action of 2,4-dichlorophenoxyacetic acid and 2,4,5-trichlorophenoxyacetic acid on bindweed. Science 100:154–155, 1944.

Klingman GC. Weed Control: As a Science. New York: Wiley, 1961.

Klingman GC. Pesticide development. In: Byrnes WR, Holt HA, eds. Herbicides in Forestry. West LaFayette, In: Department of Forestry and Natural Resources, Purdue University, 1975, pp. 15–25.

Kraus EJ, Mitchell JW. Growth regulatory substances as herbicides. Bot Gaz 108:301–350, 1947.

Martin H, Worthing CR. Pesticide Manual. 4th ed. Worcester, England: British Crop Protection Council, 1974.

Pokorny R. Some chlorophenoxyacetic acids. J Am Chem Soc 63:1768, 1941.

Weed Science Society of America. Herbicide Handbook. 7th ed. Champaign, IL: Weed Science Society of America, 1994.

Weed Science Society of America. Herbicide Handbook. 2nd ed. Geneva, NY: W. Humphrey Press, 1970.

Young DW, Fisher CE. Treatments to the foliage of mesquite (*Prosopis juliflora*). North Central Weed Control Conf Res Rept 6:147, 1949.

Zimmerman PW. The formative influence and comparative effectiveness of various plant hormone-like compounds. Torreya 43:98–115, 1943.

Zimmerman PW, Hitchcock AE. Substituted phenoxy and benzoic acid growth substances and the relation of structure to physiological activity. Contr Boyce Thompson Inst 12:321–343, 1942.

Zimmerman PW, Hitchcock AE. Proc Am Soc Hort Sci 45:353, 1944.

4

Chemistry and Properties of Herbicides Used

I. INTRODUCTION

Herbicides are an important means of woody plant control. Compared to mechanical practices, herbicides usually are less expensive, less damaging to the environment, and often more effective. Herbicide sprays, however, are subject to drift that may damage susceptible crops or valuable vegetation on nearby areas if improperly applied. A variety of herbicides and herbicide combinations are available commercially. It is essential that the user understand the properties and the effects of herbicides to use them safely and effectively. Herbicides commonly used for weed and brush control include 2,4-D, dicamba, bromacil, picloram, hexazinone, triclopyr, amitrole, and tebuthiuron. Combinations of herbicides such as 2,4-D + triclopyr, 2,4-D + picloram or dicamba, and triclopyr + clopyralid also have been used (Bovey et al., 1984; Weed Science Society of America, 1994).

II. HERBICIDE PROFILES

A. Benzoics

Zimmerman and Hitchcock (1942) showed that certain β-naphthoxy, phenoxy, and benzoic acid derivatives caused formative effects in plants. They indicated that whenever the acid form was active the esters, salts, and amides also were active, and suggested that changing the structure of some 40 known active substances could produce new compounds of growth regulator interest. Zimmerman and Hitchcock (1952) indicated the hormonelike properties of 2,3,6-trichlorobenzaldehyde and its corresponding acid, 2,3,6-trichlorobenzoic acid (2,3,6-TBA). During the 1950s and 1960s 2,3,6-TBA was tested and used as a highly effective temporary soil sterilant for controlling deep-rooted perennial weeds (Crafts and Robbins, 1962). Several chemical companies were involved in its development (Weed Science Society of America, 1967).

Another highly successful hormonelike benzoic herbicide is dicamba (3,6-dichoro-2-methoxybenzoic acid), invented by S. B. Rechter. United States patent 3,013,054 was awarded in 1958. After 40 years, it is still a highly successful herbicide. It is known commercially as Banvel.

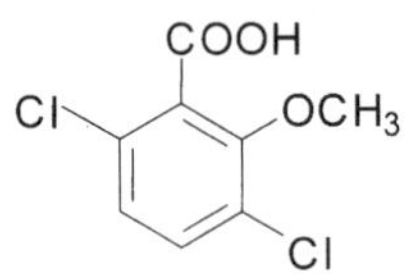

dicamba

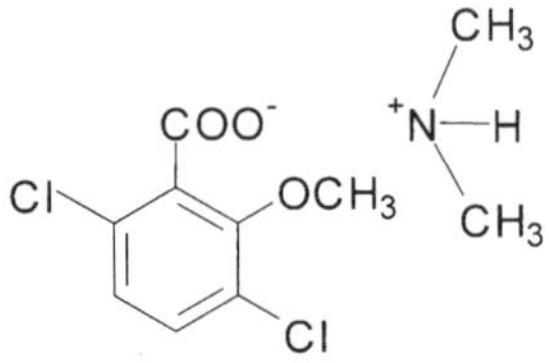

dicamba, dimethylamine salt

Dicamba is a selective translocated herbicide, that controls many broadleaf weeds in pastures and cropland and some woody plants. Similar to the phenoxy herbicides in activity and use, it is absorbed through roots as well as through foliage (Table 1). Dicamba can be applied either by ground or aerial sprays or as granules, depending on the weeds to be controlled and their proximity to susceptible crops. Dicamba can be applied in mixtures with 2,4-D or other herbicides to increase activity and/or broaden the spectrum of weeds controlled.

Dicamba has a low order of toxicity to wildlife, fish, livestock, and humans. It rapidly degrades and does not accumulate in the environment. Dicamba has a low corrosion hazard to spray equipment. It is formulated as a dimethylamine or sodium salt and is sprayed in a water carrier. In granular formulation it is an acid (Bovey, 1995; Crafts and Robbins, 1962; Weed Science Society of America, 1994).

Spray drift of dicamba to sensitive crops, conifers, and certain woody plants should be avoided. Granular formulation may be preferred where drift of sprays may present a hazard to crops. Methods of production and analysis are given elsewhere (Martin and Worthing, 1974).

It is important to read the herbicide label on the container since it provides information on proper application methods and rates and environmental safety precautions regarding spray drift or contamination of water sources.

B. Bipyridiniums

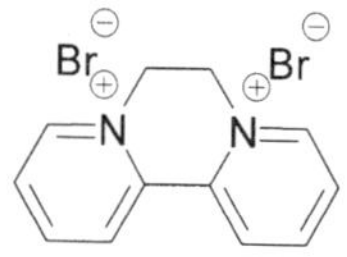

diquat dibromide

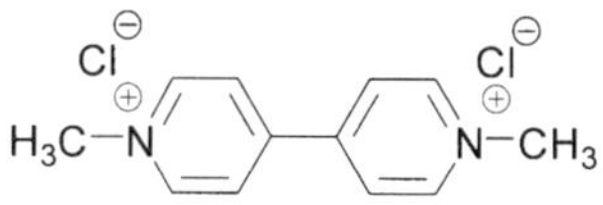

paraquat dichloride

Diquat {6,7-dihydrodipyrido [1,2-α:2′,1′*c*]pyrazinediium ion} and paraquat (1,1′-dimethyl-4,4′-bipyridinium ion), commonly called Gramoxone Extra, are desiccant and defoliant herbicides used for general contact activity against weeds and brush. In some situations they are used as selective herbicides. Woody species usually will resprout from foliar sprays of diquat and paraquat.

These compounds are water-soluble and are inactivated by soil contact. Paraquat is registered for suppressing existing sod and emerged undesirable weeds and grasses to permit pasture and range seeding. Paraquat can be used to control winter weeds in dormant warm-season grasses. Protective clothing and respirators should be used when applying paraquat because it is highly toxic when ingested.

Diquat and paraquat were first made by the Dyestuffs Division of I.C.I., Ltd. in Great Britain. Diquat was first used as a growth regulator in 1955 and parquat in 1959 (Weed Science Society of America, 1994). Production methods are available (Martin and Worthing, 1974).

C. Phenoxys

The phenoxy herbicides include 2,4-D, 2,4-DB, dichlorprop (4-chloryl-2-methylphenoxy)acidic acid (MCPA), 4-(4-chloryl-2-methylphenoxy)butanoic acid (MCPB), (±)-2-(4-chloryl-2-methylphenoxy)propionic acid (mecoprop), silvex, and 2,4,5-T. Silvex and 2,4,5-T uses have been cancelled by the U.S. Environmental Protection Agency (EPA) (Bovey et al., 1984; Burnside, 1996). Dichlorprop and 2,4-D are still used in woody plant management, especially in mixtures with other herbicides (Table 1). The herbicide 2,4-D has probably had more influence on the development of modern weed control practices worldwide than any other compound. Even though new herbicides have been developed, 2,4-D remains essential to world agriculture because of its low cost, safety, and effectiveness on many weeds (Burnside, 1996). The phenoxy herbicides have an excellent track record of over 50 years of use (Bovey, 1996).

Production of 2,4-D [(2,4 dichlorophenoxy) acetic acid] is made by the interaction of 2,4-dichlorophenol and sodium monochoroacetate to produce the sodium salt of 2,4-D. The sodium salt is then acidified to produce 2,4-D.

TABLE 1 Herbicides Used for Woody Plant Management on Rangelands and Other Areas

Herbicide	Chemical name	Commercial formulation	Mode of entry	Major uses	Rates (kg/ha)	Water solubility (ppm)
Amitrole	1 *H*-1,2,4-triazol-3-amine	Wettable powder or liquid	Foliage	Noncrop areas, poison ivy, and oak	2 to 12	280,000
Bromacil	5-bromo-6-methyl-3-(1-methylpropyl)-2,4(1*H*,3*H*) pyrimidinedione	Wettable powder or liquid	Root	Certain woody plants, herbaceous weeds	1 to 6	815
Clopyralid	3,6-dichloro-2-pyridinecarboxylic acid	Monoethanol amine salt	Foliage and root	Broadleaf weeds and brush, effective on honey mesquite	0.06 to 4.5	1000
2,4-D	(2,4-dichlorophenoxy)acetic acid	Various salts and esters	Foliage	Broadleaf weeds, certain woody plants	0.25 to 2 (4 to 5 kg/1000 L on certain woody plants)	900
Dicamba	3,6-dichloro-2-methoxybenzoic acid	Dimethylamine salt	Foliage and root	Weeds and brush	0.5 to 11	4,500
Dichlorprop	(+) 2(2,4-dichlorophenoxy) propanoic acid	Potassium salt and esters	Foliage	Certain woody plants	0.5 to 2	710
Diesel oil, kerosene			Foliage and stems	Carrier for herbicide sprays on weeds and brush, basal pours on woody plants	Adequate coverage	Insoluble
Diquat	6,7-dihydrodipyrido[1,2-α:2′,1′*c*]pyrazinedi-ium ion	Dibromide salt	Foliage contact	General weed control, noncrop areas	0.5	718,000
Fosamine	Ethyl hydrogen (aminocarbonyl) phosphonate	Miscible liquid	Foliage	Woody plant control, growth regulator	7 to 14	1,790,000
Glyphosate	*N*-(phosphonomethyl) glycine	Isopropylamine salt	Foliage	Broadspectrum nonselective weed control	0.3 to 2	900,000

Hexazinone	3-cyclohexyl-6-(dimethylamino)-1-methyl-1,3,5-triazine-2,4(1*H*,3*H*)-dione	Soluble powder, miscible liquid or pellet	Foliage and root	Weed and brush control on crop and noncrop areas, forest sites	2 to 12	33,000
Imazapyr	(±)-2-[4,5-dihydro-4-methyl-4-(1-methylethyl)-5-oxo-1*H*-imidazol-2-yl]-3-pyridinecarboxylic acid	Isopropylamine salt	Foliage and root	Weed and brush control on noncrop areas	0.5 to 4	11,272
Metsulfuron	methyl2[[[[(4-methoxy-6-methyl-1,3,5-trazin-2-yl)amino]carbonyl]amino]sulfonyl]benzoate	Dry fowable	Foliage and root	Weeds and certain woody plants in crop and noncrop areas	4 to 12 g	548 at pH 5, 2790 at pH 7, 213,000 at pH 9
Paraquat	1,1′-dimethyl-4,4′-bipyridinium ion	Dichloride salt	Foliage contact	Weed control during grass establishment, pasture renovation, and weed control on noncrop areas	0.5 to 1	Highly soluble
Picloram	4-amino-3,5,6-trichloro-2-pyridinecarboxylic acid	Potassium or amine salts	Foliage and root	Weed and brush control on crop and noncrop areas	0.14 to 9	430
Tebuthiuron	*N*-[5-(1,1-dimethylethyl)-1,3,4-thiadiazol-2-yl]-*N*,*N*,′-dimethylurea	Wettable powder and pellets	Root	Weed and brush control on rangelands, total vegetation control	0.6 to 4.2	2,500
Triclopyr	[(3,5,6-trichloro-2-pyridinyl)-oxy]acetic acid	Triethyamine salt and butoxyethyl ester	Foliage and root	Weed and brush control on rangelands and noncrop areas	0.28 to 10	430

Note: Many herbicides listed are also commercially available in mixtures with other herbicides. Some mixtures useful in weed and brush control include: 2,4-D + dichlorprop; 2,4-D + triclopyr; 2,4-D + picloram; 2,4-D + dicamba; and amitrol + ammonium thiocyanate.
Source: Weed Science Society of America, 1994.

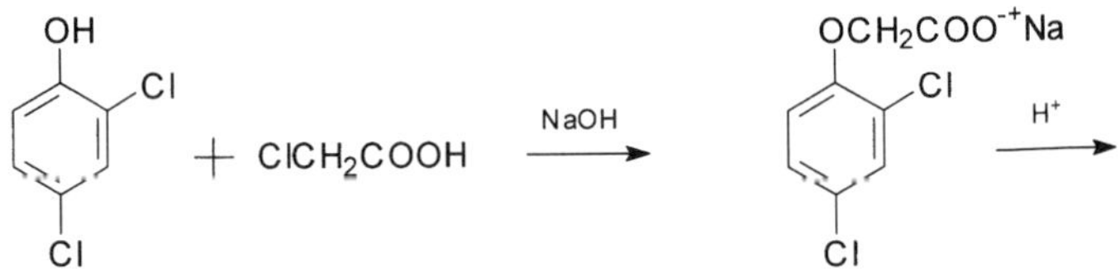

dichlorophenol chloroacetic acid sodium salt of 2,4-D

The 2,4-D acid can be reacted with a variety of alcohols to produce a large variety of esters and amine salts by reacting 2,4-D acid with the appropriate amine.

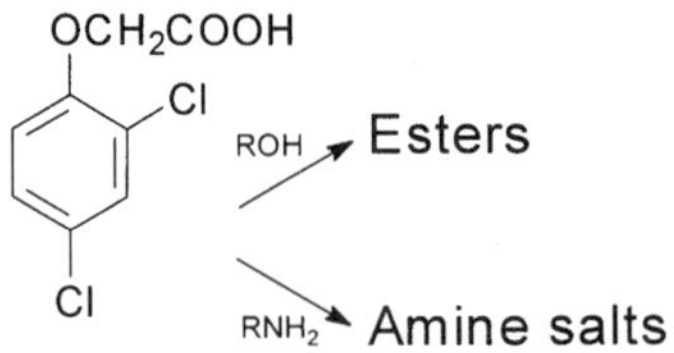

2,4-D

Both ester and amine formulations of 2,4-D are registered for broadleaf weed and brush control on pastures and rangeland. Common formulations include the diethylamine salts or the isooctyl ester of 2,4-D (Bovey, 1996). Dichlorprop (2,4-DP) may be synthesized either by condensation of 2-chloropropronic acid with 2,4-dichlorophenol or by the chlorination of 2-phenoxypropronic acid (Martin and Worthing, 1974). Dichlorprop is commonly formulated as the butoxyethyl or isooctyl ester or the dimethylamine for use in weed and brush control (Weed Science Society of America, 1994).

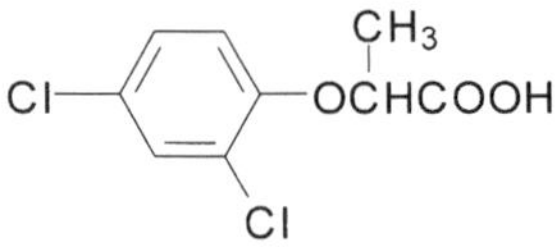

dichlorprop

Herbicide 2,4,5-T [(2,4,5-trichlorophenoxy) acetic acid] is made by an interaction of sodium chloroacetate with 2,4,5-trichorophenol (Martin and Worthing, 1974). Silvex is the propionic form of 2,4,5-T and is made by the interaction of 2,4,5-trichlorophenol with sodium chloropropionate (Spencer, 1973).

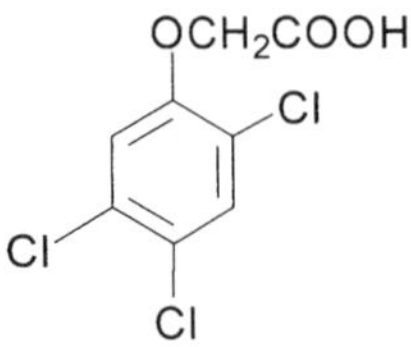

2,4,5-T

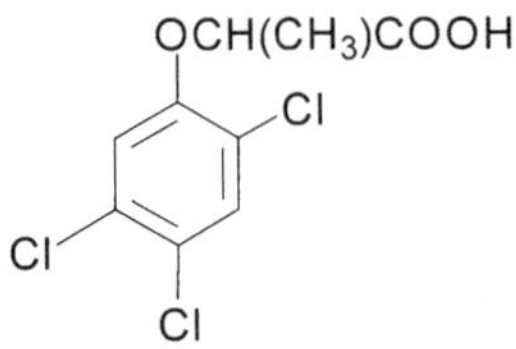

Silvex

Like 2,4-D, 2,4,5-T was formulated in various ester and mine formulations. Silvex was formulated as the potassium salt or propylene glycol butyl ether esters (Spencer, 1973). Information on MCPA, MCPB and mecroprop can be found in other sources (Weed Science Society of America, 1994; Martin and Worthing, 1974; Spencer, 1973).

Phenoxy herbicides produce ecological changes and shifts that are beneficial for livestock production and wildlife habitat when used as recommended. The phenoxy herbicides are moderately toxic to laboratory animals by oral or dermal intake and are only slightly toxic by inhalation. In the field, they are not toxic to livestock at dosages required for weed control, and they dissipate rapidly from soil, vegetation, and water sources. They do not accumulate in the food chain.

The phenoxys are practically nontoxic to soil organisms, and the soil microbial population is responsible for their rapid breakdown. They are moderately toxic to some species of fish as the ester formulation and only slightly toxic to lower aquatic organisms, birds, and wild mammals under laboratory conditions.

As indicated earlier, susceptible vegetation can be damaged from spray drift or from volatilization (especially high-volatile esters). Following label instructions and making applications during favorable weather should prevent drift and volatilization problems.

Breakdown of the phenoxys also occurs by sunlight and nonbiological means, as well as by soil microorganisms and metabolism by higher plants. Mobility of the phenoxys in the environment and leaching into the soil is very low because of the short-lived presence of the phenoxys. Extensive use worldwide indicates few if any health hazards to man. Few organic herbicides have a longer period of proven safety to animals and man than the phenoxy herbicides.

The phenoxy herbicides are considered selective in that they can control broadleaf weeds in grasslands or grass crops. Small amounts (0.25 to 2.24 kg/ha) are effective on many kinds of broadleaf plants. Extremely high rates will injure grasses.

The phenoxy compounds are readily absorbed by the leaves of most plants and are translocated to other plant parts with movement of food materials within the plant. Consequently, these compounds are effective in suppressing many "hard-to-kill" perennial broadleaf species, including woody plants. Although oil-soluble amines are available, the oil soluble formulations (usually esters) applied in kerosene or diesel oil will penetrate the bark of most woody plants and can be used as basal sprays (see Chapter 5) or foliar sprays to individual plants. The phenoxy herbicides, however, are commonly broadcast to large areas containing dense stands of weeds and brush. These herbicides should be sprayed on above-ground parts and foliage since they are not effective at economical rates as soil-applied herbicides.

The phenoxy compounds are relatively inexpensive and easy to apply. They are usually marketed as liquid concentrates as salts or esters that differ somewhat in their ability to kill plants. The ester formulations are usually most effective as foliar sprays on trees and brush, compared to the salts. The types of phenoxy herbicides commonly available as salts (amines) include dimethylamine, triethylamine, diethanolamine, trimethylamine, and triethanolamine.

As indicated earlier, amine salts can be formed by the reaction of 2,4-D acid with an organic compound containing an amino group ($-NH_2$). Other inorganic salts of the phenoxys that have been sold commercially include the ammonium, sodium, potassium, and lithium salts. They are formed by the reaction of the parent acid with an alkaili. Amines are sprayed in water carriers.

Esters are formed by combing the acid of 2,4-D with one of many different alcohols. The result is an oily liquid soluble in oil. Esters are classified as high-volatile or low-volatile, depending upon how readily they vaporize or give off fumes from the liquid phase. Esters will form an emulsion with water or water–oil carriers when properly formulated. The emulsion gives a milky appearance when mixed with water. The ester is identified by the alcohol used in the reaction. The less expensive and more abundant alcohols, such as methyl, ethyl, isopropyl, and butyl can be used, but are highly volatile when applied under high air temperatures. Such phenoxy compounds are identified on the label found on the herbicide container or package. The long-chain alcohols with an ether linkage have a lower volatility hazard than the short-chain alcohols. For this reason, they should be used in areas in which sensitive crops or vegetation are grown. The tendency for a herbicide to vaporize (volatilize) is important, since it may be carried from the target area and damage valuable vegetation under windy conditions or on hot days. Some low-volatile esters of 2,4-D or 2,4,5-T include the propylene glycol butyl ether, butoxyethanol, and isooctyl esters. Detailed information on the phenoxys is available (Bovey, 1980).

The concentrations of the active ingredient (ai) and the "acid equivalent" (ae) are indicated on the label as pounds per gallon. Acid equivalent refers to that portion of the formulation that theoretically can be converted to the acid. If an herbicide concentrate has an acid equivalent of 4 lb per gal, then 1 gal of the concentrate contains 4 lb by weight of the parent herbicidal acid, regardless of formulation. Usually the most concentrated formulations are most economical to use and usually cost less per lb than weaker concentrates. (See Chap. 12 for specific uses for each phenoxy herbicide on woody plants.) On most woody species, however, 2,4,5-T was more effective than other phenoxys, but 2,4,5-T has been replaced by other growth regulator-type herbicides even though they are not as economical to use.

Pokorny (1941) first synthesized 2,4-D and 2,4,5-T, and Zimmerman and Hitchcock (1942) were the first to report that 2,4-D and its derivatives were highly active compounds in inducing cell elongation (growth) and in causing formative effects in plants at extremely low dosages. Amchem Products Inc. introduced 2,4-D and 2,4,5-T as "weed killers" in 1944. MCPA, MCPB, and MCPP were developed in Great Britain in the 1940s and 1950s (Spencer, 1973; Martin and Worthing, 1974; Bovey, 1980; Burnside, 1996).

D. Pyridine Herbicides

Pyridine herbicides are growth-regulator types. Their chemical structures have pyridine rings. These include three very important herbicides for weed and brush control. They are clopyralid, picloram, and triclopyr. They are usually applied as foliar sprays, although they also have root-uptake activity.

1. Clopyralid (3,6-dichloro-2-pyridinecarboxylic acid) is the ai of a liquid formulation containing 3 lb ae/gal. It is formulated as a monoethanolamine salt, and is called Stinger, Reclaim, or Transline commercially. The structure of the monoethanolamine salt follows:

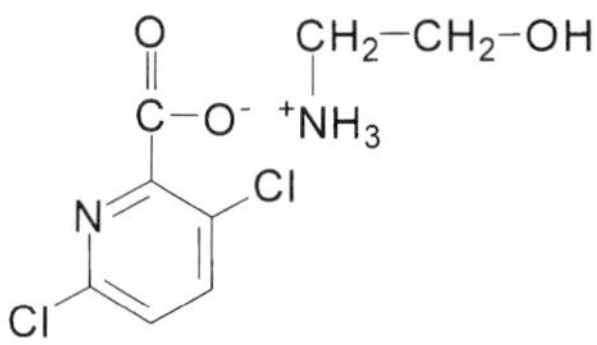

clopyralid monoethanolamine salt

Clopyralid is used for selective postemergence weed control in sugar beets, field corn, small grains, fallow, pastures and rangeland, Christmas tree plantings, and noncrop areas (Anderson, 1996). It is especially effective for honey mesquite (*Prosopis glandulosa* Torr.) control alone or combined with triclopyr (Bovey and Whisenant, 1992). Reclaim (clopyralid) is used in the southwest for mesquite, acacia, and other woody plants on rangeland. Clopyralid is moderately persistent and mobile in soil, but less persistent than picloram. (See Table 1 in Chapter 6.) It has a low order of toxicity to animals. It was discovered in 1961 and was first marketed in Europe (Weed Science Society of America, 1994).

2. Picloram (4-amino-3,5,6-trichloro-2-pyridinecarboxylic acid) is a selective, translocated herbicide that effectively controls many weed and brush species of grasslands and other areas. It is effective for soil or foliage application and can be effective as injection/cut-surface or cut-stump treatments on undesirable trees.

   Picloram can be applied in liquid sprays and as granules to brush in the spring and fall months, depending upon the species to be controlled. It is both foliar and root-absorbed. Most perennial grasses are resistant. Its high activity against many woody plants at moderate rates makes it a desirable brush killer.

Care must be taken to prevent drift of picloram to desirable plants and water sources, as was discussed with the phenoxy herbicides. Picloram is relatively persistent in the soil, especially in cooler climates. Since it is water-soluble, care must be taken to prevent its movement into water used for irrigation. It should not be applied where it can be leached or moved in significant amounts to sensitive crop areas by rainfall. Picloram has a low mammalian toxicity. It is only slightly corrosive to spray equipment (Bovey, 1995).

A common trade name is Tordon 22K (2 lb ae/gal liquid). It is usually formulated as a potassium salt. Picloram is a restricted-use pesticide because of its phytotoxicity to susceptible nontarget plants. The chemical structure picloram is

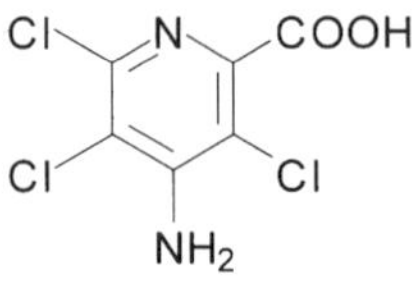

picloram

Picloram was discovered in 1960 by Dow Chemical Co. and was introduced commercially in 1963 (Weed Science Society of America, 1994).

3. Triclopyr {[(3,5,6-trichloro-2-pyridinyl)oxy] acetic acid} is a selective postemergence herbicide for use on rights-of-way, pastures, rangeland, industrial locations, and forestry sites. Triclopyr has been approved for application on honey mesquite as foliar sprays or basal stem treatments in an oil carrier. It also can be used to kill trees and brush by injection/cut-surface treatments. Mixtures of triclopyr + clopyralid (1:1) readily control honey mesquite and 1:1 mixtures of triclopyr + picloram have been used on mixed brush species, including honey mesquite, in the southwest United States. Triclopyr is readily translocated in plants and is reportedly more effective than 2,4,5-T on some species. It is moderately toxic to warm-blooded animals, and degrades moderately rapidly in the soil (Bovey, 1995). The chemical structure is

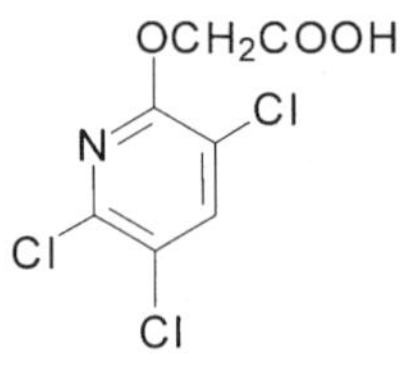

triclopyr

The butoxyester form, Grazon ET, Garlon 4 (4 lb ae/gal), and the triethylamine salt Garlon 3 or Grandstand (3 lb ae gal) are used. Like clopyralid and picloram, triclopyr is an auxin-type herbicide. It was introduced by Dow Chemical in 1975.

E. Sulfonylureas

Sulfonylurea herbicides were first introduced in the early 1980s and were unique, since very low rates—often less than 0.25 oz ai per acre—were effective on weeds. They inhibit amino acid synthesis in plants, especially in shoots and roots. They inhibit the enzyme acetolactate (ALS) necessary for the biosynthesis of valine, leucine, and isoleucine, and deprive weeds of these essential amino acids. They have a very low mammalian toxicity because they act on a single enzyme not present in mammals (Weed Science Society of America, 1994; Anderson, 1996).

The sulfonylureas have the general structure indicated. Sulfonylurea herbicides are used to control broadleaf weeds and a few annual grasses in certain crops, and to provide nonselective control of annual and perennial grass and broadleaf weeds in noncrop and industrial areas (Anderson, 1996). The sulfonylureas are readily absorbed by both the roots and shoots of plants and are translocated via xylem and phloem to meristematic tissue. The sulfonylurea of most interest to woody plant control is metsulfuron, commonly known as Ally or Escort.

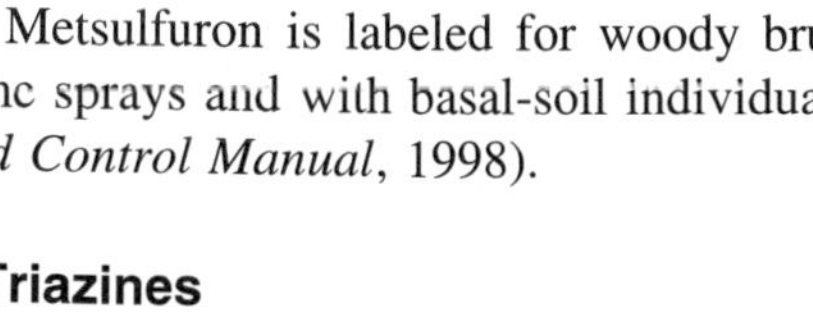

metsulfuron-methyl

Metsulfuron is labeled for woody brush and vines with low- and high-volume sprays and with basal-soil individual plant treatment on multiflora rose (*Weed Control Manual*, 1998).

F. Triazines

Hexazinone [3-cyclohexyl-6-(dimethylanino)-1-methyl-1,3,5-triazine-2,4-(1*H*, 3*H*)-dione] is the active ingredient of a wettable powder (Velpar), 90% ai weight per weight (w/w), a water-dispensable liquid (Velpar L) containing 2 lb ai/gal and a water-soluble granule (Velpar ULW) with 75% ai w/w. Hexazinone is used for selective weed control in dormant alfalfa, Christmas tree planting, pineapple, and sugarcane. It is also used for weed control in forestry sites and weed and brush control in noncrop areas (Anderson, 1996). It is registered on rangeland for individual woody plant control. Hexazinone is highly water-soluble, and can move on the soil surface from the treated site after heavy rainfall to nontarget areas (Table 1). Hexazinone has a moderately long residue in soil and is mobile on the soil surface. It has low toxicity to mammals. Its structure has been indicated (Weed Science Society of America, 1994).

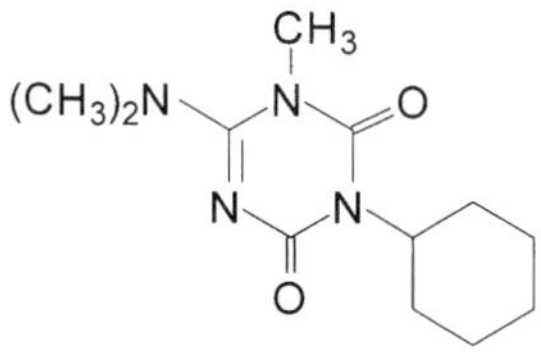

hexazinone

It is readily absorbed by roots and translocated upward in the xylen, and it inhibits photosynthesis by binding to the Q_B-binding niche on the D1 protein of the photosystem II complex in chloroplast thylakoid membranes and blocking electron transport from Q_A to Q_B. The biological properties of hexazinone were first described in 1975 by DuPont patents (Weed Science Society of America, 1994).

G. Ureas and Uracils

These compounds include bromacil, diuron, fenuron, monuron, and tebuthiuron. Ureas and uracil-type herbicides can be selective at low rates and nonselective at high rates. They usually are formulated as wettable powders for water sprays or as granules or pellets for dry application.

Fenuron, fenuron-TCA, monuron, and monuron-TCA are no longer used commercially in the United States, but fenuron and monuron are apparently registered for use in some countries. These compounds will therefore not be discussed.

Bromacil [5-bromo-6-methyl-3-(1-methylpropyl)-2,4 (1*H*, 3*H*) pyrimidinedione] is a uracil that controls a variety of woody plants. If rates above 5.6 kg/ha are used it also kills many desirable grasses and forbs on grazing lands. The chemical structure follows.

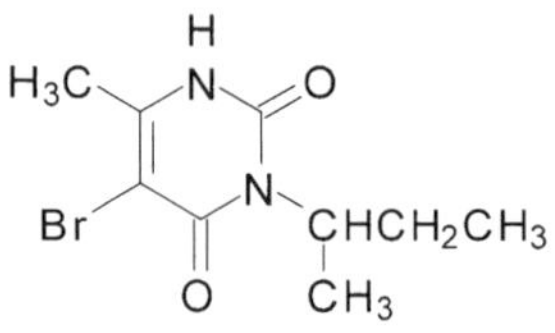

bromacil

Bromacil is used primarily for weed and brush control on noncrop areas. It controls perennial grasses and is used for selective weed control in pineapple and citrus. It is formulated as a granular water-soluble liquid and wettable powder. It has a low mammalian toxicity and is nontoxic to bees, but has an LC_{50} at 28 ppm to rainbow trout and LC_{50} at 71 ppm to bluegill (*Farm Chemicals Handbook*, 1997). It is moderately persistent and mobile in soil. Bromacil is readily absorbed by roots and translocated in the xylem to the leaves. It inhibits photosynthesis by binding onto the pigment protein of the photosystem II complex in the thylakoid membrane of the chloroplasts, thus interfering with normal electron transport into the plastoquinone pool (Weed Science Society of America, 1994; Anderson, 1996). Substituted uracils were first described in 1962 (Weed Science Society of America, 1994). Bromacil is manufactured by the bromination of 3-*sec*-butyl-6-methyluracil (Martin and Worthing, 1974).

Tebuthiuron {*N*-[5-(1,1-dimethylethyl)-1,3,4,-triadiazol-2-yl]-*N*,N′-dimethylurea} is a substituted urea and has the following structure.

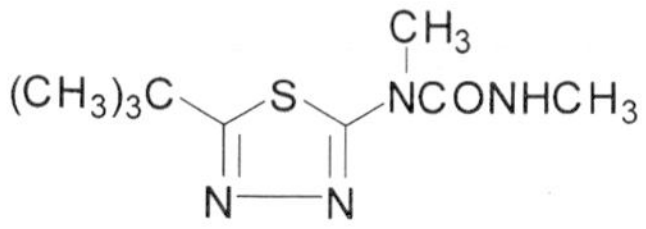

tebuthiuron

Tebuthiuron is available in a pellet (Spike 20P) that contains 20% ai w/w. Tebuthiuron is used for woody plant control. It is extremely active and will injure or kill woody plants with roots that extend into treated areas (Anderson, 1996). Tebuthiuron is a root-active, soil-applied herbicide, and sufficient rainfall is required to move the herbicide into the root zone. It controls a large number of woody plants and also injures or controls herbaceous vegetation, especially at rates exceeding 4.5 kg/ha. It can be applied in pellet form in South Texas for live oak (*Quercus virginianae* Mill) control at any time during the year, with dramatic increases in grass cover at rates of 2.2 and 4.5 kg/ha. Herbaceous cover will usually increase by the second growing season (Bovey et al., 1985).

The absorption, translocation, and mode of action of tebuthiuron are similar to those for hexazinone and other ureas and uracils. The LD_{50} for technical-grade tebuthiuron acute toxicity for rats is 644 mg/kg. Its water solubility is 2500 mg/L at 25°C and is stable to ultraviolet (UV) light. It persists in the soil for a considerable amount of time, and that helps make it an effective woody plant killer. Persistence can be considerable in areas of low rainfall (>1 year). Surface mobility of tebuthiuron is limited due to its solubility and low K_d values, which allow rapid movement into the soil (Weed Science Society of America, 1994). Synthesis and analytical methods are available (Weed Science Society of America, 1994). Tebuthiuron was discovered in 1970 by Air Supply, Inc., and was first marketed in Brazil by Eli Lilly & Co. under British patent 1,266,172 (Weed Science Society of America, 1994).

The ureas and uracils mentioned are absorbed primarily through the roots of woody plants. They may be applied in the spring or fall, when weeds and brush are actively growing and when adequate rainfall leaches them into the root zone.

Fall, winter, and early spring applications of materials such as tebuthiuron can be made, which reduces injury to forage plants and eliminates hazards of drift or movement in surface runoff water to susceptible crop areas. These compounds may kill trees a great distance from the point of application, depending upon the size of the root system and whether or not it extends into the treated area. The compounds are nonvolatile and do not corrode equipment.

Most of the urea and uracil herbicides can be injurious to some forage species when applied broadcast, especially as sprays. The method of application, whereby the materials are applied as pellets, balls, or spot-basal treatments to confine the herbicide to fewer spots in the treated area, reduces exposure to desirable forage plants. Application can be made with aerial or ground equipment. Also, application of granules, pellets, or sprays to the soil surface and spray subsurface in bands spaced 2 to 3 meters apart reduces herbicide injury to forages (Bovey et al., 1976; Flynt et al., 1976).

Most of the ureas and uracils may persist in the soil for several months at rates used for brush control and are low in toxicity to warm-blooded animals.

H. Other Organic Herbicides

1. Amitrole (3-amino-1,2,4-triazole) is a restricted use pesticide because it has been proposed to cause cancer in laboratory animals. Its chemical structure is as follows:

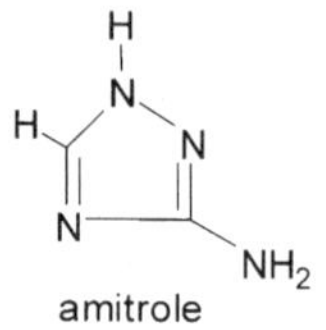

amitrole

Amitrole is generally used as a nonselective systemic herbicide to control annual grasses, annual and perennial broadleaf weeds, and certain woody plants and vines in industrial and other noncrop areas (Anderson, 1996) under the trade names Amizol (water-soluble powder—90% ai w/w) and Amitrol-T (aqueous—2 lb ai/gal). Amitrole is effective against poison ivy, poison oak, and certain other woody plants. Amitrole is effective through the roots and tops of plants, but cannot be used where there is any possibility of residues on food or feed crops. Amitrole decomposes rapidly in warm, moist soils in about 1 week. It was first introduced by Union Carbide in 1954. It is synthesized by condensation of formic acid with aminoguanidine (Weed Science Society of America, 1994).

2. Fosamine [ethyl hydrogen (aminocarbonyl) phosphonate)], commonly known as Krenite, has the following chemical structure:

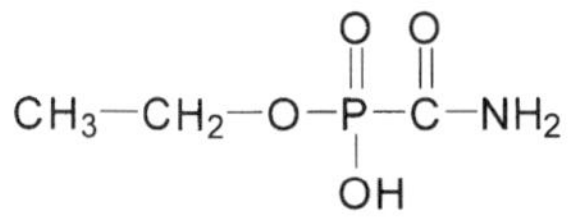

fosamine acid

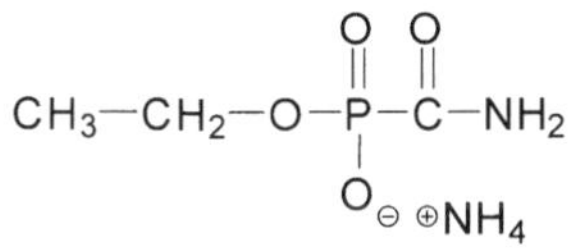

fosamine ammonium salt

It is formulated as an ammonium salt and is applied foliar for brush control on noncrop land or for site preparation for conifer planting. In most woody plants, fosamine injury is not apparent before normal leaf senescence in the fall. Leaf and bud development may be severely or completely inhibited the following spring, however. Fosamine translocation occurs both symplastically and apoplastically, and accumulation occurs rapidly (within 1 day after treatment). Complete coverage of the plant is necessary for the best results. Fosamine is synthesized by reacting triethyl phosphate with methyl chloroformate and ammonia. Fosamine was first described by DuPont in 1974.

3. Glypholsate([*N*-(phosphonomethyl) glycine] has the following structure:

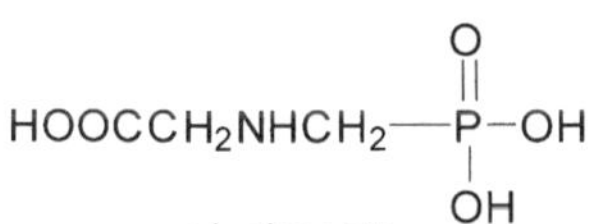

glyphosate

It is commonly formulated as an isopropyl-amine salt at 3 lb ae/gal. Trade names include Accord, Honcho, PondMaster, Protocol, and Roundup (Weed Science Society of America, 1994).

Glyphosate, a nonselective herbicide, is effective against both grasses and broadleaf plants. It is readily translocated from leaf and stem tissue to roots, killing a high percentage of many weeds. Glyphosate is inactivated by soil contact and should not injure newly seeded forages planted in treated soil. It is sprayed in a water carrier and is not corrosive to equipment. Glyphosate is registered for noncrop use and pretill weed control and as a directed spray for orchards, plantations, Christmas trees, and many other crops. Broadcast sprays over woody species will damage desirable forage plants, so applications should be made to individual plants on noncrop areas or areas to be renovated. Glyphosate has a low order of mammalian toxicity (Bovey, 1995).

Herbicidal activity was first reported in 1971 by Monsanto Co. Trimethylsulfonium salt was introduced in Spain in 1989 by ICI Agrochemicals (Weed Science Society of America, 1994), and is called Touchdown.

4. Imazapyr {(±)-2-[4,5-dihydro-4-methyl-4-(1-methylethyl)-5-oxo-1[*H*-imidazol-2-yl]-3-pyridinecarboxylic acid}, commonly known as Asenal, is an imidazolinone herbicide first introduced by American Cyanamid in the late 1970s. Imazapyr has the following chemical structure:

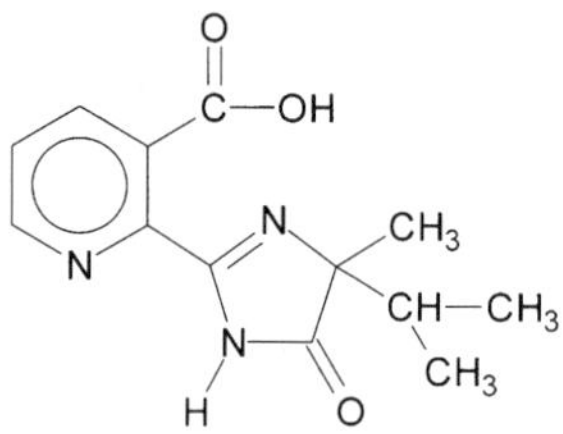

imazapyr

Imazapyr controls a wide variety of annual and perennial weeds, deciduous trees, vines, and brambles in noncrop situations. Research is being conducted to develop a granular product for use in site preparation and pine release in forestry. Imazapyr is available as an oil-soluble formulation for cut-stump, basal-bark, and frill application called Chopper (Bovey, 1995).

The imidazoline herbicides inhibit ALS, which interferes with the biosynthesis of the branched-chain amino acids isoleucine, leucine, and valine, as described for the sulfonylurea herbicides. Imazapyr has a low order of toxicity since it acts on an enzyme not present in mammals (Anderson, 1996; Weed Science Society of America, 1994). Imazapyr is persistent in soil, making it a valuable perennial weed and woody plant control agent.

5. Diesel oil and kerosene are commonly used to control honey mesquite and hiusache. From 0.25 to 4 L of oil are used per tree, depending on the trees' size. The oil is applied around the base of the tree during

dry weather when the soil is pulled away from the trunk. Application at this time enables oil to penetrate to the lower buds on the stem (Bovey, 1996).

Oils alone are not very effective herbicides when applied to the foliage of woody plants. Diesel fuel is commonly used as a diluent and carrier for oil-soluble herbicides for individual plant treatments. It is also used with and without water as a carrier in aircraft spraying (Bovey, 1995). Among the advantages of using oil as an herbicide carrier is that it may increase foliar retention, absorption, and spray drift loss (Anderson, 1996).

III. HERBICIDE COMBINATIONS

Herbicide combinations have become popular in both agronomic and grazing land herbicide use. Herbicide combinations sometimes offer the advantages of improved weed control, a greater spectrum of weeds controlled, reduced cost, and reduced residue problems. One of the first successful combinations of organic herbicides for brush control included a 1:1 mixture of 2,4-D + 2,4,5-T. The 2,4-D was useful in the mixture to reduce costs on some weeds without sacrificing effectiveness.

The picloram + 2,4,5-T (1:1) combination was more effective on honey mesquite as a foliar spray than either herbicide alone at comparable rates (Robison, 1967; Bovey et al., 1968). Bovey et al. (1968) found that 2,4,5-T sometimes could be added in equal amounts to picloram to increase control or reduce picloram rates proportionately on huisache, honey mesquite, and live oak.

Scifres and Hoffman (1972) found that combinations of dicamba + 2,4,5-T were no more effective on honey mesquite than either herbicide alone at equal rates. Dicamba, however, could be substituted for 2,4,5-T in combination with picloram. Meyer and Bovey (1973) found that mixtures of dicamba + picloram were more effective on honey mesquite that 2,4,5-T + picloram. Huisache was more effectively defoliated with picloram or picloram + dicamba than 2,4,5-T, dicamba, or other mixtures. On Macartney rose, however, some mixtures of picloram + 2,4-D or 2,4,5-T were no more effective than picloram alone. On live oak, the most effective treatments generally contained at least 1.12 kg/ha of picloram, either alone or in combination with dicamba or 2,4, 5-T. Picloram alone killed as many or more whitebrush plants than MCPA, dicamba, 2,4,5-T, or 2,4-D or mixtures of these herbicides.

These data indicate that different woody species may be most effectively controlled by a specific herbicide or herbicide combination. Nevertheless, the 2,4,5-T + picloram (1:1) mixture was effective on a number of woody plants, including huisache, catclaw, blackbrush, speckled alder, pricklypear and tasajillo cactus, winged elm, hackberry, honey mesquite, post oak, sand shinnery oak, wait-a-minute bush, and yaupon (Bovey, 1977).

More recently, a 1:1 mixture of triclopyr + clopyralid provided excellent control of honey mesquite when applied as broadcast foliar treatments (Bovey and Whisenant, 1992). Other herbicide mixtures for honey mesquite management include triclopyr + picloram, clopyralid + picloram, dicamba + picloram, triclopyr + dicamba, 2,4-D + triclopyr (Crossbow), 2,4-D + dicamba (Weedmaster), and 2,4-D + picloram (Grazon P + D) (Welch, 1997).

When a number of brush species are present, including mesquite in South Texas, triclopyr + picloram, clopyralid + picloram, or dicamba + picloram are considered.

Combinations of 2,4-D + picloram (4:1) are effective as sprays on some woody plants, but are more effective as injection/cut-surface treatments on such species as blue ash, green ash, flowering dogwood, parsley hawthorn, black hickory, mockernut hickory, American hornbeam, sweetbay magnolia, red maple, blackjack oak, post oak, water oak, common persimmon, and sweetgum (Bovey, 1977).

Bromacil (Hyvar-X) may be tank-mixed with dicamba, 2,4-D amine, MSMA, or paraquat to improve control on emerged weeds. Bromacil and diuron are available as premixes in granular form. Krovar I DF is 40% bromacil and 40% diuron, whereas Krovar II DF contains 53% bromacil plus 27% diuron by weight (Anderson, 1996; *Farm Chemicals Handbook*, 1997). Krovar is a broad-spectrum herbicide used in citrus crops and noncrop areas.

IV. READING THE HERBICIDE LABEL

The herbicide label on each herbicide container or package deserves special attention since it represents the cumulation of much research effort of the combined recommendations and thinking of competent industry, university, and government scientists. Taking time to thoroughly read the label will result in safer pesticide applications and may save much embarrassment and expense from misuse. Some important information found on the pesticide label includes the following.

1. Chemical name
2. Chemical formula
3. Active ingredients
4. Inert ingredients
5. Suggested uses
6. Directions for use
7. Precautions and warnings
8. Antidotes
9. Manufacturer

V. HERBICIDE EVALUATION

A. Woody Plants

The effectiveness of an herbicide treatment in the greenhouse or field can be accurately measured. Visual estimates of individual plants for a percentage of canopy reduction are made. Usually 10 or more plants per replication are evaluated. Ratings can be made a few months after treatment to determine the relative effectiveness of an herbicide, but in order to obtain the percentage of dead plants in the field, ratings are usually made at least 2 to 3 years after application. Plants showing no living foliage or sprouts at that time are considered dead. The percentage of canopy reduction and the percentage of dead plants can then be calculated from the values obtained. If individual plants cannot be distinguished (multistemmed brush), an overall percentage of canopy reduction of an entire research plot or subplot can be taken. The percentage of stem kill ratings of individual plants or groups of plants can also be made in treated and untreated areas. Checks or untreated areas (plots) are always included for comparison to treated plots to determine herbicide efficacy.

B. Herbaceous Vegetation

The effect of an herbicide treatment on grasses and forbs is very important since we usually select this type of vegetation for livestock consumption and wildlife habitat. Visual estimates can be made of the composition of the vegetation and approximate yield (oven-dry weight in kg/ha). Such herbicide-treated areas have to be protected from grazing animals by fencing or cages. Many methods are available for measuring and characterizing herbaceous vegetation (Phillips, 1959); however, clipping vegetation on a treated area, separating major species or vegetation types, weighing the samples, and determining oven-dry forage yield (kg/ha) of each component is the usual method.

VI. FACTORS AFFECTING RESULTS

When herbicides are applied as foliar sprays, they usually are effective only during certain periods of the year, commonly in the spring, after the leaves of woody plants have fully expanded and plants are growing actively. Also, some weeds may be susceptible for only a short time during their life cycle (seedling stage); they may be virtually unaffected later. Foliar damage from insects, hail, or drought also may render the treatment unsuccessful due to poor absorption and translocation of the chemical. Consequently, foliar sprays of herbicides should be applied when plants have developed under favorable soil moisture and temperature and environmental conditions. Each weed or brush species, however, may respond differently. Consult the proper reference or agricultural authority before wasting money on an improper treatment.

Spraying the proper rate of herbicide onto the foliage at the right time is the first step in successful application. Proper application equipment is essential. One of the most important factors is application at low wind velocity—below 16 km/hr for ground equipment and below 10 km/hr for aerial equipment. Applying herbicide when winds are above these velocities may result in a loss of the chemical from the target area and damage to adjacent areas from spray drift. If some wind is encountered, be certain the wind direction is away from susceptible crops or valuable vegetation.

Good coverage of foliage with herbicides is essential for satisfactory results. Small spray droplets of 100 μm in diameter or less give better coverage and results than do large spray droplets of 300 to 400 μm. Small droplets are subject to drift; therefore, a balance between small and large droplets is desired. Spray droplet size can be increased by modifying spray solution viscosity, nozzle type, and spray pressure.

Once on the stems and leaves, the herbicide must be absorbed in large enough amounts to be translocated throughout the plant. Barriers to absorption include the waxy outer covering of leaf surfaces and the bark on woody plants. Absorption of plant surfaces sometimes can be improved by oil-in-water carriers or by the use of small amounts of surfactants or wetting or penetrating agents in water carriers. Plants differ in their ability to absorb herbicides; absorption usually is more rapid in young plants.

Translocation of the herbicide throughout the plant or to the site of action (stem and root tissue) is necessary for control. Most herbicides move best within plants during favorable environmental conditions when the plant is growing actively. In some foliar-applied herbicides, such as the phenoxys, rates that are too high may reduce translocation and overall control. These herbicides can disrupt the tissues responsible for transport in the plant.

Rainfall before and soon after treatment is desirable to leach pelleted herbicides into the root zone of the weeds and brush as well as to stimulate growth. Herbicides that are absorbed by roots move up through the plant to kill it. Tebuthiuron applied to the soil as pellets can be applied in the Southwest at any time of year for controlling certain woody species. Tebuthiuron is not readily broken down by sunlight and will remain intact on the soil surface until adequate rainfall leaches it into the soil. Picloram and dicamba are broken down by sunlight; avoid application during hot, dry months.

Esters of triclopyr or similar herbicides in diesel oil carrier commonly are used as basal sprays. These treatments kill trees up to 13 cm in diameter. Treatments can be applied any time of the year. If the bark or soil is wet, effectiveness is reduced. For trees larger than 13 cm in diameter the bark often is too thick to penetrate with sprays, therefore the herbicide can be applied to the sapwood through frills or notches cut into the bark. In winter, species that are difficult to control may require higher herbicide concentrations or closer spacings of cuts around the tree.

VII. THE FUTURE OF WOODY PLANT MANAGEMENT WITH HERBICIDES

The search will continue for new herbicides and formulations that are safe, effective, and economical. Development has been slowed by high production costs and application and legal restrictions. Herbicides such as the sulfonylureas are desirable for weed and brush control on grazing lands, since lower amounts in the environment reduce the risk of residues in forages, livestock, wildlife, water sources, or soil.

Methods and techniques to enhance the efficacy, safety, and economics of existing herbicides may result from new formulations, herbicides, and mixtures, along with improved placement. New herbicides are needed for grasses and forb establishment on grazing lands.

Improved delivery systems are required to apply new herbicides at extremely low rates per ha. Aircraft can now apply small amounts of pelleted herbicides (Bouse et al., 1982). Low-volume application from ground sprayers with controlled droplet atomizers and electrostatic sprayers is now feasible (Bode and Butler, 1983). Electronic sensors, monitors, and control systems for broadcast or spot application assure more precise and less costly herbicide application. Ropewick applicators and wiping devices assure better placement and reduce the amount of herbicide used. These and future developments will cut costs and improve herbicide efficacy and safety.

REFERENCES

Anderson WP. Weed Science Principles and Applications. 3rd ed. New York: West, 1996.

Bovey RW. Response of Selected Woody Plants in the United States to Herbicides. USDA-ARS, agric. handbook 493, 1977.

Bovey RW. Chemistry and production of 2,4,5-T and other phenoxys. In: The Science of 2,4,5-T and Associated Phenoxy Herbicides. New York: Wiley, 1980, pp. 29–47.

Bovey RW. Weed management systems for rangelands. In: Smith AE, ed. Handbook of Weed Management Systems. New York: Marcel Dekker, 1995, pp. 519–552.

Bovey RW. Use of 2,4-D and other phenoxy herbicides on pastureland, rangeland, alfalfa forage and noxious weeds in the United States. In: Burnside OC, ed. Biologic and Economic Assessment of Benefits from Use of Phenoxy Herbicides in the United States. USDA-NAIAP rept. no. 1-PA-96, 1996, pp. 76–86.

Bovey RW, Whisenant SG. Honey mesquite (*Prosopis glandulosa*) control by synergistic action of clopyralid:triclopyr mixtures. Weed Sci 40:563–567, 1992.

Bovey RW, Davis FS, Morton HL. Herbicide combinations for woody plant control. Weed Sci 16:332–335, 1968.

Bovey RW, Meyer RE, Bouse F, Carlton JB. Seasonal response of woody plants to tebuthiuron pellets. Weed Sci 33:551–554, 1985.

Bovey RW, Flynt TO, Meyer RE, Baur JR, Riley TE. Subsurface herbicide application for brush control. J Range Mgt 29:338–341, 1976.

Bovey RW, Wiese AF, Evans RA, Morton HL, Alley HP. Control of Weeds and Woody Plants on Rangelands. AB-BU-2344. St. Paul: USDA and University of Minnesota, 1984.

Burnside OC. The history of 2,4-D and its impact on development of the discipline of weed science in the United States. In: Burnside OC, ed. Biologic and Economic Assessment of Benefits from Use of Phenoxy Herbicide in the United States. USDA-NA PIAP rept. no. 1-PA-96. 1996, pp. 5–15.

Crafts AS, Robbins WW. Properties and function of herbicides. In: Weed Control. New York: McGraw-Hill, 1962, pp. 235–241.

Farm Chemicals Handbook. Willoughby, OH: Meister Publishing, 1997.

Flynt TO, Bovey RW, Meyer RE, Riley TE, Baur JR. Granular herbicides applicator for brush control. J Range Mgt 29:435–437, 1976.

Martin H, Worthing CR, eds. Pesticide Manual. 4th ed. Worcester, England: British Crop Protection Council, 1974.

Meyer RE, Bovey RW. Control of woody plants with herbicide mixtures. Weed Sci 21: 423–426, 1973.

Phillips A. Methods of Vegetation Study. Henry Holt, 1959.

Pokorny R. Some chlorophenoxy acetic acids. J Am Chem Soc 63:1768, 1941.

Robison ED. Response of mesquite to 2,4,5-T, picloram and 2,4,5-T/picloram combinations. Proc South Weed Sci Soc 18:293–298, 1967.

Scifres CJ, Hoffman GO. Comparative susceptibility of honey mesquite to dicamba and 2,4,5-T. J Range Mgt 25:143–146, 1972.

Spencer EY. Guide to the Chemicals Used in Crop Protection. Pub. 1093. Ontario, Canada: Res. Branch, Agric. Canada, 1973.

Weed Control Manual. Willoughby, OH: Meister Publishing, 1998.

Weed Science Society of America. Herbicide Handbook. 1st ed. Geneva, NY: Weed Science Society of America, WF. Humphrey Press, 1967.

Weed Science Society of America. Herbicide Handbook. 7th ed. Champaign, IL: Weed Science Society of America, 1994.

Welch TG. Chemical Weed and Brush Control, Suggestions for Rangeland. B-1466. College Station, TX: Tex. Agric. Ext. Serv. Texas A&M University, 1997.

Zimmerman PW, Hitchcock AE. Substituted phenoxy and benzoic acid growth substances and the relation of structure to physiological activity. Contrib Boyce Thompson Inst 12:321–343, 1942.

Zimmerman PW, Hitchcock AE. Growth regulating effects of chloro substituted derivatives of benzoic acid. Contrib Boyce Thompson Inst 16:419–427, 1952.

5

Herbicide Application Technology

I. INTRODUCTION

As discussed in Chapter 4, formulated herbicides are of little value if the methods of application and the equipment are not available to treat weed problems in a safe and practical manner. The equipment for applying the chemicals must disperse the material in sometimes very small amounts at a uniform rate per ha (acre). Spraying is the most common method of applying herbicides. Sprays are useful since small quantities of herbicides can be diluted to permit even coverage of plant or soil surfaces.

Woody plants can be killed with herbicides in different ways. Herbicides can be applied by

- Spraying onto foliage
- Wiping onto stems and foliage
- Spraying basal bark or stumps
- Injecting into the sapwood of trees by mechanical devices or through frills or notches cut into the tree
- Using soil application

II. METHODS OF APPLICATION

A. Broadcast Sprays

Spray applications are classified as low-volume or high-volume sprays (Klingman, 1961). Low-volume sprays are usually between 9.4 and 280 L/ha (1 to 30 gal of total spray per acre). Ultra-low-volume spray signifies a total spray volume of 4.7 L/ha (1/2 gal per acre) or less, which may be undiluted herbicide. High-volume sprayers apply from 280 to 4,675 L/ha (30 to 500 gal or more per acre).

1. Low-Volume Sprays

As discussed in Chapter 3, several herbicides can be applied as low-volume broadcast foliar treatments for brush control. Low-volume spraying is usually quicker, easier, and less expensive to apply than high-volume sprays. The best control of large woody plants or resistant species may require complete coverage of the plant with a high-volume spray, however.

Ground equipment is practical for applying low-volume sprays to low-growing brush and for regrowth on land that has been cleared mechanically. Low-volume sprays applied in swaths up to 15 meters wide with either a spray boom with several nozzles or a large boomless nozzle have produced satisfactory coverage of plant foliage. Spray booms on ground sprayers should be mounted to clear the tallest brush by 1 meter. A spraying pressure of 207 to 275 kPa (30 to 40 lb per square inch) is used. Boomless nozzles should be mounted to clear all brush by 1 meter. Spray pressures from 207 to 414 kPa (30 to 60 lb psi) are used.

Aerial spraying is best for treating large areas or tall and dense stands of brush on rough terrain. Both fixed-wing (airplanes) and rotary-winged (helicopters) aircraft are used for aerial spraying. Helicopters are particularly useful for spraying rough terrain and small, irregular areas, and they often are used for spraying rights-of-way. Aerial spraying can give good coverage in most brush. The spray that penetrates to the lower levels in tall brush is often inadequate to kill understory plants, however. A second aerial spraying may be necessary a year or two after the first. When applying spray by aircraft, flying as close to the top of the brush as is safe is recommended. Using experienced flagmen or guides on the ground to mark individual flight swaths assures precise application. Proper swath width should not exceed 1-1/2 times the wingspan or bladespan of the aircraft.

2. High-Volume Sprays

High-volume sprays involves the application of herbicides to kill brush along roads, rights-of-way, fence rows, and other similar areas, or individual plants on pastures and rangeland. All foliage and twigs should be thoroughly wetted. High-volume sprays are usually applied to foliage with power equipment. The power sprayer should be capable of maintaining pressures up to 690 kPa (100 lb psi). This pressure is enough to force the spray through the foliage and to the tops of taller trees. Pressure higher than 690 kPa tends to form fine spray droplets that may drift and damage nearby susceptible vegetation.

Adjustable hand-operated sprayers are suitable for applying sprays to low-growing brush. The best coverage is obtained with a fan or cone-type nozzle that has a spray angle of about 40°. The nozzle should be attached at a 45° angle to an extension tube generally about 60 to 90 cm (24 to 36 in.) long. To reach the tops of tall trees or brush too far away for a wide-angle spray, the gun can be adjusted to deliver a narrow-angle stream. Adequate pressure for hand-operated equipment varies from 172 to 275 kPa (25 to 40 lb psi). Because of the large volume of spray necessary for this method, it is impractical for treating areas in which the water supply is limited, on large areas, or in areas into which truck- or tractor-mounted sprayer tanks cannot be taken.

B. Individual Plant Treatments

Individual plant treatment is especially useful for the control of undesirable, hard-to-kill species that sometimes occur in scattered stands, such as woody plants. Individual plant treatment includes foliage sprays, basal sprays, cut-surface and injection, and wiper and soil treatments (Bovey, 1977).

1. Foliar Sprays

Foliar sprays of the growth-regulator herbicides, such as 2,4-D, dicamba, picloram, and triclopyr, are usually applied in the spring and summer to individual plants or groups of plants when they are actively growing and after the leaves have reached full size. Low- or high-volume sprays can be used. Ester formulations are the least likely to be washed off should rainfall occur soon after application and are usually more effective on woody species than amine formulations. If complete coverage is necessary to kill a given weed species, high-volume (drench) sprays that wet all foliage, twigs, and terminal stems may be desirable.

2. Basal Sprays

Basal sprays are used to treat bark at the base of individual plants to the point of runoff. Mixtures of ester formulations of 2,4-D or triclopyr with diesel oil or kerosene carriers or oil alone commonly are used. Basal sprays can be used to kill brush and trees with main stems up to about 13 cm in diameter any time of the year. The effectiveness of the treatment may be reduced if the bark or soil is wet. The stem should be sprayed from the soil line to 45 cm above ground. Four L of spray will treat 50 trees 5 cm in diameter. Compressed air sprayers, knapsack sprayers, or power sprayers can be used.

Low-volume and streamline basal applications of herbicides for the control of brush were introduced to Texas rangeland in late 1987 by the extension service (Welch, 1997). Since then, many result demonstrations have been established utilizing these two techniques. Most of the demonstrations have had mesquite as the target species, although other brush species have been treated.

Low-volume basal uses a solution containing 25% herbicide and 75% diesel fuel oil. The solution is applied to the lower 30 to 46 cm of the stem with a fan or hollow-cone nozzle to wet the stem (but not to the point of runoff), thus less total volume is used per plant than in the conventional basal treatment that applies sufficient solution to allow runoff with puddling at the base of the plant. With low-volume basal, the herbicide solution must be applied completely around the stem. Because the mixture is applied only to wet the stem, the herbicide must penetrate the bark of the plant.

Streamline basal application utilizes a mixture of 25% herbicide, 10% penetrant, and 65% diesel fuel oil, or 25% herbicide and 75% diesel fuel oil. The penetrant that has been tested extensively is d,1 limonene. It is sold under the product names of Cide-Kick II and Quick Step II. It is a naturally occurring chemical derived from by-products of the citrus industry and the pine industry. The mixture is sprayed in a band (7–10 cm wide) with a straight stream nozzle to one or two sides of the stem near the ground level or at the line dividing young (smooth) and mature (corky) bark. The material must go completely around the stem. The best results have been obtained on stems that are less than 10 cm in diameter and that have smooth bark.

Streamline basal and low-volume basal application of herbicides are very useful methods for brush control. They may be used for treating a thin stand of brush, for spot treatment, and for controlling brush near desirable plants. Thinline basal bark treatment can be achieved with applications of undiluted herbicide (triclopyr) in a thin stream to all sides of the stems about 15 cm above the base of the plants. From 2 to 15 mL of chemical are required for single stems and from 25 to 100 mL for the treatment of clumps of stems.

When treating woody plants during the dormant season, 3 to 6 qt (2838 to 5676 mL) of triclopyr are added to diesel fuel to make 100 gal of spray. A knapsack or power sprayer with low spray pressure is used when the brush is dormant and after leaf drop. The upper parts of the stem or trunk are thoroughly wet to about 30 cm above the ground to the point of runoff. For oil–water emulsions 6 qt of triclopyr are mixed with 25 gal of oil and 1.5 gal of emulsifier (Triton X-100) in water to make 100 gal of solution. Consult the herbicide label for specific applications methods and mixing instructions for specific woody plants and conditions.

3. Brush Busters Technique

The most recent technology for honey mesquite control in the Southwest includes the Brush Busters leaf or stem spray (McGinty and Ueckert, 1995). The leaf or stem spray involves mixing clopyralid (Reclaim) or triclopyr (Remedy) at 0.5% by volume of the spray solution (water). A surfactant and a dye (HiLite blue dye) to mark sprayed plants make up 0.25% and 0.25 to 0.5% of the spray solution, respectively. Diesel fuel oil at 5% with an emulsifier can be substituted for the surfactant. Honey mesquite is sprayed in the spring, when the soil reaches about 24°C or higher, and can be treated until late summer. The foliage is sprayed for good coverage until the leaves glisten, but not to runoff. Herbicides are applied by garden, backpack, or cattle sprayers, or sprayers mounted on small tractors or all-terrain vehicles.

Applying herbicides to stems (basal sprays) is most cost-effective using a small-orifice nozzle, such as the Conejet 5500-X1 from Spraying Systems Company (McGinty and Ueckert, 1995). This nozzle can reduce spray quantity by 80% over standard nozzles. The spray is applied on all sides of the stem or trunk up to 30 cm from the ground, but not to runoff. Smooth-bark mesquite requires 15% triclopyr in diesel fuel oil on stems less than 4 cm in diameter and 25% triclopyr in the mix if stems are 4 to 10 cm in diameter. Rough-bark mesquite requires the 25% triclopyr spray in diesel oil diluent.

Pricklypear and other cacti are sprayed is a similar way to honey mesquite, except picloram (Tordon 22K) is used at 1% in the tank mix with 5% diesel fuel, emulsifier (28.4 mL/3785 mL diesel fuel), and dye (Ueckert and McGinty, 1997). The spray can be applied throughout the year in water diluent. Pads or stems are sprayed until nearly wet, but not to runoff. Good coverage of the plants gives the best results.

Redberry and blueberry cedar control can utilize the Brush Busters technique (McGinty and Ueckert, 1997). Leaf sprays utilize picloram (Tordon 22K) at 1% with 0.25% surfactant and 0.25 to 0.5% dye by volume. All spray solutions are mixed in water. The treatment works best on cedar less than 1 meter tall and in spring and summer when cedar is actively growing.

Soil spot sprays can be made in late winter to midspring before expected rainfall (McGinty and Ueckert, 1997). Hexazinone (Velpar L) is recommended. It is applied undiluted to the soil surface midway between the cedar stem and the canopy edge. Velpar L is applied at 2 mL for every 1 meter of plant height or every 1 meter of plant canopy diameter.

4. Cut-Surface and Injection Treatments

Trees larger than about 13 cm in diameter often have bark too thick for basal sprays to penetrate. Herbicide can be applied to the sapwood of such trees through frills or notches cut into the bark with an ax, or it can be injected mechanically into the tree.

Frills are cuts made into the sapwood. They encircle the tree and act as cups to hold the herbicide. Make the frill by ringing the trunk of the tree with ax cuts (spacing depends upon the species controlled, the herbicides used, and the time of year) that penetrate the sapwood by at least 0.5 cm. The frill can be filled with the same type of solution as is used for basal sprays. Undiluted herbicide can also be used. A notch may also be cut into one side or more at the base of the tree and sprayed with the appropriate herbicide.

The same spray equipment can be used for frills and notches as is used for basal sprays. The solution also can be applied with a plastic squeeze bottle, such as a dishwashing soap container or a small can that has a pouring spout or lip. The container can be discarded after use.

Injection treatments are very effective in killing unwanted woody plants, but are expensive because of labor requirements. Also, to be effective, correct amounts of liquid herbicide must be placed in properly spaced cuts. In the winter, resistant species may need higher concentrations or closer spacings of herbicide. Follow the directions on the herbicide label.

There are mechanical injection tools available that make cuts in tree bark and inject the herbicide in one operation. Directions for using these tools are furnished by the manufacturers.

Mechanical injection equipment should be made of well-constructed, corrosion-resistant materials. The blade should be heavy and hardened so that it can be driven into trees repeatedly without becoming dull, and so that it can be resharpened successfully. (See Chapter 12 for herbicides used in cut-surface/injection treatments.)

5. Cut Stump

If trees are felled, the freshly cut surface of the stump should be treated with herbicide to prevent sprouting of certain species. It is more efficient to prevent sprouting than killing the sprouts. Use a solution such as the ester of triclopyr in oil, as for basal sprays. If oil solutions are used, apply the spray to the cut surface and also drench the bark thoroughly from the cut to the ground. Consult the herbicide label for guidance.

6. Soil Treatment

Certain soil-active herbicide pellets or sprays can be applied to individual plants, can be placed in grids or bands, or can be broadcast to kill plants. Rainfall soon after application may be required for satisfactory results. Dry herbicide materials commonly are formulated as extruded pellets 0.13 to 0.38 cm in diameter, spherical granules from 0.13 to 1.3 cm in diameter, or tablets of varying in size up to 2.5 cm long. The use of soil-active herbicides became common in the 1970s with the development of new herbicides that were also effective for brush control (Flynt et al., 1976; Mutz and Scifres, 1976; Stritzke, 1976; Bovey et al., 1984).

The smaller herbicide granules can be broadcast or applied in bands in brush-infested areas. The larger spherical pellets and tablets are effective when applied in grids spaced on 1.8-to-2.7-meter centers. The various species respond to grid applications according to the differences in their root systems. Gambel oak (*Quercus gambelii*), which has a deep root system, does not respond well to grid-pattern applications. Because of their shallow, widespread lateral roots, shrub live oak (*Quercus germinata*) and Utah juniper (*Juniperus osteosperma*) respond well to a grid pattern in which the herbicide pellets are spaced on 0.9-, 1.8-, or 2.7-meter centers (Bovey et al., 1984). Damage to forage plants is minimized using grid or band placement. Snakeweed control is best with broadcast application of granules. Pellets can be broadcast readily by aircraft. Ground equipment can apply herbicide pellets in bands spaced at 1.8 to 2.7 meters apart (Flynt et al., 1976).

A tractor-drawn experimental sprayer has also been designed to apply soil-active herbicides in bands to the soil surface or subsurface supporting stands of brush (Bovey et al., 1976). (The ground applicators for applying herbicide pellets and liquid herbicides subsurface are described in Chapter 15.) The spacing of the bands (usually 1.2, 1.8, or 3 meters apart) and the herbicide rate depend upon the kind and size of brush being treated. This sprayer has the advantage over conventional broadcast sprays by eliminating spray drift and reducing injury to forage by placement of herbicide in bands. It can be used near sensitive vegetation, and it improves brush control on some species. The main disadvantage is the slow operation compared to application by aircraft, and it is sometimes less effective than foliar sprays with some herbicides on some woody species.

7. Wipers and Special Techniques

Herbicides can be applied to weeds and brush using specially designed tractor-mounted rope-wick or carpeted applicators. Herbicide solutions are conducted through a porous medium and wiped onto the unwanted plants. Present designs allow treatment of weeds and brush not exceeding 1.8 to 2.1 meters tall. The ropes or carpets are saturated with herbicide either undiluted or diluted in water and wiped onto unwanted plants. Hand-carried wipers for small jobs have also been constructed. These wiping devices are available commercially and enable better placement of the herbicide on weeds with less exposure to valuable plants. Mayeux and Crane (1983) suggested that the brush roller, constructed of household carpet stretched over a 25-cm-diameter-by-2-meter-long polyvinyl chloride (PVC) cylinder, can be used effectively to treat weeds and brush on rangeland. The carpeted roller is tractor mounted, saturated with herbicide as needed, and slowly and continuously rotated against the direction of travel. Treatment of small honey mesquite as well as many other weeds and brush has been highly successful.

C. Preparation of Herbicides for Application

Commercial formulations of herbicides are sold as liquid or solid concentrates and are usually diluted in water, oil, or oil–water carriers for application in the field. As indicated in Chapter 4, amine salts of the phenoxy and other herbicides are soluble in water; esters are soluble in oils. Esters of 2,4-D, or triclopyr that have oillike properties, however, can be formulated with emulsifiers and mixed with water. The dispersed oil droplets are suspended in the water by the action of the emulsifying agent. The emulsified 2,4-D or triclopyr spray solution appears milky when mixed with water diluent. To prepare spray for the field, the herbicide concentrate is mixed with the diluent (water) in the proper proportions. For this, the gal per acre (L/ha) discharges spray at the rate of 93.5 L/ha (10 gal per acre), and if the acid equivalent (ae) of the concentrate is 480 g/L (4 lb per gal), 0.95 L (1 qt; 454 g, 1 lb) of herbicide is mixed with each 93.5 L of water (oil or oil–water emulsion), to provide a rate of 1.12 kg/ha (1 lb per acre).

Oils of varying chemical and physical properties have been used as herbicidal sprays, and oils such as diesel oil have been used with 2,4-D and triclopyr esters as carriers for spray solutions. Many times oil carriers with 2,4-D and triclopyr improve herbicidal activity by increasing wetting and penetration of leaf surfaces of the weed species, especially woody plants. Basal sprays and cut-stump treatments to woody plants are prepared by mixing esters of triclopyr or similar herbicides in kerosene or diesel oil at the rate of 3.6 to 14.5 kg (8 to 32 lb) of herbicide per 378.5 L (100 gal) of oil. Injection and cut-surface treatments to unwanted trees involve the use of diluted (water) or undiluted amine forms of 2,4-D and triclopyr or the esters as prepared for basal sprays. Oil–water emulsion carriers are commonly used to reduce the amount of oil required in broadcast sprays, but such mixtures maintain the herbicidal advantages of the oil carriers. In oil–water emulsions, water is the continuous phase, and oil—dispersed in the water in small droplets—is referred to as the discontinuous phase. Aerial sprays may contain oil–water emulsion carriers for triclopyr ester for honey mesquite control, and commonly contain 1 part oil to 4 parts water, with a total spray volume of 28.1 to 46.8 L/ha (3 to 5 gal per acre). Care must be taken to prevent the separation of oils and water or producing an invert emulsion in their preparation, which is explained later in this chapter.

D. Preparation of Broadcast Sprays

1. Foliage Sprays

High-volume or low-volume spray preparations can be made with selective or nonselective herbicides. Always double check label instructions (Table 1).

2. Low-Volume Sprays

Low-volume spray preparations are made with 2,4-D, dicamba, picloram, triclropyr, or certain mixtures of these herbicides. Table 1 shows the amount of various herbicide concentrates that should be used for a 2.2 kg/ha (2 lb per acre) application.

To prepare the spray, mix the concentrate with the carrier (water, oil, oil-in-water, water-in-oil) in the proper proportions. Since the rate of treatment is 2 lb per acre, the per-acre discharge rate of the sprayer must be known. For example, if the sprayer discharges at the rate of 93.5 L/ha (10 gal per acre) and

TABLE 1 Guide for Preparing Sprays

Acid equivalent of concentrate in lb per gal	Amount (fluid measure) of concentrate: Basal spray (mix with 1 gal of kerosene or diesel fuel)	2,4-D drench (add to each 10 gal of water)	Low-volume spray (to use for 2 lb per acre application)
2.0	10 oz	1 pint	1 gal
2.6	8 oz	12 oz	3 qt
3.3	6 oz	10 oz	5 pints
4.0	5 oz	8 oz	2 qt

Source: Bovey. 1977.

if the acid equivalent of the concentrate is 240 g/L (2 lb per gal), then mix 1 gal of concentrate with enough carrier to make 10 gal of spray. Follow the herbicide label instructions.

3. Oil-in-Water Emulsions

When preparing oil-in-water emulsions for low-volume spraying observe the following steps:

Use an ester formulation whose label specifies compatibility with oil-in-water emulsion carriers.
Add about ½ of the required volume of water to a clean, dry spray tank.
In a separate tank, mix the required volume of herbicide and oil with agitation.
Add the herbicide–oil mixture while agitating the water.
With continued agitation, add the remaining water.
During use, agitate the emulsion frequently to prevent the separation of the oil and water.

4. High-Volume or Drench

To prepare high-volume sprays containing 2,4-D, for example, mix spray concentrate with water in the proportions shown in Table 1.

To prepare a high-volume spray containing wettable powders or liquids such as bromacil or hexazinone, follow label instructions.

If increased wetting and penetration of the spray is desired, add 4 oz of surfactant (surface active agent) to each 100 gal of solution.

E. Preparation for Individual Plants

1. Basal Sprays

Basal sprays are prepared by mixing esters of 2,4-D, triclopyr, or similar herbicides with either kerosene or diesel fuel oil. Triclopyr is effective for many woody plants. Mix basal sprays in the proportions given in Table 1. See the Brush Busters technique (Sec. II.B.3) for the most recent information on certain woody plants.

2. Injection/Cut-Surface Treatments

Preparations for injection/cut-surface treatments can be made from amines and esters of 2,4-D, triclopyr, or other herbicides. Treatments can be made with concentrated formulations of the herbicides. For the best results follow the directions on the herbicide label. (See Chapter 12 for specific herbicides and species controlled by injection/cut-surface treatments.)

Research has shown that the amine salts of 2,4-D are more readily translocated from injection/cut-surface treatments than are the esters. The amines can be applied undiluted or diluted in water. The esters can be diluted in either oil or oil–water emulsions.

F. Soil Treatments

Some herbicides are formulated as granules, pellets, balls, tablets, or wettable powders for application to the soil. Pelleted materials may be spread by hand, tractor, or aerial equipment. Wettable powders are usually mixed with water and sprayed directly on woody plants or on the soil under or near the brush. If individual plant treatment is made, the amount of chemical needed will depend on the size and kind of plant to be killed. Consult the herbicide label and the latest suggestions from the nearest extension specialists or research personnel of the state agricultural experiment station or the U.S. Department of Agriculture (USDA). Do not apply within 15 to 30 meters of desirable trees.

The granule or powder formulations of different herbicides may vary in the percentage of active ingredient, from 2% to more than 50%. The herbicide concentration is indicated on the label as a percentage of the active ingredient (ai) in the formulation, therefore the required amount of formulation to use per acre (kg/ha) must be determined. For instance, if a 2-lb-per-acre of herbicide application is desired with a formulation containing a 10% ai herbicide concentration, the required amount of formulation to use per acre would be 20 lb.

G. Drift Control

Invert emulsions or water–oil emulsions are the opposite of oil–water emulsions, in which the oil is the continuous phase and water droplets are the discontinuous phase (water in oil). The invert emulsion is a thick, viscous mixture—like mayonnaise—and is useful in the dispersal of herbicides where drift control is important. The spray droplets produced are larger than conventional sprays and are less subject to evaporation and movement by air currents. Reduced weed control effectiveness may result, however, from inadequate spray coverage due to the large spray droplets, and the cost of operation may increase due to the special handling and spray equipment required. Mixing instructions on the herbicide label should be followed closely.

Other commercial preparations used in ecologically sensitive areas to control herbicidal drift include spray-thickening agents with trade names such as 38-F, 41-A DF, Air Drop, Arborchem 38-F, Armix 300, Chem-Trol, Deep Six, Direct Drifgon, Driftgard, Exacto Formula 358, Exact-Trol, Get-Down, Mist-Control, More, Nalco-Trol, Nalco-Trol II, Poly Control 2, Precision Spray Control, RNA Get Down, Sta-Put, Suspen-Der, Submerge, Super Oil V, WFSI Drift Not, Windbrake, and Wind-Fall (Harvey, 1993). They are produced by several companies and are added in small amounts to spray solutions (oz/100 gal). They reduce drift by reducing fine droplets when sprayed. They can be mixed with most prepared water-soluble and emulsifiable herbicides. They are commonly polyacrylamides (polymers).

Although drift control may be improved with the various drift-control agents, reduced weed-control effectiveness may result from reduced spray coverage by minimizing small spray particles and increasing large spray droplets. Herbicidal spray preparation may become more involved, and special spray equipment may be required. Special drift control techniques may be required in areas adjacent to sensitive vegetation, water sources, and urban areas, however.

H. Adjuvants and Surface-Active Agents

Adjuvants are substances added in small amounts ($<5\%$) to an herbicide spray solution to improve its activity. Such substances are of low mammalian toxicity, but may enhance phytotoxicity, assist emulsification, increase spreading properties on leaves, promote penetration into leaves, or perform other functions. Adjuvants that reduce interfacial tension are known as surface-active agents or surfactants. They can bind together two or more incompatible phases, such as water and oil, by modifying the interface forces between them. Numerous adjuvants and surface-active agents are commercially available. Specific adjuvants are compatible with specific herbicides, as determined by research and experience. Wetting agents reduce interfacial tension and bring a liquid (spray droplets) into intimate contact with a solid (leaf surface). Waxy leaf surfaces tend to repel water, therefore the addition of a wetting agent may increase its adherence and spreading properties. Emulsifiers are usually added by the manufacturer to maintain the stability of an emulsified product. Emulsifiers occupy the interface between oil and water surfaces, coupling them together. As a result, tiny dispersed droplets in a stable emulsion are prevented from coalescing. Many wetting agents have emulsifying properties. Thickening agents are natural or synthetic polymers that are soluble in water and increase its viscosity. Many thickening agents are also sticking agents. Some thickening agents were discussed under Drift Control (Sec. II.G). Sticking agents, however, do not have to be viscosity inducers as long as they improve spray retention on leaves. Many commercial ''spreader-sticker'' agents are available. Penetrating agents are substances that increase plant absorption of an herbicide. Such substances may act to solubilize waxy leaf surfaces on the lipoidal portion of the cell wall or membrane of the plants so that penetration is enhanced (National Academy of Science, 1968).

Other agents include activators or synergists, which are chemicals that may not be toxic to plants by themselves but may in combination with an herbicide increase the herbicide's phytotoxicity. As discussed previously, many adjuvants are also activators or synergists. Herbicides used together may also provide greater phytotoxicity than either alone. For example, the addition of dicamba, picloram, or triclopyr to 2,4-D may increase its effectiveness on certain weed species.

I. Application Equipment

Various types of equipment are available for applying herbicides as solutions, emulsions, wettable powder, and granules (Ozkan, 1996). The ideal applicator distributes the herbicides uniformly on plant and soil surfaces. Spraying is the most common application method. Sprays may be applied from small hand sprayers to large power-driven units. Spraying permits direct application to the target area or plant species, as well as providing uniform application. Of the hand-operated types, the backpacked sprayer is the most popular. It employs compressed air to force liquid out of the tank through a single- or multiple-nozzle hand-carried boom. The tank capacity of hand-operated sprayers varies from 1.9 to 18.9 L (1/2 to 5 gal). They are the most useful in spraying small, weedy areas and individual plants sometimes inaccessible to power equipment.

Power-driven sprayers range from small wheel-mounted sprayers useful in home gardens and lawns to large field sprayers. Field sprayers may be mounted on trailers, tractors, or trucks, or may be self-propelled. Sprayers are classified as low-volume or high-volume, and spray output may range from less than 1 gal to several hundred gal per acre where saturation of a large mass of vegetation is required. Most weed sprayers, however, are of the low-volume type of 280 L/ha or less (30 gal or less of total spray output per acre).

Basic components of the sprayer consist of a tank to hold liquid, some type of nozzle(s) to direct the spray, and a pump to force spray through the nozzles. In addition, for accurate application, they should have a pressure regulator, pressure gauge, shut-off valve, and filter and strainer. Details of sprayer components, spraying systems, and operation can be obtained from several manufacturers and other sources.

Power-driven or ground sprayers may consist of handguns or wands with booms of one or more nozzles to treat individual plants, or boom-type sprayers that spray swaths up to 15 meters (50 ft) wide in a single pass. Boomless sprayers with a cluster of flat spray nozzles mounted in the proper assembly produce relatively uniform spray patterns in swath widths up to 9 meters (30 ft). Drift possibilities with all types of sprayers are reduced at lower spray pressures (69 to 207 kPa) (10 to 30 lb psi). Hooded sprayers are designed for low-volume applications and control spray drift. Mist blower-type sprayers can be used in some remote areas, and provide excellent coverage and weed control. These sprayers consist of nozzles placed in a high-speed blast of air. They are effective in applying insecticides in orchards, but in many situations the drift potential is too great for application of growth regulator-type herbicides.

J. Calibration

Accurate application of herbicides is extremely important to obtain the desired weed control and prevent environmental damage. A typical field-type boom sprayer may be calibrated by several methods. The most accurate method, however, is to actually spray an area of known size as follows:

1. Fill the spray tank with a known quantity of spray solution (water).
2. Refill the tank after spraying an area of known size.
3. Calculate L (gal) used and determine sprayer output in L per ha (gal per acre).

For example, if the swath width of the sprayer was 6.1 meters (20 ft) and 7.6 L (2 gal) of spray were used after traveling 66.5 meters (218 ft, $6.1 \times 66.5 = 405$ m^2 or .04 ha or $20 \times 218 = 4{,}360$ ft^2 or 0.1 acre), sprayer output would be 197 L/ha (20 gal per acre). Spray volume per acre can be changed by varying the speed of the spray vehicle, spray pressure, types of nozzles, and spacing or width of the boom. Hand-carried or aerial sprayers can be calibrated in a similar manner.

In modern agriculture, the ability to treat large acreages in a short period of time with pesticides, fertilizers, or seeds is paramount. Aerial application is sometimes the only means of applying agricultural chemicals on wet soils, rough terrain, or tall vegetation.

The components of aircraft sprayers are similar to those of the ground equipment discussed, although the speed of application may be 129 to 145 km (80 to 90 miles) per hr, whereas ground sprayer speed is usually from 4.8 to 8 km/hr (3 to 5 mph).

Spray pumps on aircraft are driven by a V-belt drive from the main engine, a small direct-connected propeller, or an electric motor. Agitation is usually provided by a hydraulic bypass from the centrifugal pump. Most booms are positioned behind the trailing edge of the wing for improved visibility and accessibility. Booms are generally less than the length of the wings to prevent drift from the wing-tip vortices. The movement of the aircraft may set up turbulence and affect drift, however.

Hydraulic pressure spray nozzles are the most popular, but various spinning brushes, disks, and screens are also used. There is no evidence that the atomizing devices (brushes, disks, screens, etc.) have an advantage over the boom-nozzle hydraulic pressure system. Spray systems include the Ziegler nozzle system, which is operated at very low pressure and employs a manually controlled cap for positive shutoff. A hole drilled through a brass plug positioned along the boom forms the nozzle tip. The spray stream formed by the nozzle strikes the flat cap, aiding in spray formation.

Another boom system uses the diaphragm check-valve nozzles. Spring tension closes the diaphragm check valve when spray pressure drops below approximately 48 kPa (7 psi) to ensure a positive shutoff. The nozzle pressure is usually operated at around 207 kPa (30 psi) during spraying. The nozzles can be spaced along the boom to give the most desirable spray pattern. Nozzle tips and orifice disks and cores can be changed to vary sprayer output and droplet size.

The spinning or centrifugal atomizer, which is powered by air (propellers) or electric motors, can produce either coarse or fine sprays. Atomization is accomplished when centrifugal force pushes liquid through the rotating screen or disk. Usually four units are used per aircraft, with two atomizers properly spaced at the trailing edge of each wing.

The bifluid nozzle system was developed to apply invert emulsions. Two liquid phases (oil and water) are mixed at the nozzle to avoid pumping the thickened material. Although the system is not widely used, it provides a well-metered system for the application of invert emulsions.

The number of nozzles (17 to 32) used with the various types of spray systems depends mainly on the plane type and operator experience. Biwing planes usually use fewer nozzles than low-wing monoplanes. For a given total flow rate, the use of fewer but higher-flow-rate nozzles produces larger spray droplets, which in turn reduces drift possibilities but increases the probability of uneven spray coverage. Using the fewest number of nozzles that will provide uniform spray coverage has been the most satisfactory. The distance between nozzles is usually equal; however, some operators place additional nozzles on the right boom to adjust for the spray pattern distortions caused by the propeller slipstream (Wiedemann et al., 1973).

The control of spray drift is especially important with aircraft, since the drift of growth-regulator-type herbicides such as 2,4-D can cause damage to nearby susceptible vegetation and may reduce the effectiveness of the treatment. Spraying should be done under proper weather conditions (low wind velocity and stable air) to deposit the chemical on the target area. The droplet size of the released spray is very important. Small droplets (10 microns in diameter) can drift up to 1.6 km (1 mile) when released at a height of 3 meters (10 ft) in a 4.8 km/hr (3 mph) wind. Large droplets (300 to 300 microns in diameter) are less subject to drift, but may not provide adequate coverage. Spray systems designed to produce a mean droplet diameter of about 200 microns is a compromise between drift control and adequate coverage. Absolute drift control is sometimes difficult, however, since present techniques do not always eliminate all small droplets sprayed. Drift control agents discussed earlier in the chapter are sometimes useful near ecologically sensitive areas to minimize the small satellite drops, however (National Academy of Science, 1968).

Ground–air communications are important in effective aerial spraying. A precise application of the chemicals is necessary to guarantee results. Flagmen or a flagging system for accurate spacing of each swath across the field is as important as any other aspect of aerial application (National Academy of Science, 1968).

K. Cleaning Application Equipment

Cleaning the sprayer immediately after use is advised. The spray tank is emptied and rinsed with water. This may be sufficient for short-time storage. Rinsing the tank–both inside and out—with kerosene or fuel oil will protect most metal parts from corrosion (Bovey, 1977).

By cleaning the sprayer, injury or killing of sensitive broadleaf plants by 2,4-D, dicamba, picloram, or triclopyr can be avoided. Sources of contamination include the tank, hose, pump, boom, nozzles, and sprayer tires.

Experience has shown that the tank is by far the most important source of contamination, probably followed by hose contamination. By thoroughly rinsing and cleaning all parts, little or no damage should be encountered.

If the sprayer is to be cleaned, it should be rinsed with a material that acts as a solvent for the herbicide. Kerosene and fuel oils remove herbicides known to be oil-soluble. Chemicals that form emulsions when mixed with water are oil-soluble. Following the oil rinse, a rinse with a wetting agent in water will help to remove the oil. Such oil-soluble herbicides as 2,4-D ester are usually the most difficult to remove. For most water-soluble herbicides, such as 2,4-D salts and amines, repeated rinsing with water is usually enough.

Some crops can be damaged or killed by traces of phenoxy herbicides, picloram, dicamba, or triclopyr that are left in the sprayer after cleaning. Before applying fungicides or insecticides to crops with equipment that has been used for herbicides, the equipment can be tested for herbicide traces. This is done by filling the tank with water and spraying a few crop plants. Sensitive plants such as tomato, cotton, and tobacco are good test plants. If the crop plants show no distorted growth after a few days, the equipment can be used safely for spraying crops. If the plants are distorted, then further cleaning is required. The equipment should be retested for cleanliness before using it on crops. For the greatest safety with sensitive crops, fungicides or insecticides should not be applied with equipment that has been used for applying herbicides.

Spray equipment can also be cleaned with a suspension of activated charcoal in water. At least one-third of a tank of water is used. For each 38 L (10 gal) of water, 114 g (1/4 1b) of activated charcoal and 57 to 114 g (1/8 to 1/4 1b) of laundry detergent is added. This mixture is agitated vigorously to distribute the charcoal through the water.

The equipment is washed for 2 min by swirling the liquid around so that it reaches all parts of the tank. Some of the liquid is pumped through the hose and nozzles. The tank is drained and the equipment is rinsed with clean water.

Spray equipment also can be cleaned with household ammonia. First, the sprayer is thoroughly rinsed with water. If an ester formulation has been used, rinsing is done with a small quantity of kerosene or diesel oil. This oil rinse is followed with clean water to which has been added about 5 g (1 t) of laundry detergent per gal. If a salt formulation of the herbicide has been used, the fuel oil rinse may be omitted.

After rinsing the sprayer, it is filled with a solution of one part household ammonia to 99 parts water (1 L ammonia in 95 L water; 1 qt ammonia in 25 gal of solution). The solution may remain in the tank, boom, and hoses for 12 to 24 hr, then dump the ammonia solution and rinse the equipment with clean water. Other commercial preparations are also available for cleaning spray equipment. Waste from cleaning spray equipment must be disposed of in an environmentally acceptable manner.

III. SUMMARY

Several herbicides are available for managing weeds and brush. Herbicides are applied to foliage (sprays) or roots (sprays or pellets), depending on the species and herbicide. Sprays can be applied to weeds and brush with hand-carried, ground, or aircraft equipment. The selection of application equipment depends on the weed and brush density, species, and area. Rough terrain and tall vegetation may limit the use of ground or hand-carried equipment. Where stands are scattered, individual plants can be treated with foliar sprays; basal stem, cut-surface, and injection treatment; or soil application. Special wiping devices also are used to treat individual plants in small or scattered stands. Calibration of application equipment and using up-to-date equipment in good repair is essential. Calibration charts and procedures are available from the chemical manufacturers, state extension and experiment stations, and other sources. Applying herbicides, cleaning equipment, and disposing of containers and waste product must be done in accordance with label instructions and EPA requirements.

REFERENCES

Bovey RW. Response of Woody Plants to Herbicides. Handbook no. 493. Washington, D.C.: U.S. Department of Agriculture, 1977.

Bovey RW, Flynt TO, Meyer RE, Baur JR, Riley TE. Subsurface herbicide applicator for brush control. J Range Mgt 29:338–341, 1976.

Bovey RW, Wiese AF, Evans RA, Morton HP, Alley AP. Control of Weeds and Woody Plants on Rangelands. AB-BU-2344. USDA and University of Minnesota Minneapolis, 1984.

Flynt TO, Bovey RW, Meyer RE, Riley TE, Baur JR. Granular herbicide applicator for brush control. J Range Mgt 29:435–437, 1976.

Harvey LT, Drift control agents/deposition and retention agents. In: A Guide to Agricultural Spray Adjavants Used in the United States. Fresno, CA: Thomson, 1993, pp. 131–151.

Klingman GC. Weed Control: As a Science. New York: Wiley, 1961.

Mayuex HS Jr, Crane RA. The brush roller—An experimental herbicide applicator with potential for range and brush control. Rangelands 5:53–56, 1983.

McGinty A, Ueckert D. Brush Busters: How to Beat Mesquite. L-5144. Tex Agric Ext Serv, College Station: Texas A&M University, 1995.

McGinty A, Ueckert D. Brush Busters: How to Master Cedar. L-5160. Tex Agric Ext Serv, College Station: Texas A&M University, 1997.

Mutz JL, Scifres CJ. Response of honey mesquite and associated range vegetation to karbutilate. Proc South Weed Sci Soc 29:244, 1976.

National Academy of Science. Herbicide formulation and application. In: Principles of Plant and Animal Pest Control, vol. 2: Weed Control. Washington, DC: National Academy of Science, 1968, pp. 233–253.

Ozkan HE. Herbicide application equipment. In: Smith AE, ed. Handbook of Weed Management Systems. New York: Marcel Dekker, 1996, pp. 155–216.

Strizke JF. Selective removal of brush by grid placement of herbicide. Proc South Weed Sci Soc 29:255, 1976.

Ueckert D, McGinty A. Brush Busters: How to Take Care of Pricklypear and Other Cacti. L-5171. Tex Agric Ext Serv, College Station: Texas A&M University, 1997.

Welch TG. Chemical Weed and Brush Control Suggestions for Rangeland. College Station, TX: B-1466. Texas Agric Ext Serv, Texas A&M University, 1997.

Wiedemann HT, Bouse LF, Haas RH, Walter JP. Spray equipment, herbicide carriers and drift control. In: Scifres CJ, ed. Mesquite, Growth and Development, Management, Economic, Control and Uses. Res monograph 1. College Station: Tex Agric Expt Sta, 1973. pp. 33–45.

6

Herbicide Residues and Impact on the Environment

I. INTRODUCTION

Herbicides used for woody plant management on pastures, rangelands, forested areas, watersheds, and noncrop areas vary in water solubility, mammalian toxicity, and persistence in the environment (Table 1). Foliar-applied herbicides, such as clopyralid, 2,4-D, dicamba, picloram, and triclopyr, may drift to nontarget sites if improperly applied. Some soil-applied herbicides may move overland or leach to unwanted sites because they are persistent and/or highly water-soluble. The potential of each herbicide to persist in soil, vegetation, or water and to move to nontarget areas in surface runoff and to groundwater will be discussed. Most herbicides used on grazing and noncrop lands are moderate to low in toxicity. When applied in compliance with label instructions, residue, toxicity, or phytotoxicity problems are rarely encountered.

While such rapid biodegradable herbicides as 2,4-D may be desirable in some environments, more persistent herbicides may be required to control persistent perennial woody plants. Very infrequent application on grazing and noncrop lands minimizes any residue or toxicity problems, however.

Herbicide loss in the environment is a very complex process (Havens et al., 1996). The primary routes of loss occur by microbial decomposition, chemical decomposition, absorption onto soil colloids, volatility, phytolysis, leaching, and surface movement and runoff. Herbicide sampling, extraction, preparation, and analytical methods are available and are presented elsewhere (Bovey, 1980; Lavy and Santelmann, 1986; Corbin and Swisher, 1986; Weber, 1986; Weete, 1986).

The main herbicide analytical tools include bioassay, spectrophotometry,

TABLE 1 Residual Activity of Herbicides Used in Woody Plant Management in Soil, Plants, and Water Sources

Herbicide	Soil	Plants	Water sources	Precautions
Amitrole	Low	Moderate	Low	Prevent residues on food crops
Bromacil	Moderate–high	High	Mobile	Prevent soil and water mobility
Clopyralid	Moderate	Moderate	Moderate	Prevent spray drift
2,4-D	Low	Low	Low	Prevent spray drift
Dicamba	Low	Low	Low	Prevent spray drift
Dichlorprop	Low	Low	Low	Prevent spray drift
Diesel oil	Low	Low	Low	Protect skin
Diquat	Very low	High	Low	Wear protective gear
Fosamine	Low	Low	Low	Prevent spray drift
Glyphosate	Very low	High	Low	Prevent spray drift
Hexazinone	Moderate	Moderate	Mobile	Prevent soil and water mobility
Imazapyr	High	Moderate	Low	Prevent spray drift
Metsulfuron	Moderate	Moderate	Low	Prevent soil particle mobility
Paraquat	Very low	High	Low	Wear protective gear
Picloram	Moderate–high	Moderate	Mobile	Protect from water sources
Tebuthiuron	High	High	Low	Avoid application near large trees
Triclopyr	Moderate	Moderate	Low	Prevent spray drift

Note: All herbicides listed are degraded by soil microbes. Diquat, metsulfuron, paraquat, and picloram are very slowly degraded by soil microbes.

thin-layer chromatography, gas–liquid chromatography, mass spectrometry, high performance liquid chromatography (HPLC), radioisotope techniques, and several miscellaneous and new methods.

Herbicide residues may also be reduced by changing the methods of application, such as by using a carpeted roller instead of broadcast sprays (Bovey, 1996). Carpeted rollers deposit herbicide on target plants instead of exposing the soil to more herbicide. Treating individual woody plants versus broadcast application may also be useful. The application of granular forms of herbicides may also reduce residue in the herbaceous cover. Better timing of herbicide application may increase efficacy while preventing loss by leaching and runoff during periods of high rainfall or periods of drought and extended sunlight that cause phytolysis.

II. HERBICIDES

A. Benzoics (Dicamba)

1. Persistence in Soil

Scifres and Allen (1973) indicated that dicamba applied at 0.28 kg/ha dissipated from grassland soils in Texas in 4 weeks and in 9 to 16 weeks at 0.56 kg/ha. Dicamba residues were generally detected no deeper than 120 cm in clay or sandy loam soils. Dicamba residues were detected at 120 cm deep 53 weeks after application of granules at 1.68 or 2.24 kg/ha to sand in semiarid grassland, however. Under moist, warm soil conditions, dicamba has a half-life of <14 days (Smith, 1973) as a result of microbial degradation (Krueger et al., 1989). Dicamba readily converts through microbial activity to 3,6-dichlorosalicyclic acid (DCSA) (Smith, 1973; 1974). Like dicamba, DCSA can undergo breakdown, but it has been reported to be slower than for dicamba (Smith, 1973; 1974). The adsorption of DCSA to the soil is significant (Murray and Hall, 1989). Although dicamba is minimally adsorbed into soils, its residues are short-lived and unlikely to become a problem in groundwater.

2. Modes of Breakdown in Soil

Microbial degradation is highly important in the disappearance of dicamba (Krueger et al., 1989; Smith, 1973; 1974). Bacteria that utilize dicamba have been isolated and identified (Krueger et al., 1989).

3. Persistence in Plants

Morton et al. (1967) studied the disappearance of 2,4-D, 2,4,5-T, and dicamba over a 3-year period from a pasture containing silver beardgrass, little bluestem, dallisgrass, and sideoats grama. No important differences were found in the persistence of different herbicide formulations. The half-life of 2,4-D, 2,4,5-T, and dicamba in green tissue was from 2 to 3 weeks after application. The half-life in grass litter was 3 to 4 weeks. The short residual of herbicides in green tissues was attributed to dilution by growth. Rainfall hastened herbicide disappearance.

Vine mesquite tolerated 0.28 kg/ha dicamba applied preemergence (Halifax and Scifres, 1972). After emergence, 'Premier' sideoats grama tolerated 0.56 kg/ha dicamba. Pre- or postemergence applications of 1.12 and 2.24 kg/ha severely retarded shoot production of all species, including 'Blackwell' switchgrass. All species germinated and grew without a reduction in shoot production in soil containing as much as 63 ppb of dicamba.

In greenhouse studies, dicamba applied preemergence and postemergence at 0.14 to 2.24 kg/ha injured seedling kleingrass (Bovey et al., 1979a). Mature plants of kleingrass, buffelgrass, King Ranch bluestem [*Bothriochloa ischaemum* (L.) Var. *Songarica* (Rupr) Celarier & Harlan], green sprangletop (*Leptochloa dubia* H.B.K.), sideoats grama, common bermudagrass [*Cynodon dactylon* (L.) Pers.], and plains bristlegrass (*Setaria macrostachya* H.B.K.) tolerated dicamba and 2,4-D at rates as much as 2.24 kg/ha. Rates of 2,4-D at 1.12 kg/ha injured buffelgrass, which tolerated dicamba (Bovey et al., 1979a).

In the field, dicamba, 2,4-D, or 2,4,5-T generally did not reduce vegetative production of common, coastal, or coastcross-1 bermudagrass when applied in the spring or fall (Bovey et al., 1974b). In central Texas, herbage production of native forage grass was increased when whitebrush [*Aloysia gratissima* (Gillies & Hook.) Troncoso] was controlled by dicamba or picloram plus dicamba (Bovey et al., 1972).

Smith and Wiese (1972) indicated that sprays of 2,4-D, dicamba, or MCPA at 1.12 kg/ha reduced lint yields of cotton from 20 to 97%. Yield losses were most severe when cotton was sprayed before blooming. These herbicides did not affect lint quality (micronaire and length), however.

4. Modes of Breakdown in Plants

Dissipation of dicamba from plants can occur by exudation through roots into the surrounding soil, by metabolism within the plant, or by loss from leaf surfaces (Weed Science Society of America, 1994). Loss by ultraviolet light is also suggested (Baur et al., 1973).

5. Persistence and Movement in Water Sources

a. Herbicides in Runoff Trichell et al. (1968) studied dicamba runoff from sloping sod plots in Texas. They found that as much as 5.5% of the applied dicamba was recovered in runoff water when 1.3 cm artificial rain was applied 24 hr after herbicide application. No dicamba was found in runoff water from a similar artificial rain application 4 months later after 21.6 cm of natural rainfall. Approximately 8% of the artificial rain was recovered as runoff.

Norris and Montgomery (1975) found maximum dicamba levels of 37 ppb about 5.2 hr after treatment at 1.3 km from the point at which the sample stream entered the treatment unit in Oregon. Dicamba residues detected the first 30 hr after application resulted from drift and direct application to the exposed surface water. By 37.5 hr, residue levels had declined to background levels; no dicamba residues were found more than 11 days after application. Dicamba levels found in streams were several orders of magnitude below threshold response levels for fish and mammals.

In 1984, Muir and Grift (1987) sampled the Ochre and Turtle Rivers, which flow into Dauphin Lake in western Manitoba, Canada, to determine levels of MCPA, diclorfop {(±)-2-[4-(2,4-dichlorophenoxy) phenoxy] propanoic acid}, dicamba, bromoxynil, 2,4-D, triallate [S-2,3,3-trichloro-2-propenyl)bis(1-methylethyl)carbamothioate], and trifluralin [2,6-dinitro-*N*,*N*-dipropyl-4-(trifluoromethyl)benzenamine], which were used widely in each watershed. Dicamba and 2,4-D were detectable throughout most of the sampling period in both rivers at low levels of <1 ppb. Levels of <6 ppb of dicamba and 2,4-D were detected in water from the Turtle River before a high-water event, possibly from sprayed ditches or rights-of way near the river. Even so, discharges of all herbicides monitored in the study were <0.1% of the amounts used in each watershed. Levels of dicamba and 2,4-D in June were still far below toxic levels for fish or fish food organisms and below levels that would affect water quality standards.

b. Impounded Water Dicamba dissipated the most rapidly from water under nonsterile, lighted conditions (Scifres et al., 1973). Pond sediment evidently contained microbial populations capable of decomposing the herbicide. Temperature was crucial in dicamba dissipation, especially in the presence of sediment. In some cases, the influence of sediment on the dissipation rate of dicamba was apparently augmented by light. Under summer conditions, dicamba at 4.4 kg/ha/surface area of ponds dissipated at about 1.3 ppm/day. Dicamba dissipated as a logarithmic function of concentration with time.

c. Influence of Dicamba in Irrigation Water on Seedling Crops Crops varied in their response to one irrigation of water containing dicamba (Scifres et al., 1973). 'Dunn' was the most susceptible cotton cultivar. Fresh weights of Dunn seedlings were reduced at 100 ppb of dicamba, whereas concentrations of 500 ppb were required for weight reduction in 'Paymaster.' 'Blightmaster' was the most tolerant cultivar studied. 'Pioneer 820' and 'RS-626' grain sorghums seedlings also tolerated all dicamba treatments. RS-626 at 500 ppb showed increased fresh weight. 'Straight-eight' cucumber seedlings tolerated irrigation water containing as much as 50 ppb dicamba, but were injured or killed by 100 and 500 ppb, respectively. Crop tolerance to dicamba in irrigation water from greatest to least were sorghum > cotton > cucumber.

B. Bipyridiniums (Diquat and Paraquat)

1. Persistence in Soil

The typical half-life of diquat and paraquat is 1000 days. They are highly persistent due to strong binding to clay (Weed Science Society of America, 1994). Diquat and paraquat residues are tightly absorbed and biologically unavailable in the soil, however.

2. Modes of Breakdown in Soil

Photochemical degradation of diquat or paraquat on the soil surface and decomposition microbiologically are probably the most important means of breakdown in soil (Calderbank and Slade, 1976).

3. Persistence in Plants

Diquat and paraquat are apparently not metabolized in higher plants, although they can be photodegraded on plant surfaces (Weed Science Society of America, 1994; Calderbank and Slade, 1976).

4. Persistence in Water Sources

Residue levels in water decline very rapidly and then disappear, due to uptake by aquatic weeds and adsorption to suspended soil particles or bottom mud (Calderbank and Slade, 1976). Photochemical degradation could account for some loss under high sunlight and clear water. Diquat and paraquat residues typically decline to 0.01 ppm or less by 4 to 14 days of treatment with initial levels at 1 ppm (Calderbank and Slade, 1976).

C. Oils (Diesel Fuel Oil or Kerosene)

Diesel fuel is a very important component of herbicide preparations for application to individual woody plants by hand-held spray equipment and application with ground or aerial sprays. Number 2 diesel fuel is commonly used to enhance the penetration of specific herbicides into the foliage and/or stems, or it can be used alone to accomplish woody plant control as a pour on basal stems or tree trunks.

Usually small amounts of diesel fuel oil are used (see Chap. 4) in herbicide mixtures, and only a small portion of the land is treated. Practical experience indicated that diesel fuel oil is short-lived in the environment, with phytotoxic effects absent in a few days to weeks.

Plants sprayed directly with straight diesel fuel (if perennial) usually produced new growth and recovery a few weeks after treatment. If diesel fuel is released in the environment, it is subject to loss by evaporation, photodegradation, and microbial breakdown (CONCAWE, 1996).

D. Phenoxy Herbicides

1. Persistence in Soil

For more than 50 years investigators have recognized that 2,4-D is rapidly inactivated in moist soil (Bovey and Young, 1980). Warm, moist soil accelerates the degradation of phenoxy herbicides by stimulating microbial activity. After application in three Oklahoma soil types, Altom and Stritzke (1973) found that the average half-life of the diethanolamine salts of 2,4-D, dichlorprop, silvex, and 2,4,5-T were 4, 10, 17, and 40 days, respectively.

In Texas, Bovey and Baur (1972) applied the propylene glycol butyl ether esters of 2,4,5-T at 0.56 and 1.12 kg/ha to soils at five locations. After 6 weeks, 2,4,5-T had disappeared from all locations. At three different locations on sandy soils in central Texas, Scifres et al. (1977) found that 2,4,5-T was reduced to trace levels of <10 ppb in 7, 28, and 56 days. Residues of 2,4,5-T were not detected below 15 cm and generally remained in the upper 2.5 cm of soil.

a. Influence of High Rates Early work by Crafts (1949) and others indicated that 2,4-D typically did not persist from one growing season to another even at high rates, largely because of microbial breakdown. Work by Bovey et al. (1968) in Puerto Rico indicated that corn, sorghum, wheat, rice, soybeans, and cotton could be grown in soils 3 months after application of a 1:1 mixture of the *n*-butyl esters of 2,4-D plus 2,4,5-T at 26.9 kg/ha without reduction in fresh weight. Except for soybeans, which were sensitive to picloram residues, similar results were obtained with these crops for a 2:2:1 mixture of 2,4-D plus 2,4,5-T plus picloram at 16.8 kg/ha. Young et al. (1974) reported that 2,4-D and 2,4,5-T were applied at massive doses to three areas at Eglin Air Force Base in Florida in the 1960s. A chemical analysis of soil cores collected in 1970 from the treated area indicated that the herbicides had degraded.

b. Effect of Repeated Treatment Gas chromatographic analysis of Canadian soils indicated no residual amounts of 2,4-D and (4-chloro-2-methylphenoxy)acetic acid (MCPA) after 40 and 34 annual treatments, respectively (Smith et al., 1989). Under laboratory conditions, the breakdown of 2 kg/ha of (^{14}C) 2,4-D or (^{14}C) MCPA was slightly faster in soils that had received continuous applications with the appropriate herbicide, which suggests that soil microbial populations adapted in response to repeated long-term use.

In two separate studies in Texas, Bovey et al. (1974a; 1975a) found that 2,4,5-T did not accumulate in soils when applied five times at 0.56 or 1.12 kg/ha every 6 months on the same area. In plots receiving 0.56 and 1.12 kg/ha of 2,4,5-T, the average concentration did not exceed 95 and 144 ppb, respectively, and most herbicide was confined to the upper 15 cm of soil and generally disappeared by the time of retreatment.

2. Modes of Breakdown in Soil

As indicated, soil microorganisms contribute greatly to the detoxification of phenoxy herbicides (Bovey and Young, 1980). Other means of degradation include chemical decomposition, thermal loss and volatilization, absorption in soils, and photodegradation. The temperature of the soil surface may easily reach 60°C in the summer. Baur et al. (1973) found 55% loss of 2,4,5-T as the free acid exposed to 60°C, but no loss at 30°C after 2 days. The K^+ salt of 2,4,5-T adjusted to pH 7 showed 30% loss at both 30 and 60°C after a 7-day exposure. Baur and Bovey (1974) exposed dry preparations of 2,4-D to 60°C, which resulted in 75% loss within 1 day after treatment. Herbicides 2,4-D and 2,4,5-T were also subject to breakdown by long-wave ultraviolet (356 nm) irradiation (Baur et al., 1973; Baur and Bovey, 1974), therefore 2,4-D or 2,4,5-T on soil and plant surfaces would be subject to losses by ultraviolet and thermal volatility in the field.

3. Persistence in Plants

Over a 3-year period, Morton et al. (1967) studied the disappearance of 2,4-D, 2,4,5-T, and dicamba from pastures containing silver beardgrass, little bluestem, dallisgrass, and sideoats grama. No important differences were found in the persistence of different herbicides. The half-life of 2,4-D, 2,4,5-T, and dicamba in green tissue was from 2 to 3 weeks after application. The half-life of grass litter was slightly longer (3 to 4 weeks) than in green tissue. The shorter half-life of herbicides in green tissue was attributed to dilution by growth. Rainfall was important in hastening herbicide disappearance.

Baur et al. (1969) applied 2.24 kg/ha of the 2-ethylhexyl ester of 2,4,5-T alone and with 0.56, 1.12, and 2.24 kg/ha of the potassium salt or isooctyl ester of picloram to pastures supporting infestations of live oak. Grass species indigenous to the site were little bluestem, brownseed paspalum (*Paspalum plicatulum* Michx.), and Indian grass (*Sorghastrum* spp.). Recovery of 2,4,5-T acid and ester from woody and grass tissues was greatest when applied with picloram. The rates of herbicide recovery in all treatments, however, were generally <10 and 0.1 ppm, 1 and 6 months, respectively, after application.

Bovey and Baur (1972) analyzed forage grasses from five locations in Texas comprising different grass species, soils, and climate, which had been treated with the propylene glycol butyl ether esters of 2,4,5-T at 0.56 and 1.12 kg/ha. Six weeks after treatment, an overall average of 98% of the 2,4,5-T had been lost from all treated areas. After 26 weeks, the herbicide levels in grass were very low, ranging from 0 to 51 ppb.

In two separate studies, Bovey et al. (1974a; 1975a) applied a 1 : 1 mixture of the triethylamine salts of 2,4,5-T and picloram at a total of 1.12 and 2.24 kg/ha to the first and second experiment, respectively, on pasture land in central Texas. Repeat treatments were made every 6 months to the same area for a total of five applications. Herbicide content on native grass was high (28 to 113 ppm) immediately after spraying, but degraded rapidly after each treatment and disappeared before new applications were made. There was no accumulation of 2,4,5-T in soils or vegetation.

Baur et al. (1969) found that most of the 2,4,5-T applied to live oak at 2.24 kg/ha as the 2-ethylhexyl ester disappeared in 6 months. They detected small amounts of both the acid (93 ppb) and ester (233 ppb) of 2,4,5-T, however. At 1 and 6 months, more 2,4,5-T was found in live oak tissue at the top of the plant than at the middle and lower stem because the top portion intercepts more spray than do lower regions. In live oak, more 2,4,5-T was found when combined with picloram than when 2,4,5-T was applied alone at equivalent rates.

Brady (1973) indicated that radioactive 2,4,5-T persisted three to seven times longer in treated woody plants as in forest soils. The half-life of 2,4,5-T was 5.5, 5.8, 6.7, and 12.4 weeks in loblolly pine, post oak, sweetgum, and red maple, respectively. All four species decarboxylated 2,4,5-T and released CO_2, with no significant difference among species or doses.

Eliasson (1973) found that 2,4-D applied as a leaf spray or a basal-bark or stem-injection treatment resulted in slight downward translocation, with some reduction in stump regrowth of aspen poplar. Persistent 2,4-D in dead leaf and woody tissue near the injection site could be noted in some cases up to 6 years after treatment.

4. Modes of Breakdown in Plants

Basler et al. (1964) established that 2,4-D and 2,4,5-T breakdown in excised blackjack oak leaves was 50% or more in 24 hr. Morton (1966) showed that approximately 80% of the 2,4,5-T absorbed by mesquite leaves was metabolized in 24 hr. Numerous other investigations also have shown the importance of metabolism in detoxification and loss of phenoxy herbicides in many plant species (Bovey and Young, 1980).

The leaves and stems of plants are the main receptors of foliar-applied herbicides. Aside from their function in decarboxylation, breakdown, and conjugation of the herbicide, leaves and plant parts may abscise or abort from the plant and fall to the soil, where the tissue and any residual herbicide may weather and decay. Aerial parts of plants may be removed by mowing machines or clipped and consumed by grazing animals. If the herbicide does not kill or stop the growth of the plant, such as happens in many grasses, the herbicide will be diluted by growth.

On plant surfaces, phenoxy herbicides are lost by photodegradation and volatilization in a manner similar to loss from soils. Rainfall is also reported as an important means of accelerating herbicide loss from litter and plant surfaces (Bovey et al., 1974a; 1975a; Morton et al., 1967).

5. Persistence and Movement in Water Sources

a. Herbicides in Runoff Using gas chromatographic and bioassay detection techniques, Trichell et al. (1968) investigated the loss of 2,4,5-T, dicamba, and picloram from bermudagrass and fallow plots of 3 and 8% slope. When determined 24 hr after application of 2.24 kg/ha, a maximum of about 2,3, and 5 ppm of picloram, 2,4,5-T, and dicamba, respectively, were found in runoff water after 1.3 cm of simulated rainfall. Losses of dicamba and picloram were greater from sod than from fallow plots, whereas 2,4,5-T losses were approximately equal. Four months after application, picloram, 2,4,5-T, and dicamba concentration in runoff water from sod plots had diminished to 0.03, 0.04, and 0 ppm, respectively. The maximum loss of any herbicide from the treated area was 5.5%, and averaged 3%.

Bovey et al. (1974a) sprayed a 1:1 mixture of the triethylamine salts of 2,4,5-T plus picloram at 1.12 kg/ha every 6 months on a native-grass watershed for a total of five treatments. Plant ''wash-off'' was the main source of herbicide detected in runoff water. Concentrations of both herbicides were moderately high (400 to 800 ppb) in runoff water if 3.8 cm of simulated rainfall were applied immediately after herbicide application. If major natural storms occurred 1 month or longer after herbicide treatment, the concentration in runoff water was <5 ppb.

Norris and Moore (1970) and Norris (1971) indicated that the concentration of 2,4-D, 2,4,5-T, picloram, and amitrole seldom exceeds 0.1 ppm in streams adjacent to carefully controlled forest spray operations in Oregon. Concentrations exceeding 1 ppm have never been observed and are not expected to occur. Chronic entry of these herbicides into streams did not occur for long periods after application.

b. Impounded Water Bovey and Young (1980) summarized the literature on the fate of 2,4-D and other phenoxys in impounded water. In general, phenoxy decomposes rapidly, especially if adapted microorganisms are present. Photodegradation of phenoxys in impounded water is also an important means of breakdown.

c. Groundwater Wiese and Davis (1964) applied 500 ml of water to wet tubes (7.6×61 cm) of dry Pullman silty clay loam topsoil to a depth of 56 cm. The diethylamine salts of 2,3,6-TBA (2,3,6-trichlorobenzoic acid) and PBA (chlorinated benzoic acid) leached to about 51 cm, while the amine salt of 2,4-D and the sodium salt of fenac (2,3,6-trichlorobenzeneacetic acid) leached to 38 cm. The amine salts of silvex and 2,4,5-T leached to approximately 23 cm. Ester of silvex, 2,4,5-T, and 2,4-D remained in the top 8 cm of soil. When excessive water (34.4 cm) was used to wet soil in the tubes, all herbicides could be detected in the leachate except monuron (*N*′-(4-chlorophenyl)-*N*,*N*-dimethylurea) and the ester formulation of 2,4,5-T.

In New Mexico, O'Connor and Wiergenga (1973) studied degradation and movement of 64 kg/ha of 2,4,5-T in lysimeter columns in the greenhouse. They concluded that pollution of groundwater from normal application rates of less than 2 kg/ha of 2,4,5-T is unlikely because of its relatively slow rate of movement in soil and its rapid biological detoxification.

Edwards and Glass (1971) applied 11.2 kg/ha 2,4,5-T (excessively high rate) to a large field lysimeter in Coshocton, Ohio. The total amount of 2,4,5-T found in percolation water intercepted at 2.5 meters deep for as long as 1 year after application was insignificant.

Bovey and Baur (1972) found little or no 2,4,5-T 12 weeks after treatment in soils at five widely separated locations in Texas after treatment with the propylene glycol butyl ether esters of 2,4,5-T at 0.56 and 1.12 kg/ha.

Bovey et al. (1975a) conducted an investigation to determine the concentration of 2,4,5-T and picloram in subsurface water after spray applications of the herbicides to the surface of a seepy area watershed and lysimeter in the Blacklands of Texas. A 1:1 mixture of the triethylamine salts of 2,4,5-T plus picloram was sprayed at 2.24 kg/ha every 6 months on the same area for a total of five applications. Seepage water was collected at 36 different dates, and one to six wells in the watershed were sampled at 10 different dates during 1971, 1972, and 1973. The concentration of 2,4,5-T and picloram in seepage and well water from the treated area was extremely low (<1 ppb) during the 3-year study. No 2,4,5-T was detected from 122 drainage samples from a field lysimeter at another site sampled for 1 year after treatment with 1.12 kg/ha of 1:1 mixture of the triethylamine salt of 2,4,5-T plus picloram. Picloram was detected in lysimeter water at only 1 to 4 ppb 2 to 9 months after treatment. Supplemental irrigation in addition to 85.5 cm natural rainfall leached 2,4,5-T and picloram into the subsoil.

6. Surveys

An extensive analysis of surface waters of Texas in 1970 for 2,4-D, 2,4,5-T, and silvex revealed zero or trace levels of these herbicides (Dupuy and Schulze, 1972). The amount of brush sprayed with 2,4,5-T annually in the 1960s was generally less than 0.4 million ha. Out of a total of 43 million ha of range and pasture lands, approximately 0.8 million ha of pasture weeds were sprayed annually with 2,4-D in Texas (Welch, 1989). Some herbicide was introduced into the environment each year, but did not contaminate surface waters.

7. Mode of Breakdown in Water

Phenoxy herbicides do not persist in water sources, and significant concentrations, if found, occur only within a short time after treatment (Bovey and Young, 1980). The loss of herbicides from treated areas by movement in runoff water is a very small percentage of the total amount applied, even under intensive natural or simulated rainfall. Phenoxy herbicides rapidly dissipate in streams and are not detected downstream from the points of application. In impounded water, phenoxys decompose rapidly, especially if adapted microorganisms are present. Even under large-scale applications to surface water, 2,4-D disappeared rapidly after application, and concentrations remained low or undetectable. In surveys of major river systems in the United States, 2,4-D appeared infrequently and in minute concentrations.

8. Spray Drift Potential

Maybank and Yoshida (1969) indicated that a typical droplet-size distribution produced by herbicide spray nozzles using water diluent contained droplets of <100 μm in diameter that were subject to drift. This could amount to 20% of the total spray volume, depending upon the types of nozzles and pressures used. Smith and Wiese (1972) found that application to cotton of 2,4-D at 0.05 to 0.1 kg/ha caused significant yield loss. The earlier the cotton was sprayed, the more severe the damage. Studies by Maybank and Yoshida (1969) indicated that drift of herbicide at 0.04 kg/ha approached those concentrations causing injury to cotton. If precautionary measures are not taken, spray droplets of <100 μm may drift several hundred meters, and application rates of 2,4-D at 0.5 kg/ha or higher may damage adjacent sensitive crops.

At four locations in Texas, Behrens et al. (1955) found that 2,4-D caused more leaf malformation in cotton than 2,4,5-T and MCPA. Silvex did not cause leaf malformations. Similarly, 2,4-D caused the greatest reduction in cotton yield, followed by 2,4,5-T and MCPA. Silvex caused the least reduction in yield. Smith and Wiese (1972) compared 2,4-D to dicamba, picloram, bromoxynil (3,5-dibromo-4-hydroxybenzonitrile), and 2,3,6-TBA. The order of damage to cotton was 2,4-D ester > 2,4-D amine >> dicamba > MCPA > picloram >> bromoxynil >> 2,3,6-TBA. Sprays of 2,4-D, dicamba, or MCPA at 0.1 kg/ha reduced lint yields from 20 to 97%. Yield losses were the most severe when the cotton was sprayed before blooming. Lint quality (micronaire and length), however, was not affected by these herbicides. 'Tamcot' cotton seedlings were injured by foliar sprays of 2,4,5-T, triclopyr, and clopyralid at 0.03 kg/ha in the greenhouse (Bovey and Meyer, 1981). No new growth occurred when cotton was treated with 0.14 or 0.56 kg/ha of 2,4,5-T or triclopyr, and only slight leaf malformations occurred at clopyralid rates of 0.03 kg/ha or less. Because clopyralid has shown excellent control of honey mesquite in Texas, damage from spray drift of this herbicide should be minimal. (Methods to control spray drift and the volatility of 2,4-D and other herbicides are discussed in Chapter 5.)

E. Pyridine Herbicides (Clopyralid, Picloram, and Triclopyr)

1. Clopyralid

Clopyralid is closely related to picloram, but reacts differently to certain weed species and is less persistent in the environment (Weed Science Society of America, 1994). Clopyralid is less effective than picloram on most broadleaf species, but is highly effective against certain broadleaf weeds, such as those in the Polygonaceae, Compositae, and Leguminosae families. It has little or no activity against grasses or crucifers. Mixtures of clopyralid with other growth regulator-type herbicides extend the spectrum of species controlled. Differing from picloram, clopyralid is resistant to photodecomposition, but is more susceptible to degradation by microbes (Weed Science Society of America, 1994). In a wide range of soils across the United States, clopyralid degrades at a medium to fast rate. It has a half-life ranging from 12 to 70 days (Weed Science Society of America, 1994).

a. Persistence in Soil and Movement in Water Sources A 1:1 mixture of the monoethanolamine salts of clopyralid and the tri-isopropanolamine salt of picloram was applied at 0.56 kg/ha each in May 1988 and June 1989 to the same area (Bovey and Richardson, 1991). Approximately 90 days after treatment, >99% of the clopyralid had dissipated, as compared with 92% of the picloram. Most herbicide was detected in the upper 30 cm of soil. Neither herbicide was detected after 1 year.

Neither herbicide was detected in subsurface water from the treated area in 1988, but concentrations of <6 ppb of clopyralid or picloram were detected in subsurface water collected 11 days and from 41 to 48 days after treatment in 1989. The study represents a worst-case scenario, because the herbicides were applied twice to bare soil and were disked into the soil to prevent loss from photodegradation. Under normal practices, the herbicides may be applied once every 5 to 20 years to weeds and brush and are not protected from photodegradation by disking.

b. Persistence in Plants See triclopyr section (Sec. II.E.3) on persistence in plants.

2. Picloram

The fate of picloram in grassland ecosystems was summarized (Bovey and Scifres, 1971), but considerable data have been generated since. Picloram can control many broadleaf weed and brush species on grasslands (Bovey and Scifres, 1971).

It has a very low order of toxicity to warm-blooded animals, but is relatively persistent in the environment. Its persistence results in its effectiveness as an herbicide. If not photodecomposed, some picloram may move laterally on the soil surface or vertically through the soil profile to a limited degree. Biological breakdown of picloram by microbes or higher plants is slow, but dilution by runoff and dissipation in impounded water are important modes of dissipation. Environmental problems with picloram are related to susceptible plant life, where contaminated runoff or irrigation water could result in damage or where spray drift injures vegetation.

a. Persistence in Soil

Subhumid Rangeland Sites: Bioassays 1 year after treatment have shown that <1 ppb of picloram was detected in soil at five Texas rangeland sites when applied at 1.1 to 4.5 kg/ha (Bovey and Scifres, 1971). The soils were sampled only once to a depth of 61 or 91 cm, however, thus the sampling did not account for possible leaching of herbicide beyond 91 cm.

In other studies, picloram was applied at 2.2 and 9 kg/ha on the Gulf Coast prairie at Victoria on a Katy gravelly sandy loam (fine-loamy, siliceous, thermic Aquic Paleudalfs) and on the post oak savannah at College Station on an Axtell fine, sandy loam (fine, montmorillonitic thermic family of Udertic Paleustalfs) (Merkle et al., 1966). The annual rainfall at both sites was 81 cm. Frequent sampling at Victoria and College Station has shown that 2.2 kg/ha of picloram disappeared from the top 61 cm of soil by 6 and 12 weeks after treatment, respectively. Picloram was not detected after 1 year, regardless of herbicide rate or sampling depth. Bioassays using field beans confirmed chemical analysis, indicating the absence of detectable residues.

Additional studies at College Station were conducted to determine picloram residue levels to a depth of 2.4 meters for 2 years after application of 1.12 kg/ha (Baur et al., 1972a). Annual rainfall was 62 and 48 cm for year 1 and 2, respectively. After 30 days, residues were 93 ppb in the top 15 cm of the Axtell fine, sandy loam soil and <5 ppb at a depth of 46 to 122 cm. After 6 months, residues were between 5 and 10 ppb to a depth of 183 cm and <5 ppb between 198 and 244 cm deep. Residues to depths of 244 cm were <5 ppb 1 year after treatment. Retreatment after 1 year did not cause picloram accumulation, and dissipation of picloram was similar for spray and granular formulations. Although picloram leached to the lower soil profile, the concentrations detected were extremely low and would not likely contaminate the soil or groundwater.

Fallowed Areas and High Rates: Much of the picloram spray is intercepted by vegetation and plant litter. To investigate the magnitude of this effect, residues and leaching of picloram applied to bare soil was determined (Bovey et al., 1969). Picloram was applied at 1.1, 3.4, and 10.1 kg/ha to an Erving clay loam and a Lakeland sand at College Station and a Nipe clay, a Fraternidad clay, and Catano sand near Mayaguez, Puerto Rico. The annual rainfall in Texas was 71 and 74 cm at the sand and clay sites, respectively, and 81, 175, and 196 cm, respectively, on the clay, Nipe clay, and Catano sand sites in Puerto Rico.

1. *Texas Studies.* Picloram was applied to dry soil that received <1 cm of rainfall during the first 6 weeks after treatment. Loss of picloram was rapid during this period (Bovey et al., 1969), presumably by photodecomposition (Merkle, et al., 1967), because picloram was exposed to sunlight on bare soil. After 3 months, picloram applied at 1.1 kg/ha disappeared from sandy soil. Picloram concentrations in the sand and clay soils treated at all rates were considerably reduced; these soils received 23 and 30 cm of rainfall, respectively.

   Six months after treatment at 1.1, 3.4, and 10.1 kg/ha, picloram was present in the upper 15, 30, and 91 cm of the clay soils, respectively (Bovey et al., 1969). Picloram was detected at nearly all levels down to 122 cm in sand, but had dissipated at the 1.1 kg/ha rate. After 18 months, a small amount of picloram (0.03 ppm) was found in only the top 15 cm of clay soil treated at 3.4 kg/ha.

   Plots treated with 10.1 kg/ha had picloram residues in the top 91 cm of clay soil at 3 and 6 months after treatment, but were found only in the top 61 cm of soil after 18 months (Bovey et al., 1969). Picloram in the sandy soil was found at 122 cm deep at the 6- and 18-month sampling dates. Bioassay with 'Black Valentine' beans (*Phaseolus vulgaris* L.) 18 months after treatment detected no picloram residues where 1.1 kg/ha was applied on clay and sandy soils and in the sandy soil at 3.4 kg/ha. Beans grown in clay soil from 0 to 15 or 15 to 30 cm and deep treated with picloram at 3.4 kg/ha were injured, however. The greatest injury to beans occurred when grown in the top 60 cm of clay soil, receiving 10.1 kg/ha of picloram. Some injury was also recorded at soil depths of 91 to 122 cm. The greatest picloram injury to beans grown in sandy soil was at 91 to 122 cm deep, probably because of leaching after receiving abundant rainfall on the site.
2. *Puerto Rican Studies.* Three months after treatment, picloram was distributed throughout the upper 51 cm of clay soils at all rates of treatment (Bovey et al., 1969). Picloram residues increased as the rate was increased. Picloram at 3.4 and 10.1 kg/ha persisted in the Fraternidad clay soil for 1 year. Disappearance of picloram was related to soil type and rainfall. Picloram was most persistent in the Fraternidad clay, where rainfall was lowest. The disappearance of picloram from the Catano sand was rapid, and no herbicide was detected 6 months after treatment in the upper 100 cm of soil. Rainfall of 122 cm on the Catano sand may have leached much of the picloram from the soil. By 1 year

after treatment in the Nipe clay, picloram was detected only where 10.1 kg/ha had been applied, but detectable concentrations were <10 ppb. The use of picloram rates >1.1 kg/ha would be uncommon. At 1.1 kg/ha, picloram residues disappeared rapidly from the sandy soil, but were detected at as much as 23 ppb in clay soils for at least 6 months. Bioassay and gas–liquid chromatographic (GLC) techniques were compared for Nipe and Fraternidad clay soils receiving 3.4 kg/ha of picloram (Bovey et al., 1969). Both methods show similar trends in picloram concentrations at most depths of sampling, but for undetermined reasons, the bioassay consistently gave higher readings for the Fraternidad clay than did gas chromatography. Bioassay procedures with 'Puerto Rico 39' cucumbers (*Cucumis sativus* L.) are accurate and sensitive to 5 ppb of picloram. Other studies also showed a close correlation between bioassay and GLC techniques (Scifres et al., 1972).

Semiarid Sites: In the rolling plains of Texas, detectable picloram residues occurred in the upper 30 to 45 cm of soil after application of 0.28 kg/ha (Scifres et al., 1971a). Picloram was applied to sandy loam soils in June or July, and then the soils were irrigated to cause leaching. After the application of as much as 23 cm of irrigation water for 15 hr within 20 days after picloram treatment, residues were typically in the upper 45 cm of soil. On seven rangeland sites in the rolling plains, picloram at 0.28 kg/ha usually dissipated from the soil profile within 1 year after treatment.

Widely Diverse Locations: In studies at five diverse locations, including east-central, south, and west Texas, picloram disappeared from soils by 1 year after application of 0.26 to 1.1 kg/ha, regardless of cumulative rainfall or location (Hoffman, 1971).

b. Factors Affecting Soil Persistence

Soil Texture: As indicated in Texas and Puerto Rico, picloram disappeared more rapidly from sandy than from clay soils (Bovey et al., 1969). Scifres et al. (1977) found that picloram disappeared within 56 to 112 days after application from two watersheds on sandy soils in east-central Texas, and most picloram was restricted to the upper 15 cm of soil. No picloram was detected deeper than 60 cm. Bovey and Richardson (1991) detected picloram as long as 181 days after application in Houston black clay, mostly in the upper 30 cm of soil. The application rate was 0.56 kg/ha in both studies.

Environment: Laboratory studies (Merkle et al., 1967) indicated that the dissipation of picloram was accelerated at high temperatures (38°C versus 4 and 20°C) and by leaching. Photodecomposition may be an important means of loss if the herbicide remains on the soil surface for several days. All these dissipation pathways have been documented (Baur et al., 1972a; Bovey et al., 1969; Bovey and Richardson, 1991; Bovey and Scifres, 1971; Hoffman, 1971; Merkle et al., 1966; Scifres et al., 1971a; Scifres et al., 1977).

Effect of Repeated Treatment: Bovey et al. (1974a; 1975a) applied a 1:1 mixture of 2,4,5-T plus picloram at a total of 1.1 kg/ha every 6 months for a total of five times on a native-grass pasture watershed and a total of 2.2 kg/ha every 6 months on an adjacent watershed. The herbicide content of the Houston black clay remained low (0 to 238 ppb) during the study. Picloram dissipated and did not accumulate at either soil site.

c. Effect of Picloram Soil Residues on Plant Growth Picloram is widely used for weed and brush control in forage grass crops (Bovey et al., 1972; Bovey and Scifres, 1971; Bovey et al., 1984b). In Puerto Rico, 'USDA-34' corn (*Zea mays* L.), 'Combine Kafir-60' sorghum (*Sorghum bicolor* L.), 'Mentana' wheat (*Triticum aestivum* L.), 'Taichung Native No. 1' rice (*Oryza sativa* L.), and 'Blightmaster' cotton (*Gossypium hirsutum* L.) could be grown without reduction in fresh weight as early as 3 months after application of 6.72 kg/ha. 'Clark' soybeans [*Glycine max* (L.) Merrill] were injured as long as 6.5 months after treatment of picloram at 6.72 kg/ha (Bovey et al., 1968). Activated carbon applied at as much as 672 kg/ha in a Toa silty clay protected oats from picloram applied at 0.56 kg/ha, but did not completely protect 'Ashly' cucumbers or 'Black Valentine' beans (Bovey and Miller, 1969). At College Station, Texas, 'Tophand' sorghum was grown in Wilson clay loam 12 months after application of 1.12 kg/ha picloram without a reduction in plant numbers, dry matter production, flowering, or germination of harvested seed (Bovey et al., 1975b). No picloram was detected in sorghum seed harvested from plants growing in picloram-treated soil as early as 6 month after application. 'Hill' soybean numbers per ha and total dry matter production were slightly depressed 14 months after picloram application (Bovey et al., 1975b).

Ryegrass (*Lolium perenne* L. and *Lolium multiflorum* Lam.) could be grown as early as 75 days (and 16 cm rainfall) after application of picloram at 1.1 or 3.4 kg/ha as a spray or granule in a Wilson clay loam soil (Baur, 1978b).

Scifres and Halifax (1972a) found that picloram did not influence germination but did affect range grass seedling growth. Radicle elongation of sideoats grama [*Bouteloua curtipendula* (Michx.) Torr.], buffalo grass [*Buchloe dactyloides* (Nutt.) Engelm.], and switchgrass (*Panicum virgatum* L.) in petri dishes was reduced by 125 ppb picloram, whereas shoot elongation was not retarded by 1000 ppb. Buffalo grass, sideoats grama, and switchgrass seedlings germinated in soil containing 500 ppb picloram and were generally not reduced in top-growth production. Top-growth production of Arizona cottontop [*Digitaria Californica* (Benth.) Henr.] and vine mesquite (*Panicum obtusum* H.B.K.), however, was reduced by 125 to 250 ppb of picloram in soil. Root production and root:shoot ratios of switchgrass seedlings were decreased when 1000 or 2000 ppb of picloram were placed on the soil surface or at 7.5 cm deep (Scifres and Halifax, 1972b). Sideoats grama root production decreased by the application of 1,000 ppb of picloram placed 2.5 cm deep, but production was increased in soil with 1000 ppb of picloram placed 15 cm deep. Root:shoot ratios in picloram-treated soil were typically no different from those in untreated soil, but root growth pattern was affected.

Grain sorghum varieties varied in their response to pre- or postemergence irrigation and postemergence spray of various picloram concentrations (Scifres and Bovey, 1970). Significant increases in dry weight of Pioneer 820, RS-625, and PAG-665 occurred when treated preemergence with irrigation water containing 500, 1000, and 2000 ppb of picloram, whereas GA-615, RS-671, Tophand, and RS-626 were retarded by these concentrations. Dry weights of Pioneer 820 and RS-625 were increased by irrigation water containing 1 to 5 ppb of picloram applied postemergence or when treated with sprays containing 0.035 to 0.7 kg/ha.

d. Persistence in Plants

Herbaceous Plants: Getzenduner et al. (1969) showed that picloram residues in grass collected from various U.S. locations generally degraded after 1 year, and residues were lower from granular than from liquid formulation. No bound form of picloram was found in grasses.

Semiarid Areas. Picloram dissipated from grass, primarily buffalo grass, at rates of 2.5 to 3% per day after applications of 0.28 kg/ha for honey mesquite control (Scifres et al., 1971b), thus more than 90% of the picloram dissipated from grasses and broadleaf herbs within 30 days after application. The dissipation of picloram from grasses was not affected by irrigation to surface runoff within 10, 20, or 30 days after application (Scrifres et al., 1971b)

Hoffman (1971) found that picloram disappeared from grasses in 3 to 6 months at more than 20 locations (mainly semiarid sites) in Texas. Picloram was applied with equal rates of 2,4,5-T for honey mesquite control. Rates of picloram varied from 0.28 to 0.84 kg/ha. If rates of picloram are from 2.24 to 4.48 kg/ha, persistence may be as long as 2 years after treatment of redberry juniper (*Juniperus pinchoti* Sudw.) with picloram granules (Bovey and Scifres, 1971). The predominant grasses in the study were little bluestem and sideoats grama.

Humid Areas. The dissipation of picloram in range grasses in humid areas was relatively rapid. Zero to small concentrations were detected 6 months after treatment (Baur et al., 1969; 1972c; Bovey et al., 1974a; 1975a). Even when treatments were repeated every 6 months for a total of five applications at both the recommended (0.56 + 0.56 kg/ha) and twice the recommended rates of picloram and 2,4,5-T, picloram dissipated rapidly and did not accumulate in the grasses or the environment (Bovey et al., 1974a; 1975a).

Woody Plants:

Semiarid Areas. Honey mesquite leaves contained about 25 ppm of picloram the day of application of 0.28 kg/ha (Scifres et al., 1971b), but contained <1 ppm 23 days later. Picloram dissipated more slowly from sand shinnery oak leaves (*Quercus havardii* Rydb.) at the same site than from honey mesquite. Nearly 2 ppm were detected in the oak leaves after 60 days. Picloram in soil surface leaf litter dissipated after 120 days.

Humid Areas. Baur et al. (1969) reported that the amounts of 2,4,5-T detected in live oak at Victoria, Texas, were greater when applied with either the potassium salt or isooctyl ester of picloram than when 2,4,5-T was applied alone. Most of the herbicide had dissipated from live oak stems 6 months after treatment. Less than 1 ppb of picloram was detected in yaupon (*Ilex vomitoria* Ait.) stems or roots 6 months after treatment or retreatment with 1.12 kg/ha of picloram (Baur et al., 1972c).

e. Effect on Plants Fourteen days after planting, aqueous picloram solutions were used to water plants growing in pots (Baur et al., 1970). Water containing 0.25 and 0.50 ppb stimulated fresh shoot weights of 'Texas No. 30' corn, 'ATX3197 MS Kafir 60' sorghum, 'Stoneville 213' cotton, 'Alabama Blackeye' cowpeas [*Vigna unguicula* (L.) Walp.], and 'Lee' soybeans 21 days after planting. It took 100 ppb to get the same effect in 'Milam' wheat. Significant stimulation in dry weight production occurred in corn, sorghum, cotton, and soybeans at 0.25 ppb and cowpeas at 1 ppb. Fresh and dry weight decreased in corn, wheat, and sorghum at 1000 ppb and in all dicot species at 100 ppb. Picloram at levels as great as 1000 ppb had no effect on the dry weights of 'Bluebell' rice and wheat or the fresh weight of rice. Picloram caused reductions in soluble protein concentrations in all monocot species and in 'HA-61' sunflowers. Significant increases in soluble protein occurred at 0.25 and 1 ppb in cowpeas and cotton, respectively.

In the field, picloram, tebuthiuron, and 2,4-D did not reduce the protein concentration in kleingrass (*Panicum coloratum* L.), buffelgrass (Cenchrus ciliaris L.), and Coastal bermudagrass [*Cynodn dactylon* (L.) Pers.], but did reduce protein concentrations in a buffel X birdwood hybrid (*Cenchrus setigerus* Vahl.) (Baur et al., 1977). Glyphosate increased the protein content in buffelgrass and kleingrass and sometimes reduced the production of buffelgrass and buffelgrass X birdwood hybrid.

f. Factors Affecting Degradation Concentrations as high as 70 ppm of picloram were detected on grasses 2 hr after foliar application of 0.56 kg/ha or 1.12 kg/ha each of picloram and 2,4,5-T (Bovey et al., 1974a; 1975a). Only 842 ppb of picloram at 0.56 kg/ha could be detected on grass after application of 3.8 cm of simulated rainfall (Bovey et al., 1974a; 1975a). Photodecomposition may also be important in the loss of picloram from treated vegetation (Baur et al., 1973). Herbicide dilution and metabolism by plant growth also influences picloram loss.

g. Persistence and Movement in Water Sources

Movement of Picloram in Surface Water: Trichell et al. (1968) determined the movement of picloram in runoff water from small plots 24 hr after application. The loss of picloram was greater from sod than from fallow. The maximum loss obtained for picloram, dicamba, or 2,4,5-T was 5.5%, and the average was approximately 3%. The time interval from picloram application to the first rainfall determined the amount of picloram that moved away from the point of application with surface runoff. Four months after application, picloram losses were <1% of that lost during the initial 24 hr after application.

Scifres et al. (1971a) indicated that picloram moved in surface runoff when 0.28 kg/ha was applied in the Rolling Plains of Texas for the control of honey mesquite. Irrigation the first 10 days after application resulted in a concentration of 17 ppb of picloram in surface runoff. Irrigation at 20 or 30 days resulted in <1 ppb of picloram residue in runoff water. No more than 1 or 2 ppb picloram was detected after dilution of runoff water into ponds.

Baur et al. (1972b) studied picloram residues from a 6.1-ha watershed treated with the potassium salt of picloram at 1.12 kg/ha near Carlos, Texas, on an Axtell fine, sandy loam soil. Samples were collected directly below the treated area and in streams below the plots after each heavy rainfall. After 3 months, concentrations of 5 ppb or less of picloram were found in runoff water. After 1.5 weeks with initial treatments in April, no picloram was found in streams from 0.8 to 3.2 km from the treated area. After a 6.1-cm rainfall 10 months after treatment, no picloram was detected in runoff water, regardless of the sampling location.

Research shows that herbicide residues can occur in surface runoff water if heavy rainfall occurs soon after treatment. When pelleted picloram was applied at 2.24 kg/ha to a 1.3-ha rangeland watershed, surface runoff of 1.5 cm from a 2.1-cm rain received 2 days after treatment contained an average of 2.8 ppm of picloram (Bovey et al., 1978b). The picloram content declined rapidly in each successive runoff event, however, and runoff water contained <5 ppb by 2.5 months after application. Loss of the potassium salt of picloram from grassland watersheds in surface runoff water was similar whether the picloram was applied as aqueous sprays or as pellets on a Houston black clay soil. Picloram plus 2,4,5-T at 0.56 kg/ha each were applied on May 4, 1970, December 4, 1970, May 4, 1971, and October 8, 1971 (Bovey et al., 1974a). No runoff event occurred until July 25, 1971, 72 days after the third herbicide treatment. The concentration of 2,4,5-T and picloram averaged 7 and 12 ppb, respectively, in runoff water and <5 ppb during subsequent runoff events. The data indicated that picloram or

2,4,5-T content was typically <5 ppb in runoff if major storms occurred 1 month or longer after treatment on the Houston black clay.

On sandy soils, Scifres et al. (1977) found only trace amounts of picloram or 2,4,5-T, which had been applied at 0.56 kg/ha each, in surface runoff water following storms about 30 days after application.

Mayeux et al. (1984) found that maximum concentrations of picloram were 48 and 250 ppb in initial runoff from an 8-ha area treated with 1.12 kg/ha in 1978 and 1979, respectively. The herbicide concentration decreased with the distance from the treated area in proportion to the size of adjacent untreated watershed subunits that contributed runoff water to streamflow. About 6% of the applied picloram was lost from the treated area during active transport.

The Movement of Picloram in Subsurface Water: Bovey et al. (1975a) conducted an investigation to determine the concentration of 2,4,5-T and picloram in subsurface water after spray applications to the surface of a seepy area watershed and lysimeter site in the Blacklands of Texas. A 1:1 mixture of the triethylamine salts of 2,4,5-T plus picloram was sprayed at 2.24 kg/ha every 6 months on the same area for a total of five applications. Seepage water was collected on 36 different dates, and 1 to 6 wells in the watershed were sampled at 10 different dates during 1971, 1972, and 1973. The concentration of 2,4,5-T and picloram in seepage and well water from the treated area was extremely low (<1 ppb) during the 3-year study. No 2,4,5-T was detected from 122 drainage samples from a field lysimeter at another site sampled for 1 year after treatment with 1.12 kg/ha of a 1:1 mixture of the triethylamine salt of 2,4,5-T plus picloram. Levels of 1 to 4 ppb of picloram were detected in lysimeter water from 2 to 9 months after treatment. In addition to a total of 85.5 cm natural rainfall, supplemental irrigation was used to leach picloram into the subsoil.

In another study, Bovey and Richardson (1991) found that picloram and clopyralid remained in the uppermost 30 cm of a Houston black clay soil. The herbicides were sprayed at 0.56 kg/ha each on the same area in 1988 and 1989 on a seepy site overlying a shallow, perched water table. No herbicide was detected in the subsurface water from the area in 1988, but concentrations of <6 ppb of both herbicides were detected in subsurface water collected 11 days and from 41 to 48 days after treatment in 1989.

Dissipation of Picloram from Impounded Water Sources: Research conducted in semiarid and subhumid environments has shown that most picloram was dissipated from impounded natural water souces within a month to 6 weeks after introduction (Haas et al., 1971). Concentrations of picloram from 1 to 2 ppb were detectable a year after application of 1.1 kg/ha to these ponds, however. In no case did treated areas adjacent to domestic water wells that were 9 to 46 meters deep result in picloram residues in wells. Once picloram moved into water catchments in the rolling plains, residues were detected for at least a year after treatment (Scifres et al., 1971a).

Dilution is important in the dissipation of picloram from impounded water. Photodecomposition is also important in reducing picloram concentrations in water. In the photolysis of picloram, certain levels of light energy are necessary for degradation of each molecule. Assuming light energy is randomly dispersed, then interception of photons by picloram molecules would be a random occurrence. In such a system, degradation of picloram would be expected to occur rapidly at first, then decrease as fewer molecules were available for light interception. Such dissipation curves were reported by Haas et al. (1971), who found that most rapid dissipation occurred within the first 3 to 4 weeks after application of picloram to impounded water. In such a concentration-dependent system, more energy must be applied for degradation of the remaining herbicide molecules than is required at higher picloram concentrations (Baur et al., 1973).

3. Triclopyr

a. Persistence in Soil In Texas, Moseman and Merkle (1977) determined that when applied in the fall, triclopyr persisted about 6 months in a Miller clay soil but dissipated 3 months after summer application. In Canada, Jotcham et al. (1989) indicated that triclopyr was slightly less persistent than 2,4,5-T, but neither herbicide was biologically active during the next season. In four different soils, triclopyr and 2,4-D had similar mobilities as determined by soil thin-layer chromatography (TLC). Schubert et al. (1980) reported that triclopyr residues in the soil decreased from a maximum of 18 to 0.1 ppm in 166 days in a West Virginia watershed.

At two sites in Oregon, Norris et al. (1987) found that triclopyr and its metabolites persisted for 1 year or more in small concentrations. They speculated that dry summers in Oregon may retard dissipation of triclopyr compared with summers in West Virginia. Triclopyr residues were confined to the top 30 cm of soil. Also in Oregon, Newton et al. (1990) found that triclopyr persisted in small amounts in the soil for 1 year after aerial application. Picloram, triclopyr, and 2,4-D residues decreased rapidly after application, leveled off 79 days after treatment, and then began a period of slow loss that continued until the following summer. however, Newton et al. (1990) found that picloram was lost more quickly than triclopyr or 2,4-D, as contrasted with results reported by Norris et al. (1987), who observed that picloram persisted longer than 2,4-D. Norris et al. (1987) worked in a nearby but drier area. Newton et al. (1990) suggested that triclopyr is similar to 2,4-D in movement and persistence.

b. Mode of Breakdown in Soils Leaching, photodegradation, and microbes degrade triclopyr (Weed Science Society of America, 1994).

c. Persistence in Plants Bovey et al. (1979b) found more picloram than triclopyr in greenhouse-grown huisache [*Acacia farnesiana* (L.) Willd.] 0, 3, 10, and 30 days after treatment with foliar sprays, soil application, or soil plus foliar treatments. In field-grown honey mesquite, more clopyralid and picloram than triclopyr or 2,4,5-T was detected in honey mesquite stem tissue (Bovey et al., 1986). Triclopyr and 2,4,5-T residues were generally <2 ppm by 3 days after application, whereas picloram and clopyralid residues were as high as 11 and 22 ppm, respectively. Concentrations of triclopyr and picloram recovered from honey mesquite stems were about 25% greater at 3 days after treatment than at 30 whereas concentrations of 2,4,5-T and clopyralid were about 50% greater at 3 days after application than at 30. Twenty months after application, concentrations of 2,4,5-T in standing dead stems were 0.2 and 0.4 ppm dry weight in upper stem phloem and upper stem xylem, respectively (Bovey et al., 1986). Phloem and xylem tissue taken from the base of dead stems had <0.01 ppm of 2,4,5-T and little or none in live resprouts. Concentrations of triclopyr in dead stems ranged from 0.06 to 0.9 ppm, but generally the herbicide could not be detected in live resprouts.

After 22 to 26 months, as much as 0.4 and 0.9 ppm of 2,4,5-T and triclopyr could be detected in dead honey mesquite stems that had fallen to the soil surface. Thorns also contained detectable concentrations of 0.1 ppm each of 2,4,5-T and triclopyr.

In comparison, concentrations of picloram ranged from 0.3 to 1.3 ppm dry weight 20 to 26 months after treatment in dead honey mesquite stems that were standing or that had fallen on the soil (Bovey et al., 1986). The concentrations of clopyralid in the same tissues ranged from 0.7 to 3.3 ppm. No clopyralid was detected in treated live stems, but concentrations of picloram ranging from 0 to 0.04 ppm were detected. Picloram and clopyralid, at 0.3 and 0.8 ppm, respectively, were detected in thorns from several dead stems.

Norris et al. (1987) found that triclopyr decreased rapidly from grasses in Oregon. Initial average concentrations of 527 ppm immediately after treatment were reduced to <0.3 ppm by 158 days after treatment. Newton et al. (1990) found that 2,4-D, triclopyr, and picloram persisted in evergreen foliage and twigs for nearly 1 year. Crowns and browse layers showed similar rates of loss, but browse layer concentrations of 2,4-D and triclopyr were only about one-third of those in crown foliage. Despite shading, picloram decreased to low levels before rainfall and remained low but detectable. Salt formulations of the herbicides were lost faster than ester formulations, and herbicide residues decreased rapidly in litter and soil.

Whisenant and McArthur (1989) showed the dissipation of triclopyr from several herbaceous and woody species in northern Idaho. Triclopyr concentrations in foliage varied among species at two sites. The highest concentration of 362 ppm occurred in shinyleaf ceanothus (*Ceanothus velutinus* Dougl. Ex. Hook.) 1 day after treatment, but by 365 days more than 98% of the triclopyr had dissipated from all species. Triclopyr residue data from the study and large herbivore toxicological data from other studies indicate that toxicity from triclopyr is unlikely under proper use.

d. Effect on Plants Triclopyr was generally more phytotoxic to seedling '5855X127C' corn, 'TAM 0312' oats, 'MS 398' grain sorghum, and 'Selection 75' kleingrass than was either 2,4,5-T or clopyralid (Bovey and Meyer, 1981). 'Caddo' wheat tolerated triclopyr at 0.56 kg/ha. Triclopyr and clopyralid caused greater injury to 'Florrunner' peanuts than did 2,4,5-T, whereas 2,4,5-T and triclopyr were more damaging to 'Tamcot' cotton and 'Liberty' cucumber than was clopyralid. At 0.14 and 0.56 kg/ha, all three herbicides killed 'Gail' soybeans. Kleingrass was not affected by any rate of clopyralid.

e. Factors Affecting Degradation Triclopyr is lost from grasses because of metabolism, growth dilution, wash-off, volatilization, and photodegradation (Norris et al., 1987).

f. Persistence and Movement in Water Sources

Surface Runoff Water: Using a helicopter, Schubert et al. (1980) treated the upper part of a watershed in West Virginia with 11.2 kg/ha triclopyr. Two streams traversed the treated area. The movement of triclopyr residues in the soil and water downslope from the treated area was insignificant.

The maximum concentration of triclopyr in stream water was 95 ppb the first 20 hr after application, similar to that observed for other herbicides applied to forest streams (Norris et al., 1987). A reduction in concentration the first 20 hr after application was attributed to photodecomposition. In September, during the first significant rains after application in May, the maximum triclopyr residues of 12 ppb were found in a small pond at the site. A 6-cm rain on November 9, causing a 6,500-L stream discharge, increased triclopyr concentrations to 15 ppb, but no more triclopyr was detected after November 11.

Groundwater: Triclopyr was applied in both ester and amine formulations on October 24, 1986, to coastal plain flatwood watersheds near Gainesville, Florida (Bush et al., 1988). Panicum grasses (*Panicum* spp. and *Dichanthelium* spp.), wiregrass (*Arietida stricta*), gallberry (*Ilex glabra*), and most herbaceous plant species were controlled by both formulations. Triclopyr applications resulted in a shift toward a bluestem-dominated understory. Triclopyr residues were detected at trace levels of 1 to 2 ppb in storm runoff during the first runoff event after application. No triclopyr residues were detected in subsequent runoff events or in any groundwater wells for 6 months after application.

Impounded Water: An examination of triclopyr and by-product 3,5,6-trichloro-2-pyridinol (TCP) residue dissipation following the application of the triethylamine salt of triclopyr at prescribed rates showed that no adverse effects should be produced on the aquatic environment (Green et al., 1989). The results showed that detectable triclopyr levels in water were variable from 3 to 21 days, residue half-life being less than 4 days. Residue accumulation in sediment, plants, and fish was negligible. Concentrations and persistence of TCP were transitory, however, results of crayfish evaluation indicated prolonged persistence of both triclopyr and TCP. Further evaluation of triclopyr and TCP accumulation in crayfish from the nonedible parts must be accomplished before a tolerance level can be established.

g. Mode of Breakdown in Water Photodegradation is a major means of triclopyr decomposition in water (Weed Science Society of America, 1994).

F. Sulfonylureas (Metsulfuron)

1. Persistence in Soil

In the laboratory in which no leaching occurred, [14C] metsulfuron methyl was lost closely following first-order kinetics, with the logarithm (base 10) of the percentage of herbicide remaining being proportional to incubation time (Smith, 1986). The half-life in clay loam, sandy loam, and clay were 70, 102, and 178 days, respectively. A soil transformation product of metsulfuron was 2-carboxymethylbenzenesulfonamide, but never exceeded 3% of the applied radioactivity in clay and sandy loam. In clay, 32% of the initial metsulfuron was transformed to the sulfonamide after 84 days. The half-life of metsulfuron in clay (pH 7.5) was more than twice that in clay loam (pH 5.2).

Walker et al. (1989) indicated that metsulfuron and chlorsulfuron adsorption was negatively correlated with soil pH and positively correlated with soil organic matter content. Degradation rates decreased with increasing soil depth and were positively correlated with microbial biomass and negatively correlated with soil pH. Walker and Welch (1989) found that triasulfuron was less mobile than metsulfuron and chlorsulfuron, but all three herbicides were largely confined to the upper 40 to 50 cm of the soil by 148 days after application. Initial rates of 32g/ha affected lettuce and sugar beet growth after 1 year.

In a field lysimeter, Bergstrom (1990) found that chlorsulfuron and metsulfuron were not recovered in leachate from normal rates (4 g active ingredient/ha) in sandy or clay soils watered in addition to natural rainfall. At rates of 8 g ai/ha over 7 months, <1% of the herbicides appeared in leachate. Anderson (1996) indicated that in areas of soil pH greater than 7.0 and with prolonged low soil temperature and low annual rainfall the sulfonylurea herbicides can remain active in soil for 2 to 4 years or more. Generally, however, the sulfonylureas are moderately persistent in soil, with a typical half-life ranging from 30 days to 6 weeks (Weed Science Society of America, 1994).

2. Mode of Breakdown in Soil

Microbial degradation is slow. Chemical hydrolysis is the major form of degradation. Degradation is increased at low pH and high soil temperature in moist soil (Weed Science Society of America, 1994; Beyer et al., 1988).

3. Persistence in Plants

In wheat, barley, and other tolerant grasses the phenyl ring of metsulfuron is hydroxylated and conjugated with glucose (Beyer et al., 1988). Several metabolites have been identified. The rapid increase in activation or detoxification of the sulfonylureas by crops has been the basis of crop tolerance. Anderson et al. (1989) found that the concentration of total radiolabeled residue in field wheat and barley treated with metsulfuron declined from 1.9 and 2.3 ppm to <0.01 and 0.03 ppm, respectively, by 9 weeks after treatment.

G. Triazines (Hexazinone)

Although hexazinone is used as a spot-soil treatment in Texas for the control of honey mesquite and other woody plants (Welch, 1997), little work has been done in Texas on its residues in soils, plants, and water sources.

1. Persistence in Soil

The mobility of hexazinone in runoff water and its leachability in soil is well documented (Allender, 1991; Bouchard et al., 1985; Feng, 1987; Feng et al., 1989; Lavy et al., 1989; Zandvoort, 1989). Hexazinone movement downslope in runoff water can sometimes injure nontarget vegetation remote from the point of application (Allender, 1991). The high water solubility of hexazinone in water (3.3 g/100 g) contributes to its leaching potential. Prasad and Feng (1990), however, found that after 1 year, hexazinone residues were reduced to 1% at the treated spot and did not move laterally beyond 0.5 meters on a sandy loam in Canada.

Greenhouse studies in silt and sandy loam soils showed that the half-life of hexazinone was 4 to 5 months (Weed Science Society of America, 1994). Under field conditions the half-life varied from 1 to 6 months, depending upon location. Microbial breakdown contributes to decomposition in soil (Weed Science Society of America, 1994). Hexazinone photodegrades on the soil surface, but volatilization losses are negligible.

2. Persistence in Plants

Hexazinone and tebuthiuron were rapidly taken up by roots of seedling winged elm (*Ulmus alata* Michx.), bur oak (*Quercus macrocarpa* Michx.), black walnut (*Juglans nigra* L.), eastern red cedar (*Juniperus virginiana* L.), and loblolly pine (McNeil et al., 1984). Four hours later, ^{14}C was detected in all parts of winged elm treated with ^{14}C-tebuthiuron and ^{14}C-hexazinone. Root and foliar absorption varied with herbicide and species. The results indicated, however, that the selectivity of tebuthiuron and hexazinone can be attributed to the amount of intact herbicide translocated to the foliage. Loblolly pine and eastern red cedar prevented accumulation of the parent compound in the foliage within 24 hr. Demethylation was the primary detoxification mechanism of tebuthiuron by eastern red cedar, loblolly pine, and bur oak. Loblolly pine, a hexazinone-resistant species, degraded hexazinone rapidly into three unknown degradation products, thereby preventing its accumulation in foliage.

Using whole-plant metabolism studies with pear [*Pyrus melanocarpa* (Michx.) Willd.] and bristly dewberry (*Rubus hispidus* L.), Jensen and Kimball (1990) found no difference in ^{14}C accumulation in leaves, but did find a greater formation of the monodemethylated metabolite B,[3-cyclohexyl-6-methylamino-1-methyl-1,3,5-triazine-2,4-dione] in the more tolerant *P. melanocarpa*.

Sidhu and Feng (1993) studied concentrations of hexazinone and two metabolites in vegetation for 2 years after broadcast application of a 10% granular formulation of hexazinone at 2 and 4 kg ai/ha in August 1986 in a boreal forest. Prewinter concentrations of hexazinone in stems of trembling aspen, Saskatoon berry, and willow ranged from 0.02 to 0.05 μg/g dry weight at 64 days after treatment. The absorption of hexazinone accelerated during the spring thaw (1987), and residues in the foliage of several woody and herbaceous species peaked during May to July. Patterns of accumulation of hexazinone and its metabolites varied with the species. Foliar concentrations diminished significantly by the end of the first growing season in 1987 (372 days after treatment), and were undetectable or extremely low at the end of the second growing season in 1988 (707 DAT). Based on the highest residue concentrations detected in several plant species, it is estimated that wildlife would ingest a maximum of 16, 28, and 24 mg of hexazinone, metabolite A, and metabolite B, respectively, for every kg of dry matter consumed. Reported LD_{50} values suggest that application of hexazinone at the 4 kg ai/ha rate or less poses no toxicological threat to wildlife.

3. Effect on Plants

Hexazinone controls many annual and biennial weeds, woody vines, and most perennial weeds and grasses, except johnsongrass (Weed Science Society of America, 1994). Hexazinone controls woody plants for site preparation and conifer release and for selective weed control in crops such as alfalfa, cacao, coffee, oil palm, pecans, pineapple, sugarcane, rubber trees, tea, and certain conifers.

4. Persistence and Movement in Water Sources

Lavy et al. (1989) found relatively small amounts of hexazinone in runoff water from a spot-gun application to a forest floor in Arkansas. Forest litter was highly effective in absorbing surface applications of hexazinone. In another study, the maximum concentration of hexazinone was 14 ppm in a stream that drained a 11.5-ha watershed treated with 2 kg/ha (Bouchard et al., 1985). Hexazinone residues of <3 ppm were detected in stream discharge for 1 year after application. The amount of hexazinone transported from the watershed in stream discharge represented only 2 to 3% of the amount initially applied.

Neary et al. (1986) found only 0.53% loss of hexazinone in streamflow of the applied herbicide in Georgia. Residues in streamflow peaked at 442 ppb in the first storm, but declined rapidly and disappeared within 7 months. Total sediment yield increased by a factor of 2.5 because of the increased runoff associated with site preparation using herbicide and salvage logging. Sediment loading remained below those produced by mechanical techniques, however, and overall water quality changes were small and short-lived. Leitch and Flinn (1983) applied hexazinone at 2 kg/ha from a helicopter to a 46.4-ha catchment. Only six of the 69 samples analyzed contained hexazinone, which was well below the maximum allowable concentration of 600 μg/L for potable water.

5. Aquatic Environment

Polyethylene exclosures were located in a typical bog lake in northeastern Ontario (Solomon et al., 1988). Triclopyr, 2,4-D, and hexazinone were applied at 0.3 and 3, 1 and 2.5, and 0.4 and 4 kg/ha, respectively. Less than 5% of the triclopyr and 2,4-D remained in water after 15 days. As much as 25% of the 2,4-D absorbed into the side of the corrals. Triclopyr could not be detected after 42 days. At 0.4 and 4 kg/ha, hexazinone could not be detected by 21 and 42 days after application, respectively. Hexazinone dissipated more rapidly than 2,4-D, and was not absorbed into sediments.

H. Substituted Ureas (Tebuthiuron)

1. Persistence in Soil

Pelleted tebuthiuron was applied aerially on duplicate plots at 2.2 and 4.4 kg/ha in the spring, summer, fall, and winter of 1978 and 1979 for mixed brush control (Bovey et al., 1982). Soil was predominantly an Axtell fine, sandy loam (Udertic Paleustalfs). Tebuthiuron persisted for more than 2 years in the Claypan Resource Area of Texas at depths of 0 to 15 cm and 15 to 30 cm as determined by 'Tamcot' cotton and 'Caddo' wheat bioassays. Tebuthiuron content ranged from 0.08 to 0.49 ppm. Deeper soil depths were not sampled. On a Houston black clay (Udic Pellustert), pellets were broadcast and applied in bands at 2.24 kg/ha (Bovey et al., 1978a). Tebuthiuron was also detected to depths of 46 to 61 cm, but not at 76 to 91 cm deep. After 6 months, most tebuthiuron was found in the 0- to 15-cm soil layer. Whether tebuthiuron leached deeper after 6 months is unknown. Tebuthiuron applied as a broadcast spray also resided mainly in the 0- to 15-cm layer. In another study, at 2.24 kg/ha, tebuthiuron persisted in the 0- to 15-cm and 15- to 30-cm layers of soil for 25 months on a Lufkin fine, sandy loam (Vertic Albaqualfs) (Meyer and Bovey, 1988). The Lufkin fine, sandy loam, however, was underlain by a claypan at 15 to 30 cm deep, whereas the Wilson clay loam was more permeable. All studies mentioned are in an area with approximately 75 to 90 cm or more annual rainfall.

In semiarid rangeland soils in north central Arizona, Johnsen and Morton

(1989) detected most tebuthiuron in the surface 30 cm of soil during the first 5 years after treatment, but small concentrations were detected as deep as 105 cm 6 and 9 years after treatment. After 9 years, from 55 to 75% of the tebuthiuron detected was at a depth of 60 to 90 cm.

2. Factors Affecting Dissipation and Leaching

Tebuthiuron has a half-life of 12 to 15 months in areas receiving 100 to 150 cm rainfall annually (Weed Science Society of America, 1994). Photodecomposition and volatilization loss from soil is negligible. Some microbial breakdown occurs, but the half-life of tebuthiuron is considerably greater in low-rainfall areas and in soils of high organic matter regardless of rainfall. Chang and Stritzke (1977) found that after six successive desorption extractions, 40% of the tebuthiuron was adsorbed into soil with 4.8% organic matter, but less than 1% was adsorbed into soil with 0.3% organic matter. The soil mobility of tebuthiuron was greater in soil with low organic matter and low clay content. Greater dissipation occurred at 15% soil moisture and at 30°C than at lower moisture and temperature levels. Baur (1978a) also found that tebuthiuron leaching was inversely related to the clay content of soil and directly related to the rate of application. Tebuthiuron is more phytotoxic in soils low in clay and/or organic matter (Chang and Stritzke, 1977; Duncan and Scifres, 1983). For these reasons, one could therefore expect greater tebuthiuron persistence in semiarid soils than in soils in humid areas, as discussed by Johnsen and Morton (1989).

3. Distribution and Dissolution of Pellets

Whisenant and Clary (1987) indicated that using a 40% active extruded pellet at 0.6 and 1.1 kg/ha left residues of 9 to 21% and 17 to 38% of the treated area, respectively. The lower percentages were from a soil with 47 g/kg soil organic carbon (OC), and higher percentages were on loam soils with 17 and 18 g/kg OC. VanPelt and West (1989) placed large tebuthiuron briquettes of 1.8 g and 13% ai beneath pinyon pines trees at the dripline, midcrown, and stem base. Residue analysis indicated that overland runoff, wind, and animals did not move the briquettes.

4. Effect on Plant Growth

a. ***Greenhouse*** Baur and Bovey (1975) compared the growth inhibition of five herbicides by applying 1.4 to 1,434 μg/plant to one unifoliolate leaf of 'Southern blackeye' cowpea and the partly unfurled true leaf of 'Topland' sorghum seedlings. The order of decreasing effectiveness for growth and herbicidal effectiveness for cowpea and sorghum was paraquat, glyphosate, tebuthiuron, 2,4-D, and endothall. Tebuthiuron and glyphosate had little inhibitory effect on the germination of sorghum, cowpeas, or 'Era' wheat.

Tebuthiuron applied preemergence or early postemergence at 0.6 kg/ha injured buffelgrass (Bovey et al., 1984a). Buffelgrass became more tolerant with age—to as old as 150 days—but plants were still injured at 1.1. kg/ha of tebuthiuron applied as foliar sprays to plants growing in pots. Buffelgrass [*Pennisetum ciliare* (L.) Link] shoot and root weights were reduced by 2 to 4 ppm of tebuthiuron placed 0 to 3, 8 to 11, or 15 to 18 cm deep in soil columns 30 days after emergence (Rasmussen et al., 1985). Plains bristlegrass seedling shoot weights were not reduced when 2 ppm of tebuthiuron were placed 8 to 11 cm deep or deeper.

b. Field Common bermudagrass and kleingrass tolerated March and April applications of tebuthiuron at 2.2 kg/ha using an 80% wettable powder formulation, but June applications reduced production (Baur et al., 1977). Buffelgrass and buffelgrass X birdwood hybrid tolerated tebuthiuron at 0.4, 1.1, and 2.2 kg/ha with March, April, and June applications. Coastal bermudagrass tolerated March but not April or June treatments. Tebuthiuron had little effect on protein concentrations of common or coastal bermudagrass, buffelgrass, and kleingrass but reduced protein concentrations in the buffel X birdwood hybrid.

Masters and Scifres (1984) reported that application of tebuthiuron pellets (20% ai) at rates as much as 2.2 kg/ha did not affect in vitro digestible organic matter of little bluestem, bahiagrass (*Paspalum notatum* Flugge), Bell rhodesgrass (*Chloris gayana* Kunth), and green sprangletop, but did increase foliar crude protein concentrations of little bluestem during the growing season of application. In the south Texas plains, tebuthiuron pellets (20% ai) at three locations at rates as much as 2.2 kg/ha did not significantly decrease buffelgrass standing crop or foliar cover compared with untreated areas (Hamilton and Scifres, 1983).

c. Natural Areas In the Texas post oak savannah during the spring aerial application of tebuthiuron pellets (20% ai) at 2.2 and 4.4 kg/ha to heavy brush cover increased grass production the second growing season after application (Scifres et al., 1981). Treated native grass stands consisted of a higher proportion of perennial species of good to excellent grazing value than stands on untreated rangeland.

Aerial application of tebuthiuron pellets at 2.2 kg/ha to mixed brush in south Texas significantly increased the grass standing crop at 1, 2, and 3 years after treatment (Scifres and Mutz, 1978). Overall grazing of the grass stand was improved, but forb production and diversity were decreased where 1 kg/ha or more of herbicide was applied. Forage stands recovered after 3 years, regardless of the herbicide rate used.

5. Reseeding on Treated Areas

In January 1976, Baur (1979) treated areas in the Texas claypan resource area near Leona, Texas, with 1.1 or 2.2 kg/ha of tebuthiuron using the wettable powder as a spray or 20% ai pellets. Tebuthiuron at 1.1 kg/ha suppressed weed cover and produced a 71% kleingrass cover. Tebuthiuron at 2.2 kg/ha prevented kleingrass establishment. In 1977, kleingrass production in plots treated with 1.1 kg/ha tebuthiuron the same year exceeded untreated areas, but 2.2 kg/ha of tebuthiuron markedly reduced kleingrass production. No coastal bermudagrass survived in tebuthiuron-treated areas on the deep sand.

In other work, Baur (1978b) showed that annual ryegrass could not be established until 261 days and 68 cm of rainfall after treatment of 1.1 kg/ha of tebuthiuron either as a spray or as granules. Rates of 3.4 kg/ha prevented revegetation by johnsongrass on 95% of the area after 499 days on the black clay loam soil.

6. Persistence in Plants

a. Herbaceous Plants The initial concentration of tebuthiuron in coastal bermudagrass was 438 ppm from spray application of 2.2 kg/ha, but never exceeded 1 ppm from broadcast- or band-applied pellets at 2.2 kg/ha (Bovey et al., 1978a). Low concentrations are desirable in forage because livestock or wildlife may graze treated areas immediately after treatment. Tebuthiuron concentrations in forage from sprays decreased rapidly with time, and residues from sprays or pellets were <2 ppm within 3 months after treatment in the Texas Blacklands prairie.

In semiarid areas, tebuthiuron or its metabolites persisted as long as 11 years after treatment (Johnsen and Morton, 1991). Tebuthiuron was detected in sideoats grama and blue grama [*Bouteloua gracilis* (H.B.K.) Lag. ex Griffiths] 10 years after application of 6.7 kg/ha. Metabolites of tebuthiuron were detected in blue grama 11 years after applications of 2.2, 4.5, and 6.7 kg/ha. The highest concentrations of tebuthiuron plus metabolites were 25 ppm in blue grama 10 years after application of 4.5 kg/ha, and 21 and 23 ppm in sideoats grama 9 and 10 years, respectively, after applications of 6.7 kg/ha. Only these three samples of 120 total samples exceeded the legal limits of 20 ppm of tebuthiuron plus metabolites in forage plants. No samples from plots treated with 4 kg/ha or less exceeded 10 ppm of tebuthiuron plus metabolites, and only 10% of them exceeded 5 ppm.

b. Woody Plants Foliage, twigs, stems, and litter from recently killed Utah juniper [*Juniperus osterosperma* (Torr.) Little] trees averaged 13.3, 0.4, 0.4, and 4.0 ppm of tebuthiuron plus its metabolites, respectively (Johnsen, 1992). Dead stems averaged 0.5 ppm in sapwood, 0.1 ppm in heartwood, and 0.4 ppm in bark 3 to 9 years after treatment. Root bark averaged 1.1 ppm, and root wood averaged 0.5 ppm. The investigator concluded that residues have little potential harm when used as firewood or fenceposts.

7. Persistence and Movement in Water Sources

a. Surface Runoff Water Pelleted tebuthiuron was applied at 2.24 kg/ha to a 1.3-ha rangeland watershed. A 2.8-cm rain 2 days after application produced 0.94 cm of runoff, which contained an average of 2.2 ppm of tebuthiuron

(Bovey et al., 1978a). Tebuthiuron concentration decreased rapidly with each subsequent runoff event. After 3 months, tebuthiuron concentration was <0.05 ppm; none was detected in runoff water 1 year after treatment. The concentration of tebuthiuron applied as a spray at 1.12 kg/ha decreased to <0.01 ppm in runoff within 4 months from a small plot receiving sumulated rainfall. On 0.6-ha plots, mean tebuthiuron concentration from sprays and pellets was 0.50 ppm or less in water when the first runoff event occurred 2 months after application. Concentrations of tebuthiuron in soil and grass from pellet applications were <1 ppm and decreased with time.

Tebuthiuron applied at 1 kg/ha as 20% ai pellets to dry Hathaway gravelly, sandy loam soil in the spring diminished by 5% at the first simulated rainfall event (37 mm) in runoff water and sediment (Morton et al., 1989). The second and third simulated rainfall events (22 and 21 mm, respectively) removed an additional 2% of tebuthiuron. When tebuthiuron was applied to wet soil in the spring, the initial simulated rainfall events, totaling 42 mm, removed 15% of the tebuthiuron. When tebuthiuron was applied to wet soil in the fall, the initial rainfall events, totaling 40 mm, removed 48% of the tebuthiuron in runoff water and sediment. No significant differences were found in the total amount of tebuthiuron within the soil profile after application to dry and wet soils. More than half of the tebuthiuron had moved into the upper 7 cm 1 day after application. Tebuthiuron was not detected below 90 cm after 165 mm of simulated rainfall and 270 mm of natural rainfall.

b. Hydrologic Effects Selected hydrologic variables were evaluated after the conversion of heavily wooded sites to open grassland with an herbicide-prescribed burning treatment sequence in east-central Texas (Lloyd-Reilley et al., 1984). Terminal infiltration rates and sediment production 3 years after aerial application of tebuthiuron pellets at 2.2 kg/ha for brush management differed little from values for untreated woody areas.

I. Substituted Uracils (Bromacil)

1. Persistence in Soil

Bromacil has more use in industrial sites and other noncrop areas than for pasture and rangeland use (Table 1). The disappearance of bromacil varies with geographic location. High rates disappear rapidly in Florida, whereas loss from soils in Oregon may take 2 years or more (Gardiner, 1975). The Half-life of 2-C^{14} labeled bromacil and terbacil was 5 to 6 months after application of 4.5 kg/ha to the surface of a Butlertown silt loam soil in the field (Gardner et al., 1969).

2. Mode of Breakdown in Soil

Microbiological degradation appears to play a very important role in the disappearance of the substituted uracils from soils (Gardiner, 1975). No 5-bromouracil or 5-chlorouracil was detected as a metabolite of bromacil or terbacil in soil or plant studies.

3. Persistence in Plants

Bromacil is readily absorbed by roots and translocated in the xylem to the leaves. Bromacil is less readily absorbed by leaves and stems. Surfactants enhance activity (Weed Science Society of American, 1994; Anderson, 1996). Orange plants grown in bromacil-treated nutrient solution for 4 weeks in sand absorbed <5% of the applied bromacil; 85% remained in the roots, and 15% was translocated to leaves and stems (Gardiner et al., 1969). Three C^{14}-labeled compounds were found in ratios of 10:5:1. The major metabolite was bromacil, followed by 5-bromo-3-*sec*-butyl-6-hydroxymethyluracil. The minor metabolite was not identified.

4. Effect on Plants

Bromacil controls many annual and perennial grasses and sedges. At rates of 5 kg/ha or more it controls many woody plants.

5. Persistence and Movement in Water Sources

Of the 13 herbicides evaluated, Weber and Best (1972) considered bromacil a persistent and mobile soil-applied herbicide. On a municipal site, bromacil and hexazinone residues were evaluated for movement after weed control around building and gas pipeline valve sites in Australia. Bromacil was found at five of seven sites, whereas hexazinone was found at four of seven sites after native trees were adversely affected up to 100 meters from the site of application (Allender, 1991).

6. Breakdown Procedure in Water

Saltzman et al. (1983) showed that uracil compounds in industrial waste effluents were rendered nonphytotoxic by a photo-oxidation procedure.

J. Other Organic Herbicides

1. Amitrole

a. Persistence in Soil Amitrole disappears rapidly from the soil (Carter, 1975). Disappearance is by soil adsorption, microbial degradation, and nonbiological reactions. The average half-life is 14 days (Weed Science Society of America, 1994).

b. Persistence in Plants The -s-triazole nucleus is very stable. Persistence in different plants varies. The primary metabolite (6 to 10%) was a serine conjugate forming β-(3 amino-1,2,4-triazol-1-yl)-α-alanine in horsetail (Weed Science Society of America, 1994). Conjugation with glycine and tannins is possible in some plants. Metabolism in plants is related to differential susceptibility to amitrole (Carter, 1975).

2. Fosamine

a. Persistence in Soil The typical half-life of fosamine is 8 days (Weed Science Society of America, 1994). Fosamine is rapidly decomposed by soil microbes.

b. Persistence in Plants Fosamine had an average half-life of 7 days in pasture grass and red clover, and none was found after 1 year (Weed Science Society of America, 1994). The primary metabolite was carbamoylphosphonic acid.

c. Persistence in Water Sources Fosamine is degraded in water exposed to artificial or natural sunlight (Weed Science Society of America, 1994).

3. Glyphosate

a. Persistence in Soil Glyphosate controls many herbaceous and woody plants (Bovey, 1985). It is recommended that spray drift or mist of glyphosate not be allowed to contact green foliage, green bark, or suckers of desirable plants.

Glyphosate has limited use on rangelands. Torstensson (1985) indicated that glyphosate is rapidly adsorbed in soil. Adsorption occurs through the phosphoric acid moiety that competes for binding sites with inorganic phosphates. Glyphosate is virtually immobile in soils. Its half-life ranges from a few days to several months and is correlated with the microbial activity of soils. The inactivation of glyphosate through soil adsorption is important.

b. Persistence in Plants Several structures indicate little or no evidence of metabolic degradation in a variety of species over several months (Duke, 1988). Initial residues of three glyphosate formulations (Vison, Touchdown, and Mon 14420) and triclopyr ester (Release) in sugar maple (*Acer saccharum*) foliage were 529,773,777 and 1630 mg acid equivalent/kg dry mass. Mean times to 90% dissipation were <16 days for glyphosate formulations, 9 days for triclopyr ester, and 33 days for triclopyr acid (Thompson et al., 1991).

c. Persistence in Water Sources Bronstand and Friestad (1985) concluded that the regular use of glyphosate in agriculture or forestry allowed only very remote chances of contaminating the aquatic environment. The compound dissipates by microbial degradation, adsorption into sediments, and photolysis.

4. Imazapyr

a. Persistence in Soil Imazapyr is generally weakly bound to soil, but adsorption increases as organic matter and clay increase. Soil adsorption increases below pH 6.5 and imazapyr generally remains in the top 50 cm of soil (Weed Science Society of America, 1994). Its field half-life ranges from 25 to 140 days, depending on soil type and environmental conditions. Weed control may occur from 3 months to 2 years, making it active for woody plant control. In Maryland, Coffman et al. (1993) found that several crop species and indigenous plants tolerated soil-applied imazapyr and triclopyr by the second growing season at rates up to 9 and 10.1 kg/ha, respectively. Microbial degradation is the principal means of degradation.

*b. **Persistence in Plants*** Susceptible weeds metabolize imazapyr slowly or not at all.

*c. **Persistence in Water Sources*** There was no detectable degradation of imazapyr in distilled water or in pH 5 and 7 buffers over 30 days (Mangels, 1991). The half-life in pH 9 was calculated to be 325 days. Simulating sunlight with a borosilicate-filtered Xenon Arc lamp indicated photolysis followed first-order kinetics. The half-lives in distilled water and pH 5 and 9 buffers were between 1.9 to 2.3, 2.7, and 1.3 days, respectively. Four degradation products were identified. Imazapyr can be rapidly degraded in aquatic systems through photolysis.

III. SUMMARY

Amitrole and fosamine are apparently lost rapidly from the soil by soil microbial activity and other means and do not become an environmental problem under labeled use (Table 1). Food crops must be free of any amitrole residue.

Diquat, paraquat, and glyphosate are strongly adsorbed to soil, making them virually biologically unavailable. Diquat and paraquat are destroyed if exposed to sunlight on plant and soil surfaces. Microbial breakdown is apparently very slow in decaying plants and soils. Glyphosate is degraded in soil and water by microbial activity. The mobility of these herbicides in most soils will not be a problem because of their strong adsorption to soil. They have a low potential for movement in runoff water.

In moist, warm soil, dicamba has a half-life of <14 days as a result of microbial degradation. Its half-life in native grasses and litter is 3 to 4 weeks. Under simulated rainfall conditions, a maximum of 5.5% of dicamba applied to a watershed was removed in runoff water. Dicamba levels found in streams after application to large watersheds were several orders of magnitude below the threshold response levels for fish and mammals.

The phenoxy herbicides such as 2,4-D and diclorprop are short-lived in the environment and have limited mobility, so movement into groundwater is unlikely. Phenoxys are rapidly decomposed by soil microbes, sunlight, and plant metabolism. Phenoxy herbicides are generally less phytotoxic to broadleaf plants than picloram. Preventing spray drift or vapors to susceptible plants such as cotton is essential, however.

Picloram degrades within 3 to 6 months in Texas soils. Persistence will be longer in cooler climates. The half-life of this herbicide varies widely, depending upon rainfall and soil temperature. Picloram tends to leach to lower soil depths, but most remains in the upper meter of soil. Picloram may move in surface runoff water, but its removal from watersheds is usually less than 5% of the total amount applied. Picloram is degraded slowly by soil microorganisms and plant metabolism, but is degraded rapidly by sunlight. Picloram is phytotoxic to a wide range of plants, especially broadleaf plants, so care must be taken to limit its movement.

Clopyralid is chemically similar to picloram, but has a shorter half-life in soil than does picloram and is subject to degradation by soil microbes. It resists degradation by sunlight. Clopyralid moves in water sources, as does picloram, but it is not phytotoxic to as many plant species, as is picloram.

In Maryland wheat, corn, and okra could be grown 8 days after application of triclopyr at 3.4 kg/ha to soil. Kidney beans, squash, and potatoes could be grown in less than 3 months after triclopyr application. In Texas triclopyr persisted for only 3 months after summer application and 6 months after fall application. The mode of breakdown in soils is by leaching, photodegradation, and microbial activity. Mobility in runoff water is similar to 2,4-D.

Metsulfuron is moderately persistent in soil, with a typical half-life of 30 days. Degradation is slow by soil microbes, and nonmicrobial hydrolysis is slow at high pH and relatively rapid at lower pH. Degradation rates increase at high soil temperature and high soil moisture.

Hexazinone is mobile in runoff water and readily leaches in some soils. Spot-gun application to brush species indicated limited movement and transport from treated watersheds in stream discharge. Its half-life varied from 1 to 6 months in soil, depending upon location.

In semiarid and humid regions, tebuthiuron may persist for long periods. Tebuthiuron is readily adsorbed in soil having high organic matter or high clay content, but may leach in soils low in organic matter or low clay content. In forage plants, however, tebuthiuron content is typically well below legal residue limits when applied as pellets. Concentrations of tebuthiuron in runoff water decreased rapidly and were <0.05 ppm after 3 months from a watershed in central Texas. Tebuthiuron resists photodecomposition and volatilization, and breakdown by microbial activity is slow.

Bromacil is degraded in soils by microbial activity. Disappearance varies with geographical location, with more rapid loss occurring in warmer soils. Bromacil is considered a moderately to highly persistent herbicide, especially at rates of 5 kg/ha and above. It is moderately mobile in the environment.

Imazapyr controls many weeds and woody plants. It is highly persistent in soils and may persist for 2 years. Microbial degradation is the principal means of imazapyr dissipation in soil. Dissipation in shallow ponds is relatively rapid.

Imazapyr is not very mobile in soil, and contamination from runoff into streams appears unlikely.

REFERENCES

Allender WJ. Movement of bromacil and hexazinone in a municiple site. Bull Environ Contam Toxicol 46:284–291, 1991.

Altom JD, Stritzke JF. Degradation of dicamba, picloram and four phenoxy herbicides in soils. Weed Sci 21:556–560, 1973.

Anderson JJ, Priester TM, Shalaby LM. Metabolism of metsulfuron methyl in wheat and barley. J Agric Food Chem 37:1429–1434, 1989.

Anderson WP. Weed Science: Principles and Applications. 3rd ed. New York: West, 1996.

Basler E, King CC, Badiei AA, Santelmann PW. The breakdown of phenoxy herbicides in blackjack oak. Proc South Weed Conf 17:651–355, 1964.

Baur JR. Movement in soil of tebuthiuron from sprays and granules. PR-3524. College Station: Texas Agri Exp Sta, 1978a.

Baur JR. Effect of picloram and tebuthiuron on establishment of ryegrass winter pasture. J Range Mgt 31:450–455, 1978b.

Baur JR. Establishing kleingrass and bermudagrass pastures using glyphosate and tebuthiuron. J Range Mgt 32:119–122, 1979.

Baur JR, Baker RD, Bovey RW, Smith JD. Concentration of picloram in the soil profile. Weed Sci 20:305–309, 1972a.

Baur JR, Bovey RW. Ultraviolet and volatility loss of herbicides. Arch Environ Contam Toxicol 2:275–288, 1974.

Baur JR, Bovey RW. Herbicidal effect of tebuthiuron and glyphosate. Agron J 67:547–553, 1975.

Baur JR, Bovey RW, Benedict CR. Effect of picloram on growth and protein levels in herbaceous plants. Agron J 62:627–630, 1970.

Baur JR, Bovey RW, and Holt EC. Effect of herbicides on production and protein levels in pasture grasses. Agron J 69:846–851, 1977.

Baur JR, Bovey RW, McCall HG. Thermal and ultraviolet loss of herbicides. Arch Environ Contam Toxicol 4:289–302, 1973.

Baur JR, Bovey RW, Merkle MG. Concentration of picloram in runoff water. Weed Sci 20:309–313, 1972b.

Baur JR, Bovey RW, Smith JD. Herbicide concentration in live oak treated with mixtures of picloram and 2,4,5-T. Weed Sci 17:567–570, 1969.

Baur JR, Bovey RW and Smith JD. Effect of DMSO and surfactant combinations on tissue concentrations of picloram. Weed Sci 20:298–302, 1972c.

Behrens R, Hull WC, Fisher CE. Field responses of cotton to four phenoxy-type herbicides. Proc South Weed Conf 8:72–75, 1955.

Bergstrom L. Leaching of chlorsulfuron and metsulfuron methyl in three Swedish soils measured in field lysimeters. J Environ Qual 19:701–706, 1990.

Beyer EM, Duffy MJ, Hay JV, Schlueter DD. Sulfonylurea herbicides. In: Kearney PC,

Kaufman DD, eds. Herbicides: Chemistry, Degradation and Mode of Action, vol. 3. New York: Marcel Dekker, 1988, pp. 117–189.

Bouchard DC, Lavy TL, Lawson ER. Mobility and persistence of hexazinone in a forest watershed. J Environ Qual 14:229–233, 1985.

Bovey RW. Efficacy of glyphosate in non-crop situations, In: Grossbard E, Atkinson D, eds. The Herbicide Glyphosate. London, U.K.: Butterworth, 1985, pp. 435–448.

Bovey RW. Weed management systems for rangelands. In: Smith AE, ed. Handbook of Weed Management Systems. New York: Marcel Dekker, 1996, pp. 519–552.

Bovey RW, Baur JR. Persistence of 2,4,5-T in grasslands of Texas. Bull Environ Contam Toxicol 8:229–233, 1972.

Bovey RW, Baur JR, Bashaw EC. Tolerance of kleingrass to herbicides. J Range Mgt 32: 337–339, 1979a.

Bovey RW, Dowler CC, Merkle MG. The persistence and movement of picloram in Texas and Puerto Rican soils. Pest Monit J 3:177–181, 1969.

Bovey RW, Hein H Jr, Meyer RE. Effect of herbicides on the production of common bufflegrass (*Cenchrus ciliaris*). Weed Sci 32:8–12, 1984a.

Bovey RW, Hein H Jr, Meyer RE. Concentration of 2,4,5-T, triclopyr, picloram and clopyralid in honey mesquite (*Prosopis glandulosa*) stems. Weed Sci 34:211–217, 1986.

Bovey RW, Ketchersid ML, Merkle MG. Distribution of triclopyr and picloram in huisache (*Acacia farnesiana*). Weed Sci 27:527–531, 1979.

Bovey RW, Meyer RE. Effects of 2,4,5-T, triclopyr and 3,6-dichloropicolinic acid on crop seedlings. Weed Sci 29:256–261, 1981.

Bovey RW, Miller FR. Effect of activated carbon on the phototoxicity of herbicides in tropical soil. Weed Sci 17:189–192, 1969.

Bovey RW, Richardson CW. Dissipation of clopyralid and picloram in soil and seep flow in the Blacklands of Texas. J Environ Qual 20:528–531, 1991.

Bovey RW, Scifres CJ. Residual characteristics of picloram in grassland ecosystems. B-1111. College Station: Texas Agri Exp Sta, 1971.

Bovey RW, Young AL. The Science of 2,4,5-T and Associated Phenoxy Herbicides. New York: Wiley, 1980.

Bovey RW, Meyer RE, Hein H Jr. Soil persistence of tebuthiuron in the claypan resource area of Texas. Weed Sci 30:140–144, 1982.

Bovey RW, Meyer RE, Holt EC. Tolerance of bermudagrass to herbicides. J Range Mgt 27:293–296, 1974b.

Bovey RW, Meyer RE, Morton HL. Herbage production following brush control with herbicides in Texas. J Range Mgt 25:136–142, 1972.

Bovey RW, Miller FR, Baur JR, Meyer RE. Growth of sorghum and soybeans in picloram-treated soil. Agron J 67:433–436, 1975b.

Bovey RW, Miller FR, Diaz-Colon J. Growth of crops in soils after herbicidal treatments for brush control in the tropics. Agron J 60:678–679, 1968.

Bovey RW, Burnett E, Meyer RE, Richardson C, Loh A. Persistence of tebuthiuron in surface runoff water, soil, and vegetation in the Texas Blacklands Prairie. J Environ Qual 7:233–236, 1978a.

Bovey RW, Richardson C, Burnett E, Merkle MG, Meyer RE. Loss of spray and pelleted picloram in surface runoff water. J Environ Qual 7:178–180, 1978b.

Bovey RW, Wiese AF, Evans RA, Morton HL, Alley HP. Control of weeds and woody plants on rangelands. AD-BU-2344. USDA and Minneapolis: University of Minnesota, 1984b.

Bovey RW, Burnett E, Richardson C, Baur JR, Merkle MG, Kissel DE. Occurrence of 2,4,5-T and picloram in surface water in the Blacklands of Texas. J Environ Qual 4:103–106, 1975a.

Bovey RW, Burnett E, Richardson C, Baur JR, Merkle MG, Baur JR, Knisel WG. Occurrence of 2,4,5-T and picloram in surface water in the Blacklands of Texas. J Environ Qual 3:61–64, 1974a.

Brady HA. Persistence of foliar-applied 2,4,5-T in woody plants. Proc South Weed Sci Soc 26:282, 1973.

Bronstad JO, Friestad HO. Behavior of glyphosate in the aquatic environment. In: Grossbard E, Atkinson D, eds. The Herbicide Glyphosate. London, U.K.: Butterworth, 1985, pp. 200–205.

Bush PB, Neary DG, Taylor JW. Effect of triclopyr amine and ester formulations on groundwater and surface runoff water quality in the Coastal Plain. Proc South Weed Sci Soc 41:226–232, 1988.

Calderbank A, Slade P. Diquat and paraquat. In: Kearney PC, Kaufman DD, eds. Herbicides; Chemistry, Degradation and Mode of Action, vol. 2. New York: Marcel Dekker, 1976, pp. 501–540.

Carter MC. Amitrole. In: Kearney PC, Kaufman DD, eds. Herbicides: Chemistry, Degradation and Mode of Action, vol. 1. New York: Marcel Dekker, 1975, pp. 377–378.

Chang SS, Stritzke JF. Sorption, movement and dissipation of tebuthiuron in soils. Weed Sci 25:184–187, 1977.

Coffman CB, Frank JR, Potts WE. Crop responses to hexazinone, imazapyr, tebuthiuron and triclopyr. Weed Technol 7:140–145, 1993.

CONCAWE. Gas Oils (Diesel Fuels/Heating Oils). Prod dossier no. 95/107. Brussels: CONCAWE'S Petroleum Products and Health Management Groups, 1996.

Corbin FT, Swisher BA. Radioisotope techniques. In: Camper ND, ed. Research in Weed Science. 3rd ed. Champaign, IL: South Weed Science Society, 1986, pp. 265–278.

Crafts AS. Toxicity of 2,4-D in California soils. Hilgardia 19:141–158, 1949.

Duke SO. Glyphosate. In: Kearney PC, Kaufman DD, eds. Herbicides: Chemistry, Degradation and Mode of Action vol. 3. New York: Marcel Dekker, 1988, pp. 1–70.

Duncan KW, Scifres CJ. Influence of clay and organic matter of rangeland soils on tebuthiuron effectiveness. J Range Mgt 36:295–297, 1983.

Dupuy AJ, Schulze JA. Selected water-quality records for Texas surface waters: 1970 Water Year. Rpt. 149. Austin: Texas Water Development Board and U.S. Geological Survey, 1972.

Edwards WM, Glass BL. Methoxychlor and 2,4,5-T in lysimeter percolation and runoff water. Bull Environ Contam Toxicol 6:81–84, 1971.

Eliasson L. Translocation and persistence of 2,4-D in *Populus tremula* L. Weed Res 13: 140–147, 1973.

Feng JC. Persistence, mobility and degradation of hexazinone in forest silt loam soils. J Environ Sci Health B22(2):221–233, 1987.

Feng JC, Sidhu SS, Feng CC, Servant V. Hexazinone residues and dissipation in soil leachates. J Environ Sci Health B24(2):131–143, 1989.

Gardiner JA. Substituted uracil herbicides. In: Kearney PC, Kaufman DD, eds. Herbicides: Chemistry, Degradation and Mode of Action, vol. 1. New York: Marcel Dekker, 1975, pp. 293–231.

Gardiner JA, Rhodes RC, Adams JB, Soboczenski EJ. Synthesis and studies with 2-C^{14} labeled bromacil and terbacil. J Agric Food Chem 17:980–986, 1969.

Getzenduner ME, Herman JL, VanGiessen B. Residues of 4-amino-3,5,6-trichloropicolinic acid in grass from application of Tordon herbicides. J Agric Food Chem 17: 1251–1256, 1969.

Green WR, Westerdahl HE, Joyce JC, Haller WT. Triclopyr (Garlon 3A) dissipation in Lake Seminole, Georgia. Final report. Misc. paper A-89-2. Washington, DC: Department of the Army, U.S. Army Corps of Engineers, 1989.

Haas RH, Scifres CJ, Merkle MG, Hahn RR, Hoffman GO. Occurrence and persistence of picloram in natural water sources. Weed Res 11:54–62, 1971.

Halifax JC, Scifres CJ. Influence of dicamba on development of range grass seedlings. Weed Sci 20:414–416, 1972.

Hamilton WT, Scifres CJ. Buffelgrass (*Cenchrus ciliaris*) response to tebuthiuron. Weed Sci 31:634–638, 1983.

Havens PL, Sims GK, Erhardt-Zabik S. Fate of herbicides in the environment. In: Smith AE, ed. Handbook of Weed Management Systems. New York: Marcel Dekker, 1996, pp. 245–278.

Hoffman GO. Practical use of Tordon 225 mixture herbicide on Texas rangelands. Down to Earth 27:17–21, 1971.

Jensen KIN, Kimball ER. Uptake and metabolism of hexazinone in *Rubus hispidus* L. and *Pyrus melanocarpa* (Michx.) Willd. Weed Res 30:35–41, 1990.

Johnsen TN Jr. Observations: Potential long-term environmental impact of tebuthiuron and its metabolites in Utah juniper trees. J Range Mgt 4:167–170, 1992.

Johnsen TN Jr, Morton HL. Tebuthiuron persistence and distribution in some semiarid soils. J Environ Qual 18:433–438, 1989.

Johnsen TN Jr, Morton HL. Long-term tebuthiuron content of grasses and shrubs on semiarid rangelands. J Range Mgt 44:249–253, 1991.

Jotcham JR, Smith DW, Stephenson GR. Comparative persistence and mobility of pyridine and phenoxy herbicides in soil. Weed Technol 3:155–161, 1989.

Krueger JP, Butz RG, Atallah YH, Cork DJ. Isolation and identification of microorganisms for the degradation of dicamba. J Agric Food Chem 37:534–538, 1989.

Lavy TL, Mattice JD, Kochenderfer JN. Hexazinone persistence and mobility of a steep forested watershed. J Environ Qual 18:504–514, 1989.

Lavy TL, Santelmann PW. Herbicide bioassay as a research tool. In: Camper ND, ed. Research Methods in Weed Science. 3rd ed. Champaign, IL: South Weed Science Society, 1986, pp. 201–217.

Leitch CJ Flinn DW. Residues of hexazinone in streamwater after aerial application to an experimental catchment planted with radiata pine. Aust For 46:126–131, 1983.

Lloyd-Reilley J, Scifres CJ, Blackburn WH. Hydrologic impacts of brush management with tebuthiuron and prescribed burning on post oak savannah watersheds. Texas Agri Ecosystems Environ 11:213–223, 1984.

Mangels G. Behavior of the imidazolinone herbicides in the aquatic environment. In: Shaner DL, O'Connor SL, eds. The Imidazolinone Herbicides. Boca Raton, FL: CRC Press, 1991, pp. 184–190.

Masters RA, Scifres CJ. Forage quality responses of selected grasses to tebuthiuron. J Range Mgt 37:83–87, 1984.

Maybank J, Yoshida K. Delineation of herbicide drift hazards on the Canadian prairies. Trans Am Soc Agri Eng 12:759–762, 1969.

Mayeux HS Jr, Richardson CW, Bovey RW, Burnett E, Merkle MG, Meyer RE. Dissipation of picloram in storm runoff. J Environ Qual 13:44–49, 1984.

McNeil WK, Stritzke JF, Basler E. Absorption, translocation and degradation of tebuthiuron and hexazionone in woody species. Weed Sci 32:739–743, 1984.

Merkle MG, Bovey RW, Davis FS. Factors affecting the persistence of picloram in soil. Agron J 59:413–414, 1967.

Merkle MG, Bovey RW, Hall R. The determination of picloram residues in soil using gas chromatography. Weeds 14:161–164, 1966.

Meyer RE, Bovey RW. Tebuthiuron formulation and placement effects on response of woody plants and soil residue. Weed Sci 36:373–378, 1988.

Morton HL. Influence of temperature and humidity on foliar absorption, translocation and metabolism of 2,4,5-T by mesquite seedlings. Weeds 14:136–141, 1966.

Morton HL, Johnsen TN Jr, Simanton JR. Movement of tebuthiuron applied to wet and dry rangeland soils. Weed Sci 37:117–122, 1989.

Morton HL, Robinson ED, Meyer RE. Persistence of 2,4-D, 2,4,5-T and dicamba in range forage grasses. Weeds 15:268–271, 1967.

Moseman TE, Merkle MG. Analysis, persistence and activity of triclopyr. Proc South Weed Sci Soc 30:64, 1977.

Muir DCG, Grift NP. Herbicide levels in rivers draining two prairie agricultural watersheds (1984). J Environ Sci Health B22:259–284, 1987.

Murray MR, Hall JK. Sorption-desorption of dicamba and 3,6-dichlorosalicylic acid in soils. J Environ Qual 18:51–57, 1989.

Neary DG, Bush PB, Grant MA. Water quality of ephemeral forest streams after site preparation with the herbicide hexazinone. For Ecol Mgt 14:23–40, 1986.

Newton M, Roberts F, Allen A, Kelpsas B, White D, Boyd P. Deposition and dissipation of three herbicides in foliage, litter and soil of brushfields of southwest Oregon. J Agri Food Chem 38:574–583, 1990.

Norris LA. The behavior of chemicals in the forest: Pesticides, pest control and safety on forest range lands. Proc., short course for pesticidal application. Corvallis, OR: Oregon State University, 1971, pp. 90–106.

Norris LA, Montgomery ML. Dicamba residues in streams after forest spraying. Bull Environ Contam Toxicol 13:1–8, 1975.

Norris LA, Moore DG. The entry and fate of forest chemicals in streams. Proceedings of Symposium or Forest Land Uses and Stream Environment. Corvallis, OR: Oregon State University, 1970, pp. 138–158.

Norris LA, Montgomery ML, Warren LE. Triclopyr persistence in western Oregon hill pastures. Bull Environ Contam Toxicol 39:134–141, 1987.

O'Connor GA, Wiergenga PJ. The persistence of 2,4,5-T in greenhouse lysimeter studies. Soil Sci Soc Am Proc 37:398–400, 1973.

Prasad R, Feng JC. Spotgun-applied hexazinone: Release of red pine (*Pinus resinosa*) from quaking aspen (*Populus tremuloides*) competition and residue persistence in soil. Weed Technol 4:371–575, 1990.

Rasmussen CA, Smith RP, Scifres CJ. Seedling growth responses of buffelgrass (*Penni-*

setum ciliare) to tebuthiuron and honey mesquite (*Prosopis glandulosa*). Weed Sci 34:88–93, 1985.

Saltzman S, Acher AJ, Brates N, Horowitz M, Gevelberg A. Removal of the phytotoxicity of uracil herbicides in water by photodecomposition. Pestic Sci 13:211–217, 1983.

Schubert DE, McKellar RL, Stevens LP, Byrd BC. Triclopyr movement on a small watershed. Abstract no. 105. Toronto: Weed Science Society of America, 1980, p. 50.

Scifres CJ, Allen TJ. Dissipation of dicamba from grassland soils of Texas. Weed Sci 21: 393–396, 1973.

Scifres CJ, Bovey RW. Differential responses of sorghum varieties to picloram. Agron J 62:775–777, 1970.

Scifres CJ, Halifax JC. Development of range grass seedlings germinated in picloram. Weed Sci 20:341–344, 1972a.

Scifres CJ, Halifax JC. Root production of seedling grasses in soil containing picloram. J Range Mgt 25:44–46, 1972b.

Scifres CJ, Mutz JL. Herbaceous vegetation changes following applications of tebuthiuron for brush control. J Range Mgt 31:375–378, 1978.

Scifres CJ, Bovey RW, Merkle MG. Variation in bioassay attributes as quantitive indices of picloram in soils. Weed Res 12:58–64, 1972.

Scifres CJ, Hahn RR, Merkle MG. Dissipation of picloram from vegetation of semi-arid rangelands. Weed Sci 19:329–332, 1971b.

Scifres CJ, Allen TJ, Leinweber CL, Pearson KH. Dissipation and phytoxicity of dicamba residues in water. J Environ Qual 2:306–309, 1973.

Scifres CJ, Hahn RR, Diaz-Colon J, Merkle MG. Picloram persistence in semi-arid rangeland soils and water. Weed Sci 19:381–384, 1971a.

Scifres CJ, McCall HG, Maxey R, Tai H. Residual properties of 2,4,5-T and picloram in sandy rangeland soils. J Environ Qual 6:36–42, 1977.

Scifres CJ, Stuth JW, Kirby DR, Angell RF. Forage and livestock production following oak (*Quercus* spp.) control with tebuthiuron. Weed Sci 29:535–539, 1981.

Sidhu SS, Feng JC. Hexazinone and its metabolites in boreal forest vegetation. Weed Sci 41:281–287, 1993.

Smith AE. Transformation of dicamba in Regina heavy clay. J Agric Food Chem 21:708–710, 1973.

Smith AE. Transformation of the herbicide dicamba and its degradative product 3,6-dichlorosalicylic acid in prairie soils. J Agric Food Chem 22:601–605, 1974.

Smith, AE. Persistence of the herbicides [14C] chlorsulfuron and [14C] metsulfuron methyl in prairie soils under laboratory conditions. Bull Environ Contam Toxicol 37:698–704, 1986.

Smith AE, Aubin AJ, Biederbeck VO. Effects of long-term 2,4-D and MCPA field applications on soil residues and their rates of breakdown. J Environ Qual 18:299–302, 1989.

Smith DT, Wiese AF. Cotton response to low rates of 2,4-D and other herbicides B-1120. College Station: Texas Agri Exp Sta, 1972.

Solomon KR, Bowhey CS, Liber K, Stephenson GR. Persistence of hexazinone (Velpar), triclopyr (Garlon), and 2,4-D in a northern Ontario aquatic environment. J Agric Food Chem 36:1314–1318, 1988.

Thompson DC, Pitt DG, Buscarini T, Staznik B, Thomas DR, Kettela EG. Initial deposits

and persistence of forest herbicide residues in sugar maple (*Acer saccharum*) foliage. Can J For Res 24:2251–2262, 1994.
Torstensson L. Behavior of glyphosate in soils and its degradation. In: Grossbard E, Atkinson D, eds. The Herbicide Glyphosate. U.K.: Butterworth, 1985, pp. 137–150.
Trichell DW, Morton HL, Merkle MG. Loss of herbicides in runoff water. Weed Sci 16: 447–449, 1968.
VanPelt NS, West NE. Above-ground dissolution of tebuthiuron particles placed beneath the crowns and small pinyon trees. J Environ Qual 18:281–284, 1989.
Walker A, Welch SJ. The relative movement and persistence in soil of chlorsulfuron, metsulfuron-methyl and triasulfuron. Weed Res 29:375–383, 1989.
Walker A, Cotterill EG, Welch SJ. Adsorption and degradation of chlorsulfuron and metsulfuron-methyl in soils from different depths. Weed Res 29:281–287, 1989.
Weber JB. Herbicide analysis and chemical property determinations using spectrophotometric methods. In: Camper ND, ed., Research Methods in Weed Science. 3rd ed. Champaign, IL: South Weed Science Society, 1986, pp. 247–264.
Weber JB, Best JA. Activity and movement of 13 soil-applied herbicides as influenced by soil reaction. Proc South Weed Sci Soc 25:403–413, 1972.
Weed Science Society of America. Herbicide Handbook. 7th ed. Champaign, IL: Weed Science Society of America, 1994.
Weete JD. Herbicide analysis by chromatographic techniques. In: Camper ND, ed. Research Methods in Weed Science. 3rd ed. Champaign, IL: South Weed Science Society, 1986, pp. 219–245.
Welch TG. Acres of Texas rangeland treated for brush and weed control. College Station, TX: Texas Agric Ext Serv, 1989 (unpublished data).
Welch TG. Chemical weed and brush control, suggestions for rangeland. B-1466. College Station: Texas Agric Ext Serv, 1997.
Whisenant SG, Clary WP. Tebuthiuron distribution in soil following application of pellets. J Environ Qual 16:397–402, 1987.
Whisenant SG, McArthur ED. Triclopyr persistence in northern Idaho forest vegetation. Bull Environ Contam Toxicol 43:660–665, 1989.
Wiese AF, Davis RG. Herbicide movement in soil with various amounts of water. Weeds 12:101–103, 1964.
Young AL, Thalken CE, Ward WE, Cairney WJ. The ecological consequences of massive quantities of 2,4-D and 2,4,5-T herbicides: Summary of a five-year field study. Abstr. no. 164. Las Vegas: Weed Science Society of America, 1974.
Zandvoort R. Leaching of fluridone, hexazinone and simazine in sandy soils in the Netherlands. Netherlands J Agric Sci 37:257–262, 1989.

7

Herbicide Toxicology and Safety

I. INTRODUCTION

Everyone working with agricultural chemicals wants to know the toxicity of the phytocide, both to himself and to other organisms. This chapter will explore herbicide toxicity and ways to minimize exposure to animals and the environment when managing woody plants. Toxicity to the test organism is given as the LD_{50} or lethal dose, expressed as milligrams (mg) of toxicant per kilogram (kg) of body weight (mg/kg), the dose that kills 50% of the test animals, usually laboratory rats. For mammalian toxicity it can be expressed as acute oral toxicity, and toxicity is determined (short-term) at the end of 24 hr (Ware, 1989).

The LC_{50} is the median lethal concentration, a concentration that kills 50% of the test organisms; expressed as mg—or cubic centimeters (cc), if liquid—per animal. It is also the concentration expressed as parts per million (ppm) or part per billions (ppb) in the environment (usually water) that kills 50% of the test organisms exposed (Ware, 1989).

Chronic toxicity is defined as the toxicity of a material determined beyond 24 hr and usually after several weeks of exposure. The "no-effect level" (NOEL) is also desirable information about an herbicide or a pesticide to determine if the application is being done at a safe level. Very toxic pesticides have an LD_{50} of 50 to 500 for a single oral dose for rats (mg/kg) and an LD_{50} of 200 to 1000 for a single dermal dose for rabbits (mg/kg) (Klaassen et al., 1986). Paraquat and diquat are the only herbicides listed in Table 1 that fit the very toxic category. Herbicides that are moderately toxic include 2,4-D, dicamba, dichlorprop, hexazinone, tebuthiuron, and triclopyr. The LD_{50} for moderately toxic compounds is 500 to 5000 mg/kg oral dose for rats (Klaassen et al. 1986). Amitrole, bromacil,

TABLE 1 Mammalian Toxicity of Herbicides Used in Woody Plant Management

Herbicide	Toxicity rating	Acute oral LD_{50} (rat; mg/kg)	Dermal LD_{50} (rabbit; mg/kg)	Inhalation LC_{50} (rat; μg/L-14-h)
Amitrole (tech)	Low	>5000	>2000	NA[a]
Bromacil	Low	5200	>5000	>4.8
Clopyralid	Low	>5000	>2000	1.3
2,4-D	Moderate	764	>2000	1.8
Dicamba	Moderate	1707	>2000	>9.6
Dichlorprop	Moderate	800	>4000	NA
Diesel fuel	Low	7400	>4100	NA
Diquat	High	230	>400	NA
Fosamine	Very low	24,400	>1638	NA
Glyphosate	Low	>5000	>5000	NA
Hexazinone	Moderate	1690	>6000	>4.5 (1-h)
Imazapyr	Low	>5000	>2000	>1.3
Metsulfuron	Low	>5000	>2000	>5.3
Paraquat	High	150	>240	Nontoxic
Picloram	Low	>5000	>2000	>0.04
Tebuthiuron	Moderate	644	>200	NA
Triclopyr	Moderate	713	>2000	NA

[a] NA = Not available.

Source: Farm Chemicals Handbook 1997; Weed Science Society of America, 1994.

clopyralid, diesel fuel, glyphosate, imazapyr, metsulfuron, and picloram are only sightly toxic at acute oral LD_{50} (rat) of 5000 to 15,000 mg/kg. Fosamine at 24,000 mg/kg is practically nontoxic (>15,000 mg/kg). If the rabbit dermal category for slightly toxic is 2000 to 20,000 mg/kg, dicamba, dichloprop hexazinone, and imayapyr can be added to the slightly toxic list.

In working with woody plant control herbicides, most are moderately to slightly toxic to mammalians, and with a few simple precautions exposure levels can be reduced to safe operational levels. The next few pages will detail acute oral, chronic, and dermal tests that determine a specific herbicide toxicity level.

II. HERBICIDES

A. Amitrole

Data on rats indicated that amitrole has an oral LD_{50} of >5000 and a dermal LD_{50} on rabbits of >2000 mg of amitrole/kg of body weight (Table 1). Sixteen-month feeding trials, however, indicated that after 90 days at 50 mg/kg/day, the male rats had enlarged thyroids (Table 2). Rats fed 500 mg/kg/day for 120 days and then returned to an amitrole-free diet 14 days before sacrifice indicated normal thyroids (Weed Science Society of America, 1994). Even though amitrole is considered only slightly toxic it is a restricted-use pesticide because of possible oncogenic effects in mammals. Data on teratogenicity, mutagenicity, and reproduction effects from amitrole are apparently not available. The LD_{50} for mallard ducks is 2000 mg/kg.

B. Bromacil

Bromacil is another herbicide that fits the slightly toxic category, with an oral LD_{50} for rats of over 5000 mg/kg (Table 1). Its dermal LD_{50} on rabbits was also >5000 mg/kg. In feeding trials the NOEL was 50 ppm for 24 months dietary in rats and 625 ppm for 12 months dietary in dogs. NOELs are given in Table 2 for teratogenic, mutagenic, and reproduction effects. Bromacil is not a teratogen or mutagen, and wildlife should not be affected by bromacil at the recommended field rates.

C. Clopyralid

Clopyralid is only slightly toxic to mammalians (Table 1). Test animals tolerated substantial amounts of clopyralid in relatively short- and long-term feeding trials (Table 2). Clopyralid is not a teratogenic, mutagenic, oncogenic, or reproductive toxin. Wildlife, including fish, birds, earthworms, and honey bees had relatively high tolerance to clopyralid. In eye irritation tests, clopyralid caused possible corneal injury in rabbits.

Table 2 Selected Data of the No Effect Level of Herbicides Used in Woody Plant Management on Several Mammalian Toxicity Evaluations and Wildlife Effects

Herbicide	Subacute toxicity day dietary—NOEL (mg/kg/day)	Chronic toxicity month dietary—NOEL (mg/kg/day)	Teratogenicity NOEL (mg/kg/day)	Reproduction NOEL (mg/kg/day)	Mutagenicity (mg/kg/d)	Wildlife species LD_{50} (mg/kg)	Wildlife species LD_{50} (mg/L)
	Mouse, rat, dog	Mouse, rat, dog	Rat, rabbit	Rat			
Amitrole	NA, NA, NA	NA, NA 16 mo-50[a], NA	NA, NA	NA	NA	Mallard duck—2000	NA
Bromacil	NA, NA, NA	NA, 24 mo—50 ppm, 12 mo—625 ppm	Maternal—20, fetal-75; maternal—100, fetal—100	250 ppm	None	NA	Bobwhite quail and mallard duck > 10,100 (8-d); bluegill sunfish—71/mg/L (48-h) rainbow trout—104 mg/L (96-h)
Clopyralid	90-d-750, 90-d-150, 180-d-250	18 mo—50, 24 mo—15, 12 mo—100	>250, 110	500	None	Mallard duck—1465	Bobwhite quail > 4640 ppm (8-d); bluegill sunfish—125 mg/L (96-h); rainbow trout—104 mg/L (96-h)
2,4-D	90-d-15, 90-d-15, NA	24 mo—5, 24 mo—5, 12 mo—1	75, 30	5	None	Bobwhite quail—500	Mallard duck > 5620 ppm (8-d); bluegill sunfish—263 mg/L (96-h); rainbow trout 377 mg/L (96-h)
Dicamba	NA, 90-d-250, NA	18 mo—115, 24 mo—125, 12 mo—60	Maternal—160, fetal—400; maternal—30, fetal—300	40	DNA-DR positive in *B. Subfillis*	Bobwhite quail—216; mallard duck—1373	Bluegill sunfish—135 m/L (96-h); rainbow trout—135 m/L (96-h)
Dichlorprop	NA, NA (no effect at 12.4, 98-d), NA	NA, NA (no effect at 12.4, 90-d), NA	NA, NA	NA		NA	Bluegill sunfish 1.1 mg/L (96-h); bluegill sunfish—160 mg/L (96-h)[b]
Diesel fuel[c]							
Diquat	NA, NA, NA	NA, 24 mo—25 ppm, 12 mo—50 ppm	NA, NA	25	NA	Hen—200–400, partridge—295	Mirror carp—67 mg/L (96-h); rainbow trout—21 mg/L (96-h)
Fosamine	NA, 90-d-10,000, NA	NA, NA, 6 mo—10,000	Maternal—350 fetal >3000, NA	5000	In vitro cytogenetics positive	NA	NA

Glyphosate	90-d-2300, 90-d > 1400, NA	24 mo—4500, 24 mo—400, 12 mo—500	1000, maternal—175, fetal >350	700	None	Bobwhite quail >4640	Mallard duck—4640 ppm (8-d); bluegill sunfish—120 mg/L (96-h); rainvow trout—86 mg/L (96-h)
Hexazinone	NA, 90-d-1000, NA	18 mo—200 ppm, 24 mo—200 ppm, 12 mo-200 ppm	100, 125	200 ppm	In vitro cytogenetics positive	Bobwhite quail 2258	Mallard duck > 10,000 ppm (8-d); bluegill sunfish—370 mg/L (96-h); rainbow trout—320 mg/L (96-h)
Imazapyr	NA, 90-d-500, 90-d-250	18 mo—1500, 24 mo—500, 12 mo—250	300, 400	500	None	Bobwhite quail and mallard duck >2150	Bluegill sunfish—100 mg/L (96-h); rainbow trout >100 mt/L (96-h)
Metsulfuron	28-d-7500 ppm, 90-d-1000 ppm, NA	18 mo—5000 ppm, 24 mo—500, 12 mo—500	40, 25	500	None	Mallard duck >2510	Bobwhite quail > 5620 ppm (8-d); bluegill sunfish > 150 mg/L (96-h); rainbow trout > 150 mg/L (96-h)
Paraquat	NA, NA, NA	NA, 24 mo—25 ppm, 24 mo—34	NA, NA	NA	NA	Bobwhite quail 981; mallard duck 4048	Brown trout—2.5 mg/L (96-h); rainbow trout—32 mg/L (96-h)
Picloram	90-d < 1050, 90-d-50, 90-d-250	24 mo-1000, 24 mo-20, 12 mo >175	>400, >400	Parental-200, reproductive 100	none	Mallard duck >2510	Bobwhite quail > 5000 (8-d), Bluegill surfish-14.5 mg/L (96-h), Rainbow trout-5.5 mg/L (96-h)
Tebuthiuron	90-d-83, 90-d-100, 90-d-12.5	18 mo-200, 24 mo-50, 12 mo-25	80, 25	20	none	Bobwhite quail and mallard duck >2500	Bluegill sunfish-87 mg/L (96-h), Rainbow trout-87 mg/L (96-h)
Triclopyr	90-d-20, 90-d-30, 6 mo-2.5	22 mo-5.3, 24 mo-3, NA	not teratogenic, not teratogenic	30	none	Mallard duck 1689	Bobwhite quail-2935 mg/L (8-d), Bluegill sunfish-148 mg/L (96-h), Rainbow trout-117 mg/L (96-h)

Note: Technical grade herbicide unless indicated otherwise. Mutagenicity included 1) gene mutation (GM), 2) structural chromosome aberration (SCA), and 3) DNA damage/repair (DNA-DR). NA = Not available. Data in columns is listed under each species from left to right.

[a] Enlarged thyroid after 90 days; rats feed 500 mg/kg/day for 120 days, then amitrole-free diet 14 days before sacrifice had normal thyroids.

[b] Dichlorprop butoxyethyl ester technical and dimethylamine salt technical, respectively.

[c] See text for information.

Source: Weed Service Society of America, 1994.

D. 2,4-D

The herbicide 2,4-D is moderately toxic to warm-blooded animals, with an acute oral LD_{50} of 764 mg/kg for male rats (Table 1). Fish show considerable tolerance to the technical acid and salt formulations (Table 2), however. For example, rainbow trout have an LC_{50} of 377 mg/L using the technical acid. The LC_{50} for rainbow trout using technical 2,4-D isooctyl ester is >5 mg/L for a 96-hr exposure. The NOELs of 2,4-D for subacute and chronic toxicity are given in Table 2. The herbicide 2,4-D is not teratogenic, embryotoxic, or mutagenic. Precautions need to be taken near water sources when using ester formulations of 2,4-D to prevent injury to fish.

E. Dicamba

Dicamba is moderately toxic to warm-blooded animals (Table 2). The NOELs are given for subacute and chronic toxicity. Mice, rats, and dogs tolerated fairly high test rates of dicamba without apparent effect, but decreased body weight and microscopic liver effect occurred at 10,000 ppm in 90-day feeding trials. Dicamba was not oncogenic, teratogenic, or mutagenic and had no adverse effects on reproduction. Most wildlife species appear tolerant of dicamba when oral LD_{50} and LC_{50} data are considered.

F. Dichlorprop

Dichlorprop is related to 2,4-D, but is the propionic acid derivative. Not as many data have been accumulated on the toxicological properties of dichlorprop as 2,4-D, but LD_{50s} for rats are similar (Table 1). Few data are available for the NOEL for subacute and chonic toxicity as well as for the teratogenic, mutagenic, and reproductive effects. Bluegill sunfish tolerated the technical dimethylamine salt (LC_{50} of 165 mg/L for 96-h), but bluegill sunfish exhibited a very low tolerance to the butoxyethyl ester technical of LC_{50} of 1.1 mg/L for 96-hr exposure.

G. Diesel Fuel

Diesel fuel has a low order of acute oral toxicity to rats (LD_{50} of 7400 mg/kg) and to dermal exposure on rabbits (LD_{50} > 4100 mg/kg, Table 1). Two-week dermal exposure to rabbits, however, indicated that 3200 and 6400 mg/kg/L caused severe skin irritation, weight loss, and high mortality (Weed Science Society of America, 1994). Histopathological examination of the treatment sites (clipped areas) reveal skin lesions, inflammation, acanthosis, hyperkeratosis, and crusting. The top-dose animals of diesel oil often showed multifocal hepatic necrosis. Groups of B6C3F1 mice treated dermally with 2000, 4000, 8000, 20,000 or 40,000 mg/kg/day of undiluted marine diesel oil for 14 days all died at the two highest doses. When treated dermally with 250, 500, 1000, 2000, or 4000 mg/

kg/day 5 days per week for 13 weeks, diesel oil showed no deaths. An increased incidence of mild chronic active dermatitis at the treatment site was reported for the high-dose group (4000 mg/kg/day). When exposed to aerosolized diesel fuel at 1.33 to 6 mg/L three times per week for 2 hr over 3 weeks (or once per week for 6 hr over 9 weeks), no neurotoxicity occurred. Significant effects were found in the lungs by decreased total capacity, however, and there was no evidence of reversal of the lung condition after 2 weeks. The frequency of exposure contributed to toxicity. In mouse skin carcinogencity tests, diesel fuel at 25 mg three times per week for 105 weeks (and at other doses and long-term exposure) caused some tumor growth (some malignant).

There is no evidence that normal conditions of storage, handling, or use of fuels will be a hazard to human health, provided that excessive skin contact is avoided. Gas oils should not be used as solvents for cleaning the skin. Accidental eye contact may cause mild stinging and redness. High exposure to the eyes of mist or vapor may cause slight eye irritation.

The toxicity of diesel fuel has been tested in the aquatic environment on fish, invertebrates, and algae (CONCAWE, 1996). Toxic effects on free swimming aquatic organisms would be expected to occur only in the immediate vicinity of a spillage. The hazard of no. 2 fuel oil to fish is determined by acute effects, with no significant additional effects resulting from long-term exposure. Number 2 fuel oil has also been tested on the eggs of mallard ducks (*Anas platyrlynchos*). An LD_{50} of 3.2 μl per egg has been reported.

If released into the environment, fuel oil will be lost by evaporation, phytolysis, and soil adsorption (CONCAWE, 1996). In aerobic soils and sediments, gas oil components will be biodegraded. The aquatic toxicity data indicated that acute LL_{50}/EL_{50} and IL_{50} values for aquatic organisms are in the 1 to 100 mg/L range.

H. Diquat and Paraquat

Diquat and paraquat are highly toxic herbicides with oral acute LD_{50s} of <500 mg/kg (Table 1). In longer-term feeding trials (at 36 ppm and 150 ppm) rats and dogs, respectively, developed cataracts (Weed Science Society of America, 1994). Chronic toxicity tests indicated that diquat was not a carcinogen or reproductive toxin. Data on its teratogenic and mutagenic effects are not available. The LD_{50s} on warm-blooded wildlife species are similar to rats, mice, and dogs. NOELs have been determined for rats and dogs in chronic toxicity trials at 25 ppm and 50 ppm, respectively.

Toxicity data for the technical dichloride salt of paraquat indicated that oral LD_{50} for male rats was 112 to 150 mg/kg, 50 mg/kg for monkeys, 48 mg/kg for cats, and 50 to 75 mg/kg for cows (Weed Science Society of America, 1994). Chronic toxicity trials indicated NOELs of 25 ppm for rats and 34 mg/

kg for dogs. Paraquat is not carcinogenic. Data on teratogenic, reproductive, and mutagenic effects are not available. Although bobwhite quail, Japanese quail, and mallard ducks are reasonably tolerant of paraquat, it is a restricted-use pesticide because of its mammalian toxicity.

I. Fosamine

According to the definition by Klaassen et al. (1986), fosamine is practically a nontoxic herbicide (acute oral $LD_{50} > 20{,}000$ mg/kg in rats). Its NOEL in subchronic toxicity feeding trials for 90 days was 10,000 ppm (the highest level tested). In chronic toxicity trials for 6 months the NOEL for dogs was 10,000 ppm. Fosamine was not teratogenic or mutagenic, and did not produce any reproductive effects. It has not been tested on wildlife (Weed Science Society of America, 1994).

J. Glyphosate

Glyphosate is another herbicide of very low toxicity to warm-blooded animals, with an oral LD_{50} of 5600 mg/kg in rats (Table 1). Subchronic toxicity tests indicate laboratory animals tolerate high concentrations of glyphosate in feeding trials, with NOEL of 10,000 ppm for mice and >20,000 ppm for rats for 90 days (Table 2). Chronic toxicity tests established that glyphosate is not a carcinogen, and laboratory animals tolerated high concentrations with no effect. Glyphosate tests indicated that it is not teratogenic or mutagenic or a reproductive toxin. Glyphosate has been tested on several wildlife species using the technical acid glyphosate and various commercial formulations.

K. Hexazinone

Hexazinone is moderately toxic to warm-blooded animals at an oral LD_{50} for rats at 1690 mg/kg (Table 1). Subchronic toxicity feeding tests for 90 days, however, indicated the NOEL for male rats at 1000 ppm. Chronic toxicity tests established a NOEL of 200 ppm for mice, rats, and dogs fed for 18, 24, and 12 months, respectively. Hexazinone was not teratogenic, and it was not a reproductive toxin. In mutagenicity tests observing structural chromosome aberration, hexazinone in vitro cytogenics was mutagenic and nonmutagenic in separate trials (Weed Science Society of America, 1994). Wildlife injury is not likely with recommended use.

L. Imazapyr

Imazapyr has a low order of toxicity, with an acute oral LD_{50} of >5000 in rats and a dermal LD_{50} of >2000 in rabbits. NOELs have been established in subchronic and chronic toxicity tests for several laboratory animals, and wildlife injury is not likely from recommended use. Imazapyr is not teratogenic or mutagenic, and it is not a reproductive toxin. The imidozolinone herbicides have a low toxicologic potential, partially because they act by inhibiting a biosynthetic process at a site present only in plants. Imazapyr is excreted rapidly by rats before it is accumulated in tissue or blood (Gagne et al., 1991).

M. Metsulfuron

Metsulfuron is considered an herbicide of low toxicity since the acute oral LD_{50} for rats is >5000 mg/kg and a dermal LD_{50} for rabbits >2000 mg/kg (Weed Science Society of America, 1994). The NOEL in subchronic and chronic toxicity trials indicated there were no histopathological or oncogenic effects and that laboratory animals tolerated high concentrations of the herbicide (up to 7500 ppm). The compound was not teratogenic, mutagenic, or embryotoxic. The NOEL for reproduction effect was 500 ppm in rats. Metsulfuron had a low toxicity on most wildlife species tested.

N. Picloram

Picloram is an herbicide of low toxicity with an acute oral LD_{50} of >5000 mg/kg on male rats (Weed Science Society of America, 1994). The NOEL has been established in subchronic and chronic tests (Table 2). Picloram is not teratogenic or mutagenic, and it is not a reproductive toxin. It has a low order of toxicity to wildlife. Restricted use is due to potential injury to susceptible nontarget plants.

O. Tebuthiuron

Tebuthiuron is considered moderately toxic to warm-blooded animals because its acute oral LD_{50} for rats is 644 mg/kg (Table 1). The dermal LD_{50} for rabbit is >200 mg/kg. The NOEL has been established for 90-day dietary trials at 83, 100, and 12.5 mg/kg/day for mice, rats, and dogs. In chronic toxicity trials the dietary NOEL for rats was 50 mg/kg/day for 24 months and 25 mg/kg/day for dogs fed for 12 months. Tebuthiuron was not teratogenic or mutagenic, and it was not a reproductive toxin (Table 2). Wildlife tolerated tebuthiuron at reasonable rates.

P. Triclopyr

The acute oral LD_{50} for rats is 713 mg/kg; 550 mg/kg for rabbits; and 310 mg/kg for guinea pigs. Dermal LD_{50} for rabbit is >2000 mg/kg (Weed Science Society of America, 1994). The NOEL has been established for triclopyr for mice, rats, and dogs at 20, 30, and 2.5 mg/kg/day in 90-day feeding trials. The dog trials were 6 months. Higher levels of triclopyr caused decreased body weight and increased mild liver and kidney effects in dogs. Triclopyr was not teratogenic, oncogenic, or mutagenic, and it was not a reproductive toxin. Wildlife should not be affected at ordinary use rates of triclopyr (Table 2).

Restricted-use pesticides were first designated by EPA in 1978 and can be applied only by certified applicators because of their inherent toxicity or potential hazard to the environment (Ware, 1989). Amitrole is restricted because it is possibly oncogenic (induces tumors), paraquat because of its mammalian toxicity, and picloram because of possible injury to nontarget plants.

Ware (1989) indicated that dose, length of exposure, and route of absorption in humans are important variables in toxicity. The amount of pesticide required to kill a human being can be correlated with the LD_{50} of the compound to rats in the laboratory. For example, the acute oral LD_{50} of 2,4-D for rats is 764 mg/kg body weight (52 grams, Table 1). Based on this, the LD_{50} for a 150-lb (68 kg) human would be 51,952 mg. If the commercial formulation of 2,4-D contained 4 lb of acid equivalent to 2,4-D per gal of concentration (1.8 kg/3785 mL), a human being would have to deliberately drink 109 mL of the concentrated product to approach the LD_{50} level. Dermal LD_{50s} are sometimes less than the acute oral LD_{50s}, which would most likely be the route of human exposure. The dermal LD_{50} for 2,4-D in rabbits is >2000 mg/kg, making lethal exposure of 2,4-D or similar products through the skin highly unlikely.

III. MINIMIZING HERBICIDE EXPOSURE

There are three ways herbicides may enter the human body to cause poisoning (Bohmont, 1989). One route includes oral exposure by the mouth and swallowing, the second involves dermal and exposure absorption through the skin and eyes, and the third inhalation or respiratory exposure.

A. Oral Exposure

The most serious exposure is through the gastrointestinal tract when the pesticide is taken into the mouth in sufficient amounts to cause injury or death. This can occur by accident or intended self-inflicted injury. Accidental oral ingestion in most cases is a result of putting the pesticide in unmarked containers such as soft drink or food containers where children or adults may consume it.

The most likely exposure to pesticides, however, probably occurs through accident or negligent handling of material, by which it is splashed or wiped onto the mouth and face during the spraying and mixing operations. Contaminated food should also be avoided by washing the hands and face before eating and keeping foods and drinks away from pesticide exposure. Oral exposure can be minimized by checking the label on the herbicide container for special instructions or warnings about oral exposure and by avoiding contaminated foods and drinks during and after spraying operations. Also, avoid touching the face and mouth with contaminated objects, hands, clothing, and masks or respirators. Although herbicides are usually of moderate to low toxicity, one should limit human exposure where possible.

B. Dermal Exposure

Most sources indicate that pesticide absorption through the skin is the most common cause of poisoning in agricultural workers (Bohmont, 1990). Those chemicals of high acute dermal toxicity such as paraquat are of the most concern. The data presented by Bohmont (1990) indicated that the human scalp, forehead, and ear canal can absorb high amounts of applied pesticide, depending upon the pesticide concentration and vapor action, degree and length of exposure, and other factors. Of the body parts listed, the scrotum area potentially is the highest in skin absorption. Protective clothing needs to fit the chemical being used, but in general a long-sleeved shirt, long-legged trousers, shoes, socks, and a hat should be used even with slightly toxic liquid or dry chemicals. When mixing slightly toxic liquid chemicals, gloves and an impermeable apron should be used.

If the pesticide is moderately toxic, water-impermeable gloves, a wide-brimmed hat, goggles, or a face shield are recommended in addition to the clothes already listed when mixing liquid spray. A respirator may also be required if the label indicates any problem with inhalation toxicity. If mixing is prolonged, water-repellent, long-legged trousers, a long-sleeved shirt, gloves, an apron, and a wide-brimmed hat are recommended in addition to a face shield and respirator.

If the chemical is highly toxic, water-impermeable clothes or disposable coveralls as listed earlier with a respirator are required. The main reason is to limit dermal, oral, and inhalation exposure as needed for safe handling and application of the pesticide.

C. Respiratory Exposure

Most herbicides used for woody plant control should not be a problem through respiratory exposure. Some individuals may be extra sensitive to nose, throat, or lung irritation, however.

A respiratory device is important to protective applicators who apply toxic pesticides. A number of commercial devices are available from farm equipment and agricultural chemical dealers. Bohmont (1990) indicated that there are two primary respiratory devices to handle toxic pesticides. One is the chemical cartridge respirator, and the other is the canister-type respirator. The cartridge respirator protects against certain pesticide dusts, mists, and fumes. The canister type (gas mask) protects the lungs as well as the eyes against chemical dust, mists, and fumes. Supplied-air respirators and self-contained breathing equipment may be needed when oxygen is deficient or highly toxic gases are so highly concentrated that the applicator must have his or her own air supply (Bohmont, 1990).

IV. HUMAN TOXICITY

Diquat and paraquat are of the most concern of the woody plant control herbicides. They may be fatal if swallowed or injurious if inhaled or absorbed through the skin (Bohmont, 1990). Lung fibrosis may develop if these materials are taken by mouth or inhaled. Prolonged skin contact will cause skin irritation. There are no adequate antidotes for diquat or paraquat, and the effects are usually irreversible.

Bovey (1980) reviewed the toxicological properties of the phenoxy herbicides. Deaths of humans by poisoning with phenoxy herbicides is rare. Authenticated case histories of sublethal effects are also very rare, but complaints of transient dizziness, sickness, and other symptoms have occurred in field workers, especially where the spray was inhaled excessively. This includes 2,4-D and dichlorprop.

Human toxicity or symptoms with any other herbicide listed in Table 1 have not been reported in applicators, mixers, or field workers to the knowledge of the author. First aid procedures for pesticide poisoning should be reviewed before application, however. Poisoning symptoms may include headache, giddiness, nervousness, blurred vision, cramps, diarrhea, numbness, or dilation of the pupils (Ware, 1989). In some cases there is excessive sweating, tearing, mouth secretions, nausea, vomiting, change in heart rate, muscle weakness, breathing problems, confusion, convulsions, coma, or death. If poisoning is suspected obtain professional help at once and call the nearest poison control center. If the poisoned person is taken to a physician or emergency ward take along the pesticide label and telephone number of the poison control center. Proper first-aid treatment immediately after recognizing poison symptoms is extremely important.

V. HANDLING HERBICIDES

The first safety rule in using an pesticide is to read the label and follow the directions and precautions listed (Ware, 1989). In selecting an herbicide make sure it controls your particular weed problem. If in doubt consult your county agent or the weed specialist in your state cooperative extension service or university.

Mix herbicides as suggested in "Minimizing Herbicide Exposure" (Sec. III). It is always recommended to wear rubber gloves and protective boots and clothing. Avoid breathing dust or fumes and splashing herbicide onto your skin.

When applying herbicides make sure the equipment is in good working order, properly calibrated, and applied during calm weather to prevent spray drift. Controlling spray drift will limit herbicide movement to nontarget water sources, plants, and animals, including man. Spray equipment and protective clothing should be properly cleaned or disposed of between spraying operations. Do not use herbicide-application equipment to apply other types of agricultural chemicals.

When stored, herbicides should be locked up and inaccessible to children and adults. Do not store herbicides with food, feed, seed, or water sources. The herbicide should be kept dry and kept from freezing. Label the storage area. Always keep herbicides in their original closed containers and labeled. Remove unwanted and outdated containers to prevent herbicide leaks and application problems with deteriorated products. Herbicides have a longer shelf life if stored in a cool, dry, shaded area at above freezing and below extremely hot temperatures (See label instructions.)

When disposing of empty herbicide containers consult the disposal regulations within your state. The state lead agency for pesticides is designated by EPA, and has EPA-approved toxic waste disposal sites and landfills (Ware, 1989). Pesticide containers that have been triple-rinsed and punctured may be disposed of as nonhazardous wastes. The water used to rinse the containers must be used again to dilute or mix with the herbicide or disposed of as hazardous waste, however. If glass, the containers are triple-rinsed and crushed and disposed of in a sanitary landfill. Paper and cardboard containers should be smashed to make them unusable and placed in a larger clean container before trashing. All such materials should be packaged so children and pets cannot contact them. Always read the pesticide label for instructions.

VI. ACTUAL HUMAN EXPOSURE TESTS

A. 2,4,5-T

Studies in animals and humans have shown that oral doses of phenoxy herbicides are rapidly absorbed and are excreted virtually completely as phenoxy acids in urine with a half-life of less than 1 day (Leng et al., 1982). The rate of absorption of 2,4,5-T into the body appeared to be slower after external exposure than after oral administration in humans. Pharmacokinetic modeling indicated that 97% of the 2,4,5-T absorbed through the skin would be cleared within 1 week. Measurement of 2,4,5-T excreted in the urine of spray crews demonstrated that the maximum absorbed dose is not likely to exceed 0.1 mg per kg of body weight per workday. Urinary excretion provided a more reliable measure of dose than analysis of patches worn by the workers (Lavy and Mattice, 1985). Exposure was highest in mixers who handled the spray concentrate and in sprayers using backpack equipment (Lavy et al., 1980). Some absorption of 2,4,5-T was apparently due to wearing contaminated clothing by field workers.

Exposure to 2,4,5-T averaged 0.0005, 0.586, and 0.033 mg/kg body weight for inhalation, patch, and internal measurements, respectively. These measurements indicated that workers excreting the highest amounts of 2,4,5-T received exposure levels below those toxic to laboratory animals (Lavy et al., 1980).

B. 2,4-D—Aerial Application

Similar to 2,4,5-T, 2,4-D human exposure from aerial or ground applicator crews is far below the human health risk level. In three helicopter spray operations, forest workers were monitored for exposure to an internal dose of 2,4-D (Lavy et al., 1982). Levels of 2,4-D were measured in the air near the breathing zone, on denim patches to estimate dermal exposure, and in the urine excreted for 2 days before and 5 days after the spraying to determine the internal dose. Each crew made two applications about 1 week apart to compare exposure from crew members wearing customary clothing and following normal precautions (T-1) with that of the same crew members wearing protective apparel and following special hygienic practices (T-2). External exposure was low, with the highest level at 0.0911 mg/kg of body weight for a batchman in T-1. The total internal dose determined by urine analyses ranged from nondetectable to 0.0557 (in T-1) or 0.0237 (in T-2) mg/kg of body weight. Those crewmen working most closely with the spray concentrate or handling spray equipment (pilots, mechanics, and batchman-loaders) showed the highest doses. Protective clothing and good hygienic practices limited exposure. On the basis of analyses of the toxic levels of 2,4-D in laboratory animals, the human exposure levels in these tests were well below that which might endanger health. Table 3 shows the safety factor as determined by Lavy and co-workers at the University of Arkansas (Lavy et al., 1981).

According to these calculations, workers with the highest levels of exposure (e.g., pilot and batchman in T_1) could have been exposed to 1200 times as much 2,4-D as they actually received before reaching the NOEL as determined by the EPA advisory panel (Lavy et al., 1981).

The table also suggests the extent to which workers were protected by special clothing and the precautions taken in T2. Exposure to 2,4-D in T1 using normal operating conditions indicates no real need for special protection from 2,4-D; however, with more toxic compounds it may be important to reduce exposure with protective clothing and careful habits.

Results from this study indicate that the application of 2,4-D does not pose a health hazard and that its use as an important method of selectively controlling broadleaf weed species need not be curtailed.

C. 2,4-D, Dichlorprop, and Picloram—Ground Application

Past studies have shown that exposure by inhalation is small compared to the amount that enters the body when the chemical comes into contact with bare skin

TABLE 3 Safety Factor for Workers and 2,4-D Levels Measured in the Study

Position of workers	Amount of 2,4-D excreted in urine (mg 2,4-D/kg of body wt)		Safe level of 2,4-D	Safety factor	
	T_1	T_2		T_1	T_2
Pilots	.0199	.0085	24	1,200	2,800
Mechanics	.00545	.0030	24	4,400	8,000
Batchmen	.0196	.0140	24	1,200	1,700
Supervisors	.0023	.00013	24	10,000	180,000
Observers	.00049	.00009	24	49,000	260,000

Note: 24 mg 2,4-D/kg body weight is the No Observable Effect Level as supplied by EPA's Scientific Advisory Panel for 2,4-D. The no effect level (24 mg/kg) divided by the amount of 2,4-D found in workers' urine in our studies = safety factor. (Example: the pilots in T1 could have received 1200 times as much 2,4-D exposures as they did and still be within the safe level.)

Source: Lavy et al., 1981; 1982.

(Lavy et al., 1984; 1987). Most herbicides—and many insecticides—that contact humans are rapidly and quantitatively excreted in the urine. Collecting and analyzing the total urine output of exposed workers allows researchers to measure the absorbed dose and to compare techniques for limiting exposure.

In the summer of 1982, 80 forest workers using Weedone 170 and Tordon 101-R in nine forest locations in Arkansas, Oklahoma, and Mississippi cooperated in a study. The objective was to determine the exposure to 2,4-D, dichlorprop, and picloram of workers using special safety procedures.

Each group of 20 workers applied herbicides by one of four methods: backpack sprayer, injection bar, hypohatchet, or hack-and-squirt. Backpack crews applied Weedone 170 in a diluted spray form. The other three crews applied undiluted Tordon 101-R (Lavy et al., 1984; 1987).

For a comparison of worker exposure under two different levels of protection, each member of each crew participated in two different tests. In the first (T1), crew members were instructed to dress and to apply the herbicide in their customary manner. In the second test (T2), they received special instructions intended to limit exposure, and were each issued new leather gloves and new boots 1 day before application. The T1 vs. T2 comparison was designed to quantify the degree of protection achieved by employing this practical set of alternatives.

Backpack sprayers received a higher dose of 2,4-D than did the other workers. A significant reduction in exposure occurred between T1 and T2 for all workers using the injection bar, hypohatchet, and hack-and-squirt methods. Those spraying 2,4-D from backpacks did not reduce their exposure level by wearing new leather gloves or new boots, apparently due to a high degree of spray contact with other parts of their bodies.

The results of this study were used in conjunction with data from other studies on toxicology and pharmacokinetics to interpret the toxicological significance of exposure. A standard toxicological benchmark is the exposure level that produces no observed effect in sensitive test animals as determined by the 2,4-D scientific advisory panel. It is well documented that at least 95% of picloram and 2,4-D is excreted in the urine within 5 days. A margin of safety was determined by dividing the NOEL by the dose absorbed by workers as determined by urine analysis. Values were adjusted for the weight of the subjects being compared (Lavy et al., 1984; 1987).

The average margin of safety for 2,4-D ranged from 245 for the backpack crew in T2 to 5,581 for the injection bar crew in T2. Picloram margins of safety ranged from 23,000 in T1 (hypohatchet) to 943,400 in T2 (injection bar). The primary reason for the large picloram margin of safety is that this herbicide does not readily penetrate the skin.

Of the 80 herbicide applicators monitored, the most exposed one—a backpack sprayer—had a margin of safety of 98. Although his clothing appeared to be nearly saturated with spray, these calculations indicate he could have received 98 times as much 2,4-D and still be below a dose level harmful to his health even if he was repeatedly exposed in a similar manner.

Results of worker-exposure studies by the University of Arkansas Pesticide Residue Laboratory using seven different herbicides and insecticides have all indicated that when used according to label specifications health-threatening levels of exposure do not occur. Since many pesticides are more toxic than Weedone 170 and Tordon 101-R, however, it is always important to follow label instructions and to limit exposure where possible. Practices to reduce exposure include washing hands before eating, using tobacco, or using the bathroom; immediately washing with soap and water any skin on which pesticide is spilled; showering as soon as possible after exposure; and wearing clean clothes, including waterproof boots and gloves.

D. Paraquat

Several Arkansas commercial grape growers operating tractor-mounted, low-boom vineyard spray rigs were monitored for potential dermal, respiratory, and internal exposure to paraquat during the 1980 and 1981 growing seasons (Forbess et al., 1982). Workers followed their usual mixing and spraying routines. Analyses by colorimetric methods revealed very low levels of paraquat exposure. Greatest dermal exposure levels, averaging 0.015 mg paraquat/kg body weight, were detected on persons operating the spray rigs. Respiratory exposure was minimal, and there was no paraquat detected in any of the urine samples collected from each worker. Those persons receiving the highest levels of paraquat exposure had measurements that were well below those found to be toxic to laboratory animals. Hazards from using this material by this method of application should be low when used in accordance to label directions and precautions.

E. Glyphosate

This study addresses the measurements of glyphosate exposure received by 14 workers employed at two tree nurseries (Lavy et al., 1992). The applicators, weeders, and scouts monitored all wore normal work clothing, which for applicators was a protective suit, rubber gloves, and boots. Measurements were made of the glyphosate that was dislodged from conifer seedlings during water rinses taken twice weekly from May through August. Only one of 78 dislodgeable residue samples contained glyphosate. Nine cotton gauze patches were attached to the clothing of each worker 1 day per week during this same period. Hand washes were taken on the same day that patches were worn. Most patches and hand washes from applicators and weeders contained measurable amounts of glyphosate. Analyses of individual patches showed that the body portions receiving the highest exposure were ankles and thighs. For scouts, only one of 23 hand washes contained glyphosate. To provide a measure of the exposure occurring via all exposure routes (dermal, ingestion, and inhalation) an analysis was made of the total urine excreted. For most workers a daily total urine collection was made for 12 consecutive weeks. Urine analysis did not reveal any positive samples. The lower limit of method validation for glyphosate in the urine samples was 0.01 μg/mL.

High rainfall—or irrigation as needed—in conjunction with normal field dissipation avenues and worker training, were cited as contributing factors for the low amounts of glyphosate exposure found. None of the exposure parameters indicated that glyphosate exposure poses a threat to human health when used under normal nursery conditions.

F. Long-Term Exposure to Pesticides

A year-long nursery worker pesticide exposure study was designed to measure and evaluate the exposure occurring to workers who had the potential for simultaneous exposure to multiple pesticides (Lavy et al., 1993). This four-state study was conducted in five nurseries (four U.S. Department of Agriculture [USDA] Forest Service and one state) involved in conifer seedling production. Primary comparisons were made among nursery workers in the Pacific Northwest and south-central United States. Worker exposure was assessed by using patches attached to clothing, hand-rinse samples, and urine excreted from potentially exposed workers. In addition, dislodgeable residue in rinsate from a water wash of pesticide-treated seedlings was also evaluated. Four different groups of field workers (designated as applicators, weeders, scouts, and packers) were included. The pesticide-absorbed dose, assessed by a urine analysis of pesticide metabolites and the deposition of pesticide on patches attached to the clothing of field workers, was monitored as they performed their duties under normal conditions (e.g., typical clothing, pesticide application). Monitoring was performed for the 14 different pesticides that were used in these nurseries. Seven pesticides were studied in more detail using biological monitoring. For these compounds, metabolities known to be excreted in the urine of exposed humans or other mammals were used to estimate the dose of pesticide absorbed by the exposed workers.

The highest percentage of positive samples came from dislodgeable residue samples (8.3%), followed by patch (3.2%), hand-rinse (2.9%), and urine samples (1.3%). To summarize the conclusions from the urinary excretion data, 12 of the 73 nursery workers in the study received a low absorbed dose of pesticide. Biological monitoring revealed that three pesticides (benomyl, bifenox, and carbaryl) were found in the urine of some of the workers. Of the 3,134 urine samples analyzed there were 42 positive; 11 urine samples were positive for benomyl, while bifenox was responsible for 13 positives and carbaryl accounted for the remaining 18. The 12-week continuous monitoring of urine showed that metabolites of these materials were rapidly excreted; thus, no build-up in the body was anticipated. Margin of safety (MOS) calculations were made to provide an assessment of the significance of the exposure. Based on the low frequency of positive urine samples in the study, the low levels of metabolites when they were found, their apparent rapid excretion rate, and the NOEL data furnished from other sources, nursery worker exposure to pesticides in these conifer nurseries is below health-threatening levels.

G. Human Herbicide Exposure from Smoke

The potential exposure to pesticide residues resulting from burning wood treated with phenoxy and pyridine herbicides was assessed (Bush et al., 1987). Wood samples from trees treated with 2,4-D, dicamba, dichlorprop, picloram, and triclopyr contained variable amounts of parent compound residues at 4, 8, and 12 months after application. At the time of the latter sampling, residues of 2,4-D, dicamba, and picloram were <2.1 mg/kg on a fresh weight basis. Mean residue concentrations of triclopyr and dichlorprop were somewhat higher, at 3.5 and 13.0 mg/kg, respectively. In a laboratory experiment, samples with known amounts of herbicide residue were subjected to either slow or rapidly burning conditions in a tube furnace. During slow combustion, relatively stable compounds such as 2,4-D, dicamba, and dichlorprop were released in significant amounts. Rapid combustion greatly enhanced the decomposition of 2,4-D, dicamba, dichlorprop, picloram, and triclopyr. A well-developed fire in a wood stove or fireplace, with active flaming combustion, where temperatures commonly reach 800 to 1,000°C, should result in greater than 95% thermal decomposition of the herbicides examined in this study. Burning of herbicide-treated wood under smoldering conditions could result in very low levels of herbicide residue in ambient indoor air. The exposure levels are less than 0.3% of the threshold limit value for 2,4-D and triclopyr, however. The exposure is also more than three orders of magnitude lower than the established acceptable daily intakes for these products.

Smoke was monitored on sites treated with labeled rates of forestry herbicides containing the active ingredients imazapyr, triclopyr, hexazinone, and picloram (McMahon and Bush, 1992). The sites were burned within 30 to 169 days after herbicide application. Tract size ranged from 2.4 to 154 ha. Personal monitors and area monitors employing glass fiber filters and polyurethane foam collection media were used. No herbicide residues were detected in the 140 smoke samples from the 14 fires conducted in this study. The sensitivity of the monitoring methods was in the 0.1 to 4.0 $\mu g/m^3$ range, which is several hundred to several thousand times less than any established occupational exposure limit for herbicides.

VII. THE PESTICIDE LABEL

The single most important safety tool for herbicide users is the herbicide label (Ware, 1989). The Federal Environment Pesticide Control Act (FEPCA) requires that herbicide use be consistent with the label. Violations can result in heavy fines, imprisonment, or both. In addition, certified applicators are required for using restricted-use pesticides. The information required by EPA for label format for unclassified and restricted-use pesticides are listed by Ware (1989), and include the following:

Product name
Company name and address
Net contents
EPA pesticide registration number
EPA formulator manufacturer establishment number
Ingredients statement
Pounds/gal statement (if liquid)
Front-panel precautionary statements
Child hazard warning (''Keep Out of Reach of Children'')
Signal word (DANGER, WARNING, or CAUTION)
Skull and crossbones and word *poison* in red
Statement of practical treatment

Referral statement
Side- or back-panel precautionary statements
Hazards to humans and domestic animals
Environmental hazards
Physical or chemical hazards
"Restricted-Use Pesticide" block
Statement of pesticide classification
Misuse statement
Reentry statement
Category of applicator
"Storage and Disposal" block
Directions for use

The label is the key information about any pesticide (Bohmont, 1990). It gives directions on how to mix and apply the pesticide and offers guidelines on safe handling, storage, and the protection of the environment. The research and development leading to the label costs millions of dollars and many years of dedicated work. It may also include the combined efforts of industry, university, and government research. Most compounds tested are discarded for a variety of reasons. Since promising pesticides are difficult to discover and develop, potential candidates must be evaluated for sales volume, widescale usage, ease and cost of manufacturing, registration with EPA, and safety. The potential pesticide is subjected to many toxicological tests (Table 2), degradation routes and residue tests, soil movement data, field performance and benefits tests, wildlife effects, and label review (Bohmont, 1990). The company petitions EPA for legal tolerances and seeks registration of labeled uses for the pesticide on pests, crops, and animals as needed to support the product. The office of pesticide programs of the EPA in Washington, D.C., reviews the pesticide for the purposes claimed and sets tolerance levels for the pesticide in the environment. A final review by EPA is done to determine if all the claims on the label are supported by the data submitted after publication in the *Federal Register*.

The label permits a manufacturer to sell the products, provides information on the distribution, storage, sale, and disposal of the product and how to use the product legally and correctly, and provides physicians with information on treatment if poisoning occurs (Bohmont, 1990).

1. Product name and ingredients. The product or trade name is the producer's or formulator's proprietary name of the pesticide. Most companies register each brand name as a trademark, and no other company is allowed to use that name. Different companies may thus have different trade names for the same chemical. The chemical name is also printed on the label, including the percentage of active ingredient. The active ingredient in liquid is the active ingredient or acid equiva-

lent of the chemical. It is also indicated in lb per gal. If in pellet or powder form it is lb of active ingredient per lb of product. The common name is usually found with the chemical name and is the generic name accepted for the active ingredient of the product. The percentage of inert (inactive) ingredients makes up the remaining 100% of the product and may include solvents, surfactants, carriers, or fillers to enhance product handling and quality. The inert ingredients need not be named.

2. Use classification. A statement will be made on the label if the chemical is a restricted-use pesticide. In this book amitrole, paraquat, and picloram are restricted-use pesticides because of toxicity or potential environmental damage and require a certified applicator to apply them.
3. Type of pesticide. The type of pesticide is listed on the front panel of the label and states in general terms what the product controls. For example: a Dow AgroSciences specimen label for Remedy (triclopyr) states that it is a speciality herbicide for the control of woody plants and broadleaf weeds on rangeland and permanent grass pastures.
4. Net contents. The front area of the pesticide label indicates how much pesticide is in the container. For dry formulations it is in lb or oz; for liquids it is in gal, qt, or pt. Also listed are the active ingredients per gal of product.
5. Name and address of manufacturer. The EPA registration number signifies that the product has been registered with EPA. The established number is a number assigned to each manufacturing plant; in case problems develop the product can be traced.
6. Signal words and symbols. Every label must contain a signal word indicating the toxicity of the product to humans. Knowing if there is a hazard helps the user protect himself and his workers with proper clothing, equipment, and exposure. ''CAUTION'' means the product is slightly toxic, ''WARNING'' means moderately toxic, and ''DANGER–POISON'' means highly toxic.
7. Precautionary statements. The statements following the signal word on the front or side of the label indicates both the route of entry (mouth, skin, lungs, eyes) and the body parts to protect.
8. Protective clothing and equipment statements. Pesticide labels vary in the statements they contain concerning protective clothing and equipment. Some labels have no statements. Guidelines for protective clothing and equipment needs can be derived from the signal-word route-of-entry statements and other information on the label. Even though the label may not specifically require them, users should wear a long-sleeved shirt, long-legged trousers, and gloves. Waterproof

clothing should be considered with prolonged pesticide exposure, such as mixing pesticides or exposure to spray.

9. Other precautionary statements. Other precautions listed deal with handling the product–prevention of contamination of food and feed, removal and cleaning of contaminated clothing, personal hygiene after pesticide use and before eating or smoking, wearing clean clothes daily, pesticide storage, and protection of children and animals from treated areas.
10. Statement of practical treatment. These statements indicate the first-aid treatments recommended in case of poisoning.
11. Environmental hazards. The pesticide used may be harmful to the environment if improperly used. If a pesticide is especially hazardous to wildlife it will be stated on the label that it is toxic to bees, fish, or certain other wildlife.

    Environmental warning statements may also alert the user to possible contaminations in runoff water, from spray drift, or from disposal of waste pesticide to the environment.
12. Physical and chemical hazards. This section of the label spells outs any special fire, explosion, or chemical hazard this pesticide may pose.
13. Directions for use. The instructions on how to use the product are printed on the label. Such directions indicate the pests controlled, the crop, animal, or site the product is to protect, what form and equipment should be used, the proper dose to use, mixing instructions, compatibility with other use products, when the materials should be applied, and other instructions.

Some pesticide labels with signal words of DANGER—POISON or WARNING contain a reentry statement that includes the amount of time that must elapse after pesticide application until it is safe to enter the treated area without wearing full protective clothing and equipment. Users must comply with the EPA's worker protection standard (WPS). The restricted-entry interval (REI) for most herbicides used in brush control is 12 hr, but this can vary with method of application, formulation, and mixtures with other herbicides, and can be extended up to 48 hr (Weed Science Society of America, 1994). Workers must be orally warned about pesticide applications, or posted warnings must be placed at treated areas according to label requirements. Notification of a pesticide application must be given to all workers who will be within 1/4 mile of the treated area during application before the REI expires. The location and description of the treated area must be given, and REI information provided with a warning not to enter during the REI. Oral and posted warnings must be given when paraquat is used. Oral warnings usually are required only for most of the materials used in woody plant control.

VIII. LAWS AND REGULATIONS

As the use of chemicals to control pests has increased in the United States, the scope of federal and state laws and regulation has also increased to protect the pesticide user, the consumer of treated products, and the environment from pesticide pollution (Bohmont, 1990). Early in this century the USDA regulated labeling and interstate distribution of agricultural chemicals for use by farmers, home owners, and industry.

The Federal Insecticide Fungicide and Rodenticide Act (FIFRA) of 1947 superseded the 1910 Federal Insecticide Act and placed the burden of proof of acceptability of a product on the manufacturer prior to marketing. It protected the user, consumer, and public from pesticides, with certain limitations. The FIFRA regulates the distribution, sale, and use of pesticides in the United States. In 1948 the Food and Drug Administration (FDA) began establishing safe levels of residue tolerances in foods. The Pesticide Chemicals Amendment (the Miller Pesticide Amendment) amended the Federal Food, Drug and Cosmetic Act in 1954 and formalized the tolerance setting by FDA. The pesticide industry was required to submit pest-control efficacy data and proof of safety of any measurable residue in food produced. In 1958 a food additives amendment included the Delaney Clause, which prohibits any residue of a carcinogen in food.

In 1959, FIFRA (1947) was amended to include nematicides, plant regulators, defoliants, and desiccants as pesticides. In 1964, FIFRA (1947, 1959) was again amended to require that all pesticide labels contain the federal registration number. It also required caution words such as WARNING, DANGER, CAUTION, and Keep Out of Reach of Children to be included on the label of poisonous pesticides.

In 1970 the responsibility for pesticide regulation by USDA was transferred to the EPA, and the responsibility to establish pesticide tolerance by FDA was transferred to EPA. The enforcement of tolerances, however, remained with FDA.

The FEPCA of 1972 (and as amended in 1975, 1978, 1980, 1981, 1988, and 1996) completely revised FIFRA. The new act regulates the use of pesticides to protect people and the environment and extends federal pesticide regulation to all pesticides, including those distributed or used within a single state (Bohmont, 1990). There are eight basic provisions.

1. All pesticides must be registered by EPA.
2. For a product to be registered the manufacturer is required to provide scientific evidence that when used as directed the product.

   a. Will effectively control the pests listed on the label.
   b. Will not injure humans, crops, livestock, or wildlife or damage the environment.
   c. Will not result in illegal residues in food or feed.
3. All pesticides will be classified into general-use or restricted-use categories.
4. Restricted-use pesticides must be applied by a certified applicator.
5. Pesticide manufacturing plants must be registered and inspected by EPA.
6. The use of any pesticide inconsistent with the label is prohibited.
7. States may register pesticides on a limited basis for local needs.
8. Violations can result in heavy fines and/or imprisonment.

As amended in 1975, 1978, 1980, 1981, 1988, and 1996, FIFRA clarifies the intent of the law and greatly influences the way pesticides are registered and used (Ware, 1989). The more important provisions of the new amendments are stated herewith.

1. Generic standards are to be set for the active ingredients rather than for each product. This provision will speed registration, since there are only about 1200 active ingredients versus 37,000 formulations in the marketplace.
2. Reregistration of older products is required. Pesticides registered prior to August 1975 must meet the new requirements to be reregistered.
3. The EPA may grant conditional registration even though certain supporting data have not been completed.
4. Efficacy data may be waived.
5. The use of data from one registrant can be used by other manufactures or formulators if paid for.
6. Trade secrets will be protected.
7. The states have primary enforcement responsibility, called ''state primacy.''
8. States can register pesticides for special local needs to meet unusual situations.
9. Users and applicators can now.
   a. Use less pesticide than labeled dosage.
   b. Use a pesticide for controlling a target pest not named on the label, providing the site or host is indicated.
   c. Apply the pesticide by any method not prohibited on the label.
   d. Mix one or more pesticide or fertilizer, providing the label does not prohibit it.

In the fail of 1988, Congress passed amendments to FIFRA, referred to as FIFRA ''Lite'' (Ware, 1989). These amendments strengthen the authority of EPA in several major areas, concentrating on: 1) reregistration, 2) fees, 3) expediting of certain types of registration applications, 4) revised responsibilities for disposal and transportation of pesticides taken off the market by EPA, and 5) limiting of the entitlement to indemnification for holders of canceled and suspended pesticides. The registration mandate requires comprehensive reevaluation of all data supporting products containing any active ingredient registered before November 1, 1984.

The 1996 amendments to FIFRA require pesticide reregistration review (Butts et al., 1997). The Food Quality Protection Act (FQPA) was signed into law on August 3, 1996, and is the first significant amendment to FIFRA since FIFRA Lite. It requires: 1) periodic review of all pesticide registration on a 15-year cycle using new food safety standards mandated under the act, 2) extension of reregistration fee authority at $14 million per year plus $2 million for 1998, 1999, and 2000, 3) emergency suspension authority, 4) enhancement incentives for the development and maintenance of minor-use registration through extensions for exclusive use of data, flexibility to waive certain data requirements, and requirement of EPA to expedite review of minor-use applications, 5) review of antimicrobial pesticides, 6) maintenance applicator and service technician training, 7) requirement of the secretary of agriculture to collect pesticide-use data on crop diseases and use of dietary significance, 8) requirement of the secretary of agriculture to support integrated pest management programs, and 9) establishment of a science review board of 60 scientists to assist the scientific advisory panel in reviews.

Under the FIFRA Lite 1988 amendments, once all the generic data requirements are met, EPA will publish a reregistration eligibility document (RED) announcing that products containing the active ingredient are eligible for reregistration. REDs for materials used in woody plant control include amitrole (Sept. 1996), bromacil (Sept. 1996), diquat (March 1995), fosamine (Dec. 1994), glyphosate (Sept. 1993), hexazinone (March 1994), paraquat (Sept. 1996), picloram (March 1995), and tebuthiuron (March 1994). Other changes or requirements under FIFRA include the following:

Proposed EPA pesticide container regulations
Expedited review for "me-too" registrations
Notifications, nonnotifications, and minor formulations amendments
Reduced-risk pesticide initiative
Minimal-risk pesticides
FIFRA good laboratory practices
Label percentage for ingredients statement
Inerts
Worker protection
Advertising of unregistered pesticides

Registration and reporting of pesticide-producing establishments
Pesticide-export policy
Minor-use crops/commodities
Minor-use or crop—third party registrations

This review cannot report all the information and requirements for pesticides. For more information consult other references (Bohmont, 1990; Butts et al., 1997; Ware, 1989).

IX. OTHER ACTS AND LAWS AFFECTING PESTICIDES

Established to protect and improve the nation's air resources, the Clean Air Act (CAA) involves the production of 2,4-D salts and esters, picloram, and other agricultural chemicals. Industrial air emission standards are being established to reduce and control air emission during the production of these chemicals.

The objective of the Clean Water Act (CWA) is the restoration and maintenance of the nation's water through prohibition of toxic pollutants (including several pesticides) into navigable waters of the United States. Navigable waters include all surface water, regardless of navigability.

The Comprehensive Environmental Response Compensation and Liability Act (CERCLA) gives EPA the authority to enforce or carry out cleanups of releases or threatened releases of hazardous substances, pollutants, and contaminants resulting from spills or hazardous waste sites where there is danger to public health, welfare, or the environment. Some pesticides are regulated as hazardous substances under CERCLA.

The Coastal Zone Management Act (CZMA) of 1990 gives states the ability to manage coastal environmental programs. The Nonpoint Pollution Control Program (NPCP) allowed the acts to concentrate on nonpoint source pollution from farm chemicals. The CZMA gives states the increased authority to implement control measures on the use of fertilizers and pesticides, and authorizes federal grants to states that establish an approved coastal zone management plan.

The Endangered Species Act (ESA) is to protect endangered and threatened species. The FIFRA requires EPA to take steps to prevent harm to these species from the use of pesticides. In 1988, ESA was amended to require EPA to work with the USDA and U.S. Department of Interior (USDI) to allow the protection of endangered species from pesticides while allowing agricultural production to continue. Congress is currently debating the ESA law for protection of the rights of land owners. The ESA will be voluntary until congressional changes are made.

As discussed earlier, the Federal Food, Drug and Cosmetic Act (FDCA) was administered by FDA. In 1970, EPA was given tolerance-setting responsibility for pesticide residues and inert ingredients in food and feed. The FDA maintains responsibility for monitoring residues and enforcing tolerances set by EPA.

Passage of the 1996 FQPA changed FDCA pesticide regulation. The EPA is currently drafting regulations to implement this new and complex law.

The Department of Transportation (DOT) is authorized under the Hazardous Materials Transportation Act (HMTA), as amended by the Hazardous Materials Transportation Uniform Safety Act of 1994, to regulate the shipment of hazardous materials in commerce. The Research and Special Programs Administration (RSPA) is responsible for promulgating, administering, enforcing and interpreting hazardous materials regulations. The Office of Hazardous Materials Safety (OHMS) within RSPA is in charge of writing regulations, granting exemptions, providing interpretations, and enforcing regulations. Hazardous materials include some agricultural chemicals. Generally, placarding of road-transport vehicles is required on each end and side. ''Poison-Inhalation Hazard'' shipping paper requirement must be placarded ''Poison'' or ''Poison Gas'' as appropriate on each side and end of transport vehicles and on portable and freight containers. Shippers must provide a 24-hr emergency response telephone number on shipping papers, and energy-response information must be readily accessible to drivers and other personnel. Individuals who perform such functions in the transportation of hazardous materials must be trained. All persons engaged in offering transport of hazardous materials must register and pay a fee to the U.S. DOT. Registrants must maintain a copy of the registration statement filed with DOT and the certificate of registration at its principal place of business for 3 years from the date of issuance. The U.S. DOT hazard registration number must be carried on all transport vehicles subject to the registration requirements. [See Butts et al. (1997) for more details on flammable and combustible materials.]

The Occupational Safety and Health Act (OSHA) was established to assure that every U.S. worker has safe and healthful working conditions and to preserve human resources. The EPA has authority under FIFRA relating to pesticide safety of farmworkers, whereas OSHA has authority over manufacturing formulating and distribution operations involving worker safety with pesticides. Many states have their own plans approved by OSHA that may be more stringent than OSHA requirements. In 1993, the American National Standards Institute (ANSI) issued a new standard for material safety data sheet (MSDS) preparation. The new MSDS standard provides 16 sections that should be included in the MSDSs. These sections include the following:

Section 1. Chemical product and company identification
Section 2. Composition, information, and ingredients
Section 3. Hazard identification
Section 4. First-aid measures
Section 5. Fire-fighting measures
Section 6. Accidental release measures
Section 7. Handling and storage

Section 8. Exposure controls, personal protection
Section 9. Physical and chemical properties
Section 10. Stability and reactivity
Section 11. Toxicological information
Section 12. Ecological information
Section 13. Disposal considerations
Section 14. Transport information
Section 15. Regulatory information
Section 16. Other information

OSHA's so-called right-to-know law requires employers with employees exposed to hazardous chemicals be provided with information, including pesticide labels, MSDSs, training, and access to written records.

The Resource Conservation and Recovery Act (RCRA) of 1976 regulates the generation, treatment, storage, transportation, and disposal of solid wastes. Pesticides can be regulated under RCRA if they meet one or more criteria of a hazardous waste. In order to manage the waste, the product must fall into one of these classes based on the amount of hazardous waste produced per month. If the material qualifies as a hazardous waste it is disposed of in accordance with RCRA requirements. Some states have RCRA requirements more stringent than federal RCRA requirements. If hazardous wastes are transported, U.S. DOT regulations may apply. The discarding of pesticides, residues, and rinsate are usually regulated under RCRA, however, disposal requirements for empty containers are mandated by EPA under FIFRA. These requirements are found on the label. [Please refer to RCRA in Title 40 of the Code of Federal Regulations, Part 260-272 (Butts et al., 1997).]

The Safe Drinking Water Act (SDWA) of 1974 (amended in 1986 and 1996) covers regulation by the EPA. On August 6, 1996, Congress amended the SDWA to give it more flexibility in addressing actual risk while at the same time reducing some of the regulatory and cost burdens placed on consumers. Specific to the use of crop-protection chemicals, the amendments to SDWA make the following changes.

Requires EPA to establish a national occurrence database of regulated and unregulated contaminants. Also, every 5 years the EPA must establish a list of 30 unregulated contaminants to be monitored.

Removed an EPA requirement to publish every 3 years a list of 25 contaminants to be regulated and replaced it with more flexible requirements to list unregulated contaminants that may require regulation every 5 years.

Gives EPA more flexibility in setting maximum contaminant levels (MCLs).

Requires EPA to use peer-reviewed science, risk assessment, and cost/benefit analysis when establishing or modifying drinking water standards.

Provides monitoring flexibility and financial assistance to water systems.

Requires development of estrogenic substances screening program for water contaminants, which may include pesticides and other chemical substances.

Creates voluntary incentive-based source water protection program, which allows local communities to petition the state for needed resources to address water quality concerns and provides financial resources to agricultural producers to address source water problems.

Requires water systems to provide extensive new information to consumers as well as EPA on both regulated and—for the first time—priority unregulated contaminants.

The Superfund Amendments and Reauthorization Act of 1986 (SARA) amended (CERCLA). One part of the new SARA provisions is Title III—the Emergency Planning and Community Right-to-Know Act of 1986. This act established new lists of "extremely hazardous substances" and "toxic chemicals" for new notification and reporting requirements. It also added new reporting requirements for the CERCLA list of "hazardous substances" and the OSHA definitions of "hazardous material." This requires that anyone handling and storing hazardous chemicals must report to state and local emergency groups, including local fire departments. An inventory of hazardous chemicals being stored must be reported. This law is under EPA jurisdiction (Bohmont, 1990). SARA Title III has four major sections:1) emergency planning, 2) emergency notification, 3) community right-to-know reporting requirements, and, 4) toxic chemicals release report. [For more information, see Butts et al. (1997).]

The emergency planning notification includes farms and is designed to identify all facilities containing any of the listed extremely hazardous substances present in excess of its threshold planning quantity (TPQ). The TPQ is based on the amount of any one of these substances that if released could present human health hazards that warrant emergency planning. If in excess of its TPQ the facility must notify its state emergency response commission, which notifies the appropriate local emergency planning committee or local fire department.

The emergency release notification is triggered if spills of any of the extremely hazardous pesticides are in excess of their reportable quantity (RQ). If the release (other than spills or other accidents) of a pesticide registered under FIFRA is used according to the label it is exempt from reporting.

Community Right-to-Know reporting pertains to the MSDS for hazardous chemicals from facilities (mainly manufacturers and importers) that are required to report and make MSDS available to wholesale and retail outlets (Bohmont, 1990).

The Toxic Chemical Release Report is limited to facilities that manufacture or import hazardous chemicals. A complete list of hazardous pesticides with TPQs is available from the USDA's Agriculture Stabilization and Conservation Service.

The SARA Title III consolidated list includes chemicals subject to reporting requirements under Title III of SARA. The list includes chemicals referenced under five federal statutory provisions as follows: 1) SARA Section 302—extremely hazardous substances, 2) CERCLA hazardous substances (RQ chemicals), 3) SARA Section 313—toxic chemicals, 4) RCRA hazardous wastes, and 5) CAA section 112(r) list of substances for accidental release prevention. Most chemicals used in brush control fall in SARA toxic chemicals Section 313, including amitrole, bromacil, 2,4-D, dichlorprop, hexazinone, picloram, tebuthiuron, and triclopyr. Dicamba, 2,4-D, and diquat are placed in CERCLA section 304 and have RQ requirements. Paraquat is listed under section 302—extremely hazardous substance (EHS)—and has an RQ of 1 lb if spilled.

Pesticides are excluded in the Toxic Substance Control Act (TSCA) of 1976 because they are regulated under FIFRA. Certain inert and raw materials are not considered pesticides until they become part of a pesticide product and may be subject to regulation under TSCA, however. The TSCA regulates production, distribution, use, and disposal of chemicals that pose unreasonable risks.

The USDA's 1996 farm bill relative to pesticides deals with record keeping. Unless records are currently prescribed by the state, commercial applicators who apply restricted-use pesticides (RUPs) may use records kept under FIFRA. Applicators maintain the following in RUP applications: 1) brand or product name and EPA registration number, 2) total RUP applied (total quantity of product), 3) location of application, size of treated area, and the crop, commodity, stored product, or site to which the RUP was applied, 4) the month, day, and year RUP applied, and 5) the name and certification number of the certified applicator who applied or supervised the application.

X. THE ENVIRONMENT AND SAFETY

In 1992, the EPA revised the WPS that protects agricultural workers from pesticide exposure. These regulations apply to agricultural workers and pesticide handlers. If the pesticide product has labeling that refers to the WPS, it must be complied with.

Even if they are not involved in actual pesticide handling or applications, agricultural workers must be provided with 1) a WPS safety poster explaining that WPS exists and offering tips on how workers can protect themselves from pesticide effects, 2) name, address, and phone number of the nearest emergency facility, and 3) information about each pesticide application from before treatment and for 30 days after the REI.

Workers must be orally warned about applications, or posted warnings must be placed at entrances to treated areas according to label requirements. Both oral and posted warnings should be used if the label requires them. Oral warnings are issued only if workers pass within 1/4 mile of the treated area. Oral warning must include: 1) location and description of the treated area and 2) REI information and a warning not to enter during REI. The EPA's compliance manual describes WPS-designated signs for posting immediately before application. They are left up during REI and removed before workers enter within 3 days of the REI's ending.

Workers must be trained in pesticide safety. They must be retrained every 5 years. The EPA developed approved training materials with rosters and verification cards. Decontamination sites must be provided by the grower within 1/4 mile of the work area free of treatment. The site must contain enough water for routine and emergency whole-body washing and eye flushing with soap and towels.

If a worker is poisoned or injured by pesticides, growers must arrange immediate medical attention. Medical personnel should be supplied with 1) the product's name, 2) the EPA registration number, 3) the active ingredient, 4) the medical information on the label, 5) a description of how it was used, and 5) any information on exposure.

Handlers and applicators must also comply with rules for agricultural workers and obtain training and information about the pesticide and its application, safety, and personal protective equipment.*

Butts et al. (1997) suggested ways to reduce pesticide waste and possible contamination. Suggested methods include: 1) purchase only needed amounts, 2) improve application accuracy, 3) eliminate leftover spray, 4) rinse containers immediately after use and reduce rinsate, and 5) modify spray equipment to improve application efficiency and choose returnable, refillable containers. Apply the amount of pesticide recommended on the label and do not overuse. Containers should be rinsed immediately after use to prevent dried and difficult-to-remove residues. The rinsate from spray equipment should be done in the field following application.

If a modern mixing and loading facility equipped with a concrete rinse pad and collection pit is used, keep the rinsate to a minimum. The rinsate can also be used when preparing the next batch of tank mix. Modern spray equipment that has in-line pesticide injection systems eliminates leftover tank solutions. With these systems the pesticide and water are kept in separate tanks. A separate metering pump feeds pesticide into the spray line. Excess water or pesticide remains in each respective tank. Other new sprayers are equipped with a small water tank to rinse the spray tank in the field. Rinse water can then be sprayed over the target field, provided registered rates do not exceed label directions. New equipment is being developed to further improve application efficiency and manage pesticide waste.

---

* For more information about EPA's WPS and other laws and regulations call your state department of agriculture or regional EPA office or EPA Headquarters, Occupational Safety Branch (H7506C), Office of Pesticide Programs, EPA, Washington, DC 20460 or phone (703) 305-7666.

Spray drift management was discussed in Chapter 5, and protective clothing and safety equipment were discussed in Section III of this chapter.

XI. CONCLUDING COMMENTS

This chapter by no means covers all the toxicology information or laws and regulations dealing with herbicides used in weed and brush control on grazing, forestry, or noncrop lands. With few exceptions, the herbicides mentioned are generally safe toxicologically and environmentally if used as suggested on the herbicide label. The more knowledge one has about the agricultural chemicals one works with as applicators or farm chemical workers, the better the agriculture and safety practices applied. Are all these regulations and laws necessary for consumer and worker safety? The consumer and users of agricultural products may indicate that the laws and regulators are not stringent enough to protect human health, wildlife, and the environment. Growers trying to make a profit to remain in business and feed the world's population have strong arguments that overregulation is a tremendous burden, however. After a brief overview of this chapter one may be rapidly overwhelmed by the extensive and complex rules applied to pesticide production and use. One can also argue that these extensive regulations have slowed and even curtailed both pesticide use and the search for new chemistry and products, as well as having added a tremendous cost to registered products. No one who has any knowledge about pesticides would argue against some regulation and training on their use, but unnecessary restrictions have depressed development and increased growers costs. Future developments will be interesting.

REFERENCES

Bohmont BL. The Standard Pesticide User's Guide. Englewood Cliffs, NJ: Prentice Hall, 1990.

Bovey RW. Toxicology of phenoxy herbicides in animals and man—General considerations. In: The Science of 2,4,5-T and Associated Phenoxy Herbicides. New York: Wiley, 1980, pp. 71–131.

Bush PB, Neary DG, McMahon CK, Taylor JW Jr. Suitability of hardwoods treated with phenoxy and pyridine herbicides for use as firewood. Arch Environ Contam Toxicol 16:333–341, 1987.

Butts ER, Myrick C, Briner WJ, Denny RL. Regulatory file, Section D. In: Farm Chemicals Handbook. Willoughby, OH: Meister, 1997, pp. D1–80.

CONCAWE. Gas Oils (Diesel Fuels/Heating Oils). Prod dossier no. 95/107. Brussels: CONCAWE'S Petroleum Products and Health Management Groups, 1996.

Farm Chemicals Handbook. Willoughby, OH: Meister, 1977.

Forbess RC, Norris JR, Lavy TL, Talbert RE, Flynn RR. Exposure measurements of applicators who mix and spray paraquat in grape vineyards. HortSci 17:955–956, 1982.

Gagne JA, Fischer JE, Sharms RK, Traul KA, Diehl SJ, Hess FG, Harris JE. Toxicology of the imidazolinone herbicides. In: Shaner OL, O'Conner SL, eds. The Imidazolinone Herbicides. Boca Raton, FL: CRC Press, 1991, pp. 180–182.

Klaassen CD, Amdur MO, Doull J, eds. Casarett and Doull's Toxicology: The Basic Science of Poisons. 3rd ed. New York: Macmillan, 1986.

Lavy TL, Mattice JD. Monitoring field applicator exposure to pesticides. In: Honeycutt RC, Zweig D, Ragsdale N, eds. Dermal Exposure Related to Pesticide Use. ACS Symp. ser. no. 273. American Chemical Society, 1985, pp. 163–173.

Lavy TL, Flynn RR, Mattice JD. Exposure of aerial applicators to 2,4-D. Arkansas Farm Research. Fayetteville, AR: Arkansas Agric. Exp. Stn, University of Arkansas, 1981, p. 7.

Lavy TL, Mattice JD, Norris LA. Exposure of forest workers using herbicides measured. Arkansas Farm Research. Fayetteville, AR: Arkansas Agric. Exp. Stn., University of Arkansas, 1984, p. 8.

Lavy TL, Shepard JS, Mattice JD. Exposure measurements of applicators spraying (2,4,5-trichlorophenoxy) acetic acid in the forest. J Agric Food Chem 28:626–630, 1980.

Lavy TL, Cowell JE Steinmetz JR, Massey JH. Conifer seedling nursery worker exposure to glyphosate. Arch Environ Contam Toxicol 22:6–13, 1992.

Lavy TL, Mattice JD, Massey JH, Shulman BW. Measurements of year-long exposure to tree nursery workers using multiple pesticides. Arch Environ Contam Toxicol 24: 123–144, 1993.

Lavy TL, Norris LA, Mattice JD, Marx DB. Exposure of forestry ground workers to 2,4-D, picloram and dichlorprop. Environ Toxicol Chem 6:2309–224, 1987.

Lavy TL, Walstad JD, Flynn RR, Mattice JD. (2,4-dichlorophenoxy) acetic acid exposure received by aerial application crews during forest spray operations. J Agric Food Chem 30:375–381, 1982.

Leng ML Ramsey JC, Braun WH, Lavy TL. Review of studies with 2,4,5-trichlorophenoxy acetic acid in humans including applicators under field conditions. In: Plimmer JR, ed. Pesticide Residues and Exposure. ACS Symp. ser. no. 192. American Chemican Society, 1982, pp. 133–156.

McMahon CK, Bush PB. Forest worker exposure to airborne herbicide residues in smoke from prescribed fires in the Southern United States. Am Ind Hyg Assoc J 53:265–272, 1992.

Ware GW. The Pesticide Book. 3rd ed. Fresno, CA: Thomas, 1989.

Weed Science Society of America. Herbicide Handbook. 7th ed. Champaign, IL: Weed Science Society of America, 1994.

8

The Phenoxy Herbicide Controversy

I. INTRODUCTION

The discovery of 2,4-D as an herbicide during World War II was the greatest single advance in weed science and a very significant advance for agriculture (Bovey, 1980c; Burnside, 1996; Peterson, 1967; Veale, 1946). In the more than 50 years since its introduction, 2,4-D continues to be the most commonly used herbicide worldwide (Burnside, 1996). In the early 1940s, Krause at the University of Chicago and others proposed to the National Academy of Sciences that the growth-regulating properties of the phenoxys should be tested for crop destruction to limit crop production for military considerations (Bovey, 1980b). After intensive testing by the U.S. Army and others, the phenoxy herbicides were never used in World War II. After World War II, however, the U.S. Army biological laboratories at Fort Detrick, Maryland, continued testing chemicals for desiccation and defoliation activity (House et al., 1967; Bovey, 1980b). In the early 1960s the South Vietnamese government requested that the U.S. Army undertake defoliation trials for use against guerrilla forces, and in 1961 and 1962 chemical agents were shipped to the Vietnamese military. These studies, in addition to those conducted in the United States, showed both that the esters of 2,4-D plus 2,4,5-T readily defoliated woody vegetation and that cacodylic acid showed promise as a fast-acting desiccant. Several thousand chemicals were tested by the U.S. Army, industry, state agricultural experiment stations, U.S. Department of Agriculture (USDA), and others for either military or agricultural use. Few chemicals were more successful than the esters of 2,4-D plus 2,4,5-T against woody vegetation (Bovey, 1980c). In addition to the phenoxys, a large body of information on picloram was also obtained concerning weed species controlled, residues in the environment, safety, and physiological behavior in plants. Assistance by industry, private, state, and federal institutions, and Fort Detrick produced basic and practical information for agriculture, as well as for military use (Bovey, 1980b).

The first involvement of the United States in spraying herbicides for military use occurred in 1962 in a Vietnam jungle along Highway 15 connecting Bien Hoa to the coastal city of Vung Tau (House et al., 1967). The herbicides had been previously tested on a large scale by the Vietnamese military in the spring of 1962 to prove their effectiveness. The U.S. Air Force defoliation program, known as Operation Ranch Hand, applied more than 81.8×10^6 L of herbicide (mainly undiluted esters of 2,4-D plus 2,4,5-T) on over 3.2 × 106 ha of forest and cropland between 1962 and 1971 (National Academy of Science, 1974). The treatments prevented enemy ambush and infiltration along roadways and waterways, exposed secret enemy caches of ammunition, food, and soldiers, and exposed troop movement and supplies from one area to another (Bovey, 1980b).

Since 1950 the Dow Chemical Company has been aware of a highly toxic impurity formed in very small amounts from the production of 2,4,5-triclorophenol, a 2,4,5-T precursor (Bovey, 1980c). The most sensitive toxic reaction of humans exposed to the impurity is chloracne, a skin disorder most prevalent on the face, neck, and back. At about the same time in Germany some workers involved in the manufacturing of trichlorophenol developed chloracne with complaints of debility, nausea, and loss of appetite (Schulz, 1957). Baurer et al. (1961) studied synthesized and isolated compounds that occurred as impurities in chlorinated phenols, and reported that 2,3,7,8-tetrachlorodibenzo-*p*-dioxin (TCDD) formed in the manufacture of 2,4,5-trichlorophenol was the cause of chloracne in the workers. These investigations also found the chlorinated dibenzo-*p*-dioxins, such as TCDD, to be 10 to 20 times more toxic than the dibenzofurans in rabbit ear tests. Further work with TCDD on human skin at extremely low dosages produced chloracne and lethal liver necrosis in rabbits. The authors indicated that production methods could be changed to overcome the toxic effect of the chlorobenzodioxins.

In 1964 several workers at the Dow Chemical Company developed chloracne, and the bioassay (rabbit ear test developed in 1944) showed a skin reaction potential from the waste oil of the 2,4,5-trichlorophenol process (Subcommittee on Energy, Natural Resources and the Environment, 1970). Exposure to waste oil and not trichlorophenol was the cause of the chloracne. The plant was immediately closed and the principal offending impurity was identified as TCDD. Dow Chemical notified various health authorities and called a meeting in 1965 of other U.S. manufacturers of 2,4,5-T to alert them to the health hazard and share analytical techniques for the impurities (Subcommittee on Energy, Natural Resources and the Environment, 1970). Dow Chemical rebuilt its plant in 1966 and did not permit any TCDD in 2,4,5-T when assayed by a method sensitive to 1 ppm.

Permissible TCDD content was lowered to 0.5 and 0.1 ppm in 2,4,5-T products in 1970 and 1971, respectively.

The primary defoliant used in Vietnam was Agent Orange, which consisted of the *n*-butyl ester of 2,4-D and 2,4,5-T active ingredient (ai) of 0.50 and 0.53 kg/L 2,4-D + 2,4,5-T; others included White (2,4-D + picloram); Blue (cacodylic acid); and Purple (n-butyl ester of 2,4-D + 2,4,5-T and the isobutyl ester of 2,4,5-T). Agent Orange was applied at high rates and undiluted at 28 L/ha (3 gal/acre) (Bovey, and Young, 1980b). Aside from killing mangrove forests in South Vietnam, representating a small percentage of the land, most woody and herbaceous vegetation recovered. Injury to animal and human populations and individuals exposed to Agent Orange was difficult to demonstrate under actual field use.

In 1979, while attempting to determine the source of dioxin in some biological samples, chemists found that a large number of combustion sources emit dioxins, including 2,3,7,8-TCDD, into the environment (Reggiani, 1988). Dioxins were found in cigarette and refuse incinerator smoke, car and truck mufflers, and charcoal-grilled steaks. These findings indicate that dioxins are uniquitous in the environment and associated with the trace chemistry of fire. This finding should change the attitude toward 2,4,5-T, since it probably has had little or no effect on the environmental load of polychlorinated benzodioxins or benzoburans produced from incineration and does not add to human health risk (Reggiani, 1988). On February 28, 1979, the U.S. Environmental Protection Agency (EPA) announced emergency suspension of 2,4,5-T and silvex on forests, rights-of-way, and pastures, and suspended additional silvex uses on aquatic, home, and recreation areas. Despite the benefits of 2,4,5-T and silvex use in agriculture these products were canceled for use by the EPA in 1985.

II. HISTORY OF THE 2,4,5-T CONTROVERSY

One of the first qualified investigators to report on the ecological effects of the defoliation program in Vietnam was F. H. Tschirley, who was assistant chief of the Crops Protection Research Branch, Crop Research Division, Agricultural Research Service, U.S. Department of Agriculture (Tschirley, 1969). Dr. Tschirley visited defoliated areas in March 1968 by aerial and ground observations with assistance from civilian and military personnel (Bovey, 1980c). He concluded that any ecological changes caused by defoliation were not irreversible, although recovery might require a long time in the mangrove-type vegetation (20 years for revegetition). Single treatments of herbicide in semideciduous forests caused little change, but repeated treatments could result in the invasion of bamboo at some sites.

Further observations of the ecology of Vietnam were made in March 1969 by Orians and Pfeiffer (1970), sponsored by the Society of Social Responsibility of Science. Their assessment indicated severe effects on vegetation from direct application of defoliants and extensive damage from herbicide drift. In their opinion the defoliation program may have adversely affected the Vietnamese attitudes toward Americans. Orians and Pfeiffer recommended that the American Association for the Advancement of Science (AAAS), in accordance with its resolutions of 1966 and 1968, establish an international research program on the long-range effects of the military use of herbicides in Vietnam.

In December 1969 the council and board of directors of the AAAS appointed M. S. Meselson, professor of biology at Harvard University, to develop a study of the effects of the military use of chemical herbicides on the ecology and human welfare of South Vietnam. Meselson appointed A. H. Westing, professor of botany at Windham College in Vermont, to direct the Herbicide Assessment Commission, the title under which the AAAS activity was known. The earlier policies, controversies, and difficulties of the AAAS in reaching its decision to sponsor an ecological investigation of Vietnam was outlined by Pfeiffer and Orians (1972).

The Herbicide Assessment Commission began its work in February 1970, including trips to Vietnam in August and September 1970 (Bovey, 1980c). The activities of the commission in dealing with the ecological damage to Vietnam and the evaluation of birth defects in the Vietnamese population from possible exposure to 2,4,5-T were reported at the 1970 annual AAAS meeting. The preliminary report considered the effects of herbicides on the mangrove forest, tropical hardwood forests, food chain, crop destruction, military considerations, and health. A vitally important aspect of the investigation studied the maternity records in a number of hospitals in various parts of the country. Because of the variation in reporting by the different hospitals, the survey approach offered little hope of proving or disproving any relation between herbicide exposure and incidence of birth defects (stillbirths, hydatidiform moles, congenital malformations) under the conditions of the study, since large populations living under similar conditions could not be compared to heavily exposed and nonexposed sprayed areas.

Monitoring birth defects in the Vietnamese population was prompted by the fact that in 1969 Bionetics Research Laboratories in Bethesda, Maryland, reported an increased incidence of abnormalities in rats and mice born of mothers exposed to 2,4,5-T during pregnancy (Bionetics Research Laboratories, 1969). Because herbicides containing this compound had been used in Vietnam, the Ministry of Health of the Republic of Vietnam and the U.S. Military Assistance Command in Vietnam cooperatively studied obstetrical records from 1960 to 1969 in 22 hospitals (Cutting et al., 1970). Their findings were reported in December 1970 as follows: in four geographical regions (capital, coastal, interior, and delta) the rates per 1000 livebirths of stillbirth and congenital malformation were

32.5 and 5.8, respectively, in the capital area, and 36.7 and 2.9 in the three remaining areas. The differences in these rates may be attributable to better maternal and neonatal care, or to more competent or thorough examination for congenital malformations in the capital area. The rates for stillbirths declined and remained unchanged for congenital malformations during this 10-year period. The only differences in these rates between the years from 1960 to 1965 and from 1966 to 1969—periods of relatively light and heavy defoliant spraying, respectively—was a downward trend (from 36.1 to 32.0 for stillbirths, and from 5.5 to 4.5 for congenital malformations).

There were no consistent differences between heavily and lightly defoliant-sprayed areas. Neither the report by Cutting et al. (1970) nor the AAAS study proved or disproved that human birth defects were increased or related to herbicide exposure.

The initial studies by Bionetics Research Laboratories, under contract with the National Cancer Institute, indicated that the offspring of mice and rats given relatively large oral doses of the herbicide 2,4,5-T during the early stages of pregnancy showed a higher than expected number of deformities. Based on this information, on October 29, 1969, Dr. Lee A. DuBridge, science advisor to the president and executive secretary of the president's environmental quality council, announced the following actions to be taken:

1. The Department of Agriculture will cancel registrations of 2,4,5-T for use on food crops effective January 1, 1970 unless by that time the Food and Drug Administration has found a basis for establishing a safe legal tolerance in and on foods.
2. The Department of Health, Education, and Welfare will complete action on the petition requesting a finite tolerance for 2,4,5-T residues on foods prior to January 1, 1970.
3. The Departments of Agriculture and Interior will stop use in their own programs of 2,4,5-T in populated areas or where residues from use could otherwise reach man.
4. The Department of Defense will restrict the use of 2,4,5-T to areas remote from the population.
5. Other Departments of the government will take such actions in their own programs as may be consistent with these announced plans.
6. The Department of State will advise other countries of the actions being taken by the United States to protect the health of its citizens and will make available to such countries the technical data on which these decisions rest.
7. Appropriate departments of government will undertake immediately to verify and extend the available experimental evidence so as to pro-

> vide the best technical basis possible for such future actions as the government might wish to undertake with respect to 2,4,5-T and similar compounds.

The announcement also stated that "although it seems impossible that any person could receive harmful amounts of this chemical from any of the existing uses of 2,4,5-T, and while the relationships of these effects in laboratory animals to effects in man are not entirely clear at this time, the actions taken will assure safety of the public while further evidence is being sought."

Following the DuBridge announcement, many parties became concerned about the quality of research and 2,4,5-T used in the Bionetics study, since it was already known that the contaminant TCDD was highly toxic when administered to laboratory animals or allowed at too high a concentration in the finished 2,4,5-T product. It was soon learned that the 2,4,5-T used in the Bionetics study was produced by a manufacturer who had since stopped production. The product contained 27 ± 8 ppm TCDD, well in excess of levels considered safe for workers involved in manufacturing, as explained earlier. The DuBridge announcement stimulated considerable toxicological and residue research and a review of 2,4,5-T by federal, state, and private industry to determine if TCDD, 2,4,5-T, or the combination could be responsible for the birth abnormalities. Dow Chemical, in consultation with scientists at the National Institute of Environmental Health Science (NIEHS) laboratories at Research Triangle, North Carolina, repeated the Bionetics studies using regular production-grade 2,4,5-T (low in TCDD) and reported their findings to the Department of Health, Education and Welfare (DHEW) and the USDA (Subcommittee on Energy, Natural Resources and the Environment, 1970). The report showed that regular production-grade 2,4,5-T did not cause birth defects in rats at 24 mg/kg/day administered orally on days 6 through 15 of gestation as determined by gross examination of fetuses. Preliminary studies with rabbits gave similar results. Scientists of the Food and Drug Administration (FDA) also observed the results of the tests.

III. OTHER CASE STUDIES

A. Globe, Arizona

To add more problems to the political controversy over 2,4,5-T, adverse comments and complaints against the U.S. Forest Service for herbicide applications near Globe, Arizona, occurred in 1969 (Bovey, 1980c). The complaints included damage to vegetation, deformed animals, and human illnesses. Applications of herbicides were made on the Tonto National Forest in 1965, 1966, 1968, and 1969. The most recent application in June 1969 consisted mostly of silvex produced by Dow Chemical. Two investigating teams of the U.S. Forest Service investigated the allegations and submitted their reports to the interested parties.

On February 13, 1970, a hearing was begun in Globe, Arizona, chaired by Congressman Richard D. McCarthy of New York State. Dr. Arthur W. Galston, professor of biology at Yale University, was a prominent witness, and assisted McCarthy in questioning other witnesses. Various local witnesses described what they believed were herbicide effects on plants, animals, and humans. The Forest Service investigations were unable to demonstrate any conclusive herbicide effects.

National television coverage of the alleged herbicide damage focused on a goat in poor condition. The goat was said to have been born in December 1968. Examination of its teeth, however, revealed it to be at least 5 years old, and its poor condition was caused by nutritional deficiency and not herbicide effect. A duck was also shown with a slipped tendon, which was probably due to a nutritional deficiency rather than exposure to herbicides.

On February 16, 1970, a third investigating team established by the Office of Science and Education, USDA, and headed by Dr. Tschirley, assessed the allegations of Kellner Canyon–Russel Gulch. The interdepartmental panel consisted of experts on herbicide effects on plants, who monitoried soil and water for pesticides, plant pathology, air pollution, fisheries and wildlife, toxicology, and tereatology. Two additional observers included a range management expert and an entomologist. After careful on-site examination of alleged damage from herbicide spraying, the investigations panel concluded in a report and press release their findings of the spray project in brief, as follows:

1. The materials used in 1965, and 1966, and 1969 included 2,4-D, 2,4,5-T, and silvex from different sources. The 4250 L of silvex produced by Dow Chemical contained less than 1 ppm TCDD.
2. There was clear evidence of drift of herbicides to a number of plants on some of the adjacent properties, which may or may not have originated from the project spraying.
3. Nine doctors serving the area of Globe were interviewed, and there was general agreement that there had been no significant increase in human illness related to the spraying.
4. Wildlife specialists reported no significant effects on birds, deer, and other wildlife.
5. Livestock owners and observers reported no increase in animal illnesses. The allegedly afflicted goat and duck were suspect, since the goat was born before the treatment and the duck was hatched approximately 6.4 km from the treated area.
6. There was evidence of woody plant mortality from root rot damage to certain yard trees from insects and woodpeckers or sapsuckers. Other plant injuries were attributed to drought, air pollution, and unusual soil properties.

7. Phenoxy herbicide persistence is usually short-lived in soil and water. Samples were taken for analysis.
8. The panel concluded that for other than minor plant damage, none of the alleged effects could be attributed to the herbicides.

B. Swedish Lapland

In the spring of 1970, Swedish newspapers reported an accumulation of sudden deaths of reindeer grazing in the Visttrask areas of Lapland (Bovey, 1980c). Approximately 30 reindeer, mainly young animals, died within a week after a heavy, wet snowfall without any previous signs of illness. It was also reported that approximately 10 reindeer cows aborted their fetuses. An examination of several reindeer by veterinarians showed inanition (empty stomachs). When given additional feed, the deaths stopped. The case was of particular interest since it was learned that the area in which the reindeer grazed had been treated with a mixture of 2,4-D plus 2,4,5-T at 0.67 + 0.34 kg/ha.

Liver and kidney samples (Erne and Nordkrist, 1970) from one cow and three aborted fetuses indicated traces of 2,4-D (0.2 to 0.5 ppm) and 2,4,5-T (0.3 to 0.1 ppm). Tree leaves contained 25 and 10 ppm of 2,4-D and 2,4,5-T, respectively. No herbicides could be found in the ground vegetation, however. The deaths of the reindeer were attributed to starvation rather than exposure to 2,4,5-T.

Subsequently, instances of congenital malformation in human infants attributed to alleged exposure of pregnant women during application of the herbicides could not be attributed to the herbicides as determined by highly competent medical scientists at the Institute of Hygiene and the Teratological Laboratories of the Karolinska Institute of Stockholm and the Institute of Human Genetics in Munster, Germany.

C. The Te Awamutu, New Zealand, Episode

In January 1972, Sare and Forbes (1972) reported the following in the *New Zealand Medical Journal* (Young, 1980):

> Two babies, born within a month of each other at our local maternity hospital, had congenital defects incompatible with life. Both had a gross myelo-meninggocele. Postmortem was performed on only one and other congenital abnormalities were brought to light.
>
> What intrigued us was that the families concerned live on adjoining hilly country farms, where for several years aerial spraying has been carried out with a chemical called 2,4,5-T, designed to kill useless vegetation. Inquires into the nature of this chemical revealed that it contains an impurity called dioxin, which is apparently one of the most powerful

poisons ever discovered. It has been investigated in the United States, partially banned in many states, and totally in others. It was likewise banned in Vietnam when its potential danger was discovered.

The suggested relationship between 2,4,5-T/TCDD and the two deformed babies quickly recieved national and international attention. Accusations that 2,4,5-T/TCDD were indeed responsible for the congenital defects soon appeared in articles in the United States (Adamson, 1974).

The circumstances surrounding these cases at Te Awamutu were thoroughly investigated by a subcommittee of the Agricultural Chemicals Board of New Zealand (Agricultural Chemicals Board, 1972). In the subcommittee report it was noted that

> The women who gave birth to deformed babies had both been exposed to 2,4,5-T during pregnancy, one person by assisting at the airstrip during spraying and the second person by helping to free the spray truck which was struck on the property and was exposed to 2,4,5-T when spraying was done to lighten the load. It was not possible to ascertain the degree of exposure in either case.
>
> The deformity common to both babies is spina bifida, caused by a failure of the end of the neural tube to close completely during early development. This deformity is one of the commonly occurring deformities, with overseas averages of about 1 per 1000 total births. In New Zealand during the period 1964–1970, 515 live births and 151 stillbirths affected with spina bifida were recorded. In the light of present embryological knowledge it may be stated that the neural tube is usually closed by the fourth week after conception and definitely by the sixth week. Medical records show that in one case exposure to 2,4,5-T during the spraying operation occurred after the neural tube would have normally closed. It is concluded that in one of these cases the reported exposure to 2,4,5-T could not have caused the birth deformity. It is not possible to state definitely in the second case whether exposure to 2,4,5-T was in any way a factor causing the deformity, and thus the subcommittee was unable to arrive at any information of value to the general topic of 2,4,5-T toxicity to human foetuses.

Shortly after publication of the subcommittee report, legal action was taken to compel the Agricultural Chemicals Board to restrict the use of 2,4,5-T herbicide (Environmental Defense Society, 1972). Although the court ruled in favor of the Agricultural Chemicals Board, it noted: ''If in its vigilant surveillance of such dangers, the plaintiff should come into possession of more convincing material than has so far been revealed, it will no doubt advise the Board in due Form'' (Dow Chemical, 1975).

In April 1977, the New Zealand television program *Dateline Monday* suggested that the occurrence of ''clusters'' of neural tube defects in the South Taranaki, Northland, and Waikato areas of New Zealand were related to the use of 2,4,5-T (McQueen et al., 1977). The New Zealand Department of Health, Division of Public Health, appointed a committee of experts to investigate the allegations. In the committee report, McQueen et al. (1977) noted that the three clusters represented 20 cases of birth defects. Seven of the cases were anencephaly (congenital defect of the crainial vault) and 13 were spina bifida (congenital defect of the bony encasement of the spinal cord). McQueen et al. (1977) noted that although this group of defects may well have occurred entirely by chance, the possibility of a common causal factor must be considered.

After a thorough investigation of each of the 20 cases reported, McQueen et al. (1977) concluded

> It is obvious from an inspection of the data for the three ''clusters'' that 2,4,5-T cannot reasonably be implicated in the causation of neural tube defects. It is true that in one or two cases there may have been some ''exposure'' to 2,4,5-T around the critical period. However, considering 2,4,5-T is the most used pesticide in New Zealand, this is not unexpected. In short, the data permit the conclusion that there is no evidence to implicate 2,4,5-T as a causal factor in human birth defects. This statement, although scientifically correct, is unfortunately commonly misunderstood by the public and even some scientists. The demand is usually for ''proof of safety'' and a statement that there is no evidence for harm is unsatisfactory to many. One rather obvious mistake that should be avoided in interpreting the results of these investigations is to assume that because there was no known family history in some cases, then genetic factors were not in any way involved. Detailed family histories were not available in most cases. Similarly, the absence of obvious chromosomal abnormalities does not exclude inherited factors. The accumulated data on 2,4,5-T and its TCDD contaminant are sufficient to give a very high assurance of safety in the normal use of this material. This belief is in accordance with the concensus of worldwide scientific opinion.

Subsequent to the investigation by McQueen et al. (1977), and in direct response to a television program describing neural tube defects in newborn infants, Lowry and Allen (1977) published a note on herbicides and spina bifida. They examined all the children referred to in the television program and found one case of an open neural tube defect (spina bifida cystica), while the other children were normal in every respect. It was possible that these apparently normal children had spina bifida occulta; however, none of them had had radiography, and it was probable that the children had nothing more than sacral dimples.

As a final note in relation to this episode, the following brief article appeared in the *New Zealand Medical Journal* (New Zealand Medical Journal, 1972):

> Publicity on certain chemicals as causation of malformations of the human fetus has been widespread. Some of the publicity has been sensation mongering and not all the remarks from the profession have been in keeping with a balanced assessment of scientific evidence. It is proper that there should be intelligent public awareness of the various environmental hazards that may come from the use of chemicals . . . in farming . . . however, those who would write of their experiences in medical journals must remember that disasters are the staple of the sensation mongers in the news media industry.
>
> Until recent publicity there had been no suggestion that 2,4,5-T, which has been used for over 20 years in New Zealand, was responsible for congenital malfunctions either in man or in farm animals. It is the duty of the physicians (and scientists) who have any concern for science to attempt to make valid observations that can be repeated. In the problem at issue, fetal malformations are nature's common mistakes which we have no desire to perpetrate or to increase, although they are the inevitable price that is paid for our place on the evolutionary scale. There are extensive gaps in our knowledge but they will be filled only by patient work. Unresolved problems of fetotoxicity can only be solved by accurate record keeping at all stages of pregnancy.

D. The Missouri Horse Arena Episode

Many of the episodes discussed in this chapter involved ''suspected'' TCDD poisoning; however, the Missouri Horse Arena episode is one in which TCDD was confirmed to be the causative agent.

In 1972, the Missouri Division of Health and the Centers for Disease Control and Prevention in Atlanta (Anonymous, 1974) reported an investigation of a horse arena in eastern Missouri, in which 54 of 57 horses exposed to the arena had died of an illness characterized by skin lesions, severe weight loss, and heptotoxicity. Birds, dogs, cats, insects, and rodents were also found dead in and around the arena, and one 6-year-old girl exposed developed hemorrhagic cystitis (characterized by blood in the urine). Analysis of urine cultures for bacterial and viral pathogens were negative. Immediately prior to the onset of illness, the arena had been sprayed with salvage oil for dust control.

The report (Anonymous, 1974) also noted that similar horse illnesses and deaths occurred in two other horse arenas in the eastern Missouri area sprayed by the same salvage oil company. The three arenas had been sprayed within 1 month of each other. Subsequent to investigation, soil from all three arenas was excavated and disposed. No further problems occurred since the excavations.

As with the earlier episodes, the report soon received national attention. Unfortunately, some newspapers misquoted the report by implying that the cause of the incident was a result of spraying contaminated 2,4,5-T herbicide (New York Times, 1974b).

Carter et al. (1975) and Kimbrough et al. (1977) have published detailed chemical, epidemiological, and pathological data of the episode. The investigations concluded that a hexachlorophene factory in southwestern Missouri had accumulated distillate residues containing 306 to 356 ppm TCDD. It was these distillate residues that were subsequently disposed of via a salvage oil company and sprayed on the horse arena. Furthermore, as Kimbrough et al. (1977) noted, the investigation demonstrated that the improper disposal of toxic chemical wastes may have serious consequences. Companies responsible for the disposal of such wastes should be aware of the toxicity of these chemical waste products and should practice proper methods of disposal.

Commoner and Scott (1976) have reviewed the Missouri Horse Arena episode in an attempt to provide consultative data to the Italian government in the wake of the Seveso, Italy, episode. Their review focused on the human reactions (symptoms) to accidental TCDD exposure and the problem of soil degradation of TCDD. They also provided an excellent chronological account of the episode.

The response of the horses to TCDD in the above episode is of special interest. In 1970, Pinsent and Lane (1970) reported on an illness in four horses and the subsequent death of one following the spraying of a brushwood killer (2,4-D and 2,4,5-T) to destroy nettles around the margin of a pasture containing the horses. The horse that died had remained on the pasture for 12 days. The postmortem examination revealed extensive lesions in the esophagus and stomach, with degenerative changes involving the liver and kidneys. The authors concluded that neither the clinical picture nor the postmortem findings supported a plant-poisoning syndrome, and thus it was not possible that the poisoning was by the herbicides. In view of the findings by Kimbrough et al. (1977), it would appear that the horses may have been exposed to TCDD.

E. The Seveso, Italy, Episode

Perhaps the most publicized chemical accident in modern times is the TCDD episode in Seveso, Italy (Young, 1980). This episode attracted worldwide interest and concern. Hundreds of scientists, physicians, and veterinarians participated in either on-site inspections, conferences, or consultations into the various facets of this episode.

Although the Seveso, Italy, episode did not involve 2,4,5-T or other phenoxy herbicides, it did involve the potential raw material required for their synthesis; that is, trichlorophenol. In relation to the phenoxy herbicides (and hexachlorophene), this episode thus represented to many people an inherent danger associated with 2,4,5-T industrial production, hence the question can we do without trichlorophenol products and thus ensure our safety from another such industrial accident?

A voluminous amount of literature has been written on this episode. Much of this material has been layman-oriented, factually confused, politically pointed, or limited primarily to humanistic interest, thus we have seen articles in newspapers and magazines with such titles as "A Deadly Saga," "The Graveyard on Milan's Doorstep," "Toxic Cloud Over Seveso," and "The Poison That Fell from the Sky." The human drama of this industrial accident cannot and must not—be minimized, however.

Parks and Sullam (1976) assembled, translated, and reviewed an impressive collection of selected articles from the Italian press for the period from July 10 to September 1976. In the introduction to this collection of articles, Parks and Sullam stated

> The social, economic, moral, medical and administrative problems which fell on Italy out of that poisonous cloud are of vast extent and complexity. They cast a very long shadow on the future, involving the rebuilding of shattered lives, the care of burned and disfugured children, health care and the shaping of policy decisions regarding industrial regulations and the operations of multinational enterprises to which this incident forces the attention of the Italian Government.

The episode of TCDD poisoning occurred on July 10, 1976, in Seveso, a small town 40 km north of Milan (Hay, 1976a; Rawls and O'Sullivan, 1976). The source of the TCDD was a chemical factory that produced trichlorophenol through the alkaline hydrolysis of tetrachlorobenzene. When the temperature in a steam-heated reaction vessel rapidly increased, a safety disk ruptured, sending a plume of trichlorophenol, TCDD, and other products 30 to 50 meters high above the factory. The cloud apparently rose into the air, cooled, and came down over a cone-shaped area approximately 2 km long and 700 meters wide. An area of 110 ha (designated as zone A) was evacuated only after hundreds of animals had died and many people had reported skin disorders. Several measurements of TCDD on vegetation in this area and areas adjacent to the factory were in the 1 to 15 ppm range, with one reading as high as 51.3 ppm.

The chemical plant involved was the ICMESA chemical plant, owned by Givaudan, a Hoffman-LaRoche subsidiary. Hay (1976b) reported that the best available estimate is that 2.3 to 3 kg dioxin (TCDD) was released in the reactor discharge. On this basis and through information obtained from a model simulating the explosive reaction at the ICMESA plant, it has been suggested that only a few hundred grams of TCDD were actually deposited on vegetation and other surfaces. Recent data (Hay, 1976b) also confirm that 98% of the dioxin was strongly adsorbed in the top 4 cm of soil.

A Giovanardi (1976), of the Institute of Hygiene in Milan, also described the magnitude of the episode. He noted that chemical analyses of zone A indicated that TCDD levels in the soil were equal to or greater than 0.001 ppm. The inhabitants of this area were evacuated in three stages, on July 6, July 28, and August 2. The area was located south–southeast of the ICMESA factory, and was downwind of the factory at the time of the accident. The area was triangular-shaped and covered approximately 100 ha (the vertex of the triangle corresponds to the ICMESA factory). Of the 700 inhabitants of this area, approximately 200 (43 families) were living in the northern part (60 ha), and 500 (138 families) in the southern part. An analysis of soil from zone B indicated that TCDD concentrations were less than 0.001 ppm. This area was not evacuated. It included two adjacent zones located south–southeast of zone A, covering approximately 200 ha. The area had a total of 4280 inhabitants (1256 families) divided between a large urban center and an extensive rural area with some small residential aggregates.

Forth (1977) reviewed the toxicology of TCDD in relation to the genesis of the Seveso episode. He reported that over 500 people in Seveso had been treated for presumed toxic symptoms. He noted that children were particularly affected by skin symptoms, but it remained an open question as to whether this involved the immediate dermatitis resulting from caustic soda and ethylene glycol or was actually chloracne from TCDD exposure. It was unclear whether or not children were particularly sensitive or whether their particular biotype (playgrounds, sports equipment, schools, etc.) involved a particular exposure.

Thirteen months after the Seveso episode, Walsh (1977) noted that the total number of confirmed chloracne cases over the past year had been 134. Most other illnesses, however, appeared to have been short-lived. The initial skin rashes noted immediately after the accident were probably due to trichlorophenol rather than TCDD. No one has attempted to assess the long-term effects associated with this TCDD exposure. As Walsh (1977) stated, however, ''A source of almost universal regret is that from the start a greater systematic effort wasn't made to maximize the scientific knowledge gained in the aftermath of Seveso.''

Whiteside (1977) viewed Seveso as another ''grim episode'' in a continuing series of episodes (e.g., Vietnam and numerous industrial accidents) attached to the handling of dioxin contamination in its manufacturing operation at ICMESA. He concluded his ''report'' by suggesting this type of neglect can only be prevented by halting all production of trichlorophenol. Moreover, in the United States, the continuing application of dioxin on millions of acres of rangeland (via the application of 2,4,5-T) can only be stopped by the U.S. EPA's taking ''swift and drastic action'' to suspend the use of dioxin-containing pesticides.

1. Personal Observations of the Contaminated Zone in Seveso, Italy

In November 1977, Young (1980) was invited to Italy to assist the Seveso Authority (a scientific commission) in the formulation of a decontamination program for the most heavily contaminated parts of zone A (areas designated A-1 through A-5). In briefings by Professor Zurlo, chairman of the decontamination program and a faculty member, Istituto Igiene Industriale della Clinica del Lavoro della, Universita di Milano, the following scenario was presented.

Approximately 180 workers had been employed at ICMESA at the time of the accident. Since 1975, the plant's daily production of trichlorophenol was approximately one metric ton. The reactor had been manufactured in 1966 and installed at the Seveso plant in 1971. It was capable of handling 10 m^3 of reactant product (tetrachlorobenzene. caustic soda, and ethylene glycol). The reactor was controlled by a closed-circuit operation. On the day prior to the accident the reaction was incomplete and hence left at a reaction temperature of 160°C for the weekend. According to thermographic records, early Saturday morning July 10, an uncontrolled rise of temperature occurred following the failure of a safety shutdown device. Subsequently the reactants were heated to 230 to 240°C for 4 to 5 hr at a pressure of 3-4 atmosphere (atm), which eventually triggered the rupture of a second safety valve. The fumes (consisting of almost the entire contents of the reactor) were ejected via a 12-cm-diameter pipe extending from the vessel 50 meters to the roof of the plant with an opening to the atmosphere. The noise of the release was heard and the 40-meter-high cloud observed by most of the surrounding population for 20 to 30 min. (The time of the accident was placed at 1230 hr, July 10, 1976.)

Professor Zurlo's comments on the levels of contamination and areas and number of families evacuated were in agreement with those reported by Giovanardi (1976). A review of the medical findings revealed discrepancies with previous reports, however (Hay, 1976b; Walsh, 1977). Zurlo stated that as of November 1977, 67 cases of chloracne had been clearly diagnosed. The majority of these cases were young children, with girls near the age of puberty more frequently and more severely affected. Medical statistics collected from Seveso and the surrounding communities and representing periods before and after the accident, did not confirm any increase in spontaneous abortion. Of the men involved in closing the ICMESA plant during the first 2 weeks following the accident, only one man subsequently developed chloracne.

Professor Zurlo indicated that over 70,000 animals had died as a consequence of the incident. Deaths related to the toxicity of TCDD, however, were less than 2000, and were primarily domesticated fowl, rabbits, or guinea pigs. An additional 1000 animals were sacrificed by the regional veterinary service following the evacuation of the residents. These animals were thought to be contaminated and thus could not be moved with their owners. These included livestock (cows, goats, sheep, and horses) and household pets (dogs and cats). The remaining dead animals that were reported were primarily rats and mice that were killed with rodenticides to prevent movement of TCDD from contaminated areas into the adjoining community. All of the dead animals were burned in a pit on the ICMESA complex.

A tour (in protective clothing) of zones A-1 through A-5 revealed both the magnitude of contamination and the efforts of the decontamination program. All vegetation from areas A-6 and A-7 had been sacked and stacked in area A-5. A massive quantity of top soil, removed from zones A-6 and A-7, was physical removal of all contaminated surfaces.

At this time the primary technique for the destruction of the TCDD in the remaining parts of zone A was by soil biodegradation. It was proposed that all of the soil and organic matter collected from A-5 and A-6 be mixed and spread on areas A-2 through A-4. The soil would be fertilized, planted with vegetation, and maintained enclosed for an indefinite period of time.

One last note—the manifestation of chloracne was not limited to the first days after the accident, but continued to occur during the following 18 months. The skin lesions were rarely severe, mostly mild or extremely mild, and were always inclined to rapid and complete healing. Recidivism of the symptomatology was rare, and occurred practically only in cases with mild or very mild lesions. In those individuals evidencing chloracne, no laboratory evidence has been observed of significant hepatotoxicity, deranged porphyrin metabolism, or abnormal neurologic findings. It should also be recorded that of the 140 families evacuated in late July 1976 from areas A-3 through A-7, 80 of them had been reestablished into their own homes in areas A-6 and A-7 by November 1977. An additional 40 families were to be moved into their homes by January 1978. The families from area A-3 were to be permanently reestablished elsewhere in the Seveso community.

2. Discussion of the Dioxin Episodes

The only episodes in which TCDD was actually confirmed as a causative agent were those involving some of the industrial accidents, the horse arena episode in eastern Missouri, and the Seveso, Italy, episode (Young, 1980). In their review of these episodes, Young et al. (1978) noted that the exposure of the people involved was from days (most industrial accidents) to weeks (Seveso) to months (Missouri). Nevertheless, no human deaths were reported, although in both Missouri and Seveso, numerous animal deaths did occur. The clinical experience from the two episodes and the industrial episodes, involving at least 1100 individuals, support the opinion that patients without chloracne were extremely unlikely to have suffered the toxic effects of TCDD. In general, only in the most severe cases of chloracne has symptomatology persisted, admittedly for many years in a few instances.

The available scientific literature suggests that the episodes in Arizona, New Zealand, and Sweden were primarily the result of emotionalism associated with zealous press coverage. Although each incident began subsequent to field applications of phenoxy herbicides, it was highly unlikely that the symptoms reported were attributable to actual pesticide or TCDD exposure. The behavior in the environment of 2,4,5-T and TCDD following normal field applications lends little credence to accusations that significant bioaccumulations occurred in humans to initiate the toxic symptoms reported. Furthermore, the absence of confirmed illness in domestic livestock or wildlife in these three episodes also addresses the issue of whether an actual toxic exposure occurred. The magnitude of the dosage required to elicit toxic symptoms in animals might be obtained only under the most extreme cases (e.g., spills or sequential repetitive applications). These extreme situations were not noted in the episodes in Arizona or New Zealand. Moreover, the episodes in Arizona, New Zealand, and Sweden all occurred in the same time period—a period when numerous articles appeared in the world press on the alleged human health effects of herbicide Orange and TCDD in South Vietnam. The effects these articles had on the actual episode can only be speculated.

F. U.S. Air Force Ecological and Disposal Studies

In 1962, the Armament Development and Test Center at Eglin Air Force Base in Florida, was charged with the responsibility of designing, developing, and testing aerial dispersal systems for the application of herbicides in Southeast Asia (Young et al., 1975). A total of 157,000 kg of herbicides was applied from 1962 to 1970 to a 3.0 km^2 area known as test area C-52A. Herbicides Orange, Purple, White, and Blue were sprayed. Some of the 2,4,5-T contained significant levels of TCDD, as indicated by later soil analysis (up to 1500 ppt) on an area that had received approximately 39,500 kg of 2,4,5-T from 1962 through 1964.

The data indicated that climatic factors rather than herbicide residue influenced vegetative establishment (Young et al., 1975). The large seeded grasses, switchgrass (*Panicum virgatum*), and wolly panicum (*P. lanuginosum*) were the first perennial species to assume dominance. Annual herbs, rough buttonweed (*Diodia teres*), and poverty weed (*Hypericum gentianoides*) rapidly occupied space between the dominant grasses. With time, other grasses, such as broomsedge (*Andropogon virginicus*) and coastal lovegrass (*Eragrostis refracta*), encroached. The areas were well-diversified in species. Areas having 5 to 20% cover contained 14 species (8 grasses, 6 herbs), whereas an area of 80 to 100% cover had 28 species (18 grasses, 20 herbs). Rodents appeared to act as dispersal agents for seed.

Although the Air Force initiated ecological studies of test area C-52A in 1967, it was not until 1973 that extensive ecological studies of the terrestrial and aquatic ecosystems of this area were conducted (Young et al., 1975; 1978). The last applications of herbicides on this test area occurred in 1970. No residues of 2,4-D or 2,4,5-T were detected (detection limit of 10 ppb) in any soil samples collected during the period for 1971 to 1972. Residues of the contaminant TCDD were still present in 1978, however.

Fifty-four soil samples were collected to a depth of 0 to 15 cm from throughout the test area during the period from 1973 to 1978. The levels of TCDD ranged from <10 to 1500 ppt. The median concentration was 30 ppt, whereas the mean was 165 ppt. The ecological survey, extending over a 5-year period, documented the presence of at least 123 different plant species, 77 bird species, 71 insect families, 20 species of fish, 18 species of reptiles, 18 species of mammals, 12 species of amphibians, and two species of mollusks. At least 170 biological samples were analyzed for TCDD, including 30 species of animals. No TCDD was found in any of the plant species examined. TCDD was found in nine species of animals, however, including two rodent species: beachmouse (300–1500 ppt, liver) and hispid cotton rat (<10–210 ppt, liver); three species of birds: meadowlark (100–1020 ppt. liver), mourning dove (50 ppt, liver), and Savannah sparrow (69 ppt, liver); three species of fish: spotted sunfish (85 ppt, liver), mosquitofish (12 ppt, whole body), and sailfin shinner (12 ppt, whole body); and one reptile, the six-lined racerunner (360–430 ppt, muscle).

Gross pathology was done on all species collected for TCDD residue analyses. Histopathological examinations were performed on over 300 beachmice (*Peromyscus polionotus*) or hispid cotton rats (*Sigmodon hispidus*) from the test area and a control field site. Examinations were performed on the heart, lungs, trachea, salivary glands, thymus, liver, kidneys, stomach, pancreas, adrenals, large and small intestines, spleen, genital organs, bone, bone marrow, skin, and brain. Initially the tissues were examined on a random basis without the knowledge of whether the animal was from a control or test area. All microscopic changes were recorded, including those interpreted as minor or insignificant. The tissues were then reexamined on a control and test basis, which demonstrated that the test and control rodents could not be distinguished histopathologically.

Laboratory and field investigations of the food-chain components, habits, and habitats of the various species contaminated with TCDD suggested that contaminated soil rather than contaminated food was the principal route of exposure to the animals. Beachmice having 540 to 1300 ppt TCDD in their livers had 130 to 140 ppt TCDD in their pelts (Young et al., 1975), thus, grooming the pelt probably resulted in the TCDD uptake observed.

The racerunner (*Cnemidophorus sexlineatus*), a reptile, contained significant levels of TCDD in the viscera (360 ppt) in the soils highest in TCDD. The reptiles showed no significant lesions or variations in visceral mass from treated or untreated areas, however.

In 1973, 5966 arthropod specimens belonging to 71 insect families were collected (Young et al., 1975) over the 3.0 km^2 test area. A similar study performed in 1971 produced 1796 specimens representing 70 insect families. Increased numbers for 1971 to 1973 are believed due to increased vegetation cover on the area. There was little change in the population diversity, however.

An aquatic area immediately adjacent to the sprayed area was drained by five streams (Young et al., 1975). A total of 22 species of fish was collected from 1969 to 1976. The results indicated that no significant changes had occurred in the ichthyofauna of either the treated or the untreated streams during the study period.

In decontamination and disposal studies, Young et al. (1976) indicated that TCDD may be degraded by soil microorganisms, especially when in the presence of other chlorinated hydrocarbons (the estimated half-life for TCDD in the presence of 2,4,-D and 2,4,5-T was 225 to 275 days). TCDD is also degraded by sunlight and is readily bound to activated charcoal. Wild animals exposed to TCDD-contaminated soils accumulate TCDD, but apparently did not exceed the levels of TCDD in the environment. In the field, animals tolerated TCDD levels of 10 to 1500 ppt without effect.

Stark et al. (1975) demonstrated that herbicide Orange (surplus supply from Vietnam) could be incorporated into the soil in amounts up to 4480 kg/ha without inhibiting the predominant soil bacteria studied. Growth of *Pseudomonas* spp. and *Bacillus* spp. was stimulated. No detrimental effects on the environment were noted.

G. Vietnam Veterans

Thousands of man hours and millions of dollars have been spent on the 2,4,5-T issue as a result of its use in Vietnam. There have been numerous individuals, committees and scientific societies, and representatives from federal, state, and private action and research groups that have studied the data and concluded that 2,4,5-T is not a threat to human health or the environment. The controversy continues, however, because there are individuals and groups who oppose its continued use. Some veterans from the Vietnam War claim serious health problems they believed are related to Agent Orange (Reggiani, 1988), even though spraying was stopped in 1970. The Veterans Administration has argued that the long-term health effects of Agent Orange are still largely unknown. In 1979 the White House established an "interagency work group" to study the possible long-term health effects of phenoxy herbicides and contaminants. The group was chaired by the general counsel of the Department of Health and Human Services and included representatives of the Department of Defense, the Department of Health, the Veterans Administration, the EPA, the Department of Agriculture, the White House Office of Science and Technology Policy, and the Congressional Office of Technology Assessment. The research efforts and the studies related to the Agent Orange issue are still in progress today. In August 1981 an "Agent Orange work group'' was established at the White House and raised to Cabinet counsel level (Reggiani, 1988). It must be remembered that the Vietnam veterans were also exposed to such toxic chemicals as insecticides and antimalarial drugs.

The scientific panel of the interagency work group gave priority to Vietnam veterans who were possibly exposed to Agent Orange by gathering medical and scientific knowledge on the long-term effects on populations associated with occupational or accidental contamination with dioxin (Regginai, 1988). The U.S. government initiated 10 epidemiologic studies of Vietnam veterans and five health surveillance projects. The evaluation of the mortality rate among air force personnel who conducted aerial herbicides spraying missions (Operation Ranch Hand) is nearly identical to that of the comparison matched group. The same conclusion was reached after data analysis of the morbidity study. There was no evidence to support a cause-and-effect relationship between herbicide exposure and adverse health findings in Operation Ranch Hand.

The possible relationship between exposure to Agent Orange and its contaminant, 2,3,7,8-tetrachlorodibenzo-p-(dioxin), during the Vietnam War and chloracne was investigated (Burton et al., 1998). The index subjects were veterans of Operation Ranch Hand, the unit responsible for aerial herbicide spraying in Vietnam from 1962 to 1971. Other air force veterans who served in Southeast Asia during the same period but who were not involved with spraying herbicides served as comparisons. None of the Ranch Hand veterans was diagnosed with chloracne; therefore, the analyses was restricted to acne. No meaningful or consistent association was found between dioxin exposure and the prevalence of acne with or without regard to anatomical location. The results suggested that exposure of Ranch Hand veterans to dioxin was insufficient for the production of chloracne, or perhaps that the exposure may have caused chloracne that resolved and was currently undetectable.

The private sector population was also studied for the mortality, morbidity, and reproductivity of several people involved in dioxin-exposure accidents in manufacturing plants of TCP, PCP, or 2,4,5-T. These studies involved long-term exposure (accidents occurring in the late 1940s to the mid-1970s) and chloracne symptoms. The data from these exposures to dioxin indicated that dioxin did not modify the mortality rate and pattern of these people. Also, there are no discernible effects of dioxin exposure on human reproduction. The biological marker of the dioxin exposure, chloracne, is still persistent in some individuals 30 years or more after exposure, however. There is also no link between soft tissue sarcoma and exposure to phenoxy herbicides or genotoxic effects.

As indicated earlier, the discovery of polychlorinated benzodioxins and benzofurans in fly ash and the flue ashes of incinerators, cigarette smoke, car and truck mufflers, and charcoal-grilled stakes has changed the Agent Orange, 2,4,5-T herbicides, and Seveso accident episodes of dioxin exposure (Reggiani, 1988). Reggiani (1988) stated that since these substances are widespread in the environment and since they can be detected at extremely low levels, we must assume that life now takes place in a minefield of risks from perhaps thousands of substances. The public cannot be guaranteed a risk-free environment.

This chapter represents only a small portion of the research and defense efforts that went into attempts to obtain EPA approval of 2,4,5-T for continued use. One attempt was the publication of a 445-page cooperative impact assessment report by the 2,4,5-T Assessment Team to the reputable presumption against registration of 2,4,5-T. It was entitled "The Biologic and Economic Assessment of 2,4,5-T" (technical bulletin no. 1671), and was developed by USDA and state experiment station personnel and submitted to EPA on February 15, 1979. A similar report was submitted for silvex. These and many other reports and committees defending 2,4,5-T were largely ignored by EPA, but represented excellent science and thousands of man hours of work.

IV. CONCLUSIONS

This chapter is an attempt to show that despite all attempts to discredit 2,4,5-T as a "bad actor" in the environment or as being toxic to animals under field use, it was not proven after 40 years of widespread use. Even when used at extremely high rates and repeated treatment (see Sec. III.F) adverse effects could not be demonstrated.

V. ATTACK ON 2,4-D

The herbicide 2,4-D has been used commercially throughout the world for more than 50 years and was first registered in the United States in 1948 (Bovey, 1980c). Since that time it has become essential and the most researched agricultural chemical. More than 40,000 scientific articles have been published on 2,4-D (Journal of the American College of Toxicology, 1992).

Questions were first raised about 2,4-D and other phenoxy acid herbicides with the publication of a series of case-control studies by Swedish investigator Lennart Hardell during the late 1970s (Journal of the American College of Toxicology, 1992). Dr. Hardell hypothesized that exposure to phenoxy herbicides as well as to dioxin contaminates in 2,4,5-T was associated with three rare forms of cancer: Hodgkin's disease, soft tissue sarcoma, and non-Hodgkin's lymphoma (Journal of the American College of Toxicology, 1992).

When 2,4,5-T was suspended in 1979 the EPA considered taking similar action with 2,4-D, but the agency did not see 2,4-D as an imminent health hazard when used according to label instructions and precautions (Josephson, 1980).

Controversy with 2,4-D surfaced again in 1986 when the National Cancer Institute (NCI) published data from a case-control study associating non-Hodgkin's lymphoma (NHL) with farming practices in the Midwest, including the use of herbicides for more than 21 days per year. Following publication of these data, the EPA commissioned the Ontario Ministry of Environment, Agriculture Canada, and the Council for Agricultural Science and Technology for independent reviews of the NCI study. They all came to the same conclusion—that the cancer risk hypothesis was not borne out by the NCI study and that 2,4-D use did not pose an unreasonable risk to public health (Journal of the American College of Toxicology, 1992).

In 1990 the NCI published data that suggested that among Nebraska farmers there was an increased risk of NHL with increasing use of 2,4-D as measured by reported days of use per year. A blue-ribbon panel convened by the Harvard School of Public Health considered the NCI data and other epidemiological studies worldwide on 2,4-D and cancer. They concluded a link between 2,4-D use and cancer had not been established (Journal of the American College of Toxicology, 1992).

In 1991, NCI published another study implicating 2,4-D as a cancer hazard of malignant lymphoma among dogs whose owners used 2,4-D. A panel commissioned by the industry task force II on 2,4-D research data concluded that due to the limitations of the study, no association between 2,4-D use and malignant lymphoma in dogs could be established.

In 1992, NCI investigators published a third case control study of farmers that reported a small risk of NHL associated with 2,4-D use, although the risk did not appear to increase with latency or failure to use protective equipment. Munro et al. (1992) indicated that cohort studies of exposed workers to 2,4-D do not generally support the specific hypothesis that 2,4-D causes cancer. A critical evaluation of the exposure data indicated that exposure to 2,4-D in user groups is intermittent and much lower than doses tested chronically in long-term animal studies that have not shown evidence of tumor induction. Moreover, the structure of 2,4-D does not suggest it would be a carcinogen. It is excreted unchanged, and there is no evidence that it is metabolized to critically reactive metabolites or accumulates in tissue. It does not possess any of the characteristics of nongenotoxic animal carcinogens, thus the available mechanistic studies provide no plausible basis for a hypothesis of carcinogencity. There is no evidence that 2,4-D adversely affects the immune system; neurotoxic and reproductive effects have only been associated with high toxic doses that would not be encountered by 2,4-D users. New label requirements indicate protective clothing and equipment to further reduce exposure, but available data indicate that the public health impact of 2,4-D, including the risk of human cancer, was negligible in the past and would be expected to be even smaller in the present and future. Reviews by Bond and Cook (1993) and Johnson and Wattenberg (1996) support conclusions made by Munro et al. (1992). Carlo et al. (1992) concluded that as a result of the numerous limitations in the design of the NCI study referred to earlier it can be concluded that it did not show an association between dog owners' use of 2,4-D and canine malignant lymphoma.

In 1981 the EPA issued a data call-in (DCI) on 2,4-D acid (Page, 1993). Thirteen manufacturers and formulators of 2,4-D formed the industry task force on 2,4-D research data (task force I) to jointly develop the data required by DCI as allowed under the Federal Insecticide, Fungicide and Rodenticide Act (FIFRA). The task force had spent $4 million when the 1988 DCI (registration standard) was issued on the acid, ester, and amine salt forms. In 1988 membership in the new task force, task force II, dropped to six members. Problems were encountered to develop a more sensitive analytical procedure for 2,4-D, causing delays and escalating costs in meeting DCI requirements. The total cost approached $30 million, and task force members dropped to four (DowElanco, Rhône-Poulenc, NuFarm, and AGRO-GOR), because the cost would exceed the expected returns from the sale of the product. The EPA reregistration DCI required over 200 new studies to support continued registration (Hammond, 1995). The new studies have incorporated both state-of-the-art and new technologies that permitted improved understanding of the fate and breakdown products in animals, crops, and the environment.

VI. CONCLUSIONS

Available studies show that 2,4-D is rapidly excreted and is not metabolized to reactive intermediates. It does not produce genotoxic or neurological effects in animals, and it is not carcinogenic. An expert EPA panel concluded that there was at most only weak evidence for an association between cancer and 2,4-D, however, so wearing more rigorous protective clothing is a label requirement (Hammond, 1995; Johnson and Wattenberg, 1996).

VII. BIOLOGICAL AND ECONOMIC ASSESSMENT

In 1988 Agriculture Canada published an evaluation of 2,4-D use in Canadian agriculture (Canada Agriculture, 1988). It concluded that 2,4-D was the lowest-cost broadleaf weed herbicide in Canada and its loss would require users to switch to higher-cost alternatives and in some cases, incur additional crop loss. The total retail value of 2,4-D sold to end users was about $81 million, or 15% of total herbicide sales. Most 2,4-D used in agriculture is in cereal crops in the prairie provinces (95%). Potential production loss would range from $42 to $55 million, and increased herbicide expenditure would range from $16 to $27 million, for a total of $58 to $82 million. For all uses in Canada, the potential cost impact ranged from $130 to $200 million per year. If all phenoxy herbicides were withdrawn the loss would range from $420 to $499 million.

Importantly, as toxicological and environmental safety data were being refined for reregistration of 2,4-D, detailed information on the biological and economic assessment benefits of phenoxy herbicide has been provided for the United States (USDA, 1996). The economic impacts of either 2,4-D or all phenoxy herbicide loss in agriculture and nonagricultural use have been summarized (Szmedra, 1996).

About 55 million lb of phenoxy herbicides were used during 1992 in the United States, with 2,4-D taking up 86% of total use or about 47 million lb of acid equivalent. The severest economic effects from banning phenoxy herbicides would be felt in major field crops, rangeland, pasture land, and alfalfa used for forage. The total loss of phenoxy herbicides in these major uses only could result in net societal losses approaching $1.35 billion, which combine producer and consumer effects of yield, cost, and price change. A total phenoxy herbicide loss scenario would result in the following yield losses:peanut (13%), flax (6.7%), sugarcane (5.7%), green peas (5.6%), alfalfa (5.2%), asparagus (4.9%), barley (3.8%), sorghum (2.4%), and wheat (2.2%). Other major impacts include a loss of $367 million in turfgrass, $284 million in small grain production, $180 million in increased costs of noxious weed control, and $111 million in the production of orchard, vineyard, soft fruit, and nut crops. The estimated aggregate economic impact of losing only 2,4-D in the United States is a loss of $1.68 billion. If all the uses of phenoxy herbicides were canceled, the net societal loss would be $2.56 billion. These estimates describe the yield and financial impacts of the initial production the year after a phenoxy herbicide cancellation. Subsequent years' losses and financial impacts probably would be less as both farmers and markets adjust to the new production situation.

Related to herbaceous and woody plant control on pasture land, the annual net societal effect of a 2,4-D ban considering increased weed control cost and forage yield loss would be $383.4 million and $384.2 million for all phenoxy herbicides (Bovey, 1996). About 8.7 million acres of pasture land (8%) are treated annually, with an average use rate of 1.1 lb/acre. On rangeland, about 4.9 million acres (1%) are treated annually, with an average use rate of 1.2 lb/acre. The 1992 net societal effect of a 2,4-D ban would be a $79.5 million loss and $80.4 million for all phenoxy herbicides.

A 2,4-D ban for noxious weed control (sometimes including woody plants) is estimated at $162 million, and for all phenoxy herbicides, a loss of $180 million (Bovey, 1996).

Newton (1996) indicated that for rights-of-way management, banning 2,4-D would cost $19.1 million, and $300,000 for substituting for dichorprop, for a total of $19.4 million in 1992. For forest applications, the annual loss of 2,4-D would be about $7.2 million and $605,000 for dichlorprop, or about $17.8 million for 1992 (Newton, 1996).

Alternatives to the phenoxys herbicides for pasture land, rangeland, noxious weeds, rights-of-way, forestry, and certain noncrop uses include primarily clopyralid, dicamba, glyphosate, picloram, and triclopyr, all of which are much more costly to use than 2,4-D or other phenoxy herbicides.

Worldwide loss data for 2,4-D and all phenoxy herbicides are not available, but would be considerable.

VIII. SUMMARY

Research on dioxin and its biological effects is ongoing; however, the conclusions about the safety of the phenoxy herbicides has not changed since stated by Bovey and Young (1980a).

The discovery of 2,4-D and 2,4,5-T as selective herbicides in 1942 precipitated the greatest single advance in the science of weed control and one of the most significant in agriculture. Development of other phenoxy herbicides, such as MCPA, 2,4,-DB, 2,4-DP, and silvex soon followed. Today a large variety of safe and effective preplant, preemergence, and postemergence herbicides of divergent chemistry and formulations are available as indispensable tools for agriculture and noncrop uses. The development of the phenoxy herbicides led the way for technology in developing other new and specific herbicides for agriculture, although the phenoxys remain essential herbicides. When used according to recommended procedures, these herbicides have had little or no detrimental effect on the environment.

Extensive research indicated the phenoxys are rapidly degraded in the environment and do not accumulate in the food chain. Market basket surveys indicated the phenoxys occur very infrequently in human food and in very small amounts. Phenoxy herbicides are lost from soil, plant, and water sources by volatility, photodegradation, rainfall, dilution, biodegradation, and other means. The phenoxys are essentially nontoxic to soil organisms at normal field rates, and the soil microbial population is responsible for their rapid breakdown.

The phenoxys are moderately toxic to laboratory animals (mammals) by oral or dermal intake, but are only slightly toxic by inhalation. They are also moderately toxic to some species of fish (ester formulation), but only slightly toxic to lower aquatic organisms, birds, and wild mammals under laboratory conditions. Under field conditions, they are not toxic to livestock at the dosages required for weed control, and few hazards to wildlife have been reported from authentic research investigations.

When used according to labeled instructions, human health hazards associated with the worldwide use of phenoxy have been insignificant (in our search of thousands of scientific articles on these herbicides [spanning a period of 50 years], we found less than two dozen documented medical reports of human intoxication). Poisoning in man has occurred by self-ingestion or when grossly misused, but such occurrences have been extremely rare.

Damage to susceptible vegetation and crops can occur from spray drift or from volatilization (especially high-volatile esters) of phenoxy herbicides. Low-volatile esters or other formulations (amino salts) should be considered. Following label instructions, using proper spray equipment, and making application during favorable weather (low wind velocity, etc.) will minimize the drift problem. In some situations, however, the risk of damage to valuable vegetation may be too great, and other weed control methods should be used if possible or feasible.

Considerable research data are available on the physiological effects and mode of action of the phenoxy herbicides in higher plants. At extremely low concentrations, phenoxys mimic natural auxin activity (hormone effects) in plants, and in excess but small amounts exhibit herbicidal properties. Mode of action is extremely complicated, since many biochemical responses are triggered that may lead to abnormal growth, epinastic manifestations, plugging of vascular channels, and ultimately death of the plants. On the other hand, many weed and crop plants are virtually unaffected by any phenoxy herbicide and therefore are useful selective weed-control agents. The phenoxy herbicides are widely used because they are more efficient and less injurious to the environment than alternative methods. Loss of these compounds would add considerably to the cost of food, forest products, electric power, transportation, and services, with the cost being borne directly by the consumer.

Ando et al. (1970) reported that 2,4-dichlorophenol, a synthetic and natural derivative of 2,4-D, was isolated from the fermented broth of a soil fungus, Penicillium species. They suggest that such derivatives are not artifact hormones, but natural growth-regulatory hormones of plants, such as gibberellins and cytokinins. In other words, phenoxy or similar compounds may occur in nature and function in plant metabolism on a wider scale than formerly indicated. Further research is needed to unlock the secret of auxin activity and other so-called synthetic hormonelike compounds such as 2,4-D to understand their basic metabolism and physiological function.

REFERENCES

Adamson L. Spray now—Pay later? Environ Action July 6: 9–13, 1974.

Agricultural Chemicals Board. Report of the Subcommittee on 2,4,5-T. Wellington, New Zealand: Agricultural Chemicals Board. 1972, pp. 1–10.

Ando K, Kato A, Suzuki S. Isolation of 2,4-dichlorophenol from a soil fungus and its biological significance. Biochem Biophys Res Commun 39:1104–1107, 1970.

Anonymous. Illness associated with TCDD-contaminated soil Missouri. Morb Mort 23(34):299, 1974.

Bauer H, Schulz KH, Spiegelberg U. Berufliche Vergiftungen bei der Herstellung van Chlorophenol-Verbindungen. Arch Gewerbepath Gewerbehyg 18:538–555, 1961.

Bionetics Research Laboratories. Evaluations of the Teratogenic Activity of Selected Pesticide and Industrial Chemicals in Mice and Rats vol. III. Bethesda, MD: Bionetics Res. Lab., 1969.

Bond GG, Cook RR. The paradox of herbicide 2,4-D epidemology. In: Wang RGM, Knaak JB, Mailbach HI, eds. Health Risk Assessment: Dermal and Inhalation Exposure and Absorption of Toxicants. Boca Raton, FL: CRC Press, 1993, pp. 471–490.

Bovey RW. Use of 2,4-D and other phenoxy herbicides on pastureland, rangeland, alfafa forage and noxious weeds in the United States. In: Biologic and Economic Assessment of Benefits from Use of Phenoxy Herbicides in the United States. USDA-NAPIAP report no. 1-PA-96. Washington, DC. 1996, pp. 76–86.

Bovey RW, Young AL. Conclusions and recommendations. In: The Science of 2,4,5-T and Associated Phenoxy Herbicides. New York: Wiley, 1980a, pp. 449–454.

Bovey RW, Young AL. Military use of herbicides. In: The Science of 2,4,5-T and Associated Phenoxy Herbicides. New York: Wiley, 1980b, pp. 371–403.

Bovey RW. The 2,4,5-T controversy. In: The Science of 2,4,5-T and Associated Phenoxy Herbicides. New York: Wiley, 1980c, pp. 1–28.

Burton JE, Michalek JE, Rahe AJ. Serum dioxin, chloracne, and acne in veterans of operation Ranch Hand. Archives Environ Health 53:199–204, 1988.

Burnside OC. The history of 2,4-D and its impact on development of the discipline of weed science in the United States. In: Biological and Economic Assessment of Benefits from Use of Phenoxy Herbicides in the United States. USDA-NAPIAP dept. no. 1-PA-96. Washington, DC. 1996, pp. 5–15.

Canada Agriculture. An Economic Assessment of the Benefits of 2,4-D in Canada: Agriculture Canada. Guelph, Ontario: Deloitte, Haskins & Sells, 1988.

Carlo GL, Cole P, Miller AB, Munro IC, Solomon KR, Squire RA. Review of a study reporting an association between 2,4-dichlorophenoxyacetic acid and canine malignant lymphoma: Report of an expert panel. Reg Toxicol Pharmacol 16:245–252, 1992.

Carter CD, Kimbrough RO, Liddle JA, Cline RF, Zack MM Jr, Barthel F. Tetrachlorodibenzodioxin: An accidental poisoning episode in horse arenas. Science 188:738–740, 1975.

Commoner B, Scott RF. Accidental Contamination of Soil with Dioxin in Missouri: Effects and Countermeasures. St. Louis: Center for the Biology of Natural Systems, Washington University, Sept. 24, 1976.

Cutting RT, Phuoc TH, Ballo JM, Benenson MW, Evans CH. Congenial Malformations Hydatidiform Moles and Still Births in the Republic of Vietnam 1960–1969. no. 903.233. Washington, DC: U.S. Army, Department of Defense, U.S. Government Printing Office, 1970.

Dow Chemical. Comments of the Dow Chemical Company on the paper by Lucile Adamson. ''Spray Now—Pay Later?'' published in Environ. Action, July 1974, pp. 9–13. Midland, MI: Dow Chemical, 1975.

Environmental Defense Society. The case against 2,4,5-T. NZ Environ 2:16–21, 1972.

Erne K, Nordkrist M. Suspected Herbicide Poisoning of Reindeer. Stockholm: report, State Veterinary Institute to the Poison Board, 1970.

Forth W. 2,3,7,8-Tetrachlorodibenzo-1-4-dioxin (TCDD). Das Ungluck von Seveso. Deut Arzt Fort Aktuelle Med 44(3):2617–2626, 1977.

Giovanardi A. (1976). Decontamination Program for the Dioxin-Contaminated Areas of

Seveso and Media. Milan: Guinta Regionale Della Lombardia, Ministero Della Sanita. Aug. 18, 1976, pp. 1–17.

Hammond LE. New perspectives on an essential product: 2,4,-D. Down to Earth 50(2): 1–5, 1995.

Hay A. Toxic cloud over Seveso. Nature 262 (5570):636–638, 1976a.

Hay A. Seveso: The aftermath. Nature 263 (5578):538–540, 1976b.

House WB, Goodson H, Gadberry HM, Dockter KW. Assessment of Ecological Effects of Intensive or Repeated Use of Herbicides. ARPA order no. 1086. Final report Kansas City, MO: Midwest Res. Int. sponsored by Adv Res Project Agency, Department of Defense, 1967.

Johnson RA, Wattenberg EV. Risk assessment of phenoxy herbicides: An overview of the epidemiology and toxicology data. In: Biologic and Economic Assessment of Benefits from Use of Phenoxy Herbicides in the United States. USDA-NAPIAP report. no. 1-PA-96. Washington, DC. 1996, pp. 16–40.

Johsephson J. Forest pesticides: An overview. Environ Sci Tech 14(10):1165–1168, 1980.

Journal of the American College of Toxicology. Preface in a comprehensive, integrated review and evaluation of the scientific evidence relating to the safety of the herbicide 2,4-D. J Am College Toxicol 11(5), 1992.

Kimbrough RD, Carter CD, Liddle JA, Cline RE, Phillips PE. Epidemiology and pathology of a tetrachlorodibenzodioxin poisoning episode. Archives Environ Health 28: 77–85, 1977.

Lowry RB, Allen AB. Herbicides and spina bifida. Can Med Assoc J 117:850, 1977.

McQueen EG, Veale MO, Alexander WS, Bates MN. 2,4,5-T and Human Birth Defects. New Zealand Department of Health, Division of Public Health, 1977.

National Academy of Science. The effects of herbicides in South Vietnam. Part A. Summary and Conclusions. Washington, DC: Committee on the Effects of Herbicides on Vietnam, Div. of Biological Sciences, Assembly of Life Sciences, National Research Council. National Academy of Science, 1974.

Munro IC, Carlo GL, Orr JC, Sund KG, Wilson RM, Kennepohl E, Lynch BS, Jablinske M, Lee NL. A comprehensive, integrated review and evaluation of the scientific evidence relating to the safety of the herbicide 2,4-D. J Amer Coll Toxicol 11(5): 559–664, 1992.

New York Times. Death of animals laid to chemicals, NY Times. Aug. 28: 36, 1974.

Newton M. Phenoxy herbicides in rights-of-way and forestry in the United States. In: Biologic and Economic Assessment of Benefits from Use of Phenoxy Herbicides in the United States. USDA-NAPIAP report no. 1-PA-96. Washington, DC. 1996, pp. 165–178.

New Zealand Medical Journal. Fetotoxicity. NZ Med J 75(480): 304–305, 1972.

Orians GH, Pfeiffer EW. Ecological effects of the war in Vietnam. Science 168:544–554, 1970.

Page DL. 2,4-D Registration Update. Belhaven, NC: Industry Task Force II on 2,4-D Research Data, 1993.

Parks M, Sullam E. The Chemical Cloud that Fell on Seveso. Washington, DC: Rachel Carson Trust for the Living Environment, 1976.

Peterson GE. The discovery and development of 2,4-D. Agric Hist 41:243–253, 1967.

Pinsent PJN, Lane JG. A case of possible 2,4-D and 2,4,5-T poisoning in the horse. Vet Rec 87:247, 1970.

Pfeiffer EW, Orians GH. The military uses of herbicides in Vietnam. In: Harvest of Death. New York: Free Press, 1972, pp. 117–176.

Rawls RL, O'Sullivan A. Italy seeks answers following toxic release. Chem Eng News 54(35):27–35, 1976.

Reggiani G. Historical overview of the controversy surrounding Agent Orange. In: Young AL, Reggiami GM, eds. Agent Orange and Its Associated Dioxin: Assessment of a Controversy. New York: Elsevier, 1988, pp. 29–76.

Sare WM, Forbes PI. Possible dysmorphogenic effects of an agricultural chemical: 2,4,5-T. New Zealand Med J 75 (476):37–38, 1972.

Schulz KH. Clinical and experimental tests for etiolopic chloracne. Arch Klein Exp Derm 206:589–596, 1957.

Stark HE, McBride JK, Orr GF. Soil Incorporation/Biodegradation of Herbicide Orange. Dugway Proving Grounds, UT: U.S. Army, doc. no. DGP-FR-C615F, 1975.

Subcomittee on Energy, Natural Resources and the Environment. Hearing before the Subcommittee on Energy, Natural Resources and the Environment of the Committee on Commerce, United States Senate, 91st Congress (Honorable P. A. Hart, presiding). Effects of 2,4,5-T on Man and the Environment. Serial 91–60. U.S. Government Printing Office: Washington, DC, 1970.

Szmedra P. Potential economic effects of banning phenoxy herbicides in the United States. In: Biologic and Economic Assessment of Benefits from Use of Phenoxy Herbicides in the United States. USDA-NAPIAP report. no. 1-PA-96. Washington, DC. 1996, pp. 41–75.

Tschirley FH. Defoliation in Vietnam. Science 163:779–786, 1969.

USDA. Biologic and Economic Assessment of Benefits from Use of Phenoxy Herbicides in the United States. USDA-NAPIAP report no. 1-PA-96. Washington, DC. 1996.

Veale PO. 2,4-D—Its future in weed control. Agric Chem: X 20–22, 1946.

Walsh, J Seveso: The questions persist where dioxin created a wasteland. Science 197(4308):1064–1067, 1977.

Whiteside T. A reporter at large: The pendulum and the cloud. New Yorker: 30–35, July 25, 1977.

Young AL. (1980). The chlorinated dibenzo-*p*-dioxins. In: Bovey RW, Young AL, The Science of 2,4,5-T and Associated Phenoxy Herbicides. New York: Wiley, 1980, pp. 133–205.

Young AL, Calcagni JA, Thalken CE, Tremblay JW. The Toxicology, Environmental Fate, and Human Risk of Herbicide Orange and Its Associated Dioxin. tech report OEHL-TR-78-92. Brooks Air Force Base, TX: USAF Occupational and Environmental Health Laboratory, Aerospace Medical Division, 1978.

Young AL, Thalken CE, Arnold EL, Cupello JM, Cocherham LG. Fate of 2,3,7,8-Tetrachoro-Dibenzo-*p*-Dioxin (TCDD) in the Environment: Summary and Decontamination Recommendations. tech. report USAFA-TR-76-18. Dept. of Chemistry and Biological Science. Colorado Springs, CO: U.S. Air Force Academy, 1976.

Young AL, Thalken CE, Ward WE. Studies of the Ecological Impact of Repetitive Aerial Applications of Herbicides on the Ecosystem of Test Area C-52A. Eglin AFB, FL. tech. report AFATL-TR-75-142. Eglin Air Force Base, FL: Air Force Armament Lab., 1975.

9

Herbicide Fate and Activity in Woody Plants

I. INTRODUCTION

Knowledge of the factors affecting the absorption, translocation, and mode of action of an herbicide in woody plant control is essential for optimal results. Lethal amounts of foliar- or soil-applied herbicide must be absorbed by leaf or root tissue for translocation within the plant to the site of action. Understanding the factors that produce optimum efficacy at minimal herbicide rates results in more efficient and safe use of the herbicide. This chapter explores many factors affecting the absorption, translocation, and mode of action of the herbicides used in woody plant control and how those factors influence their efficacy on individual woody species and woody plants in general.

II. HERBICIDES

A. Benzoics (Dicamba)

Dicamba has been used for nearly 40 years, and while considerable data on its absorption, translocation, and mode of action have been investigated for herbaceous plants, the body of data for woody plants is not nearly as extensive. Dicamba is rapidly absorbed by either roots or foliar tissue (Frear, 1976). Foliar-absorbed dicamba is translocated basipitally and acropetally, and is mobile in both xylem and phloem. Movement out of mature leaves and continuous redistribution into younger leaves and meristematic tissues support the concept of an active source-to-sink, phloem-mediated translocation system. Mobile forms of the herbicide may include unchanged dicamba, 5-OH-dicamba and its conjugate. The rate of absorption, movement, and distribution of assimilates may vary with species and stage of growth.

Dicamba is relatively stable in higher plants, but it undergoes hydroxylation or demethoxylation with hydroxylation (Ashton and Crafts, 1981). Dicamba produces symptoms similar to the phenoxy herbicides with epinasty of young shoots and proliferation of growth. [For more details consult the *Herbicide Handbook* (Weed Science Society of America, 1994).]

B. Bipyridyliums (Diquat and Paraquat)

Bovey and Davis (1967) found that plant species vary in their rates of paraquat absorption. Washing yaupon (*Ilex vomitoria*) and live oak (*Quercus virginana*) 1 hr after application of paraquat reduced its effectiveness regardless of air temperature, but washing winter peas (*Pisum sativum*) 10 min after application had little effect on phytotoxicity. Field-grown mesquite (*Prosopis glandulosa*) showed extensive leaf necrosis when leaves were washed 20 min after application, but live oak showed no injury. Winged elm (*Ulmus alata*) response was intermediate. The temperatures of 5 to 6°C delayed leaf necrosis of greenhouse-grown oats (*Avena sativa*), winter peas, husache (*Acacia farnesiana*, mesquite, live oak, and yaupon for at least 48 hr after paraquat treatment compared to air temperatures of 24 to 28°C. From paraquat absorption and concentration studies Bovey and Miller (1968) found that the phytotoxic differences between white, variegated, and green leaves of hibiscus (*Hibiscus rosa sinensis*) were not due to the amount or rate of absorption or protoplasmic susceptibility but to the differences in the chlorophyll content of treated leaves. The same reduced activity of paraquat on white leaf tissue was demonstrated for sorghum (*Sorghum bicolor*) and alpina (*Alpina purpurata*).

Davis et al. (1968b) studied the uptake and transport of picloram and 2,4,5-T both alone and in combination, as well as in the presence of paraquat by honey mesquite, huisache, yaupon, and beans. Paraquat reduced the transport of picloram by mesquite, huisache, and beans. Paraquat increased the uptake of picloram by yaupon but did not affect transport. Brown and Nix (1975) studied the uptake and movement of paraquat in the transpiration stream of 25-year-old slash pine (*Pinus elliotti*) trees with methyl-^{14}C-paraquat (10μ Ci/tree). The data indicated that paraquat had a strong affinity for cellulose, resulting in slow movement both vertically and horizontally.

Baur et al. (1969a) showed that paraquat caused rapid disintegration of the plasmalemma of mesquite mesophyll cells followed by rupturing of the chloroplast membranes and loss of chloroplast turgor. No changes were noted in the mitochondria, Golgi bodies, endoplasmic reticulum or nucleus, or in the composition of the cytoplasm or cell walls of treated tissue. Merkle et al. (1965) concluded that light and O_2 are essential for a rapid bleaching of the pigment system of broadleaf beans by paraquat. This bleaching appeared related to the destruction of a protective system that normally prevents photo-oxidation. Light (but not 0_2) was essential for changes in membrane permeability by paraquat in mesquite, honeysuckle (*Lonicera saponica*), and broadleaf beans (*Phaseolus vulgaris*).

The mechanism of action of diquat and paraquat involves the formation of the free radical by reduction of the ion and subsequent auto-oxidation to yield the original ion. The free radical itself does not appear to be the primary toxicant. During the auto-oxidation of the paraquat free radical to the ion, H_2O_2 (hydrogen peroxide), O_2-(superoxide radical), + OH (hydroxyl radical), and O_2 (single oxygen) are formed. Each by-product is potentially phytotoxic. Research suggests that the hydroxyl radical is responsible for paraquat-induced lipid peroxidation and related phytotoxic symptoms. The photosynthetic apparatus, light, and molecular oxygen are required cofactors for these reactions (Ashton and Crafts, 1981).

In herbaceous plants, absorption into the foliage is rapid (30 min), but it is slower in some woody plants (Bovey and Davis, 1967). Paraquat translocates only in the apoplast (including the xylem), and foliar-applied paraquat usually remains in the treated leaves. The fate of diquat in plants is similar to paraquat's (Weed Science Society of America, 1994).

C. Primary Plant Growth Regulator Herbicides (Phenoxys, Clopyralid, Picloram, and Triclopyr)

1. Foliar Absorption

a. ***Herbicide Configuration and Type*** The progressive chlorination on the absorption of phenoxyacetic acid (POA) by bean leaf discs using 2-, 4,- 2,4-, 2,6-, 3,5-, 2,4,5- and 2,4,6,-chloroderivatives of POA leads to increased absorption in light and dark, with some exceptions (Sargent and Blackman, 1969). Further results on the effect of chlorination of POA on uptake by fronds of *Lemma minor* L. support this absorption theory (Kenney-Wallace and Blackman, 1972).

In 1949 it was concluded that the amine salt of 2,4,5-T in a water–oil emulsion was absorbed and translocated more consistently than other herbicides in honey mesquite, including the amine or esters of 2,4-D (Fisher and Young, 1948; 1949; Young and Fisher, 1949a, b; 1950). Plant response was measured by the percentage of tissue killed, and may not represent actual leaf absorption. The herbicide 2,4,5-T is generally more effective on woody plants than 2,4-D. Some exceptions exist but may not be related to absorption efficiency.

In contrast, Norris and Freed (1964; 1966a) found that 2,4-D is absorbed more readily by big leaf maple (*Acer macrophyllum*) than 2,4,5-T as the 2-ethyl hexyl esters. They also found that diclorprop and silvex (propionic acids) as the 2-ethyl hexyl esters were absorbed more readily than 2,4-D or 2,4,5-T esters.

Davis et al. (1968a) studied the uptake of picloram and 2,4,5-T in leaves of 10 woody species, including honey mesquite, and found that in most species, picloram entered faster and accumulated at higher concentrations than 2,4,5-T 14 and 48 hr after application. In other studies, honey mesquite leaves also absorbed picloram more rapidly and extensively than 2,4,5-T, but moisture stress reduced foliar uptake of picloram, whereas absorption of 2,4,5-T was unaffected (Davis et al., 1968b). In winged elm moisture stress did not affect absorption of picloram or 2,4,5-T. Bovey and Mayeux (1980) found higher concentrations of clopyralid than 2,4,5-T, triclopyr, or picloram in honey mesquite stems and roots 3, 10, and 30 days after application to the soil or to foliage in the greenhouse.

More clopyralid than 2,4,5-T, triclopyr, and picloram was usually detected in upper and basal stem phloem and xylem in field-grown honey mesquite (Bovey et al., 1986; Meyer and Bovey, 1986). Gas chromatographic analysis of field-grown honey mesquite indicated that more than twice as much clopyralid was absorbed by leaves than picloram by 4 hr after treatment (Bovey et al., 1988a). After 1, 3, and 8 days, more than three times as much clopyralid was transported to the upper stem phloem than picloram.

The leaves of spiny asters (*Aster spinosus* Benth.) absorbed less 2,4-D and picloram than sunflowers at 2, 4, and 6 hr after exposure (Mayeux et al., 1979b). Picloram concentration was usually less than 2, 4-D except after 4 hr of exposure in spiny aster leaves. Foliar absorption may be a limiting factor in spring after response to herbicides. The potassium salt of 2,4-D and picloram were used. The absorption of 2,4-D and picloram by undisturbed stem tips compared to regrowth stem tips was greater at 2 hr after treatment, but was usually no different at 4 and 6 hr after treatment (Mayeux et al., 1979b). The main stems absorbed less herbicide at 2 and 4 hr than undisturbed stem tips but were no different at 6 hr after treatment.

When 2,4-D and picloram absorption was monitored in spiny asters from field applications of 2,4-D or picloram, peak concentrations occurred in the leaves at 24 to 48 hr after treatment and declined up to 120 hr after treatment (Mayeux et al., 1979b). Peak concentrations from the broadcast of applications of 2,4-D at 2.24 kg/ha and 1.12 kg/ha of picloram resulted in about 15 to 16 μg/g of herbicide per gram fresh weight of spiny aster leaves. There were usually no significant differences between herbicide concentrations in shredded and non-shredded terminal stems. No 2,4-D or picloram was detected in the rhizomes of spiny asters.

In tissue culture, the greatest absorption of picloram and dicamba in soybeans and cottonwood (*Populus deltoides* Marsh.) occurred from agar the first 24 hr after treatment (Diaz-Colon et al., 1972). Absorption remained nearly static thereafter for 14 days. More dicamba was absorbed by soybean and cottonwood tissue cultures than either picloram or 2,4,5-T.

More picloram than triclopyr was absorbed by greenhouse-grown huisache leaves by 3, 10, and 30 days following broadcast sprays applied at 1.12 kg/ha (Bovey et al., 1979b). The potassium salt of picloram and the triethyamine salt of triclopyr was used.

Detached leaves of Drummond's goldenweed [*Isocoma drummonlii* (T. & G.) Greene.] absorbed the potassium salt of 2,4-D from aqueous solutions more slowly than did sunflower, and both species absorbed less of the potassium salt of picloram than 2,4-D (Mayeux and Scifres, 1980). There was no difference in the absorption of herbicides by leaves 6 hr after exposure to solutions containing 0.5% (v/v) surfactant. Attached Drummond's goldenweed leaves absorbed about 50% of the available 2,4-D (diethylamine) and 25% of the available picloram (potassium salt) within 5 days after spraying in the field during July or November.

b. Herbicide Formulation Morton et al. (1968) found larger amounts of 2,4,5-T in honey mesquite leaves treated with the butoxyethyl esters of 2,4,5-T than with the ammonium salts. The concentration of 2,4,5-T translocated to the stems, however, was similar. Most data suggest that ester formulations of the phenoxy herbicides penetrate leaf surfaces more readily than amine salts (Hull, 1970). This may or may not result in greater accumulation of herbicide in the roots, since the esters are not translocated as readily as the amine salt formulations (Bovey and Young, 1980; Hull, 1956). Norris and Freed (1964; 1966a) showed that the 2-ethyl hexyl ester of 2,4-D and 2,4,5-T were absorbed by big leaf maple seedlings to a much larger extent than the acid or triethanolamine form. Harvey (1990), however, found that the uptake of the amine salt was greater than the acid or ester form either as foliar or basal bark treatments in groundsel bush (*Baccharis halimifolia*).

The leaves of greenhouse-grown honey mesquite rapidly absorbed triclopyr from both ester and amine formulations (Bovey et al., 1983). After 4 hr about 23% of both formulations were absorbed from an application of commercial-grade triclopyr. By 24 hr after treatment more triclopyr acid was recovered from the application of the amine versus the ester. A large proportion of the herbicide was absorbed by the treated leaf, especially the amine formulation, but not all the triclopyr was recovered after transport because of possible degradation and lack of analysis of the ester fraction. The triethylamine salt or the ethylene glycol butyl ether ester of triclopyr were applied at 54 μg and 62 μg, respectively, to one leaf/plant/pot, on five plants/replication, with four replications.

In another study (Bovey et al., 1983), the analytical grade of the triethylamine salt and the *n*-butoxy ethyl ester of triclopyr were applied to greenhouse-grown honey mesquite leaves at 65 μg and 62 μg, respectively, with five plants/replication and three replications in a randomized complete design. After 0, 4, and 24 hr, the absorption of both forms was about equal when both the acid and ester were analyzed and added together for the ester. Most of the ester was rapidly hydrolyzed to the acid by 4 and 24 hr after treatment.

In field-grown honey mesquite (Bovey et al., 1983), more triclopyr acid (131 μg) was detected in the leaves after application of the amine than the ester (39 μg) at 0 hr, but by 4 hr (105 vs. 110 μg) or 24 hr (54 vs. 58 μg) no differences in triclopyr formulations were apparent. The loss of absorbed triclopyr from leaves was largely due to transport to the stem.

An application of the monoethanolamine salt and the 2-ethylhexyl ester of clopyralid to greenhouse-grown honey mesquite leaves with a pipet indicated that about twice as much clopyralid was absorbed within 15 min from the ester form (26%) as from the amine form (12%) of the total recovered (Bovey et al., 1989). After 24 hours, however, absorption of the ester was less than of the amine. In another study (Bovey et al., 1990a), treated leaves of greenhouse-grown honey mesquite absorbed more clopyralid within 15 min (0 hr) after pipet application of the oleylamine salt compared to the monoethanolamine salt or the 1-decyl ester. After 24 hr, treated leaves absorbed and transported more clopyralid into the plant of the salt formulations compared to the 1-decyl ester.

c. ***Herbicide Mixtures*** In greenhouse-grown mesquite, Davis et al. (1968b) found that the uptake and transport of 2,4,5-T decreased in the presence of picloram, but the uptake and transport of picloram increased in the presence of 2,4,5-T. Increasing ratios of 2,4,5-T: picloram up to 16:1 continued to increase uptake and transport of picloram. The inverse effect occurred for 2,4,5-T when picloram: 2,4,5-T ratios were increased.

Paraquat reduced the absorption and translocation of picloram in greenhouse-grown honey mesquite, huisache, and beans in 1:1 mixtures at 0.012 molar (M) each (Davis et al., 1968b). In the field, paraquat increased the uptake of picloram by yaupon but did not effect transport.

Turner (1972) showed that DEF (S,S,S-tributyl phosphorotrithioate) increased picloram, 2,4,5-T, and mecoprop salts efficacy on four woody species, privet (*Ligustrum ovalifolium*), poplar (*populus gelrica*), bluegum (*Eucalyptus globulus*), and guava (*Psidum guajava*). Mixtures of DEF with the esters of these herbicides are not synergestic and are often antagonistic. The mode of action of DEF is not clear, but it facilitates entry of water-soluble growth-regulator herbicides into leaves with little or no effect on herbicide movement in the plant. Hinshalwood and Kirkwood (1988) found that 2,4-D increased absorption of ^{14}C-asulam in bracken ferns, but neither 2,4-D nor ethephon significantly affected total basipetal translocation. Ethephon decreased ^{14}C-asulam accumulation in the rhizomes.

Baur et al. (1971) studied the absorption of picloram and 2,4,5-T in detached live oak (*Quercus virginiana* Mill.) leaves immersed in aqueous solutions for up to 4 hr. Herbicide concentration ranged from 10^{-3} to 10^{-6} M; solutions were adjusted to pH 4, 6, 7, or 8. The absorption of picloram in the presence of equimolar concentrations of 2,4,5-T exceeded that for picloram alone. Picloram had no effect on 2,4,5-T absorption.

Baur et al. (1969b) found that recovery of 2,4,5-T as acid and ester was greater in live oak tissues treated with mixtures of picloram (ester or salt) than in tissue treated with 2,4,5-T as the 2-ethylhexyl ester applied alone.

Bovey et al. (1988a) found that the addition of picloram or triclopyr to clopyralid at equal rates in field-grown mesquite (0.28 + 0.28 kg/ha) increased clopyralid concentrations in field-grown honey mesquite leaves by 4 hr after treatment compared to clopyralid applied alone.

d. Carriers and Adjuvants Fisher et al. (1956) evaluated a wide range of oils and oil–water emulsions as well as water as 2,4,5-T spray carriers for controlling honey mesquite. The 1:3 diesel fuel oil–water emulsion was considered equally effective and more economical to use than specially formulated oils. In some instances, the use of water alone as the carrier reduced the effectiveness of the 2,4,5-T applications. Hull (1956) indicated similar results on velvet mesquite. A nontoxic oil in a 1:4 oil–water emulsion as a carrier for 2,4,5-T resulted in considerably greater injury to the nontreated distal foliage than diesel oil as a carrier. Behrens (1957) found that when diesel fuel alone was used as the carrier on greenhouse-grown plants at spray volumes of 117 and 299 L/ha, the effectiveness was reduced, compared to 37 L/ha. The reduced effectiveness was attributed to the phytotoxicity of the diesel fuel, which caused rapid killing of the leaves, limiting 2,4,5-T translocation. Bovey et al. (1979a), however, found that no differences occurred in the canopy reduction of honey mesquite when water, diesel oil, or diesel oil:water carries at ratios of 1:3, 1:9, 1:18 were used with the 2-ethylhexyl ester of 2,4,5-T in 187 L/ha diluent. Scifres et al. (1973a) found that absorption of 2,4,5-T ester was more rapid in a paraffin-oil carrier than in diesel fuel, water, or emulsions of the oils in water carriers. No significant differences in the percentage of mesquite control have resulted from foam carriers (McCall et al., 1974; Scifres et al., 1974) compared with conventional sprays or the addition of sufactants to the spray solution (Bovey et al., 1979a). Most commercial herbicide formulations have sufficient surfactant and wetting properties for wetting plant surfaces, and the addition of more surfactant or emulsifier to the spray solution may have only a limited effect. Hull and Morton (1971) found that 2,4,5-T, picloram, or a 1:1 mixture of both were more effective on honey and velvet mesquite seedlings treated in dimethyl sulfoxide, ethylene glycol, phytobland oil, and water 50:25:15:10 by volume than in aqueous carriers.

The addition of surfactant I (trimethylnonylpolyethoxyethanol) or surfactant II (4-isoprophenyl-1-methyl-cyclohexane) at 1 and 0.6% (v/v) of the spray solution enhanced the phytotoxicity of the triethylamine salt of triclopyr, picloram, and the butoxyethanol ester of 2,4,5-T in greenhouse-grown but not field-grown honey mesquite (Bovey and Meyer, 1987). Adjuvants, however, did enhance clopyralid activity (Bovey and Meyer, 1987; Bovey et al., 1988a). Surfactants I and II at 0.5% by volume enhanced clopyralid absorption in field-grown honey mesquite leaves by 4 and 24 hr after the application of broadcast sprays with 0.28 kg/ha of clopyralid (Bovey et al., 1988a). The enhanced uptake of clopyralid with surfactants may be partially due to enhanced spray retention of clopyralid on honey mesquite leaves due to the presence of surfactant (Bovey et al., 1987; Bovey and Meyer, 1987).

Organosilicone surfactants Sylgard 309 [2-(3-hydroxypropyl)-heptamethyl-trisiloxane, ethoxylated, acetate EO glycol,-allyl,-acetate] or Silwet L-77 (polyalkyleneoxide modified polymethylsiloxane copolymers) added to the spray solution at 0.1, 0.25, or 0.5% by volume did not increase spray deposition, absorption, translocation, or phytotoxicity of clopyralid in greenhouse-grown honey mesquite (Bovey et al., 1994). Mayeux and Scifres (1980) found that 0.5% surfactant by volume in the spray solution did not enhance the absorption of 2,4-D or picloram by detached Drummond's goldenweed leaves. Mayeux and Johnson (1989) indicated that the addition of surfactant had little effect on the absorption of picloram in Lindheimer prickly pear (*Opuntia lindheimeri* Engelm.).

Meyer et al. (1972a) found that field-grown honey mesquite leaves were functional in herbicide uptake for about 4 days after application. The maximum absorption apparently occurred the day of spraying, however thus any agent or force that causes leaf removal too quickly after spraying reduces control. In most cases it is important to use a carrier that will penetrate the waxy surface of the leaf but will not kill the leaves or cause abscission soon after spraying (Scifres et al., 1973b).

Hall (1973) found that adding surfactants to the iso-octyl ester of 2,4,5-T and the dimethylamine of dicamba generally did not increase the absorption or translocation of the herbicides when they were applied to the foliage of five hardwood species and loblolly pine. Adjusting the pH of the formulations to 3, 6, or 9 with a buffering solution had small and inconsistent effects upon absorption and translocation.

Sharma and Vanden Born (1970) found that Atlox 210 (a nonionic blended surfactant containing polysorbate, mono- and diglycerides, butylated hydroxyonisole, butylated hydroxytoluene, and propylene glycol in a water–isopropanol medium at 1% by vol.) and high relative humidity enhanced the penetration of the potassium salt of picloram and the dimethylamine of 2,4-D into detached aspen poplar leaves. The surfactant was more effective with the amine formulations than for the ethyl or butoxyethanol ester of 2,4-D. Abaxial leaf surface penetration was greater than adaxial (upper surface). More herbicide penetrated in spring versus late summer and fall applications. Although balsam poplar is more resistant than aspen poplar to picloram, more herbicide was absorbed by balsam poplar.

Sands and Bachelard (1973a) found that surfactants markedly increased the uptake of picloram by *Eucalyptus viminalis* and *E. polyanthemos* leaf discs. The surfactant effect depended largely on the degree of wetting by the applied solution on the leaf surface. Sands and Bachelard (1973b) concluded that stomata played a major role in herbicide uptake in eucalpytus leaf discs.

Tan and Crabtree (1994) studied the impacts of pH and sorption-desorption of 'Pegosperse' 100-O (PEG 100-O: diethylene glycol mono-oleate, containing 15% diester) surfactant by apple (*Malus pumila* M.) leaf cuticles on surfactant-enhanced cuticular penetration of 2,4-D. Glass cylinders were affixed to enzymatically isolated adaxial apple leaf cuticles after the cuticle segments had been soaked in 10 ml/L PEG 100-O solution and washed for 20 and 120 min, respectively. Quantities of [^{14}C] 2,4-D in the glass-cuticle chambers passing through the cuticles at pH values from 1 to 6.5 were determined. PEG 100-O significantly increased cuticular penetration of dissociated 2,4-D at pH 4.5; the surfactant had no effect on the penetration of undissociated 2,4-D at pH 1.0. Surfactant-enhanced penetration of 2,4-D occurred only when the surfactant was in the cuticles, while the process of surfactant sorption-desorption alone had no effect on penetration. These results support a "hydrophilic channel" hypothesis that surfactants may create hydrophilic channels or increase the area of the channels in the cuticle, and consequently enhance the passing of polar molecules such as dissociated 2,4-D through the cuticle.

Brady (1970a) found that the addition of ammonium nitrate greatly increased the absorption of the isooctyl ester of 2,4,5-T by sweetgum (*Liquidambar styraciflua*), post oak (*Quercus stellata*), red maple (*Acer rubrum*), and loblolly pine (*Pinus taeda*) leaves. Phosphoric acid alone or mixed with ammonium nitrate caused less increase.

Wilson and Nishimoto (1975a) discovered that ammonium sulfate increased picloram activity on seedling guava (*Psiduim guajava*), strawberry guava (*Psiduim cattleianum*), and dwarf beans (*Phaselus vulgaris*) when added to the spray solution. The absorption of ^{14}C-picloram by strawberry guava leaves was increased about fivefold by 0.5% and 10% ammonium sulfate and increased ^{14}C-picloram about four times in the upper stem and attached leaves compared to picloram alone. Wilson and Nishimoto (1975b) found the ammonium sulfate was the only one of six sulfate salts that increased picloram absorption in detached leaves of strawberry guava.

e. Spray Characteristics Spray droplet size affects phytotoxicity, depending upon the species studied. In some species, herbicidal efficiency decreases as droplet size increases above 500 micrometers in diameter (Hull, 1970). Behrens (1957) reported that droplet size, spray volume, and herbicide concentration had no direct influence other than minor effects on the response of honey mesquite or cotton to 2,4,5-T, but that droplet spacing of 465 droplets per cm^2 was considered the maximum spacing that would maintain a high level of herbicidal effectiveness.

The addition of surfactant WK (trimethynonylpolyethoxyethanol) at 0.5% by volume of spray solution caused an increased uptake of clopyralid by the upper canopy of greenhouse-grown honey mesquite (Bovey et al., 1987). Enhanced uptake after 24 hr was probably a result of a twofold increase in the deposit of clopyralid on the plant (Bovey et al., 1987; 1991). A greater deposit of clopyralid on plant surfaces after the addition of surfactant was associated with a reduced liquid surface tension and a greater percentage of spray volume in small droplets (<204 μm diam.). The addition of surfactant WK at 0.5% by volume of spray solution caused a twofold increase in the deposition of the monoethanolamine salt but not the oleylamine salt of clopyralid (Bovey et al., 1991). There were no differences in spray deposit between spray droplet size spectrums of 160 or 330 μm Dv.5 or spray solution application of 47 or 187 L/ha.

An air-assist spray nozzle at 9.4 L/ha by volume resulted in greater initial clopyralid deposit and detection in the upper canopy of greenhouse-grown honey mesquite than application by a conventional hydraulic nozzle at 9.4 or 187 L/ha (Bovey et al., 1994). Air-assist application did not increase phytotoxicity compared to hydraulic nozzles. In the field, honey mesquite mortality and canopy reduction 16 months after aerial applications of clopyralid were significantly less in the 624 μm droplet treatment in two of four experiments when compared to plots treated with smaller droplet sizes (325 and 475 μm Dv.5) (Whisenant et al., 1993). Mortality increased with larger spray volumes (19, 37, and 75 L/ha), particularly with 625 μm droplets. Mortality data showed that larger droplet sizes require larger spray volumes for greatest efficiency.

Hand-carried spray equipment with 800067,8001 and 8015 Spraying Systems flat fan nozzles were used to obtain 37,187, and 935 L/ha in a diesel oil–water 1:3 (v/v) ratio (Bovey et al., 1979a). Diluent of 187 L/ha produced greater canopy reduction than 37 or 935 L/ha in honey mesquite, winged elm, and Macartney rose (*Rosa bracteata* Wendl.) using the ester form of 2,4,5-T at 0.56 and 2.2 kg/ha on honey mesquite and winged elm and 2.2 kg/ha of 2,4-D ester on Macartney rose. When 1.1 kg/ha of (4-chloryl-2-methylphenoxy)acetic acid (MCPA) was used on white brush there were no differences in 37,187, or 935 L/ha diluent, but 187 L/ha of diluent was superior to other spray volumes on live oak using 2.2 kg/ha of 2,4,5-T. Diluent 37 L/ha may have resulted in insufficient coverage for maximum herbicide uptake for most species using ground equipment, whereas 935 L/ha may have resulted in the loss of the herbicide from plant surfaces in excessive runoff, except with live oak.

f. Concentration of Applied Herbicide It is commonly thought that there are optimum concentrations for maximum effectiveness of translocated herbicides. Above these concentrations it is thought that the effectiveness of the herbicide is reduced by contact injury to the conducting tissues, which results in reduced amounts of material being transported out of the leaf. The effect of concentration of 2,4,5-T on uptake has been measured by Davis et al. (1968a) in an experiment on detached leaves of live oak (*Quercus virgininana*). They found that the quantity of herbicide absorbed increased as the concentration was increased. Morton (1966) similarly demonstrated a linear relationship between the amount of 2,4,5-T applied to the leaves of honey mesquite seedlings and the quantity absorbed. The amount applied ranged from 5 to 100 μg/leaf, and absorption was from 2.5 to 52.7 μg/leaf over 72 hr. Badiei et al. (1966) also found that the amount of 2,4,5-T butoxyethanol ester absorbed by blackjack oak leaves (*Quercus marilandica*) increased as the amount applied increased, although the percentage absorbed decreased.

Swanson and Baur (1969) found that picloram absorption by discs of potato (*Solonum tuberosum*) tuber tissue immersed in buffered (pH 4.0 to 8.0) solutions increased with concentration (5×10^{-4} M to 5×10^{-3} M) and with time up to 24 hr. Both absorption and leakage were temperature-dependent. Wilson and Nishimoto (1975b) found that the magnitude of ammonium sulfate-induced increase in detached strawberry guava leaves in ^{14}C-picloram absorption was not affected by picloram concentration in the range of 250 to 2000 ppm or by leaf age.

Richardson and Grant (1977) found that more 2,4,5-T was absorbed by blackberry leaves at high concentrations that at low concentrations, although the percentage absorbed in 24 hr decreased from 13% to 6% as the concentration was increased.

g. pH Effects Discs of potato tuber tissue were immersed in buffered (pH 4.0 to 8.0) solutions of picloram (5×10^{-4} M to 5×10^{-3} M) for 1 to 36 hr (Swanson and Baur, 1969). The uptake of picloram during incubation and the leakage after return of the discs to untreated buffer were determined by gas chromatographic analysis of extracts of the tissue and ambient buffer. Picloram absorption increased with concentration and with time up to 24 hr. The maximum uptake occurred at pH 4.0, and very little picloram was absorbed at pH 7.0 and 8.0. Both absorption and leakage were temperature-dependent. The rate and extent of leakage was greatest at the highest concentration. Typically, more than 90% of the picloram absorbed from 5×10^{-3} M was lost to fresh buffer within 12 hr. Sterling and Lownds (1992) also found the most favorable absorption of picloram by broom snakeweed (*Gutierrizia sarothrae*) in solution at pH 4.

Baur et al. (1971) measured the absorption from a solution of the potassium salt of 2,4,5-T by detached leaves of live oak. They found that absorption was highest from solutions of pH 4, but not significantly different at pH 6, 7, or 8. In further experiments Baur et al. (1974) measured the penetration of 2,4,5-T acid into leaves of honey mesquite from solutions of pH 3.5 to 9.5. The 2,4,5-T was applied by immersing the leaves in solution, applying droplets under high humidity so they would not evaporate, or applying droplets at ambient conditions so they would evaporate within 45 min. They found that uptake of 2,4,5-T was favored when leaves were immersed in a solution of pH 3.5 and when leaves were treated with droplets of the same solution and then maintained in a moist condition. Leaves on which the droplets were allowed to dry only showed some enhancement of uptake at pH 3.5 up till 60 min. There was reduced uptake at pH 9.5 compared with lower pH values under the latter conditions. The quantity of herbicide absorbed was highest for leaves under the high humidity treatment at pH 3.5 to 5.5.

Mayeux and Johnson (1989) found that absorption decreased with increased pH of the picloram solution in detached prickly pear pads, indicating that picloram diffused through the cuticle as the undissociated molecule. Under laboratory conditions, weak acids penetrate best at low pH values where the molecules are largely in the undissociated form (Crafts, 1961a; 1964). In this state, they more readily penetrate the lipoidal phases of the cuticle and leaf cells. Under field conditions, however, little benefit of improved control has been shown by adjusting the pH of the spray solution.

h. Temperature Absorption of foliar-applied herbicides generally increases with increased temperature. This has been confirmed by Sharma and Vanden Born (1970), who demonstrated increased absorption of the dimethylamine of 2,4-D by detached leaves of aspen poplar when the temperature was increased. Morton (1966) described experiments in which significantly greater amounts of 2,4,5-T acid were absorbed by leaves of mesquite seedlings with increasing temperature. Differences between 21°C and 29°C were not detected, but increased absorption did occur at 38°C. The uptake of 2,4,5-T isooctyl ester by six hardwood species and loblolly pine was examined by Brady (1970b). He found that all species absorbed more 2,4,5-T at 35°C than at 13°C, although for some species there was no difference between 13°C and 24°C, and for water oak (*Quercus nigra* L.) there was no difference in absorption between 24°C and 35°C. Wills and Basler (1971) found some correlation between the uptake of 2,4,5-T butoxyethanol ester in winged elm and temperature, although they did not eliminate possible interference from other environmental factors. Richardson (1975) found that the rate of absorption of 2,4,5-T butylester by detached leaves of blackberry (*Rubus procerus* P.J. Muell.) increased when the ambient temperature was increased from 18°C to 30°C.

In their review on the mechanisms of foliar penetration, Robertson and Kirkwood (1969) discussed the possible reasons for the temperature effect on foliar absorption. The mechanisms for increased penetration of NAA (naphthaleneacetic acid) through isolated pear leaf cuticles (*Pyrus communis* L. cv. Bartlett) have been discussed by Norris and Bukovac (1969) in relation to the lipoidal nature of cuticular membranes. These authors have suggested that temperature influences the rate of penetration of lipophilic molecules through the cuticle by altering the physical state of the cuticle (e.g., the viscosity of the fatty molecules). The effect of temperature on cellular metabolism would probably also have some effect on absorption.

*i. **Humidity*** Conditions of high humidity favor stomatal opening and prolong drying time of spray deposits. They may also influence the degree of hydration of the cuticle. Foliar penetration under conditions of high humidity thus favors increased herbicide absorption.

Using detached leaves of aspen poplar (*Populus tremuloides*) Sharma and Vanden Born (1970) found that uptake of 2,4-D as the dimethylamine was doubled under a high humidity level when compared with a low humidity level. Repeated wetting of the treated spots under low humidity resulted in increased uptake, but still not to the extent of that under high humidity. This indicates that the lower rate of drying under high humidity is only partly responsible for the increased penetration of the herbicide. Presumably, as the treated adaxial surface was astomatous, hydration of the cuticle played some part in increasing absorption.

In experiments examining the uptake of 2,4,5-T acid by honey mesquite, however, Baur et al. (1974) found that rewetting the treated spots did not stimulate any further uptake under conditions of low humidity. Sharma and Vanden Born (1970) claimed that penetration of 2,4,-D ethyl ester was not affected by humidity. In their experiments the droplets of 2,4-D ester were covered to minimize volatility losses, however which would have resulted in a saturated atmosphere around each droplet. It is therefore possible that a very localized humidity effect may have influenced uptake.

Pallas (1960) has demonstrated consistently greater absorption and translocation of 2,4-D triethanolamine salt at humidities from 70% to 74% than at humidities between 34% to 48%. This corresponded with increased stomatal opening. The drying time of the applied droplets was similar and was not thought to cause the differences observed. The author did not distinguish between absorption and translocation. In experiments on wolftail (*Carex cherokeenis* Schwein) Burns et al. (1969) found enhanced absorption of the sodium salt of 2,4-D at high humidity. The increased uptake may have been caused partly by the temperature of the high humidity treatment being higher than that of the low humidity treatment in this experiment.

The results of Morton (1966) are in disagreement with the above. In experiments on mesquite seedlings Morton measured the foliar penetration of 2,4,5-T acid at humidities ranging from 35% to 100% and found no response. He considered this due to the relatively xerophytic nature of mesquite, which may have enabled it to adapt to changes in humidity more readily than some other plants. Later work by Baur et al. (1974), however, demonstrated that 2,4,5-T acid was absorbed to a greater extent by leaves of mesquite under conditions of high humidity (saturated atmosphere) than by leaves under a low humidity treatment (60% ± 5% RH).

Foliar absorption of picloram was investigated in broom snakeweed, a rangeland shrub (Sterling and Lownds, 1992). Absorption was dependent on the relative humidity and temperature, with the greatest uptake at 94% relative humidity and 35°C, respectively. Absorption was also greatest at pH 4 and least at pH 8. These data suggest that picloram is absorbed via simple diffusion, and absorption is dependent on favorable environment and solution pH.

j. Moisture Stress Merkle and Davis (1967) showed that foliar absorption of 2,4,5-T and picloram in beans (var. Black Valentine) was unaffected by extreme moisture stress. Moisture stress reduced foliar uptake of picloram in honey mesquite but not in winged elm (Davis et al., 1968c). Moisture stress did not affect absorption of 2,4,5-T in honey mesquite or winged elm. Wills and Basler (1971) also measured the absorption of 2,4,5-T acid by winged elm seedlings and found no significant differences over a range of moisture stress levels. In contrast, the absorption of 2,4,5-T butoxyethanol ester by blackjack oak (*Quercus marilandica* Muenchh.) was significantly reduced by increasing moisture stress, although the reduction in absorption was small (Badiei et al., 1966).

Bovey and Clouser (1998) found in preliminary studies that water stress (−1.3 to −2.8 MPa) did not affect the absorption and translocation of clopyralid in greenhouse-grown honey mesquite 4 or 24 hr after herbicide treatment. The addition of triclopyr (synergistic to clopyralid) increased clopralid uptake at low-water stress (−1.3 MPa), but at high-water stress (−2.8 Mpa) triclopyr decreased the clopyralid uptake.

In two separate studies in Canada, triclopyr absorption was not significantly affected in water-stressed water oak and southern red oak (*Q falcata* Michx.) seedlings (Seiler et al., 1993) or in red maple (*Acer rubrum* L.) seedlings (Bollig et al., 1995). Triclopyr translocation was reduced in both studies, however.

k. Light Light assists herbicidal penetration by stimulating stomatal opening in most species (National Academy of Science, 1968). Measurement of herbicide absorption by honey mesquite as influenced by quality and intensity of light has not been determined. Brady (1969), however, found that the absorption of the isooctyl ester of 2,4,5-T increased as light intensity increased up to 2680 foot candles, but it decreased thereafter in post oak (*Quercus stellata* Wangenh.) and water oak. Absorption of 2,4,5-T increased as light intensity increased up to 4000 foot candles in longleaf pine (*Pinus palustris* Mill.) and American holly (*Ilex opaca* Ait.). Davis et al. (1968a) found that the uptake of picloram by live oak leaves decreased as light intensity increased.

Scifres et al. (1973c) found that honey mesquite seedlings that developed under shade were more easily killed by 2,4,5-T sprays than seedlings grown in sunlight. The increased effectiveness under shade may have been due to limited cuticle development (Hull et al., 1975), and hence greater herbicide uptake. Baur and Swanson (1968) found that honey mesquite grown during short days was more susceptible to 2,4,5-T or picloram than that grown during long days. The reason for this difference is not clear, but may be related to cuticular development.

l. ***Rainfall*** Bovey and Diaz-Colon (1969) found that oil-soluble formulations (esters) of 2,4-D, 2,4,5-T, and picloram were less affected by artificial rainfall than water-soluble herbicides, such as paraquat and cacodylic acid on guavas (*Psidium guajava* L.) and mangoes (*Mangifera indica* L.) The oil-soluble phenoxy herbicides usually retained their effectiveness even when the leaves were washed within 15 min after treatment. Field-grown honey mesquite leaves showed complete leaf necrosis even when leaves were washed 20 min after treatment with paraquat, indicating rapid absorption (Bovey and Davis, 1967). Winged elm and live oak showed little injury under the same conditions, but both showed about 40% leaf necrosis when washed 60 min after treatment.

In North Carolina, the shoots of naturally established turkey oak (*Quercus laevis* Walt.) and red maple (*Acer rubrum* L.) were treated with aqueous dilutions of three herbicidal products at three rates each (Upchurch et al., 1969). Simulated rainfall was applied at 1.3 or 2.5 cm at 5, 15, 60, or 120 min after herbicidal application. Responses measured 10 and 13 months after herbicidal application were the percentage control of original shoots, percentage control of new shoots, shoot height, and number of live stems per plant. Neither the ester nor amine derivatives of 2,4,5-T were reduced as to their action on woody plants by the simulated rainfall applied. The action on woody plants of picloram plus 2,4-D was markedly reduced by the application of all of the simulated rain treatments at any of the intervals at which they were applied following herbicidal treatment.

The foliar activity of the amine salts of glyphosate, dicamba, picloram, clopyralid, and triclopyr was decreased on greenhouse-grown huisache when simulated rainfall was applied up to 240 min after herbicide treatment (Bovey et al., 1990b). The effectiveness of the butoxyethyl ester of triclopyr or 2,4,5-T was not reduced by rainfall washoff within 15 min after application. In natural huisache stands, injury from triclopyr ester or amine salts of picloram or clopyralid was not reduced by simulated rainfall at 60 min after herbicide treatment. In the greenhouse and field, honey mesquite leaves rapidly absorbed most herbicides, and triclopyr, 2,4,5-T, picloram, and clopyralid were highly phytotoxic even when simulated rainfall was applied within 15 min after herbicide treatment (Bovey et al., 1990b).

m. ***Time of Day*** No absorption data are available for huisache or Macartney rose relative to time of day or season of application. The data suggested, however, that 1 : 1 ratios of picloram plus 2,4,5-T as the triethylamine salts were more effective when applied to these woody plants in the evening (6:00 P.M.)

than in the morning (6:00 A.M.) or at midday (1:30 P.M.) (Bovey et al., 1977). The best control of huisache was in June, while the best control of Macartney rose occurred in September and October. The poorest control occurred when the internal water stress was highest.

n. ***Leaf Structure and Development*** As indicated earlier, the best time for application of foliar herbicides for honey mesquite control is during a 50- to 90-day period after the first leaves emerge in the spring. By May 20, leaflets of honey mesquite in Brazos County have usually attained full maturity (Meyer et al., 1971). The upper cuticle is usually 5 to 8 microns thick, and the lower cuticle is usually 2 microns thick. Penetration of cuticle by herbicides appears sufficient for herbicidal effect and translocation to other parts of the plant, however. In most plants there is a relationship between cuticular development and both the composition and foliar absorption of herbicides (Hull et al., 1975). The more mature the leaf, the greater the cuticular development, and that may partially explain the resistance of honey mesquite to herbicide sprays applied late in the growing season, even though limited stomatal penetration can occur when the cuticules become very thick (Hull et al., 1975). Mayeux and Jordan (1984) found that the amount of epicuticular wax per unit leaf area on honey mesquite leaves was least in April and May, increased until July, and remained stable thereafter at several locations in Texas. Population means ranged from around 4 mg dm^{-2} in east-central Texas during April and May to a maximum of over 10 mg dm^{-2} on leaves of trees growing in north-central Texas in October. Honey mesquite growing in arid west Texas and semiarid south Texas had no more wax per unit leaf weight or area than those in humid east-central Texas. Mayeux and Wilkinson (1990) also determined the chemical composition of epicuticular wax on honey mesquite. Jacoby et al. (1990) studied wax on the leaves of honey mesquite in northeastern Texas. They found trends similar to Mayeux and Jordan (1984) in wax accumulation on honey mesquite leaves, but found considerable variation in wax accumulation among individual trees. Jacoby et al. (1990) further stated that increasing amounts of epicuticular wax on the leaves of honey mesquite during early summer may contribute to increasing resistance to foliar-applied herbicides.

Wax thickness on honey mesquite may act as an herbicide barrier, but the data indicate that most herbicides used are rapidly absorbed and transported by honey mesquite leaves (Bovey et al. 1983; 1986; 1987; 1988a,b; Hull, 1956; Morton, 1966; Scifres et al., 1973a). Bovey et al. (1986) found concentrations of 2,4,5-T, triclopyr and picloram in upper stem phloem and xylem 3 days after treatment to be greater in early May or late April than in late May, June, July, August, or September, but the differences were not always significant. Control with 2,4,5-T, picloram, and triclopyr declined in July, August, and September (Meyer and Bovey, 1986), but is related to anatomical and physiological changes in honey mesquite plants instead of leafwax accumulation (Meyer et al., 1971). Meyer and Bovey (1986) and Jacoby et al. (1991) reported successful late-season application with clopyralid on honey mesquite. The data on clopyralid concentrations in honey mesquite in upper stem phloem and xylem 3 days after treatment over a 2-year period showed no differences among dates of sampling (Bovey et al., 1986). Higher than normal rates were used (1.1 kg/ha versus 0.28 or 0.56 kg/ha). Mayeux et al. (1979b) indicated that poor herbicide uptake by spiny aster top growth was not adequately explained by cuticle thickness, and that the chemical characteristics of the epidermal covering may be more responsible for high resistance to herbicide penetration than cuticle thickness. Large quantities of viscous, noncrystilline epicuticular waxes were observed on the leaves of three goldenweed (*Isocoma*) species (Mayeux et al., 1981). Leaves of field-grown plants of the least herbicide-suspectable species, common goldenweed [*I. coronopifolia* (Gray) Greene.], increased from 71 mg dm^{-2} in March to 286 mg dm^{-2} in October. This was two to four times greater than the amount present on other goldenweed species or Drummon's goldenweed [*I. drummondii* (T. & G.) Greene.] and jimmyweed [*I. Wrightii* (Gray) Rydb.]. Greenhouse-grown plants produced quantitics similar to field-grown plants, but maximum production occurred during the summer. Wilkinson and Mayeux (1990) determined that the chemical composition of the epicuticular wax of *I. coronopifolia* and *I. drummondii* was 85% to 95% free fatty acid and alcohols. Alkane (<5%), ester (<2%), and ketone (<3%) concentrations were low. The short-chain, free fatty acids and alcohols suggested that they are hydrophilic compared to other plants and helps explain both the observed loss of epicuticular waxes in rainfall (Mayeux and Jordan, 1981; 1987) and the variation in responses of these weedy shrubs to herbicide sprays (Mayeux et al., 1979a; Mayeux and Jordan, 1980; Mayeux and Scifres; 1981). The control of *I. coronopifolia* and *I. drummondii* with broadcast sprays of translocated herbicides strongly depends upon substantial rainfall prior to treatment (Mayeux and Jordan, 1981). The removal of leaf waxes increased picloram accumulation in detached leaves of both species by a factor of eight, demonstrating that these deposits effectively reduce herbicide entry (Mayeux and Jordan, 1980). Mayeux and Johnson (1989) found that removing the epicuticular wax from mature pads (cladophylls) of Lindheimer prickly pear cactus increased picloram absorption by four- to sixfold, while the addition of surfactant had little effect on absorption. Picloram entered detached pads at the areoles more readily than through the surrounding cuticle. New pads absorbed more picloram than the old pads. Most of the picloram remained in the waxy surface of old and new pads. About 2% of the applied picloram was recovered from within the epicuticular wax after 30 days. Little picloram was absorbed by the roots. Wilkinson and Mayeux (1990) determined that 97% of the total epicuticular wax of *O. engelmanii* was alkanes and esters on greenhouse-grown cladoyphylls. The most prevalent carbon number of alkanes from buds were C_{15} through C_{25}, while those of fully expanded cladophylls were C_{29} through C_{35}. Both odd- and even-numbered ester components were present.

The abaxial surface of the leaf usually absorbed more herbicide than the adaxial (upper) surface using picloram or 2,4,5-T on 10 species of woody plants (Davis et al., 1968a). Leaf surfaces vary from abaxial to adaxial on the same species and between species. Meyer and Meola (1978) have provided data on the leaf and stem surfaces of many Texas woody plants.

o. Other Seasonal Studies Dalrymple and Basler (1963) have found that the absorption of 2,4,5-T butoxyethanol ester by blackjack oak was the highest in early spring, decreased to a minimum in June (early summer), then increased again until August (late summer). The respiration rate of the tissue was similar for the absorption of the butoxyethanol ester of 2,4,5-T. Sharma and Vanden Born (1970) measured the effect of stage of growth on the penetration of 2,4-D dimethylamine and ethyl ester into leaves of aspen poplar from June to September. There were no differences in the absorption of 2,4-D ester, although 2,4-D amine had a period of maximum absorption in July. It was thought that the increased absorption in July may have been due to cracks and punctures in the cuticle, which would favor absorption of polar compounds by the aqueous route.

The effect of the spray date on the absorption of the propylene glycol butyl ether ester of 2,4,5-T was measured by Brady (1971). He found that sweetgum (*Liquidamber styraciflua* L.) absorbed the most 2,4,5-T in June and a minimum in May. There was markedly reduced absorption in July and September. A similar pattern was established for green ash (*Fraxinus pennsylvanica* Marsh.) and loblolly pine. Water oak had a maximum absorption in May, a slightly reduced absorption in June and July, and a greatly reduced absorption in September. The ambient temperature at the times of treatment was not noted in this experiment. Later work by Brady (1972) showed a similar pattern, in that more 2,4,5-T isooctyl ester was absorbed in May than in August by detached leaves of five out of six woody species. Similar results were obtained by Wills and Basler (1971) when they measured the absorption of the butoxyethanol ester of 2,4,5-T by winged elm. They noted, however, that there was some correlation between absorption and the temperature at the time of spraying. Absorption of the propylene glycol butyl ether ester of 2,4,5-T by creosote bush [*Larrea tridentata* (DC.) Cov.] was shown to be higher after May than earlier in the year (Schmutz, 1971). The leaf wiping and autoradiographic method used here to estimate absorption gave only qualitative results. Detached blackberry leaves treated with 2,4,5-T butyl ester at an ambient temperature of 26 to 28°C, had a higher rate of absorption if treated in the autumn than in the summer (Richardson, 1975).

p. Metabolism, Degradation, and Mode of Action Morton (1966) found that approximately 80% of the 2,4,5-T absorbed by the leaves of honey mesquite seedlings was metabolized after 24 hr. Metabolism was completely inhibited at 10°C, and a lower rate of metabolism was noted at 38°C than at 21°C and 29°C. Picloram, however, is more resistant to degradation in plants than 2,4,5-T (Bovey and Scifres, 1971).

The effects of picloram on protein synthesis in bean (var. Astro) hypocotyl and hook tissues were studied by (Baur and Bowman) (1972). Picloram (10^{-4}M) was shown to have a stimulatory effect on ^{14}C-1-DL-leucine uptake in hook but not in hypocotyl tissues. Maximum leucine incorporation and maximum total protein concentration occurred in hook tissues treated with 10^{-4} M picloram. Inhibition of protein synthesis with cycloheximide (CH) and erythromycin (ERY) indicated that endogenous and picloram-stimulated protein synthesis is a function of the 80S cytoplasmic ribosomes rather than 70S chloroplast or mitochondria ribosomes.

Gas chromatographic and radioisotopic analyses were made of cell wall, chloroplast mitochondria, and the remaining cytoplasm fractions of cowpea (*Vigna sinensis* Endl. var. Southern Blackeye) trifoliates acropetal to primary leaves treated with the growth regulator picloram (Baur and Bowman, 1973). Most of the picloram was recovered from the remaining cytoplasm. Concentrations in chloroplast and mitochondria were consistently low.

Pretreatment of potato (var. Russet) tuber discs in pH 5.5 buffer significantly reduced the uptake of picloram (10^{-3}M) (Baur and Bovey, 1970). Tissue pretreated in buffer at 7°C subsequently absorbed more picloram than tissue pretreated at 25°C. Inclusion of cetyl trimethyl ammonium bromide (CTAB, 2×10^{-4} M) in the treating solution caused a significant increase in picloram uptake in tissues that were not pretreated in buffer. The reduction in uptake caused by buffer pretreatment was effectively reversed when CTAB was included in the treating solution. The results suggest that picloram uptake by potato tissue is related to the availability of the quaternary ammonium binding sites provided by membrane phosphatides.

In nutrient agar comparative concentrations (10^{-3} to 10^{-5} M) of 2,4,5-T were generally more inhibitory to the growth of tissue cultures of soybeans (var. Acme) and cottonwood than were either picloram or dicamba (Diaz-Colon et al., 1972). Compared to untreated tissue, dicamba or picloram at 10^{-6} M in the nutrient agar resulted in a 200% increase in the growth of soybean tissue. At 10^{-5} and 10^{-6} M dicamba also produced an increase in the growth of cottonwood tissue. The greatest absorption of picloram and dicamba by tissue cultures from agar occurred during the first 24 hr after treatment. Absorption remained nearly static thereafter for 14 days, however. More dicamba was absorbed by soybean and cottonwood tissue cultures than either picloram or 2,4,5-T. In another study (Bovey et al., 1974), mixing equimolar solutions of 2,4,5-T + dicamba, 2,4,5-T + picloram, or picloram + dicamba at 10^{-4}, 10^{-6}, and 10^{-8} M did not increase phytotoxicity over that of the most phytotoxic herbicide of the pair.

Changes were studied in the concentration of picloram with time in roots, stems, and leaves of 20-day-old seedlings of huisache and honey mesquite (Baur and Bovey, 1969). Exposing the root systems to aqueous solutions of picloram (1.0 ppm huisache and 10.0 ppm honey mesquite) for 24 hr killed approximately 60% of the treated plants. In honey mesquite, picloram was redistributed and eventually lost over a 5-day period, whereas neither redistribution nor loss occurred in huisache.

Honey mesquite leaves absorbed high amounts of clopyralid as foliar sprays, as indicated by concentrations of 10 μg/g fresh weight or more in basal stem phloem by 4 days after treatment (Bovey et al., 1988b). Small quantities of clopyralid (<1 μg/g) were detected in basal stem phloem after spray applications of clopyralid to defoliated plants or roots of foliated plants treated by soil application. When applied to foliated plants, the 0.56 kg/ha of clopyralid killed 60% or more plants, but none was killed when clopyralid sprays were applied to defoliated plants or when 2.2 kg/ha of clopyralid was applied to the soil. Water, diesel oil plus water, or water plus surfactant were equally effective as clopyralid carriers as they were as foliar sprays.

2. Translocation

Once an herbicide is absorbed by leaves and stems a key factor in killing woody plants is translocation of the phytocide to the lower stem and roots. The phloem is the principal food-conducting tissue in vascular plants. Compounds such as 2,4-D are translocated through the phloem from regions of carbohydrate synthesis (leaves) to sugar-importing tissues, such as roots, buds, shoot tips, seeds and fruit, and other leaves. The direction of herbicide movement is determined by the patterns of food distribution and utilization within the plant, since translocation of food may also occur from roots to leaves or between other plant parts (Crafts, 1961a, b; 1964; Crafts and Yamaguchi, 1964). Ideally, at least for the phenoxy herbicides, it is best to apply foliar sprays when food transport is occurring from the leaves (basipetal) to other plant parts (roots) so as much herbicide as possible is translocated to the base of the stem and roots. In the case of honey mesquite and many woody plants, this occurs under springtime conditions after the foliage is mature enough to export sugars and the plant is rapidly growing radially. If the herbicide is applied at other times during the year the results may be unsatisfactory since assimilate (food) movement may be limited. For successful chemical control of woody plants, movement of phytotoxic materials to regenerative tissues (buds) may be necessary to eliminate their growth potential. The greatest concentration of buds on honey mesquite occurs on the trunk in the first 30 cm below the soil line (Fisher et al., 1946; 1956; 1959; Meyer et al., 1971).

a. Herbicide Configuration and Type Fisher and Young (1948) reported that sodium arsenite, sodium arsenate, sodium chlorate, ammonium sulfamate, sulfamic acid, ammonium thiocyanate, 2,4-D, and 2,4,5-T were the only chemicals out of several hundred tested that were absorbed by the foliage and translocated in sufficient amounts to kill dormant buds on the underground stem. Increasing chemical concentrations did not improve translocation. Young and Fisher (1949) reported that the ester of 2,4,5-T was a more effective treatment than several formulations of 2,4-D or other chemicals.

Rapid transport of 2,4,5-T was determined by plant response after spraying an exposed branch of honey mesquite and observing the effect on the leaves of sprayed versus shielded or protected plant parts (Young and Fisher, 1949a). Fisher and Young (1949) also used a tip-immersion method to test translocation by immersing the tip of a branch into a container of herbicide for a period of from 1 to 24 hr and observing translocation by plant response. These methods were very useful in determining the most effective transported herbicides but did not quantitively measure herbicide content in plant tissue.

Norris and Freed (1966a) have shown that molecular configuration can alter the amount of herbicide translocated to the root system in big leaf maple (*Acer macrophyllum*). They found that 2,4,5-T and dichlorprop were transported to a greater extent than 2,4-D and silvex. They also demonstrated (Norris and Freed, 1966b) that 2,4-DB is translocated to a much larger extent than the above chemicals even though it was less well absorbed. It is obvious that the molecular configuration of these chemicals alters the way they affect plant systems, and thus influences the degree to which they are translocated.

Merkle and Davis (1966) recovered 76% of the 2,4,5-T and 94% of the picloram from bean plants (Black Valentine) 4 hr after application to a primary leaf. Most of the 2,4,5-T and picloram was found in the treated leaf and leafwash, but small quantities of picloram were detected throughout the plant, including the roots. Probably considerable metabolism of 2,4,5-T occurred, since recovery was only 76%. The data are similar in honey mesquite seedlings, in that the mesquite absorbed the picloram more readily and extensively than 2,4,5-T (Davis et al., 1968c). After 4 hr the apex contained both herbicides, but only picloram occurred in the roots. After 24 hr the apex and roots contained more picloram than 2,4,5-T. The amount of picloram and 2,4,5-T absorbed and transported in winged elm were similar.

In the field, Davis et al. (1972) found that the 2,4,5-T content in honey mesquite stem phloem 48 hr after treatment was higher in stems within 20 cm of the foliage than those near the soil line. Similar levels of 2,4,5-T occurred from application of either 0.56 or 1.12 kg/ha, whereas three times as much picloram occurred in plants treated with 1.12 kg/ha versus 0.56 kg/ha. Herbicide concentration was highest in June and lowest in August.

Meyer et al. (1972a) indicated the time required by herbicides to be retained on the plant after spraying to give maximum canopy reduction or mortality varied among greenhouse-grown honey mesquite, huisache, and whitebrush [*Aloysia gratissima* (Gillies & Hook.) Troncoso] and field-grown honey mesquite, huisache, whitebrush, live oak, Arizona ash (*Fraxinus velutina* Torr.), and winged elm. In most species, however, herbicide absorption and transport were complete within a 4-day period or less as compared to undefoliated treated plants.

Bovey and Mayeux (1980) studied the effectiveness and transport of 2,4,5-T, picloram, triclopyr, and clopyralid in greenhouse-grown honey mesquite. Higher concentrations of clopyralid than 2,4,5-T, triclopyr, or picloram usually were found in honey mesquite stems and roots 3, 10, and 30 days after application to soil, foliage, or both. This may be one reason why clopyralid and picloram in upper-stem phloem were not different 4 hr after treatment but clopyralid concentrations were significantly higher than picloram at 1, 3, and 8 days after treatment in greenhouse-grown honey mesquite (Bovey et al., 1988a).

Bovey et al. (1986) found comparisons of triclopyr, 2,4,5-T, picloram, and clopyralid similar to studies by Davis et al. (1972) using picloram versus 2,4,5-T on honey mesquite. More clopyralid than 2,4,5-T, triclopyr, or picloram was usually detected in upper and basal stem phloem and xylem. More herbicide tended to be detected in stems when herbicides were applied early (May and June) than late (August and September) in the season. Concentrations of triclopyr and picloram diminished about 25% from 3 to 30 days after treatment, whereas concentrations of 2,4,5-T and clopyralid diminished about 50% for the same time period.

Bovey et al. (1967) found that soil applications of picloram were more effective on greenhouse-grown huisache than foliar treatments. Soil and foliar applications at 0.56 kg/ha, however, were lethal on foliated plants, whereas hand-defoliated plants showed considerable regrowth. Plants treated with 0.28 kg/ha required 24 hr before leaf removal for maximum herbicide effectiveness. Concentration of picloram in roots from soil and foliar application was similar. Absorption and movement studies showed that 24 hr were required to move lethal amounts of picloram into stem and root tissues of huisache after foliar treatment.

Further research by Bovey et al. (1979b) indicated that more picloram than triclopyr was found in greenhouse-grown huisache up to 30 days after treatment as soil, foliar, or soil plus foliar treatments. Picloram also showed greater herbicidal activity.

Mayeux et al. (1979b) found spiny asters absorbed little herbicide relative to that in annual sunflowers. Spiny asters absorbed and transported more 2,4-D than picloram applied as the potassium salts of both herbicides by 4 and 6 hr after treatment. In general, 2,4-D behaved similarly when applied to leafless spiny asters in June and to foliated spiny asters in March. The concentration of 2,4-D in terminal stems of regrowth 74 days after shredding rose rapidly to almost 19 μg/g fresh weight at 24 hr after application and gradually fell to about 5 μg/g after 120 hr. Maximum 2,4-D concentration in terminal stems of undisturbed spiny asters was less than 14 μg/g. Picloram concentrations in terminal stems were highly variable. There were no differences in herbicide concentrations in shredded and nonshredded terminal stems. No 2,4-D or picloram was detected in rhizomes.

Drummond's goldenweed leaves absorbed about 50% of available 2,4-D (diethylamine) and 25% of available picloram (potassium salt) within 5 days after spraying in the field during July or November (Mayeux and Scifres, 1980). The herbicide accumulation in Drummond's goldenweed taproots after spray applications was generally slow, regardless of the season, but translocation to taproots was substantially greater after application in November than after treatment in March or July. Accumulation of picloram in taproots was faster and more extensive than accumulation of 2,4-D. Based on mortality at 6 months after treatment, however, both herbicides were translocated in quantities adequate for control. The greater effectiveness of picloram, despite its lower foliar uptake than 2,4-D, is attributed to its greater mobility and root uptake in Drummond's goldenweed after broadcast sprays.

Picloram was translocated basipetally from treated new pads to untreated old pads rather than in the opposite direction, but concentrations in untreated pads were low (<1 μg/g) (Mayeux and Johnson, 1989). Little picloram was absorbed by the roots compared to the pads, and little was translocated into or out of the roots. These results conflict with the view that the effectiveness of picloram for prickly pear control is attributable to extensive root uptake and acropetal transport. Observations of plants 6 months after treatment indicated that soil applications were more effective than sprays in the glasshouse, however.

b. Herbicide Formulation Norris and Freed (1964; 1966a) examined the effect of formulation on the translocation of 2,4-D and 2,4,5-T in big leaf maple seedlings. More 2,4-D was translocated to the roots when applied as an acid than as an ester or amine, and more 2,4,5-T was translocated to the roots when applied as an ester rather than as an acid or amine. The efficiency of translocation of the absorbed material was least for both esters, and high for both 2,4,5-T amine and 2,4-D amine and acid. Morton et al. (1968) showed that there are no differences between the amount of 2,4,5-T translocated in honey mesquite when applied as the butoxyethyl ester or the ammonium salt, even though more ester is absorbed than the ammonium salt.

Hull (1956) reported in velvet mesquite that when carried in a nontoxic oil emulsion, the free acid, the triethylamine, and the sodium salts of 2,4,5-T all demonstrated a greater tendency to be translocated to more distant portions of the plant than did ester formulations. Tschirley and Hull (1959), however, found the ester of 2,4,5-T consistently more effective than the amine formulation on velvet mesquite under field conditions. The research data of Fisher et al. (1956) on honey mesquite agree with that of others (Tschirley and Hull, 1959) in that the low-volatile esters and suspended acids were more consistent in killing mesquite than either the high-volatile esters or the amine formulations. The reasons for the superior performance of the ester of 2,4,5-T over the amine formulations has not been clearly established, but the ester formulations probably penetrate the wax and cuticle on the leaf more readily than the amine. Beck et al. (1975a, b), however, showed little difference in effectiveness between the ester and amine formulations of 2,4,5-T on honey mesquite. Differences in formulation may have been masked by a high rate of application or by the fact that sprays were applied to the base of the plants.

Bovey and Mayeux (1980) found no differences between the ethylene glycol butyl ether esters or triethylamine salt of triclopyr in greenhouse-grown honey mesquite stems and roots 3, 10, and 30 days after aqueous applications to the soil, foliage, or soil plus foliage.

In other detailed studies, triclopyr was rapidly transported from the treated leaf to other plant parts. Triclopyr concentrations recovered 4 hr after treatment in the upper canopy, lower canopy, and roots averaged 0.12, 0.19, and 0.09 μg, respectively (Bovey et al., 1983). Concentrations of triclopyr recovered after 24 hr were not significantly different than after 4 hr in the canopy. No ester from ester application was recovered in the canopy, other than that in the treated leaf. The uptake and transport of triclopyr applied as either the ester or the amine 4 and 24 hr after treatment were similar. Triclopyr recovered from the stems of honey mesquite in the field ranged from 0.16 to 0.72 μg/g in phloem and from 0.04 to 0.20 μg/g in xylem from a broadcast spray application of the butoxyethanol ester at 1.12 kg/ha. Concentrations of triclopyr were usually not significantly different in either the upper or lower stems, whether sampled 3 or 30 days after treatment.

Foliar sprays of the monoethanolamine salt, potassium salt, free acid, and 1-decyl ester of clopyralid were more effective in killing greenhouse-grown honey mesquite than the 2-ethylhexyl ester at rates of 0.28 kg/ha or less (Bovey et al., 1989). More clopyralid was transported to the lower canopy from application of the monoethanolamine salt and potassium salt than the 2-ethylhexyl ester of clopyralid at 4 hr or 1, 3, or 8 days after treatment. Application of the monoethanolamine salt and the 2-ethylhexyl ester to leaves with a pipet indicated that about twice as much clopyralid was absorbed within 15 min from the ester form (26%) than from the amine form (12%) of the total recovered. After 24 hr, however, absorption of the ester was less than the amine. More than twice as much clopyralid was transported from the treated leaf after application of amine than after application of the ester. Only the acid form of clopyralid was transported away from the site of application of either the original ester or amine forms.

Foliar sprays of the monoethanolamine salt, oleylamine salt, and 1-decyl ester of clopyralid were about equally effective in killing greenhouse-grown honey mesquite (Bovey et al., 1990a). The treated leaves absorbed more clopyralid within 15 min after pipet application of the oleylamine salt compared to the other formulations. After 24 hr, treated leaves absorbed and transported more clopyralid into the plant after application of the salt formulations compared to that of the 1-decyl ester. There were no consistent differences among clopyralid formulations in the transport of clopyralid from foliar sprays at 4 hr or 1, 3, or 8 days after treatment. Only the acid form of clopyralid was transported from the site of application of either the ester or the amine formulation.

c. ***Herbicide Mixtures*** The picloram: 2,4,5-T combination was particularly useful in honey mesquite control (Bovey et al., 1968; Robison, 1967). Davis et al. (1968b) found that the transport of picloram to the lower stem in greenhouse-grown honey mesquite was increased in the presence of 2,4,5-T, whereas the uptake and transport of 2,4,5-T was decreased in the presence of picloram. Increasing ratios of 2,4,5-T: picloram in mixtures up to 16:1 continued to increase uptake and transport of picloram; the reverse effect occurred for 2,4,5-T when 2,4,5-T: picloram ratios were decreased. More total herbicide was transported when the 2,4,5-T: picloram combination was used than either herbicide used alone at equal rates. This may help to explain the greater effectiveness of the herbicide combination in controlling honey mesquite. When paraquat was combined with picloram on honey mesquite, huisache, and beans, the transport of picloram to the lower stem was reduced because of damage of the transport system by paraquat.

In field studies, Davis et al. (1972) found that the highest concentrations of 2,4,5-T, picloram, or combinations of 2,4,5-T: picloram in phloem were associated with the dates of the best control of honey mesquite established by numerous investigations. Adding 2,4,5-T to picloram caused an increase in the amounts of picloram in the phloem in four of five dates of application. These data agree with the laboratory and greenhouse investigations described above (Davis et al., 1968b), therefore the combination of picloram and 2,4,5-T was generally more effective than either herbicide applied alone.

The recovery of 2,4,5-T as the acid and ester was significantly greater in live oak tissues treated with mixtures of the 2-ethylhexyl ester of 2,4,5-T (2.2 kg/ha) plus the potassium salt or isooctyl ester of picloram (0.56, 1.1 and 2.2 kg/ha) than in tissues treated with 2,4,5-T ester alone (Baur et al., 1969b). The recovery of 2,4,5-T as the ester was noted in the middle and lower-stem tissues. Between 90 and 99% of the herbicide recovered 1 month after treatment was gone 6 months after treatment. Evaluation of brush reduction 2 years after treatment indicated that the mixtures of picloram salt and 2,4,5-T resulted in greater reduction of brush than mixtures of picloram ester and 2,4,5-T or 2,4,5-T alone.

Bovey et al. (1988a) found that the addition of picloram or triclopyr to clopyralid at equal rates (0.28 + 0.28 kg/ha) increased the clopyralid concentrations in field-grown honey mesquite by 1 day after treatment compared to clopyralid applied alone.

d. Carriers and Adjuvants A large number of diluents and adjuvants for herbicides have been evaluated for control of woody plants with some success, as discussed in Carriers and Adjuvants under Foliar Absorption (see Sec. II.C) (Bovey et al., 1979a; Bovey and Meyer, 1981; Fisher et al., 1956; Hull, 1956).

An April application of 1.12 kg/ha of picloram plus 5.0% X-77 plus 10.0 or 25.0% of dimethylsulfoxide (DMSO) produced more effective canopy reduction of yaupon 1 year after treatment than did other herbicide treatments or dates of treatment (Baur et al., 1972). The increased canopy reduction by retreatment by the second year suggest that X-77 rather than DMSO was the effective component in the spray mixtures.

Scifres et al. (1973a) compared water, diesel oil, diesel oil: water: (1:4) emulsion, paraffin oil, and paraffin oil: water: (1:4) emulsion for carriers of 0.56 kg/ha of the butyl ether ester of 2,4,5-T in greenhouse-grown honey mesquite. No differences related to carrier occurred in the amount of 2,4,5-T translocated to the stem and roots, although greater amounts of herbicides were absorbed by leaves treated with diesel or paraffin oil carriers.

The concentration of clopyralid in basal stem phloem of field-grown honey mesquite sampled 4 and 30 days after spraying 0.56 kg/ha to the foliage showed no differences in water, diesel oil plus water 1:4 (v/v), and water plus surfactant 0.5% (v/v) diluents (Bovey et al., 1988b). Surfactant I (trimethylnonylpolyethoxyethanol) or surfactant II (4-isoprophenyl-1-methyl-cyclohexane) at 0.5% by volume of the spray solution enhanced clopyralid absorption and transport to the upper stem phloem in field-grown honey mesquite by 1 day after spraying 0.28 hg/ha clopyralid. Enhanced absorption and transport of clopyralid may be partially due to enhanced spray retention of clopyralid on the leaves with surfactant (Bovey et al., 1987; 1988a; 1991).

Organosilicone surfactants Silgard 309 or Silwet L-77 added to the spray solutions at 0.1, 0.25, or 0.5% by volume did not increase spray deposition, absorption, translocation, or phytotoxicity of clopyralid in greenhouse-grown honey mesquite (Bovey et al., 1994).

e. Spray Characteristics The addition of surfactant WK at 0.5% (v/v) caused an increased uptake of clopyralid by the upper canopy of greenhouse-grown honey mesquite (Bovey et al., 1987). Enhanced uptake after 24 hr was probably a result of a twofold increase in the deposit of clopyralid on the plant (Bovey et al., 1987; 1991). A greater deposit of clopyralid on plant surfaces after the addition of surfactant was associated with reduced liquid surface tension and a greater percentage of spray volume in small droplets (<204 μm diam.). The addition of surfactant WK at 0.5% (v/v) of the spray solution caused a twofold increase in the deposition of the monoethanolamine salt of clopyralid but not the oleylamine salt of clopyralid (Bovey et al., 1991). There were no differences in spray deposit between spray droplet-size spectrums of 160 or 330 μm Dv. 5 or spray solution application of 47 or 187 L/ha.

An air-assist spray nozzle at 9.4 L/ha by volume resulted in a greater initial clopyralid deposit and detection in the upper canopy of greenhouse-grown honey mesquite than application by conventional hydraulic nozzle at 9.4 or 187 L/ha (Bovey et al., 1994). Air-assist application did not increase phototoxicity compared to hydraulic nozzles. In the field, honey mesquite mortality and canopy reduction 16 months after aerial application of clopyralid were significantly less in the 624-μm-droplet treatment in two of four experiments when compared to plots treated with smaller droplet sizes (325 and 475 μm Dv. 5) (Whisenant, 1993). The mortality increased with larger spray volumes (19, 37, and 75 L/ha), particularly with the 625-μm droplet. The mortality data show that larger droplet sizes require larger spray volumes for the greatest efficacy.

f. ***Concentration of Applied Herbicide*** Fisher et al. (1956) indicated that a rate of 0.56 kg/ha of low-volatile ester of 2,4,5-T in a 1-to-3 ratio of oil–water emulsion effectively controlled honey mesquite. Increasing the amount of 2,4,5-T did not increase the percentage of mesquite killed. Davis et al. (1972) found that 2,4,5-T butoxyethanol ester applied at 0.56 and 1.12 kg/ha did not result in significant differences in the concentration of the herbicide in the phloem tissue of mesquite. Translocation of 2,4,5-T applied as the butoxyethanol ester from the treated leaves of blackjack oak was found to increase as the amount of herbicide applied was increased, however (Badiei et al., 1966). This paralleled a similar increase in absorption.

g. ***pH Effects*** See pH Effects under Foliar Absorption (Sec. II.C).

h. ***Temperature and Relative Humidity*** Translocation of 2,4,5-T in honey mesquite seedlings was primarily basipetal (downward) from the point of application at 21°C, both acropetal (upward) and basipetal at 29°C, and only a short distance acropetal at 38°C (Morton, 1966). The quantities of 2,4,5-T translocated into untreated tissues at 38°C were less than at 21°C and 29°C. The highest concentrations of 2,4,5-T were found in tissues with the highest soluble sugar concentrations. From 3 to 27% of the 2,4,5-T absorbed by honey mesquite leaves was subsequently detected in untreated stem, leaf, and root tissues. The total amounts of C^{14} (carboxyl-labeled 2,4,5-T) detected in the untreated tissues of the seedlings tended to increase with increasing humidity, particularly in the roots and lower stems.

Radosevich and Bayer (1979) found that 2,4,5-T, triclopyr, and picloram transport was greater in periods of warm temperatures (29°C and 13°C, day and night) and long days (16-hr photoperiod) than cool temperatures (13°C and 2°C, day and night) and a 12-hr photoperiod in five plant species as revealed by autoradiographs. They found little metabolism of any herbicide, and each herbicide moved readily in the symplast (phloem); however, root application revealed limited apoplastic (xylem stream) mobility. Among the herbicides tested, ^{14}C associated with triclopyr was more mobile than 2,4,5-T or picloram in tanoak (*O. densiforus*), snowbush ceanothus (*Ceanothus velutinus*), bigleaf maple (*Acer macrophyllum*), and beans and barley (*Hordeum vulguve*).

Translocation of 2,4,5-T to the roots of four woody species, when applied as the isooctyl ester of 2,4,5-T, was found to increase with increasing temperature (Brady, 1970b). With three other species translocation was highest at the lowest temperature. Wills and Basler (1971) found that translocation of 2,4,5-T applied as the butoxyethanol ester was the highest in winged elm trees when the ambient temperature was high.

i. ***Rainfall*** See Rainfall under Foliar Absorption (Sec. II.C).

j. ***Time of Day*** See Time of Day under Foliar Absorption (Sec. II.C).

k. ***Light*** Light intensity affected translocation of 2,4,5-T to the roots of woody plants (Brady, 1969). There was a negative linear relationship between light intensity and the 2,4,5-T content of post oak roots. In water oak roots, however, herbicide levels increased as light intensity increased. In longleaf pine and American holly, translocation was not significantly influenced by light intensity. Badiei et al. (1966) found that more 2,4,5-T applied as the butoxyethanol ester was translocated away from the treated leaf of blackjack oak in the light than in the dark.

l. ***Moisture Stress*** Moderate moisture stress in beans did not have a significant effect on the translocation of picloram, but did on the translocation of 2,4,5-T (Merkle and Davis, 1967). Advanced stress significantly reduced the translocation of both herbicides. Translocation of 2,4,5-T was apparently more sensitive to changes in moisture stress than was translocation of picloram, however. Picloram was more mobile than 2,4,5-T at all moisture stress levels studied. After 4 hr, as much picloram was translocated to the apex and central stem of bean plants from a 24-μg application, as was 2,4,5-T at 8 hr after a 50-μg application. This agrees with studies on honey mesquite, in which both herbicides were detected in the apex only 4 hr after treatment, but only picloram occurred in the roots (Davis et al., 1968c). After 24 hr, the apex and roots contained more picloram than 2,4,5-T. The phloem-cortex accumulated greater quantities of picloram than the xylem-pith, indicating major transport via the symplast. After 90 hr, herbicide concentrations in most tissues were either unchanged or higher than after 24 hr. These data support observations by Meyer et al. (1972a), which indicated that a period of 3 to 4 days was required for honey mesquite to absorb and translocate herbicide for maximum killing of stems. Moisture stress sufficient to slow growth markedly reduced the transport of picloram and 2,4,5-T into untreated tissues.

Bovey et al. (1977) found in field studies that a 1:1 combination of the triethylamine salts of picloram: 2,4,5-T was more effective on huisache and Macartney rose when applied in the evening than in the morning or at midday. The internal water stress of the plants was less at night after the 6:00 P.M. treatment than after the 6:00 A.M. or 1:30 P.M. treatment, allowing a more favorable environment for absorption and translocation of the herbicide.

Bovey and Clouser (1998) found in preliminary studies that water stress of −1.3 to −2.8 Mpa did not affect absorption and translocation of clopyralid in greenhouse-grown honey mesquite 4 or 24 hr after treatment. The addition of triclopyr (synergistic) to clopyralid increased clopyralid uptake at low-water stress (−1.3 Mpa), but decreased clopyralid uptake at high-water stress (−2.8 Mpa).

Bollig et al. (1995) found moisture stress reduced the transport of triclopyr in red maple (*Acer rubrum*) seedlings into shoots and roots but had no apparent effect on leaf absorption. Similar data were noted in plant moisture stress studies with water oak and southern red oak seedlings, where triclopyr translocation to stems and roots was 62 and 48% lower, respectively, in water-stressed plants (Seiler et al., 1993).

m. Growth Stage and Seasonal Effects Most woody plants are most susceptible to herbicides at certain times of the year when the plants are growing in a favorable environment. This could correlate with spring and/or fall treatment or other times of the year. The best time for treatment depends upon the species and environment, and usually occurs when woody plants have enough mature leaf cover to intercept, absorb, and transport lethal amounts of the appropriate herbicide or are able to absorb and transport soil-applied herbicides.

Meyer et al. (1972b) sprayed honey mesquite in the field with three herbicides at 14 different dates during 1969 and 1970. The most effective control of honey mesquite occurred from treatments applied between April 30 and July 6. Picloram and a picloram: 2,4,5-T (1:1) mixture were the most effective herbicides. Plant characteristics most closely associated with control included the widest translocating phloem thickness, the most rapid rate of new xylem ring radial growth, and the lowest predawn leaf moisture stress. The environmental variables most clearly associated with honey mesquite control were lower maximum air temperatures of 25°C to 36°C 1 week before treatment, a maximum soil temperature of 17°C to 26°C at a depth of 91 cm 1 week before treatment, and a decreasing percentage of soil moisture from 25 to 18% at a depth of 61 to 91 cm 1 week before treatment. In subsequent studies (Meyer, 1977) of responses to spraying on 36 dates from March to October during a 4-year period, the percentage of honey mesquite canopy reduction was directly correlated with the total phloem thickness, the rate of new xylem ring radial growth, and the rate of upward methylene dye movement in the xylem, and was inversely correlated with minimum leaf moisture stress. The rate of new xylem ring radial growth and the thickness of the translocating phloem appeared most often in the equations.

Measurements of seasonal effects on translocation of 2,4,5-T in blackjack oak have been made by Badiei et al. (1965). They showed that the amount of 2,4,5-T in the roots paralleled the free sugar content of the roots, and that maximum kill was obtained when translocation was high. Later experiments (Badiei et al., 1966) showed that translocation of 2,4,5-T applied as the butoxyethanol ester to blackjack oak is highest in June. Dalrymple and Basler (1963) found that the translocation of the same chemical in blackjack oak was the highest in early spring and decreased until July. They also found that respiration rates of the leaf tissue followed a similar pattern to the amount of herbicide translocated.

Brady (1971) found that the maximum translocation of 2,4,5-T, applied as the propylene glycol butyl ether ester, occurred during May and June in green ash (*Fraxinus pennsylvanica*) and during May in water oak and loblolly pine, while there was no significant seasonal effect on translocation in sweet gum. Translocation of the same chemical applied to creosote bush grown in the field was at a maximum when applied about 30 days after effective summer rains, while movement was minimal during the cool, dry winter and spring (Schmutz, 1971). This indicates that translocation of the herbicide probably correlates with moisture stress. Wills and Basler (1971) also found that the movement of 2,4,5-T was highest in June and July in winged elm. This, however, also correlated with herbicide absorption and higher temperatures. Davis et al. (1972) found that the concentration of 2,4,5-T in the phloem of honey mesquite reached a maximum in June, when the air temperature was high.

Bovey et al. (1986) investigated the concentrations of triclopyr, 2,4,5-T, picloram, and clopyralid application to natural stands of honey mesquite on seven different dates from May to September in 1980 and 1981. The herbicides were all applied broadcast to separate replicated plots at 1.12 kg/ha. The concentrations of 2,4,5-T and triclopyr were less than 2 μg/g fresh weight regardless of application date. The concentrations of picloram and clopyralid were as high as 11 and 22 μg/g fresh weight, respectively, in upper stem phloem at some dates of application. Higher concentrations of all herbicides were detected in upper stem phloem than in the upper stem xylem or basal stem phloem or xylem. More herbicide was likely to be detected in stems when applied early in the season (May and June) rather than late (August and September) in the season. Concentrations of triclopyr and picloram recovered from honey mesquite stems were about 25% greater at 3 than at 30 days after treatment, whereas concentrations of 2,4,5-T and clopyralid were about 50% greater at 3 than at 30 days after application.

3. Root Penetration and Translocation

*a. **Laboratory and Greenhouse Studies*** In the greenhouse, huisache and honey mesquite can be killed by soil-applied herbicides and show as much or more picloram, triclopyr, or clopyralid content in root tissue up to 30 days after treatment as foliar spray content (Bovey et al., 1967; 1979a; Bovey and Mayeux, 1980). In the field, however, huisache and especially honey mesquite are difficult to kill with soil-applied herbicides in heavy clay soils (Bovey and Meyer 1978; Meyer and Bovey, 1979).

Baur and Bovey (1969) studied changes in the concentration of picloram in roots, stems, and leaves of 20-day-old huisache and honey mesquite plants exposed for different lengths of time. Exposing roots to aqueous solutions of picloram for 24 hr killed about 60% of the treated plants. It took 10 times more herbicide to give the same response in honey mesquite (10 ppm) as huisache (1 ppm). In honey mesquite, picloram was redistributed and eventually lost from the plant into the rooting solution over a 5-day period, whereas huisache, a more susceptible plant, showed no redistribution or loss of picloram.

Mayeux and Johnson (1989) found in Lindheimer prickly pear that picloram concentrations within pads treated in the glasshouse were greater when the herbicide was applied to new pads (4.6 μg/g) after 30 days. More picloram was translocated basipetally from treated new pads to untreated old pads than in the opposite direction, but the concentrations in untreated pads were low (<1 μg/g). Little picloram was absorbed by roots compared to pads and little was translocated into or out of roots. These results conflict with the view that the effectiveness of picloram for prickly pear control is attributable to extensive root uptake and acropetal transport. Observations of plants 6 months after treatment, however, indicated that soil applications were more effective than foliar sprays in the glasshouse.

*b. **Field Studies*** As indicated earlier, controlling honey mesquite with soil-applied herbicides has not been very effective in the field. Triclopyr, picloram, and clopyralid, however, are highly effective when applied to soil in pots supporting honey mesquite under greenhouse conditions. Possibly the extensive root system of honey mesquite and impermeable heavy clay soils in some areas may partially preclude effective control under field conditions.

When applied as soil treatments for honey mesquite control, herbicides such as picloram, dicamba, karbutilate, bromacil, tebuthiuron, and prometon (Bovey et al., 1969; Meyer and Bovey 1979; Scifres et al., 1973b; 1978) have generally been ineffective at economical rates. Honey mesquite was more effectively controlled in the field when liquid formulations of karbutilate and tebuthiuron were applied subsurface than on the soil surface, however, (Meyer and Bovey, 1979).

Leaves absorbed high amounts of clopyralid as foliar sprays on honey mesquite, as indicated by concentrations of 10 μg/g fresh weight or more in basal stem phloem by 4 days after treatment (Bovey et al., 1988b). Small quantities of clopyralid (1 < μg/g) were detected in basal stem phloem after spray applications of clopyralid to defoliated plants or roots treated by soil application. When applied to foliated plants, the 0.56 kg/ha of clopyralid killed 60% or more plants, but none was killed when clopyralid sprays were applied to defoliated plants or when 2.2 kg/ha of clopyralid were applied to the soil.

III. MODE OF ACTION

A. Phenoxy Herbicides

Although the phenoxy herbicides have probably received more detailed investigations than any other class of herbicides relative to mode of action, their exact mechanism remains unknown. A long list of physiological responses are induced in plants after treatment by the phenoxys, which may act alone or in many combinations to affect or kill the plant. The initial action involves the absorption and penetration of plant surfaces, absorption into the symplast, migration to the vascular system, and translocation from the leaves to the stem, the roots, and other plant parts. After distribution within the plant, biochemical responses are triggered that may lead to abnormal growth such as apinastic manifestations (twisting and curvature of leaves, stems, or roots) and plugging of vascular channels, ultimately starving the plant. Abnormal growth is believed to be caused by abnormal nucleic acid metabolism. Herbicides 2,4-D, 2,4-DB, dichlorprop, 2,4,5-T, silvex, and MCPA are similar in mode of action (Bovey and Young, 1980).

B. Pyridine Herbicides

Picloram and clopyralid are readily absorbed by roots and foliage and translocated throughout the plant (Weed Science Society of America, 1994). They are degraded very slowly within the plant. Translocation is both symplastic and apoplastic. Phytotoxic symptoms and mode of action are similar to the phenoxys and other hormonelike herbicides. Triclopyr manifests symptoms on weeds similar to picloram and the phenoxy herbicides.

C. Benzoics

Dicamba is included here since it produces symptoms similar to the phenoxy herbicides with epinasty of young shoots and proliferation of growth. It is readily translocated in both the symplast and apoplast and accumulates in areas of high metabolic activity, such as meristems. Dicamba is relatively stable in higher plants, but undergoes hydroxylation or demethylation with hydroxylation (Ashton and Crafts, 1973).

D. Sulfonylureas (Metsulfuron)

1. *Symptomology*: Few data are available on the fate of metsulfuron in woody plants. Metsulfuron has minor use for woody plant control. In herbaceous plants, meristematic areas gradually become chlorotic and necrotic, followed by foliar chlorosis and necrosis.
2. *Absorption/translocation*: Rapid foliar and root absorption. Metsulfuron translocates extensively in the xylem following root absorption, and less so in the phloem after foliar application. It accumulates in meristematic areas (Weed Science Society of America, 1994).
3. *Mechanism of action*: Inhibits acetolactate synthase (ALS), also called acetohydroxyacid synthase (AHAS), a key enzyme in the biosynthesis of the branched-chain amino acids isoleucine, leucine, and valine (Beyer et al., 1988). Plant death results from events occurring in response to ALS inhibition, but the actual sequence of phytotoxic processes is unclear.
4. *Metabolism in plants*: Wheat and barley rapidly metabolize metsulfuron by hydroxylating the benzene ring at the #4 carbon followed by glucose conjugation at the ring hydroxyl group. Also detected were low levels of hydroxylation of the methyl group at carbon #6 on the triazine ring (Beyer et al., 1988).

E. Triazines (Hexazinone)

1. *Symptomology*: Foliar chlorosis followed by necrosis.
2. *Absorption/translocation*: Seedling winged elm, bur oak, black walnut (*Juglans nigra*), eastern red cedar (*Juniperous virginiana*), and lobolly pine were treated in nutrient solution with ring-labeled ^{14}C-tebuthuiron and hexazinone. In 4 hr ^{14}C was detected in all sections of the winged elm, implying rapid apoplastic movement (McNeil et al., 1984). After 24 hr the older leaves contained significantly more ^{14}C than at 4 hr. In the other woody plants the major site of radioactive accumulation was in the roots. Loblolly pine accumulated more tebuthiuron and hexazinone than black walnut, bur oak, and eastern red cedar. Absorption of ^{14}C-hexazinone from most to least was loblolly pine > black walnut > bur oak = eastern red cedar. The presence of the three metabolites of hexazinone in loblolly pine suggests that it may be resistant to hexazinone since it degrades hexazinone rather than its ability to limit uptake.
3. *Mechanism of action*: Inhibits photosynthesis by binding to the Q_B-binding niche on the D1 protein of the photosystem II complex in chloroplast thylakoid membranes, thus blocking electron transport from Q_A to Q_B (Weed Science Society of America, 1994). This stops CO_2 fixa-

tion and production of ATP and $NADPH_2$ (all needed for plant growth), but plant death occurs in most cases by other processes. The inability to reoxidize Q_A promotes the formation of triplet-state chlorophyll, which interacts with ground-state oxygen to form singlet oxygen. Both triplet chlorophyll and singlet oxygen can abstract hydrogen from unsaturated lipids, producing a lipid radical and initiating a chain reaction of lipid peroxidation. Lipids and proteins are attacked and oxidized, resulting in a loss of chlorophyll and carotenoids and in leaky membranes, which allow cells and cell organelles to dry and disintegrate rapidly.

4. *Metabolism in plants*: Hexazinone was converted to several hydroxylated and/or demethylated triazinone metabolites in *Pyrus melanocarpa* and *Rubus hispidus* following root or foliar uptake. Accumulation in *P. melanocarpa* of higher levels of the monodemethylated metabolite (3-cyclohexyl-6-methylamineo-1-methyl-1,3,5-triazine-2,4-dione) compared to *R. hispidus* may contribute to the greater tolerance in *P. malamocarpa* (Jensen and Kimball, 1990). Metabolism contributed to hexazinone selectivity in red and jack pine (Wood et al., 1992) and probably in loblolly pine (McNeil et al., 1984).

F. Ureas and Uracils (Bromacil and Tebuthiuron)

1. Bromacil

1. *Symptomology*: Foliar chlorosis and necrosis.
2. *Absorption/translocation*: Readily absorbed by roots and translocated in the xylem to the leaves (Gardner, 1975), bromacil is less readily absorbed by the leaves and stems. Surfactants enhance foliar activity. Orange plants grown on bromacil-treated nutrient solution in sand, however, absorbed $<5\%$ of the applied bromacil; 85% of the absorbed remained in the roots and 15% was translocated to the stem and leaves (Gardner et al., 1969).
3. *Mechanism of action*: Inhibits photosynthesis by binding to the Q_B-binding niche on the D1 protein of the photosystem II complex in chloroplast thylakoid membranes, thus blocking electron transport from Q_A to Q_B (Weed Science Society of America, 1994). This stops CO_2 fixation and production of ATP and $NADPH_2$ (all needed for plant growth), but plant death occurs in most cases by other processes. The inability to reoxidize Q_A promotes the formation of triplet-state chlorophyll, which interacts with ground-state oxygen to form singlet oxygen. Both triplet chlorophyll and singlet oxygen can abstract hydrogen from unsaturated lipoids, producing a lipid radical and initiating a chain reaction of lipid peroxidation. Lipids and proteins are attacked and oxi-

dized, resulting in the loss of chlorophyll and carotenoids and in leaky membranes, which allow cells and cell organelles to dry and disintegrate rapidly.

4. *Metabolism in plants*: Most of the applied bromacil was not metabolized in orange seedlings, but some 5-bromo-3-*sec*-butyl-6-hydroxymethyluracil and an unknown minor metabolite were produced. No 5-bromouracil was detected as a metabolite of bromacil in plants (Weed Science Society of America, 1994).

2. Tebuthiuron

1. *Absorption/translocation*: Readily absorbed into roots, less so into foliage (Steinert and Stritzke, 1977). Tebuthiuron is readily translocated following root absorption (McNeil et al., 1984).
2. *Mechanism of action*: Inhibits photosynthesis by binding to the Q_B-binding niche on the D1 protein of the photosystem II complex in chloroplast thylakoid membranes, thus blocking electron transport from Q_A to Q_B (Weed Science Society of America, 1994). This stops CO_2 fixation and production of ATP and $NADPH_2$ (all needed for plant growth), but plant death occurs in most cases by other processes. The inability to reoxidize Q_A promotes the formation of triplet-state chlorophyll, which interacts with ground-state oxygen to form singlet oxygen. Both triplet chlorophyll and singlet oxygen can abstract a hydrogen from unsaturated lipids, producing a lipid radical and initiating a chain reaction of lipid peroxidation. Lipids and proteins are attacked and oxidized, resulting in a loss of chlorophyll and carotenoids and in leaky membranes, which allow cells and cell organelles to dry and disintegrate rapidly.
3. *Metabolism in plants*: Primarily degraded by N-demethylation and hydroxylation of the terbutyl side chain (Eaton et al., 1976).

G. Other Organic Herbicides (Amitrole, Fosamine, Glyphosate, and Imazapyr)

1. Amitrole

1. *Symptomology*: The primary symptom is bleaching (albinism) in leaves and shoots, and is most evident in meristems and developing leaves. Bleached tissues eventually wilt and become necrotic.
2. *Absorption*: Amitrole absorption into roots or penetration across the leaf cuticle is extremely rapid. Over 90% of foliar-applied amitrole penetrated the shoot of field horsetails 1 day after application (Coupland and Peabody, 1981; Carter, 1975). The foliar uptake rate of amitrole was considerably faster than that of glyphosate and fosamine.

3. *Translocation*: Amitrole is polar and nonionized, is strongly mobile in the apoplast (including the xylem), and is largely retained in the phloem once it has crossed the plasma membrane. Consequently, amitrole is considered ambimobile, translocating in both apoplasm and symplasm. Based on the injury pattern, the herbicide appears to accumulate at the growing points (Carter, 1975).
4. *Mechanism of action*: Inhibits accumulation of chlorophyll and carotenoids in the light (Carter, 1975), although the specific site of action has not been determined. Research indicates that the histidine, carotenoid, and chlorophyll biosynthetic pathways probably are not the primary sites of amitrole action. Instead, amitrole may have a greater effect on cell division and elongation than on pigment biosynthesis (Weed Science Society of America, 1994).
5. *Metabolism in plants*: The primary metabolite (6–10%) was a serine conjugate forming β-(3-amino-1,2,4-triazol-l-yl)-α alanine. Amitrole may also conjugate with glycine (Carter, 1975) in some plants, and may complex with tannins. Differential susceptibility to amitrole has been associated with varying rates of herbicide metabolism.

2. Fosamine

1. *Symptomology*: In most woody plants, fosamine injury is not apparent before normal leaf senescence in the fall. Leaf and bud development is severely or completely inhibited the following spring, however. Leaves that emerge appear abnormally small and spindly. Certain other plants, such as pines and bindweed, may show a response soon after application. Suppression of terminal growth may occur in moderately susceptible to resistant species.
2. *Absorption*: The penetration of fosamine into leaf tissue is generally slow, with less than 50% of the applied herbicide absorbed 7 to 8 days after application in blackberries (Richardson, 1980), field horsetails (Coupland and Peabody, 1981), and multiflora roses (Mann et al., 1986). Retention and penetration is further reduced when fosamine is applied to very hairy leaf surfaces. Good penetration occurs, however, when fosamine is applied to young stems. The slow rate of fosamine absorption increases the potential for washoff by rainfall. Surfactants may increase the fosamine penetration rate.
3. *Translocation*: Fosamine translocation occurs both symplastically and apoplastically, but is limited in most species examined. In blackberries, multiflora roses, and field horsetails, the herbicide initially translocated somewhat rapidly, accumulating to near maximum concentration in nontreated tissues 1 day after treatment. No significant increase in fosamine accumulation occurred thereafter, however. Field experience

as well as limited translocation of fosamine in radioisotope studies supports the view that complete coverage of the plant is required for adequate control under field conditions.

4. *Mechanism of action*: Not well understood. Fosamine strongly inhibited mitosis in tissues of mesquite, but this may be a secondary effect (Morey and Dahl, 1980).
5. *Metabolism in plants*: Fosamine had an average half-life of 7 days when applied POST to pasture grasses and red clover (Chrzanowski, 1983). The primary metabolite was carbamoylphosphonic acid (CPA), which reached a maximum concentration after 2 to 4 weeks. Carboxyphosphonic acid also was detected as a metabolite. No fosamine or its metabolites were found in pasture turf or clover 12 months after treatment. Late-summer application of fosamine to pin oak did not result in rapid decline of the herbicide because the plant went into dormancy shortly after treatment. In contrast, the half-life of foliar-applied fosamine in apple seedlings grown under greenhouse conditions was 2 to 3 weeks.

3. Glyphosate

1. *Symptomology*: Growth is inhibited soon after application, followed by general foliar chlorosis and necrosis within 4 to 7 days for highly susceptible grasses and within 10 to 20 days for less susceptible species. Chlorosis may appear first and be most pronounced in immature leaves and growing points. Foliage sometimes turns reddish-purple in certain species. Regrowth of treated perennial and woody species often appears deformed, with whitish markings or striations. Multiple shoots (sometimes called a witch's broom) may develop at the nodes.
2. *Absorption*: Moderately absorbed across the cuticle when POST-applied (Boerboom and Wyse, 1988). The isopropylamine salt of glyphosate is more readily absorbed than glyphosate acid, and surfactant and ammonium sulfate further increase absorption of the isopropylamine salt. Glyphosate transport across the plasmalemma is slower than most herbicides (especially nonpolar herbicides), probably because of its negative charge at physiological pH. A phosphate transporter may contribute to glyphosate movement across the plasmalemma (Duke, 1988).
3. *Translocation*: Primarily translocated in the symplast with accumulation in underground tissues, immature leaves, and meristems. Apoplastic translocation has been observed in tall morning glories (Dewey and Appleby, 1983) and quackgrass (Klevorn and Wyse, 1984), but most results suggest little to no apoplastic movement. Glyphosate may interfere with its own translocation from treated leaves by interfering with carbon partitioning and metabolism (Geiger and Bestman, 1990). Lob-

lolly pine and yaupon (*Ilex vomitoria*) tolerate glyphosate and absorb significantly less than red maple and white oak (*Q. alba*), indicating the importance of foliar absorption as a glyphosate entry barrier (Green et al., 1992). Although absorption was similar in the sensitive white oak and the tolerant red maple, white oak accumulated more glyphosate in roots than red maple, indicating that translocation patterns also contribute to glyphosate tolerance in some woody species.

4. *Mechanism of action*: Inhibits 5-enolpyruvylshikimate-3-phosphate (EPSP) synthase, which produces EPSP from shikimate-3-phosphate and phosphoenolpyruvate in the shikimic acid pathway. EPSP inhibition leads to depletion of the aromatic amino acids tryptophan, tyrosine, and phenylalanine–all needed for protein synthesis or for biosynthetic pathways leading to growth. The failure of the exogenous addition of these amino acids to completely overcome glyphosate toxicity in higher plants (Duke, 1988) suggests that factors other than protein synthesis inhibition may be involved. Although plant death apparently results from events occurring in response to EPSP synthesis inhibition, the actual sequence of phytoxic response is unclear.
5. *Metabolism in plants*: not appreciably metabolized when applied at phytotoxic rates. Glyphosate is slowly metabolized to amino methyphosphonic acid (Weed Science Society of America, 1994).

4. Imazapyr

1. *Symptomology*: Growth is inhibited within a few hr after application, but injury symptoms usually do not appear for 1 to 2 weeks or more. Mersitematic areas gradually become chlorotic, followed by a slow foliar chlorosis and necrosis.
2. *Absorption/translocation*: Foliar absorption usually is rapid (within 24 hr), but may vary with plant species. Imazapyr also is absorbed by roots. It moves readily in both xylem and phloem when root- or shoot-absorbed. On some woody plants, such as glyphosate, imazapyr can be washed from leaf surfaces for several hr after treatment.
3. *Mechanism of action*: Inhibits ALS (also called acetohydroxyacid AHAS), a key enzyme in the biosynthesis of the branched-chain amino acids isoleucine, leucine, and valine (Stidham and Singh, 1991). Plant death results from events occurring in response to ALS inhibition, but the actual sequence of phytotoxic processes is unclear. Some secondary effects may include disruption of photosynthate translocation, hormone imbalance due to interruption of source/sink relationships, and interference in DNA synthesis and cell growth.
4. *Metabolism in plants*: Tolerance is due to rapid metabolism, initiated by hydroxylation of the imidazolinone ring to form 2-carbamoylnico-

tinic acid. In this hydrolysis-mediated metabolism, an imidazopyrrolopyridine derivative also is formed. Susceptible weed species metablize imazapyr slowly or not at all (Lee et al., 1991).

REFERENCES

Ashton FM, Crafts AS. Benzoics. In: Mode of Action of Herbicides. New York: Wiley, 1981, pp. 139–163.

Badiei AA, Basler E, Santelmann PW. A study of phenoxy herbicides in blackjack oak. Proc South Weed Conf 18:603, 1965.

Badiei AA, Basler E, Santelmann PW. Aspects of movement of 2,4,5-T in blackjack oak. Weeds 13:302–305, 1966.

Baur JR, Bovey RW. Distribution of root-absorbed picloram. Weed Sci 17:524–528, 1969.

Baur RR, Bovey RW. The uptake of picloram by potato tuber tissue. Weed Sci 18:22–24, 1970.

Baur JR, Bowman JJ. Effect of 4-amino-3,5,6-trichloropicolinic acid on protein syntheisis. Physiol Plant 27:354–359, 1972.

Baur JR, Bowman JJ. Subcellular distribution of picloram. Physiol Plant 28:372–373, 1973.

Baur JR, Swanson CR. Effect of nutrient level and daylength on growth and susceptibility of mesquite and huisache to 2,4,5-T and picloram. In: Brush Research in Texas. Tex. Agric. Exp. Stn. Consol. PR-2583-2609, College Station, TX: Texas A&M University, 1968, pp. 35–38.

Baur JR, Bovey RW, El-Seify Z. Effect of paraquat on the ultrastructure of mesquite mesophyll cells. Weed Res 9:81–85, 1969a.

Baur JR, Bovey RW, Smith JD. Herbicide concentrations in live oak treated with mixtures of picloram and 2,4,5-T. Weed Sci 17:567–570, 1969b.

Baur JR, Bovey RW, Riley I. Effect of pH on foliar uptake of 2,4,5-T-1 ^{14}C. Weed Sci 22:481–486, 1974.

Baur JR, Bovey RW, Smith JD. Effect of DMSO and surfactant combinations on tissue concentrations of picloram. Weed Sci 20:298–302, 1972.

Baur JR, Bovey RW, Baker RD, Riley I. Absorption and penetration of picloram and 2,4,5-T into detached live oak leaves. Weed Sci 19:138–141, 1971.

Beck DL, Sosebee RE, Herndon EB. Control of mesquite regrowth of different ages. J Range Mgt 28:408–410, 1975a.

Beck DL, Sosebee RE, Herndon EB. Control of honey mesquite by shredding and spraying. J Range Mgt 28:487–490, 1975b.

Behrens R. Influence of various components on the effectiveness of 2,4,5-T sprays. Weeds 5:183–197, 1957.

Beyer EM, Duffy MJ, Hay JV, Schlueter DD. Sulfonylurea herbicides. In: Kearney PC, Kaufman DD, eds. Herbicides: Chemistry, Degradation and Mode of Action. 2nd ed. vol 3. New York: Marcel Dekker, 1988, pp. 117–189.

Boerboom CM, Wyse DL. Influence of glyphosate concentration on glyphosate absorption and translocation in Canada thistle (*Cirsuim arvense*) Weed Sci 36:291–295, 1988.

Bollig JJ, Seiler JR, Zedaker SM, Thompson JW, Lucero D. Effect of plant moisture stress and application surface on uptake and translocation of triclopyr with organosilicone surfactant in red maple seedlings. Can J For Res 25:425–429, 1995.

Bovey RW, Clouser AR. Unpublished data. 1998.

Bovey RW, Davis FS. Factors affecting the phototoxicity of paraquat. Weed Res 7:281–289, 1967.

Bovey RW, Diaz-Colon JD. Effect of simulated rainfall on herbicide performance. Weed Sci 17:154–157, 1969.

Bovey RW, Mayuex HS Jr. Effectiveness and distribution of 2,4,5-T, triclopyr, picloram, and 3,6-dichloropicolinic acid in honey mesquite (*Prosopis juliflora* var. *glandulosa*). Weed Sci 28:666–670, 1980.

Bovey RW, Meyer RE. Control of huisache with soil applied herbicides. J Range Mgt 31:179–182, 1978.

Bovey RW, Meyer RE. The response of honey mesquite to herbicides. B-1363. College Station, TX: Tex Agric Exp Stn, 1981.

Bovey RW, Meyer RE. Influence of adjuvants and plant growth regulators on herbicide performance in honey mesquite. J Plant Growth Reg 5:225–234, 1987.

Bovey RW, Miller FR. Phytotoxicity of paraquat on white and green hibiscus. sorghum and Alpina leaves. Weed Res 8:128–135, 1968.

Bovey RW, Scifres CJ. Residual Characteristics of Picloram in Grassland Ecosystem. B-1111. College Station, TX: Tex Agric Exp Stn, 1971.

Bovey RW, Young AL. Physiological effects of phenoxy herbicides in higher plants. In: The Science of 2,4,5-T and Associated Phenoxy Herbicides New York: Wiley, 1980, pp. 217–238.

Bovey RW, Baur JR, Diaz-Colon JD. Phytotoxicity of 2,4,5-T, picloram and dicamba alone and in mixtures in tissue culture. Weed Sci 22:191–192, 1974.

Bovey RW, Davis FS, Merkle MG. Distribution of picloram in huisache after foliar and soil applications. Weeds 15:245–249, 1967.

Bovey RW, Davis FS, Morton HL. Herbicide combinations for woody plant control. Weeds 16:332–335, 1968.

Bovey RW, Franz E, Whisenant SG. Influence of organosilicone surfactants and spray nozzle types on the fate and efficacy of clopyralid in honey mesquite (*Prosopis glandulosa*) Weed Sci 42:658–664, 1994.

Bovey RW, Haas RH, Meyer RE. Daily and seasonal response of huisache and Macartney rose to herbicides. Weed Sci 20:577–580, 1977.

Bovey RW, Hein H Jr, Keeney FN. Phytotoxicity, absorption and translocation of five clopyralid formulations in honey mesquite (*Prosopis glandulosa*). Weed Sci 37: 19–22, 1989.

Bovey RW Hein H Jr, Meyer RE. Absorption and translocation of triclopyr in honey mesquite (*Prosopis iuliflora* var. *glandulosa*) Weed Sci 31:807–812, 1983.

Bovey RW, Hein H Jr, Meyer RE. Concentration of 2,4,5-T, triclopyr, picloram and clopyralid in honey mesquite (*Prosopis glandulosa*) stems. Weed Sci 34:211–217, 1986.

Bovey RW, Hein H Jr, Meyer RE. Phytotoxicity and uptake of clopyralid in honey mesquite (*Prosopis glandulosa*) as affected by adjuvants and other herbicides. Weed Sci 36:20–23, 1988a.

Bovey RW, Hein H Jr, Meyer RE. Mode of clopyralid uptake by honey mesquite (*Prosopis glandulosa*). Weed Sci 36:269–272, 1988b.

Bovey RW, Ketchersid ML, Merkle MG. Distribution of triclopyr and picloram in huisache (*Acacia farnesiana*). Weed Sci 27:527–531, 1979b.

Bovey RW, Meyer RE, Morton HL. Use of a woody plant nursery in herbicide research. B-1216. College Station, TX: Tex Agric Exp Stn, 1979a.

Bovey RW, Meyer RE, Whisenant SG. Effect of simulated rainfall on herbicide performance in huisache (*Acacia farnesiana*) and honey mesquite (*Prosopis glandulosa*) Weed Tech 4:52–55, 1990b.

Bovey RW, Stermer RA, Bouse LF. Spray deposit of clopyralid on honey mesquite (*Prosopis glandulosa* Torr.). Weed Tech 5:499–503, 1991.

Bovey RW, Hein H Jr, Keeney FN, Whisenant SG. Phytotoxicity and transport of clopyralid from three formulations in honey mesquite. J Plant Growth Reg 9:65–69, 1990a.

Bovey RW, Hein H Jr, Meyer RE, Bouse LF. Influence of adjuvants on the deposition, absorption and translocation of clopyralid in honey mesquite (*Prosopis glandulosa*) Weed Sci 35:253–258, 1987.

Bovey RW, Morton HL, Baur JR, Diaz-Colon JD, Dowler CC, Lehman SK. Granular herbicides for woody plant control. Weed Sci 17:538–541, 1969.

Brady HA. Light intensity and the absorption and translocation of 2,4,5-T by woody plants. Weed Sci 17:320–322, 1969.

Brady HA. Ammonium nitrate and phosphoric acid increase 2,4,5-T absorption by tree leaves. Weed Sci 18:204–206, 1970a.

Brady HA. High temperature boosts 2,4,5-T activity in woody plants. Proc South Weed Sci Soc 23:234–236, 1970b.

Brady HA. Spray date effects on behaviour of herbicides on brush. Weed Sci 19:200–202, 1971.

Brady HA. Drop size affects absorption of 2,4,5-T by six hardwood species. Proc South Weed Sci Soc 25:282–286, 1972.

Brown CL, Nix LE. Uptake and transport of paraquat in slash pine. Forest Sci 21:359–364, 1975.

Burns ER, Buchanan GA, Hiltbold AE. Absorption and translocation of 2,4-D by wolftail (*Carex cherokeensis* Schwein). Weed Sci 17:401–404, 1969.

Carter MC, Amitrole. In: Kearney PC, Kaufman DD, eds. Herbicides: Chemistry, Degradation and Mode of Action. 2nd ed. vol 1. New York: Marcel Dekker, 1975, pp. 377–398.

Chrzanowski RL. Metabolism of [^{14}C] fosamine ammonium in brush and turf. J Agr Food Chem 31:223–227, 1983.

Coupland D, Peabody DV. Absorption, translocation and exudation of glyphosate, fosamine and amitrole in field horsetail (*Equisetum arvense*). Weed Sci 29:556–560, 1981.

Crafts AS. The Chemistry and Mode of Action of Herbicides. New York: Wiley Interscience, 1961a.

Crafts AS. Translocation in Plants. New York: Holt, Rinehart and Winston, 1961b.

Crafts AS. Herbicide behavior in the plant. In: Audus LJ, ed. The Physiology and Biochemistry of Herbicides. New York: Academic, 1964, pp. 75–110.

Crafts AS Yamaguchi S. The autoradiography of plant materials. Manual 35. Davis, CA: California Agric Exp Stn, 1964.

Dalrymple AV, Basler E. Seasonal variation in absorption and translocation of 2,4,5-trichlorophenoxyacetic acid and respiration rates in blackjack oak. Weeds 11:41–45, 1963.

Davis FS, Bovey RW, Merkle MG. The role of light, concentration, and species in foliar uptake of herbicides in woody plants. Forest Sci 17:164–169, 1968a.

Davis FS, Bovey RW, Merkle MG. Effect of paraquat and 2,4,5-T on the uptake and transport of picloram in woody plants. Weeds 16:336–339, 1968b.

Davis FS, Merkle MG, Bovey RW. Effect of moisture stress on the absorption and transport of herbicides in woody plants. Bot Gaz 129:183–189, 1968c.

Davis FS, Meyer RE, Baur JR, Bovey RW. Herbicide concentrations in honey mesquite phloem. Weed Sci 20:264–267, 1972.

Dewey SA, Appleby AP. A comparison between glyphosate and assimilate translocation patterns in tall morning glory (*Ipomoea purpurea*) Weed Sci 31:308–314, 1983.

Diaz-Colon JD, Bovey RW, Davis FS, Baur JR. Comparative effects and concentration of picloram, 2,4,5-T and dicamba tissue culture. Physiol Plant 27:60–64, 1972.

Duke SO. Glyphosate. In: Kearney PC, Kaufman DD, eds. Herbicides: Chemistry, Degradation and Mode of Action. 2nd ed. vol. 3. New York: Marcel Dekker, 1988, pp. 1–70.

Eaton BJ, Magnussen JD, Rainey DP. Metabolism of tebuthiuron in soil and plants. Abstr no. 199. Denver, CO: Weed Science Society of America, 1975, p. 83.

Fisher CE, Young DW. Some factors influencing the penetration and mobility of chemicals in the mesquite plant. Proc North Central Weed Control Conf 6:197–202, 1948.

Fisher CE, Young DW. Tip immersion studies on mesquite (*Prosopis juliflora*). North Central Weed Control Conf res rept 6:214–215 (abstract), 1949.

Fisher CE, Fults JL, Hopp H. Factors affecting action of oils and water-soluble chemicals in mesquite eradication. Ecol Monographs 16:109–126, 1946.

Fisher CE, Meadors CH, Behrens R. Some factors that influence the effectiveness of 2,4,5-trichlorophenoxyacetic acid in killing mesquite. Weeds 4:139–147, 1956.

Fisher CE, Meadors CH, Behrens R, Robinson ED, Marion PT, Morton HL. Control of mesquite on grazing lands. Bull. 935. College Station, TX: Tex Agric Exp Stn, 1959.

Frear DS. The benzoic acid herbicides. In: Kearney PC, Kaufman DD eds. Herbicides: Chemistry, Degradation, and Mode of Action. 2nd ed vol. 2. New York: Marcel Dekker, 1976, pp. 541–607.

Gardner JA. Substituted uracil herbicides. In: Kearney PC, Kaufman DD, eds. Herbicides: Chemistry, Degradation and Mode of Action. 2nd ed. vol. 1. New York: Marcel Dekker, 1975, pp. 293–321.

Gardner JA, Rhodes RC, Adams JB Jr, Soboczenski EJ. Synthesis and studies with 2-C^{14} labeled bromacil and terbacil. Agr Food Chem 17:980–986, 1969.

Geiger DR, Bestman HD. Self-limitation of herbicide mobility by phytotoxic action. Weed Sci 38:324–329, 1990.

Green TH, Minogue PJ, Brewer CH, Glover GR, Gierstad DH. Absorption and translocation of [^{14}C] glyphosate in four woody species. Can J For Sci 22:785–789, 1992.

Hall O. Limitations of surfactant and pH effects on herbicide behavior in woody plants. Weed Sci 21:221–223, 1973.

Harvey GJ. Uptake and distribution of acid, ester and amine salt formulations of ^{14}C-2,4-D by groundsel bush (*Baccharis halimifolia*) with time. Plant Protec Q 5:156–159, 1990.

Hinshalwood AM, Kirkwood RC. The effect of simultaneous application of ethephon of 2,4-D on the absorption, translocation and biochemical action of Asulam in bracken fern (*Pteriduim aquilinum* (L.) Kuhn). Can J Plant Sci 68:1025–1034, 1988.

Hull HM. Studies on herbicidal absorption and translocation in velvet mesquite seedlings. Weeds 4:22–42, 1956.

Hull HM. Leaf structure as related to absorption of pesticides and other compounds. In: Gunther FA, Gunther JD, eds., Residue Reviews, vol. 31. New York: Springer-Verlag, 1970, pp. 45–93.

Hull HM, Morton HL. Morphological response of two mesquite varieties to 2,4,5-T and picloram. Weed Sci 19:712–716, 1971.

Hull HM, Morton HL, Warrie JR. Environmental influences on cuticle development and resultant foliar penetration. Bot Rev 41:421–452, 1975.

Jacoby RW, Ansley RJ, Meadors CH. Late season control of honey mesquite with clopyralid. J Range Mgt 44:56–58, 1991.

Jacoby PW, Ansley RJ, Meadors CH, Huffman AH. Epicuticular wax in honey mesquite: Seasonal accumulation and intraspecific variation. J Range Mgt 43:347–350, 1990.

Jensen KI, Kimball ER. Uptake and metabolism of hexazinone in *Rubus hispidus* L. and *Pyrus melanocarpa* (Michx.) Willd. Weed Res 30:35–41, 1990.

Kenney-Wallace G, Blackman GE. The uptake of growth substances. XIV. Patterns of uptake by Lemma minor of phenoxyacetic and benzoic acids, following progressive chlorination. J Exp Bot 23:114–127, 1972.

Klevorn TB, Wyse DL. Effect of leaf girdling and rhizome girdling on glyphosate transport in quackgrass (*Agropyron repens*). Weed Sci 32:744–750, 1984.

Lee A, Gatlerdam PE, Chiu TY, Mallipudi NM, Fiali RR. Plant metabolism. In: Shaner DL, O'Connor SL, eds. The Imidazolinone Herbicides. Boca Raton, FL: CRC Press, 1991 pp. 151–165.

Mann RE, Witt WW, Rieck CE. Fosamine absorption and translocation in multiflora rose (*Rosa multiflora*). Weed Sci 34:830–833, 1986.

Mayeux HS Jr, Johnson HB. Absorption and translocation of picloram by Lindheimer pricklypear (*Opuntia lindheimer*) Weed Sci 37:161–166, 1989.

Mayeux HS Jr, Jordan WR. Epicuticular wax content and herbicide penetration of goldenweed leaves. Proc South Weed Sci Soc 33:289 (abstract), 1980.

Mayeux HS Jr, Jordan WR. Rainwash and water stress alter epicuticular wax content of goldenweed leaves. Proc South Weed Sci Soc 34:264 (abstract), 1981.

Mayeux HS Jr, Jordan WR. Variation in amounts of epicuticular wax on leaves of *Prospis glandulosa*. Bot Gaz 145:26–32, 1984.

Mayeux HS Jr, Jordan WR. Rainfall removes epicuticular waxes from *Isocoma* leaves. Bot Gaz 148:420–425, 1987.

Mayeux HS Jr, Scifres CJ. Foliar uptake and transport of 2,4-D and picloram by Drummond's goldenweed (*Isocoma drummondii*). Weed Sci 28:678–682, 1980.

Mayeux HS Jr, Scifres CJ. Drummond's goldenweed and its control with herbicides. J Range Mgt 34:98–101, 1981.

Mayeux HS Jr, Wilkinson RE. Composition of epicuticular wax on *Prosopis glandulosa* leaves. Bot Gaz 151:240–244, 1990.

Mayeux HS, Drawe DL Jr, Scifres CJ. Control of common goldenweed with herbicides and associated forage release. J Range Mgt 32:271–274, 1979a.

Mayeux HS Jr, Scifres CJ, Meyer RE. Some factors affecting the response of spiny aster to herbicide sprays. B. 1197. College Station, TX: Texas Agric Exp Stn, 1979b.

Mayeux HS Jr, Jordan WR, Meyer RE, Meola SM. Epicuticular wax on goldenweed (*Isocoma* spp.) leaves: Variation with species and season. Weed Sci 29:389–393, 1981.

McCall HG, Scifres CJ, Merkle MG. Influence of foam adjuvants on activity of selected herbicides. Weed Sci 22:384–388, 1974.

McNeil WK, Stritzke JK, Basler E. Absorption, translocation and degradation of tebuthiuron and hexazinone in woody species. Weed Sci 32:739–743, 1984.

Merkle MG, Davis FS. The use of gas chromatography for determining the translocation of picloram and 2,4,5-T. Proc South Weed Conf 19:557–561, 1966.

Merkle MG, Davis FS. Effect of moisture stress on absorption and movement of picloram and 2,4,5-T in beans. Weeds 15:10–13, 1967.

Merkle MG, Leinweher CL, Bovey RW. The influence of light, oxygen and temperature on the herbicidal properties of paraquat. Plant Physiol 40:832–835, 1965.

Meyer RE. Seasonal response of honey mesquite to herbicides. B-1174 USDA, ARS and Tex Agric Exp Stn, 1977.

Meyer RE, Bovey RW. Control of honey mesquite (*Prosopis juliflora* var. *glandulosa*) and Macartney rose (*Rosa bracteata*) with soil-applied herbicides. Weed Sci 27: 280–284, 1979.

Meyer RE, Bovey RW. Influence of environment and stage of growth on honey mesquite (*Prosopis glandulosa*) response to herbicides. Weed Sci 34:287–299, 1986.

Meyer RE, Meola SM. Morphological characteristics of leaves and stems of selected Texas woody plants. ARS tech. bull. no. 1564. Washington, DC: USDA, 1978.

Meyer RE, Morton HL, Haas RH, Robison ED. Morphology and anatomy of honey mesquite. Tech. bull. no. 1423. Washington, DC: USDA, ARS and Tex Agric Exp Stn, 1971.

Meyer RE, Bovey RW, Riley TE, Mckelvy WT. Leaf removal effect after sprays to woody plants. Weed Sci 20:498–501, 1972a.

Meyer RE, Bovey RW, Riley TE, Mckelvy WT. Influence of plant growth stage and environmental factors on the response of honey mesquite to herbicides. B-1127. Washington, DC: USDA, ARS and Tex Agric Exp Stn, 1972b.

Morey PR, Dahl BE. Inhibition of mesquite (*Prosopis juliflora* var. *glandulosa*) growth by fosamine. Weed Sci 28:251–255, 1980.

Morton HL. Influence of temperature and humidity on foliar absorption, translocation, and metabolism of 2,4,5-T by mesquite seedlings. Weeds 14:136–141, 1966.

Morton HL, Davis FS, Merkle MG. Radioisotopic and gas chromatographic methods for measuring absorption and translocation of 2,4,5-T by mesquite. Weed Sci 16:88–91, 1968.

National Academy of Science. Physiological aspects of herbicidal action. In: Principles of Plant and Animal Pest Control vol. 2. Weed Control. National Academy of Science publ. 1597. Washington, DC: National Academy of Science, 1968, pp. 146–163.

Norris RF, Bukovac MJ. Some physical-kinetic considerations in penetration of naphthaleneacetic acid through isolated pear leaf cuticle. Physiol Plant 22:701–712, 1969.

Norris LA, Freed VH. Effects of formulation and molecular configuration on absorption and translocation of some phenoxy herbicides in big leaf maple seedlings. In: Western Weed Conference Research Progress Report. 1964, pp. 25–27.

Norris LA, Freed VH. The absorption and translocation characteristics of several phenoxyalkyl acid herbicides in big leaf maple. Weed Res 6:203–211, 1966a.

Norris LA, Freed VH. The absorption, translocation and metabolism characteristics of 4-(2,4-dichlorophenoxy) butyric acid in big leaf maple. Weed Res 6:283–291, 1966b.

Pallas JE. Effects of temperature and humidity on foliar absorption and translocation of 2,4-dichlorophenoxy-acetic acid and benzoic acid. Pl Physiol Lanc 35:575–580, 1960.

Radosevich SR, Bayer DE. Effect of temperature and photoperiod on triclopyr, picloram, and 2,4,5-T translocations. Weed Sci 27:22–27, 1979.

Richardson RG. Foliar penetration and translocation of 2,4,5-T in blackberry (*Rubus procerus* P.J. Muell). Weed Res 15:33–38, 1975.

Richardson RG. Foliar absorption and translocation of fosamine and 2,4,5-T in blackberry (*Rubus procerus* P.J. Muell). Weed Res 20:159–163, 1980.

Richardson RG, Grant AR. Effect of concentration on absorption and translocation of 2,4,5-T in blackberry (*Rubus procerus* P.J. Muell). Weed Res 17:367–372, 1977.

Robertson MM, Kirkwood RC. The mode of action of foliage applied translocated herbicides with particular reference to the phenoxy-acid compounds. I. The mechanism and factors influencing herbicides absorption. Weed Res 9:224–240, 1969.

Robison ED. Response of mesquite to 2,4,5-T, picloram, and 2,4,5-T/picloram combinations. Proc South Weed Conf 20:199 (abstract), 1967.

Sands BR, Bachelard EP. Uptake of picloram by eucalyptus leaf discs I. Effect of surfactants and nature of the leaf surfaces. New Phytol 72:69–86, 1973.

Sands BR, Bachelard EP. Uptake of picloram by eucalyptus leaf discs II. Role of stomata. New Phytol 72:87–99, 1973b.

Sargent JA, Blackman GE. Studies on foliar penetration. IV. Mechanisms controlling the rate of penetration of 2,4-dichlorophenoxyacetic acid (2,4-D) into leaves of *Phaseolus vulgaris*. J Exp Bot 20:542–555, 1969.

Schmutz EM. Absorption, translocation, and toxicity of 2,4,5-T in creosotebush. Weed Sci 19:510–516, 1971.

Scifres CJ, Baur JR, Bovey RW. Absorption of 2,4,5-T applied in various carriers to honey mesquite. Weed Sci 21:94–96, 1973a.

Scifres CJ, Kienast CR, Elrod DJ. Honey mesquite seedling growth and 2,4,5-T susceptibility as influenced by shading. J Range Mgt 26:58–60, 1973c.

Scifres CJ, McCall HG, Fryrear DW. Foam systems as herbicide carriers for range improvement. MP-1156. College Station, TX: Tex Agric Exp Stn, 1974.

Scifres CJ, Mutz JL, Meadors CH. Response of range vegetation to grid placement and aerial application of karbutilate. Weed Sci 26:139–144, 1978.

Scifres CJ, Bovey RW, Fisher CE, Baur JR. Chemical control of mesquite, In: Scifres CJ, ed. Mesquite. Res monograph 1. College Station, TX: Tex Agric Exp Stn, 1973b, pp. 24–32.

Seiler JR, Cazell BH, Schneider WG, Zedaker SM, Kreh RE. Effect of plant moisture stress on adsorption and translocation of triclopyr in oak seedlings. Can J For Res 23:2213–2215, 1993.

Sharma MP, Vanden Born WH. Foliar penetration of picloram and 2,4-D in aspen and balsam poplar. Weeds 18:57–63, 1970.

Steinert WG, Stritzke JF. Uptake and phytotoxicity of tebuthiuron. Weed Sci 25:390–395, 1977.

Sterling TM, Lownds NK. Picloram absorption by broom snakeweed (*Gutierrezia sarothrae*) leaf tissue. Weed Sci 40:390–394, 1992.

Stidham MA, Singh BK. Imidazolinone-acetohydroxyacid synthase interactions. In:

Shaner DL, O'Connor SL, eds. The Imidazolinone Herbicides. Boca Raton, FL: CRC Press, 1991 pp. 72–90.

Swanson CR, Baur JR. Absorption and penetration of picloram in potato tuber discs. Weed Sci 97:311–314, 1969.

Tan S, Crabtree GD. Cuticular penetration of 2,4-D as affected by interaction between a diethylene glycol monooleate surfactant and apple leaf cuticles. Pestic Sci 41:35–39, 1994.

Tschirley FH, Hull HM. Susceptibility of velvet mesquite to an amine and an ester of 2,4,5-T as related to various biological and meteorological factors. Weeds 7:427–435, 1959.

Turner DJ. The influence of additives on the penetration of foliar growth regulator herbicides. Pestic Sci 3:323–331, 1972.

Upchurch RP, Coble HD, Keaton JA. Rainfall effects following herbicidal treatment of woody plants. Weed Sci 17:94–98, 1969.

Weed Science Society of America. Herbicide Handbook. 7th ed. Champaign, IL: Weed Science Society of America, 1994.

Whisenant SG, Bouse LF, Crane RA, Bovey RW. Droplet size and spray volume effects on honey mesquite mortality with clopyralid. J Range Mgt 46:257–261, 1993.

Wilkinson RE, Mayeux HS Jr. Composition of epicuticular wax on *Opuntia engelmanii*. Bot Gaz 151:342–347, 1990.

Wills GD, Basler E. Environmental effects on absorption and translocation of 2,4,5-T in winged elm. Weed Sci 19:431–434, 1971.

Wilson BJ, Nishimoto RK. Ammonium sulfate enhancement of picloram activity and absorption. Weed Sci 23:289–296, 1975a.

Wilson BJ, Nishimoto RK. Ammonium sulfate enhancement of picloram absorption by detached leaves. Weed Sci 23:297–301, 1975b.

Wood JE, Stephenson GR, Hall JC, Horton RF. Selective phytotoxicity of hexazinone in *Pinus resinosa* and *Pinus banksiana*. Pestic Biochem Physiol 44:108–118, 1992.

Young DW, Fisher CE. Shield tests on mesquite (*Prosopis juliflora*). North Central Weed Control Conf Res Rept 6:200 (abstract), 1949a.

Young DW, Fisher CE. Treatments to the foliage of mesquite (*Prosopis juliflora*) with ground equipment. North Central Weed Control Conf Res Rept 6:147 (abstract), 1949b.

Young DW, Fisher CE. Toxicity and translocation of herbicides in mesquite. Proc North Central Weed Control Conf 7:95–99, 1950.

10

Herbicide Spray Drift, Vapor, and Residue Effects on Crop Plants

I. INTRODUCTION

An added benefit of woody plant control is the simultaneous control of the herbaceous weeds (nonwoody) associated with woody plants. Most of the herbicides discussed in this book readily control herbaceous weeds. Selection of the herbicide may depend upon the degree of selectivity needed. For example, in pasture and rangeland situations to control woody plants and leave desirable grasses and forbs, low rates of growth-regulator-type herbicides such as 2,4-D or dicamba may be desired. If total vegetation control is desired, more nonselective herbicides, such as bromacil, paraquat, or glyphosate—or combinations of contact plus soil-applied materials—may be desired for quick burndown and residual control. Selectivity may also be controlled by the herbicide rate and the timing of application, depending upon the species' susceptibility to the herbicide.

Some crop plants are extremely sensitive to woody plant herbicides, and extremely small amounts cause injury or death. Great care must be taken during application to prevent spray drift or volatilization from treated to nontreated areas. This chapter will explore herbaceous plant susceptibility to woody plant herbicides and factors affecting the results.

II. WHERE TO GET HELP

Most state agricultural experiment station or state extension service systems have publications or specialists that provide suggested herbicides for specific herbaceous weeds in forage, agronomic and horticultural crops, pastures, and forestry and noncrop areas. If your state does not have such information many times neighboring states have similar weed problems and can provide information. The *Weed Control Manual* (Meister Publishing Co., 37733 Euclid Ave., Willoughby OH 44094-5992) lists a large number of weeds and the approved herbicides used in various crop and noncrop situations, including many of the herbicides discussed in this book. Other sources include Klingman et al. (1983), who list about 520 herbaceous weeds and their response to phenoxy herbicides, dicamba, picloram, amitrole, and glyphosate. The *Handbook of Weed Management Systems*, edited by A. L. Smith, includes weed control in various oil seeds; grains; horticulture, pasture, and hay crops; turfgrass; rangeland; and forestry (Smith, 1995). Noxious weeds on rangeland have been thoroughly researched and discussed (James et al., 1991). Weed and brush control in rights-of-way and forestry in the United States has also been discussed (Newton, 1996). Extension and research specialists of state universities, private consultants, and private industry personnel can help by providing oral information or published data and herbicide labels.

III. SAFE USE AND PRECAUTIONS

Spray drift is the physical movement of spray particles from the spray nozzle. It can move from the treated area to nontarget areas, where it could cause damage if susceptible vegetation is present. Since herbicide drift depends upon wind speed, spray droplet size, and boom height, wind speeds in excess of 10 mph (16 km/hr) are too high for the safe application of herbicides—especially hormone-type herbicides—when susceptible crops are in the area. In addition to the hazards of spray drift, weed control is reduced, since less spray reaches the weeds. Wind velocity can be determined by using an accurate, inexpensive wind gauge.

Volatility is the loss and movement of an herbicide as fumes or vapor in the air. Some herbicides may volatilize in temperatures greater than 32°C and low relative humidity (especially the ester formulations of the phenoxy herbicides), and nonvolatile forms should used. Injury and yield loss can result in susceptible plants if air currents move volatile herbicides to nontarget areas. The amount of fumes or vapors given off is related to the vapor pressure of the chemical (Mullison and Hummer, 1949). 2,4-D acid, amine salts, and sodium salts have very low volatility and cause little or no volatility hazard, whereas the methyl, ethyl, and isopropyl esters of 2,4-D are very volatile. Low-volatile esters of 2,4-D are made using high-molecular-weight alcohols (Klingman, 1961). In addition, an ether linkage (-0-) further reduces the volatility. The iso-octyl ester (8 carbon) and the butoxy ethyl ester are both low-volatile esters. High-molecular-weight alcohols are relatively expensive, so their corresponding 2,4-D esters are more expensive to make than volatile forms.

IV. HERBICIDE VAPOR EFFECTS

A. Phenoxy Herbicides

The potential for high-volatile characteristics to cause injury to nontarget vegetation has long been recognized (Marth and Mitchell, 1949; Baskins and Walker, 1953). Some ester formulations of 2,4-D and 2,4,5-T were of concern because of the damaging effects of their vapors on plants (Hitchcock et al., 1953). When a phenoxy herbicide is to be used near susceptible crops, legal use may depend upon its low volatility.

The relative volatilities of 2,4-D esters have been measured by exposing selected test plants in cellophone bags or bell jars to vapors of the herbicide for a designated time and air temperature (Marth and Mitchell, 1949; Zimmerman et al., 1953). Leafy epinasty and stem curvature were used to evaluate initial response 24 hr after treatment. Leaf modification, stem proliferation, and growth inhibition were considered at the end of 7 to 10 days (Zimmerman et al., 1953). These studies established that the magnitude of response increased with the concentration and volatility of the chemicals. Temperature increases from 24°C to 47°C increased plant response, especially with the isopropyl (high-volatile) versus the propylene glycol butyl ether ester (low-volatile) 2,4-D forms. A progressively increased length of exposure caused an increased magnitude of test plant response with either the high- or low-volatile esters. Based on 1- and 9-day evaluations, lowest to highest, low-volatility 2,4-D formulations were alkanolamine < isooctyl < ethoxyethoxy propanol < butoxy propyl < butoxy ethyl < tetrahydrofurfuryl < propylene glycol butyl ether. The isopropyl form was slightly more volatile than the pentasol form of the high-volatility 2,4-D esters (Zimmerman et al., 1953).

Using Rutgers tomato plants (*Lycopersicon esculentum* Mill.) exposed to 2,4-D vapors, Baskins and Walker (1953) found that the lower alkyl esters (isopropyl, butyl, alkyl, amyl, and/or pentyl) were volatile at normal temperatures, while the amine forms (dimethylamine, ethanolamine, diethanolamine, triethanolamine, and alkanolamine) were low in volatility. Increased temperature from 22°C, 34°C, and 48°C increased plant response to the butyl ester, but some esters, such as the propylene glycol and tetrahydrofurfurl, had low volatility until exposed to 48°C. The data of Baskins and Walker (1953) agree with those of Zimmerman et al. (1953).

To better quantify the volatility of 2,4-D, Flint et al. (1968) determined the vapor pressures of four common commercial low-volatile esters and a high-volatile ester of 2,4-D by gas–liquid chromatography. The order of increasing volatility and the vapor pressure of these esters in mmHg at 187°C are as follows: isooctyl-2.7; 2-ethylhexyl-3.0; butoxy ethanol-3.9; propylene glycol butyl ether-3.9; and reference isopropyl-16.7.

Grover (1975) studied the effect of time, flow rate, and temperature on the volatilization of technical grade *n*-butyl ester of 2,4-D in a closed-air-flow system. Gas chromatography was used to quantify 2,4-D. The amount of ester volatilized was linear with time and at constant temperature and air flow, but the volatility increased about eightfold when the temperature was increased from 30°C to 50°C. Que Hee and Sutherland (1974) indicated that the rate of volatilization of a given compound is usually considered to be directly proportional to its vapor pressure, which is measured at the thermodynamic equilibrium at a given temperature. In the field, however, volatilization is under kinetic control, since the atmosphere acts as an infinite reservoir and the wind as a dispersive agent.

The degree and rate of volatilization and the partial vapor pressures of the methyl, n butyl, and octyl esters of 2,4-D were determined at ambient temperatures using a kinetic system approximating field-air conditions (Que Hee and Sutherland, 1974). The initial surface area/applied mass ratio Q, as well as the temperature and the type of compound, determined the volatility. Only when the Q values were kept constant did the volatility decrease with increasing chain length and decreasing saturated vapor pressure as expected. A mathematical model was constructed. Similar results were obtained for the amine salts that had saturated vapor pressures of approximately 10^{-10} mmHg at 38°C. Vapor drift can be essentially eliminated by using the amine salts.

Noble and Hamilton (1990) indicated that in Australia a CIPAC/AO AC test with tomato plants is used to specify the volatility ratings of herbicide ester formulations. This work compares the tomato plant test with an alternative chemical one. The concentrations of esters and the effective molecular weight and density of each formulation were used with the ester vapor pressures to calculate its herbicide vapor pressure both as complete and evaporated formulations. The range was from 28.8 mPa (at 25°C) for a mixture of 2,4-D esters to 0.07 mPa (at 25°C) for a 2,4,5-T-(isooctyl) formulation as complete formulations, and 35.5 and 0.16 mPa (at 25°C) as evaporated ones. A value of 0.6 mPa (at 25°C) was selected on the basis of the tomato plant test as the cutoff area for low-volatile esters and is recommended to be included in specifications for herbicide esters. Formulations with an herbicide vapor pressure above 3.3 mPa (at 25°C) are high volatility ones according to the tomato plant test, while between 0.6 to 3.3 mPa (at 25°C) is a borderline region, in which the test gives mixed results. Levels of 2,4-D-ethyl and methyl were added to pure 2-ethylhexyl esters of 2,4-D and a 2,4,5-T-(isooctyl) formulation to find what level of contamination would change the rating of these esters from low to high volatility. Formulations of 2,4-D-(isooctyl) should not contain more than 1.1 g/L 2,4-D as a methyl ester or 2.0 g/L 2,4-D as an ethyl ester. Formulations of 2,4,5-T-(isooctyl) should not contain more than 2.6 g/L 2,4-D as a methyl ester or 4.7 g/L 2,4-D as an ethyl ester.

Phytotoxicity from herbicide vapor is readily demonstrated in the laboratory, but appears to be observed infrequently in the field (Breeze, 1994). This may be because the visible symptoms are often hard to ascribe to herbicides with certainty, and in many cases damage may be only slight. Unfortunately, there is also a lack of reliable field data, in part because it is difficult to do phytotoxicity experiments and to measure vapor concentrations. Whereas vapor of the free acid of 2,4-D is phytotoxic in the laboratory and the dimethylamine (DMA) salt of dicamba releases toxic vapor, no such effect has been noted from 2,4-D DMA, even though it has a higher pK_a (an indication of the potential of the free acid to hydrate and thus give the nonvolatile anion) than dicamba. Fluroxypyr vapor appears to be toxic in laboratory conditions, but no damage has been reported from the field. Possible reasons for discrepancies between laboratory and field effects include plant size; it has been shown that the response of young tomato plants to 2,4-D is very sensitive to age. Laboratory experiments are often carried out on small plants. Certain combinations of plant species and herbicides show growth promotion at low doses, but no such response has been noted in the field.

The phytotoxic effects of 2,4-D vapors have been demonstrated on a number of plants in addition to tomatoes, including lettuce (*Lactuca sativa* L.) (Bennet, 1989; Breeze and West, 1987a,b; Breeze, 1990; Breeze and van Rensburg, 1991; 1992; van Rensburg and Breeze, 1990), barley (Breeze et al., 1992), tomatoes, lettuce, cabbage (*Brassica oleracea* L.), sunflowers (*Helianthus annuus* L.), field beans (*Vicia faba* L.), and white clover (*Trifolium repens* L.) (Breeze and West, 1987a). Cotton (*Gossypium* sp.) is also very sensitive to 2,4-D vapors (Weise et al., 1992).

The fact that volatile 2,4-D compounds readily cause vapor drift damage on a wide variety of plants is therefore well documented, especially under laboratory conditions, in which plants are confined to chambers or containers with the herbicide vapor. Most of the research has been done with high-volatility esters, which are not widely used and have generally been replaced by the less volatile long-chain compounds. In controlled-environment studies, Bennett (1989) showed that tomato plants subjected to the isooctyl ester/ioxynal applied at 40 μg, 100 μg, and 1000 μg per plant killed all the plants in 35 days. None of the treatments killed plants applied in the vapor phase, however, although a variety of sublethal plant responses developed.

Gile (1983) studied the airborne losses of seven commercial 2,4-D formulations in a simulated wheat field. The butyl and isooctyl esters had the greatest airborne losses, followed by the propylene glycol butyl ether esters. The amines, ethanol/isopropyl amine, and two dimethyl amines were nearly nonvolatile. Two to 4 days after application the 2,5-dichloro-4-hydroxyphenoxyacetic acid metabolite predominated in the plant samples, with only traces of the parent esters. No detectable 2,4-D residues were found in soil.

Simpson et al. (1980) found no high-volatility or low-volatility esters of 2,4-D after monitoring three counties in California despite numerous 2,4-D applications. The isooctyl ester of 2,4-D was not detected in air samples taken from

September 1987 to March 1988 in the Tala Valley in Natal, despite its widespread use in sugarcane (Smit et al., 1992). Polar forms of 2,4-D were found, however. Vogel (1998) found detectable levels (<7 ppb) of 2,4-D in lettuce grown in the Tala Valley during the summer (January 1991), while that grown in winter and spring did not contain detectable 2,4-D. The butoxyethyl ester of triclopyr was not found in Tala Valley lettuce, but it is not used as extensively as the 2,4-D esters.

B. Picloram

Gentner (1964) found that in a closed system the vapors from the potassium salt of picloram caused both severe damage and death in pinto beans (*Phaseolus vulgaris* L). Vapors from the potassium salt of picloram caused more injury to pinto beans than vapors from the same application rates of the propylene glycol butyl ether esters of 2,4-D or the dimethylamine salt of dicamba.

C. Dicamba

Factors influencing dicamba drift, especially vapor drift, were examined in field and growth-chamber studies (Behrens and Lueschen, 1979). In field experiments, potted soybeans [*Glycine max* (L.) Merr.] exposed to vapors rising from corn (*Zea mays* L.) foliarly treated with the sodium (Na), dimethylamine (DMA), diethanolamine (DEOA), or *N*-tallow-*N*, N^1, N^1-trimethyl-1, 3-diaminopropane (TA) salts of dicamba developed dicamba injury symptoms. Dicamba volatilization from treated corn was detected on soybeans for 3 days after the application. Dicamba vapors caused symptoms in soybeans placed up to 60 meters downwind of the treated corn. When vapor and/or spray drift caused soybean terminal bud kill, the yields were reduced. In growth-chamber studies, dicamba volatility effects on soybeans could be reduced by lowering the temperature or increasing the relative humidity. Rainfall of 1 mm or more on treated corn ended dicamba volatilization. The dicamba volatilization was greater from corn and soybean leaves than from velvetleaf (*Abutilon theophrasti* Medic.) leaves and blotter paper. The volatilization of dicamba formulations varied in growth-chamber comparisons, with the acid being the most volatile and the inorganic salts being the least. Under field conditions, however, using less-volatile formulations did not eliminate dicamba symptoms on soybeans. The volatile component of the commercial DMA salt of dicamba was identified by gas chromatography-mass spectrometry as free dicamba acid.

The diglycolamine salt of dicamba appeared to be less volatile than the dimethylamine salt of 2,4-D and the butoxyethyl ester of triclopyr in some instances when soybeans were used as indicator plants (Sciumbato, 1999). The volatility of the DGA salt of dicamba, however, was generally equal to that of the DMA salt of 2,4-D when cotton was used as the test plant. Although both salt formulations of 2,4-D and dicamba proved to be less volatile than the ester of triclopyr, there were no conclusive results that revealed the DGA salt of dicamba to be unequivocally less volatile than the DMA salt of 2,4-D.

V. SIMULATED DRIFT AND SUBLETHAL DOSE EFFECTS ON CROP PLANTS

The phenoxy and other hormonelike herbicides such as picloram (Tordon), dicamba (Banvel), triclopyr (Remedy), and clopyralid (Reclaim) are extremely important for the control of weeds and brush on grazing lands. These herbicides, however, can sometimes cause injury to forage, agronomic, and horticultural crops if spray drift is allowed from treated areas to nontarget areas. Precautions are printed on the label and applications should be made during low wind velocity with air currents moving away from sensitive vegetation.

A. Cotton (*Gossypium hirsutum* L.)

1. 2,4-D

It was recognized early that drift from 2,4-D could cause serious injury to cotton (Brown et al., 1948; Dunlap, 1948; Ergle and Dunlap, 1949). Arle (1954) reported that injury to cotton also could occur from volatilization of 2,4-D ester, use of 2,4-D-contaminated sprayers, 2,4-D carried in irrigation water, smoke from burning weeds and brush treated with 2,4-D, and reuse of 2,4-D containers.

In Mississippi, Carns and Goodman (1956) indicated that applications of 0.001 kg/ha of 2,4-D at the seedling stage delayed cotton maturity but had little effect on seed yield. The application of 0.01 kg/ha delayed maturity and reduced yield. At the squaring stage, 0.001 kg/ha had no clear-cut effect on yield or earliness, but 0.01 kg/ha reduced yield. At flowering, applications of 0.001 and 0.01 kg/ha had little effect, but 0.1 kg/ha reduced both yield and earliness. The boll stage was most resistant to 2,4-D.

In Arizona, Arle (1954) found that seed yields from 'Acala 44' cotton were reduced when spray applications of the amine salt of 2,4-D were made at low rates on June 7 (a few squares on each plant) and picked on September 30. At later harvests, however, the trend was reversed, and the yield from treated areas was consistently higher than from untreated areas. An application made on July 7 (numerous squares and blooms) caused greater yield reduction at the first and second picking than June treatments, but late-season recovery on December 28 yielded about 350% over the control. The August 17 treatment had little or no adverse effect on the quantity of cotton picked. Arle's experiments also indicated that cotton could tolerate quantities of up to 1.1 kg/ha of 2,4-D in irrigation water without adversely affecting yields.

Differential tolerance of cotton cultivars to 2,4-D was noted when drift from neighboring fields caused injury in a yield trial (Regier et al., 1986). Under these conditions, 'Paymaster 145' cotton did not show injury symptoms and out-yielded 11 other cultivars. Yield of the 12 cultivars was negatively correlated with an injury rating where 1 was no leaf injury, and 5 was all leaves having 2,4-D injury symptoms. 'McNair 307' and 'G&P 3774' cotton were injured the most by the 2,4-D drift. In an effort to explain the tolerance of 'Paymaster 145,' two studies were conducted to determine if leaf hairiness affected resistance to 2,4-D (Dilbeck et al., 1987). In the first study, lines used 'Paymaster 145' along with 'Texas marker-1' isolines 'pilose' and 'smooth leaf.' A single spray of 2,4-D at 0.56 kg/ha was allowed to drift across the three lines. In a second study, five lines were sprayed with 2,4-D at 0.001, 0.01, and 0.056 kg/ha. Visual ratings and lint yield indicated an advantage to leaf hairiness, but 'Paymaster 145' yielded the most, showing the presence of an additional genetic mechanism for tolerance to 2,4-D.

2. 2,4-D Versus Other Phenoxys

Most research indicates that 2,4-D is more phytotoxic to cotton than other phenoxys (Behrens et al., 1955; Goodman, 1953; Miller et al., 1963; Porter et al., 1959; Watson, 1955). Goodman (1953) indicated that 2,4-D reduced the yield of 'Coker 100W' cotton more than 2,4,5-T. Goodman et al. (1955) found that at equal rates 2,4-D reduced cotton yield 10 times more than 2,4,5-T. [(4-chloryl-2-methylphenoxy)acetic acid] (MCPA) caused more leaf modification in cotton than 2,4,5-T, but the yield response was similar. Cotton exposed to 2,4-D generally was the most sensitive at the seedling and square stages, as compared to the flowering or boll stages. Cotton response to MCPA and 2,4,5-T was similar to 2,4-D, except the magnitude of injury was less. The butyl ester of silvex was more injurious than the butyl ester of 2,4,5-T. Dichlorprop had little or no effect on cotton yield. Using 'Coker 100W' cotton, Watson (1955) generally agreed with Goodman (1953) and Goodman et al. (1955), except that 2,4-D and MCPA were about equal in reducing the yields of cotton. Watson (1955) indicated the amine salt of silvex was more injurious than the amine salt of 2,4,5-T, and that the sodium salt and ester of silvex were more injurious than the amine salt.

At four locations in Texas, Behrens et al. (1955) found that 2,4-D caused more leaf malformation in cotton than 2,4,5-T and MCPA. Silvex resulted in no appreciable leaf malformations. Similarly, 2,4-D caused the greatest reduction in cotton yield, followed by 2,4,5-T and MCPA, with silvex causing the least reduction in yield.

Porter et al. (1959) evaluated the low rates of several herbicides at four stages of cotton growth. The eight- to 10-leaf stage was the most sensitive to 2,4-D when measured by yield reduction. Silvex and 2,4,5-T were nonspecific in yield reduction relative to the stage of cotton growth. Rates of 2,4-D as low as 0.0001 kg/ha did not stimulate yield of cotton, as reported by other investigators. Miller et al. (1963) reported that cotton seed yields were the most drastically reduced when 0.01 and 0.1 kg/ha of 2,4-D were applied during the flowering and fruit-setting stages. Seed quality was reduced by 2,4-D treatment, and fiber quality was reduced by foliar application of 2,4-D at 0.1 kg/ha.

3. Other Hormone Herbicides

Smith and Wiese (1972) compared 2,4-D to dicamba (Banvel), picloram (Tordon), bromoxynil (Buctril), and 2,3,6-TBA. The order of damage to cotton was 2,4-D ester > 2,4-D amine >> dicamba > MCPA > picloram >> bromoxynil >> 2,3,6-TBA. Sprays of 2,4-D, dicamba, or MCPA at 0.1 kg/ha reduced lint yields from 20 to 97%. Yield losses were the most severe when the cotton was sprayed before blooming. Lint quality (micronaire and length) was not affected by these herbicides, however.

'Tamcot' cotton seedlings were injured by foliar sprays of 2,4,5-T, triclopyr (Garlon), and clopyralid (Reclaim) at 0.03 kg/ha acre in the greenhouse (Bovey and Meyer, 1981). No new growth occurred when cotton was treated with 0.14 or 0.56 kg/ha of 2,4,5-T or triclopyr. Clopyralid was less injurious to cotton than triclopyr and 2,4,5-T, and only slight malformations occurred in the leaves at 0.03 kg/ha. Since clopyralid has shown to give excellent control of honey mesquite in Texas, damage from the spray drift of this herbicide should be minimal.

Bovey et al. (1968) found that 'Blightmaster' cotton could be planted in a tropical soil 2 months after applications of the potassium salt of picloram at 6.7 kg/ha, the butyl ester of 2,4-D + 2,4,5-T at 13.5 kg/ha each, or a 2:2:1 mixture of the isooctyl esters of 2,4-D + 2,4,5-T + picloram at a total of 16.8 kg/ha.

4. Triclopyr

Fratesi et al. (1988) found that triclopyr amine was less injurious than both 2,4-D amine and bromoxynil in the field. At midseason (sixth node) triclopyr and bromoxynil were significantly less injurious than 2,4-D amine. When treated at first bloom, the injury from triclopyr amine was increased. It was equivalent to bromoxynil, but less than 2,4-D. Triclopyr was applied at 0.28, 0.028, and 0.0028 kg/ha; bromoxynil at 0.56, 0.056, and 0.0056 kg/ha; and 2,4-D at 0.14, 0.014, and 0.0014 kg/ha.

Snipes et al. (1991) found that triclopyr applied at 0.06 kg/ha at pinhead square reduced cotton height in 1987 but not in 1988. Triclopyr applied at pinhead square and early bloom reduced flowering initially, and reduced the total bloom in 1987 but not in 1988. Cotton maturity was delayed by triclopyr applied during early bloom and reduced cotton yield. The greatest yield reduction was at 0.06 kg/ha of triclopyr.

5. Glyphosate

During the late 1970s, researchers tried glyphosate (Roundup) as a postemergence topical herbicide in cotton and soybeans (Banks and Santlemann, 1977; Wills, 1978). In order to increase cotton tolerance to glyphosate, Jordan and Bridge (1979) evaluated the resistance of 405 genotypes to sprays 0.84, 1.7, and 3.4 kg/ha during early bloom. There was considerable difference in tolerance, and six of the most resistant lines were sprayed the next year at 0.56 and 0.84 kg/ha. None of the treated lines yielded as much as untreated comparisons; however, genotypes 'DES 04-11' and 'DES 04-606' yielded more than untreated controls when glyphosate at 0.5 kg/ha was directed at the base of the cotton plants. Although there was considerable tolerance of some cotton lines to glyphosate, it was not safe to spray the many varieties that were planted. Because the crop is quite tolerant, glyphosate can be used in ropewick applicators or recirculating sprayers, or for spot treatments to control perennial weeds in cotton (Keeley et al., 1984a, b). Using a ropewick applicator to control johnsongrass markedly increased the yield of cotton compared to cultivating and hoeing (Keeley et al., 1984b).

6. Diquat

Bovey and Miller (1968) found that diquat at the equivalent of 0.63 kg/ha killed all leaves on 1-month-old seedling Blightmaster cotton. It was the lowest rate that killed all leaves.

7. Paraquat

Scifries and Santlemann (1966) found that fiber quality was not affected by paraquat at 0.14 and 0.28 kg/ha. Tall cotton was less susceptible to paraquat than smaller plants. Paymaster 101A and Lankart 57 were more susceptible to paraquat than the Parrot, Pima 5-2, Acala 4-42 (glanded, glandless), and Verden varieties. Lankart 57 was not damaged by 0.14 kg/ha, but at 0.28 kg/ha reduced yield and fiber coarseness. At 0.28 kg/ha Verden boll weight was reduced with some stem burn, but the yield was not reduced. The other varieties were not affected in yield, fiber quality, or boll weight by paraquat at 0.14 or 0.28 kg/ha.

8. Physiological Effects of 2,4-D and Picloram

a. 2,4-D All seed from bolls (1 to 15 days old) set prior to treatment with 0.0000036 or 0.000036 oz (0.1 or 1.0 mg) of 2,4-D per plant produced seedlings that exhibited symptoms of 2,4-D injury in 60-day-old 'Stoneville 2B' cotton (McIlrath et al., 1951). Some seedlings from seed produced in bolls initiated 8 weeks after application of 0.0000035 to 0.00035 oz (0.1 to 10 mg) 2,4-D per plant exhibited 2,4-D injury, but seed formed 14 weeks after application did not produce malformed seedlings. Plants treated with 0.00000035 or 0.0000012 oz (0.01 or 0.04 mg) of 2,4-D at seedling or floral primordia stages showed no transmission of the stimulus into seed embryos, but seed embryos formed in bolls initiated 5 weeks after anthesis showed significant 2,4-D injury. McIlrath and Ergle (1953) showed that 2,4-D symptoms persisted in 'Stoneville 2B' and 'Marie Galante' cottons up to 6 months. The 2,4-D was extracted 80 days after application from plants showing leaf malformation characteristic of 2,4-D. Morgan and Hall (1963), however, indicated that cotton decarboxylated the side chain of 2,4-D several times faster than sorghum, and in young leaves and bolls slowly converted 2,4-D to a chromatographically different material. After the cotton recovered from 2,4-D treatment, it could not be detected in subsequent vegetative or reproductive growth.

b. Picloram Picloram increased the soluble protein concentration of cotton when applied at 1 ppb into the roots of seedlings (Baur et al., 1970). At 100 ppb picloram, soluble protein tended to be reduced but was not different from that for untreated plants. At 0.25, 0.50, and 1 ppb of picloram, dry and fresh weights of cotton were increased compared to the untreated plants.

B. Soybean [*Glycine* max (L.) Merr.]

1. 2,4-D

a. Foliar Application Slife (1956) found that Hawkeye soybeans tolerated 2,4-D amine in the early stage of growth (8 to 13 and 18 to 23 cm tall), but became more sensitive as they matured (46 to 51 and 76 to 81 cm tall). Seed yield was not affected at 0.07 and 0.14 kg/ha of 2,4-D amine at the two early stages of growth. Yield was slightly affected when soybeans were sprayed at 46 to 51 cm tall at 0.07 and 0.14 kg/ha of 2,4-D. The 0.28 and 0.56 kg/ha rates reduced yields at all stages, but were less severe when soybeans were 8 to 13 cm tall.

Hart et al. (1991) found that William seedling soybeans were injured 80 and 100% with 0.3 and 2.2 kg/ha of 2,4-D alkanolamine salt, respectively. *Glycine* accessions, however, had 15% or less injury 4 weeks after 2,4-D application. The greatest 2,4-D tolerance occurred with *G. latifolia* and *G. microphylla*. Recovery or absorption of ^{14}C-2,4-D up to 14 days after treatment was similar in all accessions as well as distribution of ^{14}C in various plant parts. Metabolism of 2,4-D in tolerant accessions (81 to 89%) was higher and more rapid than in susceptible accessions (about 50%). Distribution of the five metabolites and parent 2,4-D differed between tolerant and susceptible accessions. More rapid metabolism of 2,4-D in treated leaves of tolerant *Glycine* accessions can explain differential 2,4-D responses.

b. Soil Applications When applied preplant at 0.56, 1.12, and 1.68 kg/ha at 0, 7, and 14 days before planting, 2,4-D ester delayed emergence and injured soybeans in 1991 and 1992, depending upon the 2,4-D rate (Krausz et al., 1993). Delayed emergence ranged from 0 to 67% and 0 to 50% in 1991 and 1992, respectively. Injury at 15 days after planting combined over the years ranged from 0 to 61%. Minimal (0 to 8%) delay in emergence and minimal (0 to 5%) injury at 15, 30, 45, and 60 days after planting occurred when 2,4-D at 0.56 or 1.12 kg/ha was applied 7 to 14 days before planting. No significant yield reduction occurred due to 2,4-D rate, application timing, or their interaction.

Bovey et al. (1968) found that Clark soybeans were sensitive to extremely high rates of 1:1 combinations of 2,4-D: 2,4,5-T butyl esters applied at 26.9 kg/ha in a tropical soil. Soybeans were more susceptible to the herbicide than cotton, corn, sorghum, wheat, and rice, which could be grown in treated soils as early as 3 months after application. When 2,4-D, 2,4,5-T, and picloram were combined at 6.1, 6.7, and 3.4 kg/ha, respectively, for a total of 16.8 kg/ha, all crops could be grown as early as 3 months after treatment (Bovey et al. 1968). Cotton was slightly injured when seeded in treated soil 3 months after treatment, but not at 6 months or more. Soybeans planted 13.5 months after herbicide treatment showed no injury.

2. Other Hormone Herbicides

a. Foliar Application In the greenhouse seedling Gail soybeans were injured or killed by all rates of triclopyr, 2,4,5-T, or clopyralid at rates of 0.002, 0.009, 0.03, 0.14, and 0.56 kg/ha (Bovey and Meyer, 1981). By 4 weeks after treatment, soybeans surviving the lower rates of 2,4,5-T were severely stunted. Triclopyr and clopyralid caused callus formulation on the stems and epinasty at the lower rates in addition to stunting.

In the field, Fratesi et al. (1988) applied triclopyr amine at 0.28, 0.028, and 0.0028 kg/ha, bromoxynil at 0.56, 0.056, and 0.0056 kg/ha, and 2,4-D amine at 0.14, 0.014, and 0.0014 kg/ha to soybeans. Triclopyr was least injurious early (second trifoliate stage), while bromoxynil was most injurious. At midseason (fourth trifoliate) and late season triclopyr was highly injurious. To minimize the possible damage by drift to soybeans, early-season application to adjacent weedy areas was suggested.

Foliar applications of 2,4-D up to 0.14 kg/ha had little effect on yield at prebloom, and only slightly reduced yield when applied at flowering (Wax et al., 1969). Dicamba and picloram were more injurious than 2,4-D at prebloom. Dicamba and picloram severely restricted soybean yield when applied at flowering. Dicamba at 0.035 kg/ha and picloram at 0.009 kg/ha reduced soybean yield about 50%.

b. Soil Application Soybeans could be grown in a tropical soil by 6.5 months after application of 6.7 kg/ha picloram without adversely affecting growth (Bovey et al. 1968). Rainfall was 134 cm. On a Wilson clay loam after application of 1.12 kg/ha of the potassium salt of picloram, soybeans showed injury for more than 1 year in Texas (Bovey et al., 1975). Soybeans planted in 1972 were killed in plots treated after 6 months or less with picloram. After 12 months soybeans per ha and total dry matter production were only slightly depressed, although herbicide symptoms (leaf epinasty) were visible. Data in 1973 were similar to 1972. The annual rainfall was about 80 cm.

Wax et al. (1969) found that soil-incorporated applications of 2,4-D at rates up to 0.56 kg/ha or dicamba at rates up to 0.28 kg/ha just before planting did not reduce soybeans yields. Picloram soil incorporated at 0.035 kg/ha reduced soybean yield almost 40%. At rates of 0.035 to 0.14 kg/ha picloram caused slight to moderate leaf malformations on soybeans planted the following year in Illinois, but did not reduce yield.

In North Dakota, Thorsness and Messersmith (1991) found that soybean height, stand, and yield were reduced by clopyralid residues after 1 year when clopyralid was applied at 0.56 kg/ha at Fargo. Soybean yield at Prosper was not affected by clopyralid residues, but plant height was reduced by 0.28 and 0.56 kg/ha applied the previous year.

C. Fieldbeans (*Phaseolus vulgaris* L.)

Lyon and Wilson (1986) investigated the effects of 2,4-D and dicamba dimethylamine salts at reduced rates on Great Northern Valley fieldbeans. Herbicides were applied at 1.1, 11.2, and 112.5 g/ha at the preemergence, second trifoliate leaf, early bloom, and early pod stages of development. All rates of 2,4-D applied preemergence or in the second trifoliate leaf did not reduce seed yield, delay maturity, or reduce germination of treated seed. Dicamba or 2,4-D applied at 112.5 g/ha in early bloom or early pod stages reduced seed yield, delayed maturity, and reduced germination percentage. Dicamba was more phytotoxic than 2,4-D to fieldbeans.

D. Alfalfa (*Medicago sativa* L.)

1. 2,4-D and Others

Vernal alfalfa response was evaluated with chlorsulfuron, thifensulfuron, 2,4-D, glyphosate, and selected combinations of these herbicides at rates simulating spray drift during the fourth trifoliate leaf stage following the first cutting in 1990 and 1991 (Al-Khatib et al., 1992c). Rates used were 0.3, 0.9, 2.6, and 8.7 g/ha for chlorsulfuron and thifensulfuron; 4, 14, 43, and 142 g/ha for bromoxynil and glyphosate; 11, 37, 112, and 374 g/ha for 2,4-D dimethylamine; and 3+3, 10+11, 29+32, and 96+105 g/ha for 2,4-D plus glyphosate, respectively.

The order of phytotoxicity was 2,4-D > chlorsulfuron > thifensulfuron > glyphosate > bromoxynil. By the end of each growing season, alfalfa recovered from injury caused by all herbicides except the highest rates of 2,4-D and 2,4-D plus glyphosate. The alfalfa stand was reduced only by 2,4-D and 2,4-D plus glyphosate.

2. Paraquat

Foy and Witt (1993) found that paraquat applied in March to alfalfa less than 1 year old sometimes decreased alfalfa yields. Paraquat applied after cuttings did not affect alfalfa yields in most experiments. Paraquat rates were 0.3, 0.6, 0.8, and 1.1 kg/ha.

3. Glyphosate

Dawson (1992) found that glyphosate at 75 to 150 g/ha did not injure flowers, impair seed set, or reduce yield or quality of seed when applied to alfalfa foliage before the bud stage. Glyphosate injured flowers, caused flower abscission, and sometimes reduced seed yields when applied to alfalfa when buds or blooms were present.

4. Metsulfuron

Moyer (1995) found that the GR_{50} (rate required to reduce growth by 50%) values for Beaver alfalfa immediately after application of chlorsulfuron, triasulfuron, and metsulfuron were <2.2 g/ha applied to a Typic Boroll soil. At 6 g/ha metsulfuron injured alfalfa 80% seeded 1-year after application.

E. Red Clover (*Trifolium pratense* L.)

Whole plant and tissue-culture experiments were conducted to determine the difference in phytotoxicity of 2,4-D and its metabolite, 2,4-DCP, to red clover (Taylor et al., 1989). At the whole plant level, the mean concentration of 2,4-DCP (10 mM) required to cause 50% growth inhibition (I_{50}) of shoot dry weight was 24 times greater than for 2,4-D (0.42 mM). Using callus tissue, the I_{50} value for 2,4-DCP (0.28 mM) was 22 times greater than for 2,4-D (0.013 mM) based on dry weights. The callus tissue was 36 and 32 times more sensitive to 2,4-DCP and 2,4-D than shoot tissue based on dry weights, respectively. These data indicate that 2,4-DCP was less phytotoxic than 2,4-D to red clover both in vitro and in vivo.

F. Rose Clover (*Trifolium hirtum* All.) and Subclover (*T. subterranum* L.)

Kay (1964) found that paraquat removed weedy annual grasses from seedings of rose clover (*Trifolium hirtum* All.) and subclover (*T. subterranum* L.), with only temporary damage to the clovers when applied at early growth stages. Paraquat applied under weed-free conditions reduced herbage yields of Lana woollypod vetch (*Vicia dasycarpa* Ten.) and Wimmera ryegrass (*Lolium rigidum* Gaud.) at three dates of application; neither herbage nor seed yield of rose or subclover was reduced. Paraquat rates were 0, 0.7, 0.14, 0.28, 0.56, 1.12, and 2.24 kg/ha.

G. Sugar Beets (*Beta vulgaris* L.)

1. 2,4-D

The alkanolamine salt of 2,4-D was applied at 0, 0.017, 0.035, and 0.07 kg/ha to 'Mono Hy D2' sugar beets (Schweizer, 1978). The greatest top growth reduction occurred when 0.07 kg/ha of 2,4-D was applied to the oldest plants (12-leaf stage). All rates of 2,4-D reduced the components of sucrose yield (percentage sucrose, percentage purity, and root weight), to cause a significant reduction in recoverable sucrose. The yields of recoverable sucrose components were reduced 6.8, 7.8, and 13.2% by the 0.017 kg/ha rates, respectively.

2. 2,4-D and Other Hormone Herbicides

Schroeder et al. (1983) found similar effects of 2,4-D on sugar beets to those found by Schweizer (1978). 2,4-D was applied at 0.035, 0.14, and 0.28 kg/ha, dicamba at 0.017, 0.07, and 0.14 kg/ha, and picloram at 0.007, 0.014, and 0.028 kg/ha on 'ACH 17' sugar beets. When applied at early growth stages, 2,4-D tended to decrease root yield, but decreased purity and extractable sucrose content as much as 54% when applied at later growth stages. Dicamba at 0.14 and 2,4-D at 0.28 kg/ha decreased extractable sucrose/ha and root yield, while picloram at 0.028 kg/ha did not. 2,4-D at all rates, dicamba at 0.14, and picloram at 0.028 kg/ha increased sucrose loses during postharvest storage, therefore sugar beets inadvertently exposed to these herbicides by spray drift during the growing seasons should be processed immediately after harvest.

3. Metsulfuron

Metsulfuron applied to soil at 6 g/ha injured 'KW344' sugar beets 47% when seeded 1 year after application in Canada (Moyer, 1995).

H. Summer Rape (*Brassica napus* L.) and Turnip Rape (*B. compestris* L.)

Field experiments were conducted to compare the effects of the diethylamine salt of 2,4-D at 0.07, 0.14, 0.28, and 0.56 kg/ha on plant growth, seed production, and seed quality on 'Zephyr' summer rape and 'Span' turnip rape (Betts and Ashford, 1976). The seed yield of turnip rape at all stages of growth was reduced significantly by each additional rate of 2,4-D. Summer rape was less affected than turnip rape by 2,4-D because turnip rape leaves retained 60% more 2,4-D spray per unit of leaf area than summer rape. Retention of spray by turnip rape is probably due to the rough pubescent leaf surface as opposed to the smooth and waxy leaf surface of summer rape.

I. Flax (*Linum usitatissimum* L.)

1. Foliar application

Nalewaja (1969) found 'Bolley' flax was susceptible very early when less than 5 cm tall and during flowering to foliar sprays of the dimethyamine salt of dicamba at 0.011, 0.014, and 0.018 kg/ha. Dicamba decreased flaxseed yield, percentage of oil and iodine number of the oil, seed germination, and plant height, and increased the number of days to maturity by application at either one or both of the susceptible stages of flax.

2. Soil Application

Flax growth was not affected by clopyralid residues 1 year after planting in soil treated with 0.07 or 0.56 kg/ha of clopyralid (Thorsness and Messersmith, 1991).

J. Tobacco (*Nicotiana tabacum* L.)

Klingman (1967) found that tobacco was extremely sensitive to picloram. Residues of 12 g/ha in soil caused serious damage to tobacco. Broadcast sprays of 1 g/ha applied shortly after setting tobacco in the field reduced dollars returned by 50%. The tobacco plant gained tolerance to 6 g/ha by the time it was 61 cm tall. Picloram applied when tobacco was 61 cm tall and also when treated after topping controlled sucker growth at rates of 1 and 6 g/ha.

K. Concord Grapes (*Vitis labruscana* L.)

Ogg et al. (1991) applied the dimethylamine salt formulation of 2,4-D up to four times per year from late April to early June at 2.5, 10, and 25 ppmw on 'Concord' grapes. In one experiment 2,4-D was applied in 1979 only. In another experiment 2,4-D was applied in 1979, 1980, and 1981. When 2,4-D symptoms in July were severe, yields were reduced by as much as 85%, but when 2,4-D symptoms were slight in July, grape yields were not reduced. Symptoms of 2,4-D on grape leaves did not persist beyond the year of treatment, but grape yields did not fully recover until the second year after the last 2,4-D treatment. In the first year of treatment, low grape yields were correlated with reduced weight and number of berries per cluster. In the second and third year of treatment, low grape yields were also correlated with fewer clusters per shoot. When 2,4-D was applied 1 year only, soluble solids in grape juice were reduced as much as 9%. When applied in three consecutive years, soluble solids in juice were higher in grapes with moderate to severe 2,4-D symptoms. In the first subsequent nontreatment year after 3 years of 2,4-D treatment, juice quality was not affected, and 2,4-D was not detected in grapes. Based on these experiments the authors concluded that grapes with slight 2,4-D symptoms would be expected to have reduced yield or reduced juice quality.

L. Wine Grapes (*Vitis vinifera* L.)

1. Repeated Exposure to Herbicides

Bhatti et al. (1996) applied several sulfonylurea herbicides or 2,4-D to 'Lemberger' wine grapes up to three times per week in 1992 and 1993 to simulate spray drift. All herbicides visually injured grapevines. Grapevines generally recovered within 40 to 60 days from single low-level doses of sulfonylurea herbicides, with multiple exposures of grapevines to 2,4-D at 1.2, 3.7, and 11.2 g/ha and chlorsulfuron plus metsulfuron at 0.022 + 0.008, 0.066 + 0.024, and 0.198 + 0.072 g/ha (1/900, 1/300, and 1/100 of the maximum rate used in cereal weed control). 2,4-D and chlorsulfuron plus metsulfuron at 1/100 of the maximum rate used for wheat caused the greatest injury, persisted through the growing season, and reduced pruning weight. Based on the research, 2,4-D was more phytotoxic than several sulfonylurea herbicides.

2. Single Exposure to Herbicides

Bhatti et al. (1997) applied chlorsulfuron, trifensulfuron, bromoxynil, 2,4-D, and 2,4-D plus glyphosate on 'Lemberger' wine grape at 1/100, 1/33, 1/10 and 1/3 of maximum rate used in wheat or fallow. All herbicides except bromoxynil and thifensulfuron cause symptoms on grapevines at the highest rate. The most severe symptoms were caused by 2,4-D and 2,4-D plus glyphosate. Shoot growth, leaf area, internode length, and dry cane weight decreased as rates of 2,4-D and 2,4-D plus glyphosate increased, and fall exposure to these herbicides can adversely affect the growth of grapevines the following spring.

VI. EFFECT OF SUBLETHAL HERBICIDE RATES ON OTHER VEGETABLE AND FIELD CROPS

A. 2,4-D

Hemphill and Montgomery (1981) applied sublethal sprays of the dimethylamine salt of 2,4-D at 0, 2.1, 10.4, 20.8, 52.5, 104, 208, and in some cases 416, 1040, and 2080 g/ha to field, grown vegetable crops. The growth stage was first bloom for beans cucumbers, peppers, potatoes, and tomatoes; 16 days prior to harvest for cucumbers (second year) with a taproot diameter of 0.6 meters for root crops; six- to eight-leaf stage for cabbage, cauliflower, and broccoli; just prior to heading for lettuce; and when onions averaged 20 cm tall. In addition, seedlings of cauliflower and broccoli were treated at the three-leaf stage. Crops and cultivars used the second year were 'Blue Lake 274' bush beans, 'Nantes' carrots, 'Pacer' cucumbers, 'Russet Burbank' potatoes, and 'Early Girl' tomatoes.

'Early Girl' tomatoes and root crops were the most sensitive to 2,4-D, with 2.1 g/ha distorting tomato fruit shape and elongating 'Cherry Belle' radishes. All root crops ('Nantes' carrots, 'American Purple Top' rutabagas 'Just Right' turnips, and radishes) were rendered unmarketable by 10.4 g/ha, and gross yields were reduced by 104 g/ha. The yield of 'Yolo Wonder' peppers was increased by 2.1 g/ha, but severely depressed by 104 g/ha. 'Victory' cucumber fruit shape was distorted at 11 g/ha, and the yield was slightly reduced. 'Buttercrunch' butterhead lettuce, 'Ithaca' crisphead lettuce, 'Grand Rapids' leaf lettuce, 'Sweet Spanish' onion, and 'Market Topper' cabbage were the least sensitive to 2,4-D. Exposure to 20.8 g/ha did not reduce lettuce or cabbage yield. Onion yields were reduced by 104 g/ha. 'Blue Lake 274' bush bean yield was decreased by 22 g/ha, but 'Russet Burbank' potato yield increased at 16 g/ha.

Martin and Fletcher (1972) indicated that 4-to-9-weeks-old 'Amanda' lettuce showed epinasty and necrotic and chlorotic areas within 24 hr after treatment with >50 μg/ plant of 2,4-D. Meinhardt et al. (1991) exposed 4-week-old 'Great Lakes' and 'Robinvale' lettuce once to 1 mm and 3 mm of simulated rain containing the acid or dimethylamine salt of 2,4-D in a growth tunnel for 3 weeks. Leaf epinasty or hyponasty and duck's foot-type leaves on lettuce seedling symptoms occurred from 10 ng dm^{-3} exposure of both 2,4-D formulations and rainfall exposure.

B. Other Herbicides

Bovey and Meyer (1981) found that clopyralid and triclopyr at 0.56 kg/ha were more injurious to 'Florrunner' peanuts (*Arachis hypogaea* L.) than 2,4,5-T. Clopyralid produced injury on peanuts similar to 2,4,5-T at rates of 0.002, .009, 0.03, and 0.14 kg/ha, but 2,4,5-T and triclopyr were more injurious to 'Liberty' cucumbers and 'Tamcot' cotton than clopyralid. All three herbicides killed 'Gail' soybeans at 0.14 and 0.56 kg/ha. Rates of 0.002 and 0.009 kg/ha of 2,4,5-T produced little or no injury to the four crops except at 0.009 kg/ha on cucumbers, where injury was 89%.

Derksen (1989) sprayed dicamba, chlorsulfuron, and clopyralid at sprayer contaminant rates on 'Cultivar 894' sunflowers (*Helianthus annus* L.), 'Common Brown' mustard [*Brassica juncea* (L.) Czern. & Coss], and 'Laird' lentils (*Lens culinaris* Medik) alone and with grass weed herbicides sethoxydim and diclofop. Dry weight production in the greenhouse and crop tolerance ratings and yield in the field indicated that the grass weed herbicides enhanced crop injury from dicamba and clopyralid. When the broadleaf weed herbicides were applied at rates simulating sprayer tank residues alone or with grass weed herbicides, yield losses ranged up to 40% in sunflowers, 70% in mustard, and 95% in lentils, compared to the untreated check.

Al-Khatib et al. (1992a) evaluated the response of roses (*Rosa dilecta* Rehd.) to different herbicides applied as simulated drift. Chlorsulfuron, thifensulfuron, bromoxynil, glyphosate, and a combination of 2,4-D and glyphosate were applied over the top of established rose plants at 1/3, 1/10, 1/33, and 1/100 of the maximum labeled rate for small-grain production. All herbicides injured roses. The greatest injury was from chlorsulfuron and 2,4-D, and the least from bromoxynil and glyphosate. Plants recovered from the injury caused by all treatments except for the highest rates of chlorsulfuron and 2,4-D, which continued to show significant injury at the end of the growing season. Although all herbicides had characteristic symptoms, some of these were very similar to those caused by other stresses, therefore because of the potential ambiguity of the visual symptoms, any allegation about herbicide drift should be based on a report of all symptoms and should be supported by residue analysis.

Al-Khatib et al. (1992b) evaluated the response of sweet cherries (*Prunus avium* L.) to different herbicides applied at rates simulating drift. Chlorsulfuron, thifensulfuron, bromoxynil, 2,4-D, glyphosate, and a combination of 2,4-D and glyphosate were applied on one side of one- and two-year-old established cherry trees at 1/3, 1/10, 1/33, and 1/100 of the maximum rate for small grain production. The order of herbicide phytotoxicity was chlorsulfuron > 2,4-D > glyphosate > 2,4-D + glyphosate > thifensulfuron > bromoxynil. The trees recovered from injury caused by all treatments except higher rates of chlorsulfuron, 2,4-D, and glyphosate. The herbicides caused characteristic symptoms, but some resembled disease, mineral deficiency, and environmental stress symptoms, therefore any allegations about herbicide drift based on chronic symptoms should be supported by an analysis of the plant tissue.

VII. CROP PLANT RESPONSE TO HERBICIDE SOIL RESIDUES

A. Conservation Tillage Systems with 2,4-D and Dicamba

Dicamba and 2,4-D are used prior to seeding in conservation tillage systems (Moyer et al, 1992). The herbicide 2,4-D ester and 2,4-D amine with dicamba or glyphosate applied 0 or 15 days prior to spring seeding damaged 'Pivot' canola (*Brassuia napus* L.), 'Century' peas (*Pisum sativum* L.), 'Laird' lentils, and 'Beaver' alfalfa. The legumes were damaged by spring-applied dicamba. There was a slight reduction in the total dry matter of 'Katepwa' wheat (*Triticum aestivum* L.) and barley (*Hordeum* sp.) by 2,4-D applied in spring prior to seeding. All but lentils were tolerant of recommended fall applications of 2,4-D. The rates used were dicamba at 0.14 kg/ha, 2,4-D ester at 0.55 kg/ha, 2,4-D amine + dicamba at 0.55 + 0.14 or 0.28 kg/ha, and 2,4-D amine + glyphosate at 0.55 + 0.33 kg/ha in spring and 2,4-D amine or ester at 0.4 or 0.8 kg/ha, dicamba at 0.14 and 0.6 kg/ha, and 2,4-D amine + dicamba at 0.4 + 0.14 kg/ha applied in the fall.

B. Metsulfuron Soil Residue

In other studies Moyer (1995) found that metsulfuron at 6 g/ha injured alfalfa, lentils, sugar beets, and peas 80, 72, 47, and 38% seeded 1 year after application. Barley, canola, corn, beans, flax, potatoes, and wheat showed no injury to metsulfuron residues.

C. Clopyralid Soil Residue

Thorsness and Messersmith (1991) indicated that 'Culbert' flax, 'Norchip' potatoes, and 'Hartman' safflowers were not affected by clopyralid residues when applied 1 year before application of 70 to 560 g/ha in Fargo, North Dakota, 'McCall' soybean yield was reduced by 560 g/ha, and 'Seed Tech 315' sunflower yield by 280 and 560 g/ha. Potatoes and sunflowers in Langdon and Prosper, North Dakota, and lentils in Langdon were not affected by clopyralid residues. Soybean yield at Prosper was not affected, but plant height was reduced at 280 and 560 kg/ha of clopyralid. The authors concluded that clopyralid applied at labeled rates of 280 g/ha or less in silty clay, clay loam, and silty clay loam soils of North Dakota would probably not adversely affect dicot crops tested in these experiments planted 11 months or more after treatment.

D. Crop Response to Hexazinone, Imazapyr, Tebuthiuron, and Triclopyr Soil Residue

Field investigations were conducted for 2 years to characterize the responses of wheat, kidney beans, field corn, squash, okra, potatoes, and bananas to soil-applied hexazinone, imazapyr, tebuthiuron, and triclopyr at the rates used for controlling woody perennial plants (Coffman et al., 1993). Test species were planted at seven selected intervals through the first and second years after herbicide applications to assess any residual herbicide activity. Tebuthiuron treatments of 2.2 kg/ha were not tolerated by any bioassay species planted 436 days after application. Only potatoes were tolerant of residual hexazinone 436 days after application. All species except bananas generally tolerated residual imazapyr and triclopyr by the second growing season. By the third growing season, indigenous plants repopulated all herbicide-treated areas except those treated with tebuthiuron. Plots treated with tebuthiuron were 90% free of indigenous plants 40 months after herbicide application.

E. Crop Responses to Picloram Residues

Baur et al. (1970) applied picloram once to the soil in water to eight species of herbaceous plants 14 days after planting in the greenhouse. An analysis of aerial portions 21 days after planting revealed a significant stimulation in fresh weight of corn, sorghum, cotton, cowpeas and soybeans treated at from 0.25 to 0.50 ppb, and of wheat at 100 ppb. Significant stimulation in dry weight was noted in corn, sorghum, cotton, and soybeans at 0.25 ppb, and in cowpeas at 1.0 ppb. Significant decreases in fresh and dry weight occurred in corn, wheat, and sorghum at 1000 ppb, and in all dicot species at 100 ppb. Herbicide treatments had no effect on the dry weights of rice and wheat, or the fresh weight of rice. Herbicide treatments caused a reduction in soluble protein concentrations in all monocot species and sunflowers. Significant increases in soluble protein occurred at 1.0 and 0.25 ppb in cotton and cowpeas respectively.

Picloram residues in soil may occur as a result of weed control practices on cropland and injure subsequent crops. Bovey et al. (1975) conducted a study to determine how soil residues of picloram affected the growth and dry matter production of 'Tophand' grain sorghum [*Sorghum bicolor* (L.) Moench] and 'Hill' soybeans [*Glycine max* (L.) Merr.] grown in soil at various time intervals after picloram application. The potassium salt of picloram was applied at 1.12 kg/ha and incorporated by disking into a Wilson clay loam soil 1.5, 3, 6, 7, 12, 14, 16, 19, and 26 months before planting sorghum and soybeans in the spring of 1972 and 1973.

Sorghum was grown on picloram-treated soils 12 months after application without a reduction in plant numbers, dry matter production, flowering, or germination. No picloram was detected in sorghum seed (immature and mature) produced by plants seeded 6 or 12 months after the application of herbicide to the soil. Because soybeans are sensitive to small amounts of picloram in the soil, they should not be grown in a picloram-treated clay soil for at least 2 years after application. Earlier planting risks a possible reduction in stand and dry-matter production in situations similar to this study.

F. Crop Response to Hexazinone

The effects of hexazinone on gas exchange and chlorophyll a fluorescence kinetics were investigated in container-grown loblolly pine seedlings (Johnson and Stelzer, 1991). Hexazinone at concentrations from 10^{-4} to 10^{-8} M was applied as soil drench. Photosynthesis in seedlings treated with hexazinone concentrations greater than 10^{-6} M was partially or completely inhibited throughout the 14-day study. Chlorophyll a fluorescence kinetics indicated that the inhibition was due to disruption of PSII electron transport. At hexazinone concentrations below 10^{-5} M, quantum yield and rates of photosynthesis and transpiration were increased on day 1 and remained elevated through day 7, whereas chlorophyll, a fluorescence, was unaffected. The results suggest that in loblolly pine, sublethal concentrations of hexazinone may function in a manner similar to cytokinins.

G. Crop Response to Imazapyr

Bovey et al. (1998) conducted greenhouse experiments to determine the persistence of imazapyr and the tolerance of a wide variety of bioassay plants to imazapyr residues in central Texas soils. Eleven bioassay crops were planted in pots 1 day before and 2, 6, and 12 months after imazapyr application at rates of 0, 0.02, 0.04, 0.07, 0.14, 0.28, 0.56, and 1.12 kg/ha. Imazapyr was phytotoxic 12 months after treatment to all crop plants in at least one rate. Forage grasses and herbs were highly susceptible to imazapyr residual carryover damage, as were corn and cabbage. Beans and squash were tolerant of imazapyr residues. No statistical differences occurred in imazapyr phytotoxicity to corn grown in leached versus nonleached pots.

H. Influence of Dicamba in Irrigation Water on Seedling Crops

Crops varied in their response to one irrigation of water containing dicamba (Scifres et al., 1973). 'Dunn' was the most susceptible cotton cutlivar. Fresh weights of Dunn seedlings were reduced at 100 ppb of dicamba, whereas concentrations of 500 ppb were required for weight reduction in 'Paymaster.' 'Blightmaster' was the most tolerant cultivar studied. 'Pioneer 820' and 'RS-626' grain sorghums seedlings also tolerated all dicamba treatments. RS-626 exposed to 500 ppb showed increased fresh weight. 'Straight-eight' cucumber seedlings tolerated irrigation water containing as much as 50 ppb dicamba, but were injured or killed by 100 and 500 ppb, respectively. Crop tolerance to dicamba in irrigation water from the greatest to the least were sorghum > cotton > cucumber.

VIII. SIMULATED DRIFT AND SUBLETHAL DOSE EFFECTS ON GRASS CROPS

Grass crops are usually much more tolerant of hormonelike herbicides at sublethal rates than dicot crop plants, but may not be to soil-applied herbicides, such as hexazinone, tebuthiuron, imazapyr, and bromacil, or contact herbicides, such as paraquat or glyphosate.

A. Wheat (*Triticum aestivum* L.)

1. Hormonelike Herbicides

In a 3-year study, Klingman (1953) found that yield reduction of 'Pawnee' winter wheat of up to 10% could result from the use of 0.56 or 1.12 kg/ha of 2,4-D ester or amine. Higher rates caused more severe injury. The results of using 2,4,5-

T were similar to 2,4-D. Fall spraying caused the greatest injury. When damaged, 2,4-D caused reduced kernels and spikelets per head, and delayed wheat maturity. Yield reductions were relatively minor at rates below 0.56 kg/ha of 2,4-D ester and 1.12 kg/ha of the amine in all stages of growth. No changes occurred in the number of culms per meter of row, heads per meter, or percentage of germination. Growth abnormalities occurred with treatments in the early stages of growth. The percentage of protein was inversely associated with yield, and when no yield changes occurred, 2,4-D had no effect on protein content.

Martin et al. (1989) applied labeled herbicides to 'Buckskin' winter wheat at three growth stages (three-leaf, fully tillered, and prejointing and midboot). Dicamba at 0.14 kg/ha, dicamba + 2,4-D at 0.14 + 0.14 kg/ha, dicamba + MCPA at 0.14 + 0.14 kg/ha, chlorsulfuron at 0.05 kg/ha, MCPA amine at 0.55 kg/ha, and picloram + 2,4-D at 0.02 + 0.4 kg/ha decreased yield at the three-leaf stage; bromoxynil at 0.55 kg/ha and bromoxynil + MCPA at 0.55 kg/ha each did not. Wheat yield was reduced by chlorsulfuron, dicamba, dicamba + 2,4-D amine, MCPA, 2,4-D amine, or 2,4-D ester at the fully tillered, prejointing stage. Wheat tolerated most herbicides at the midboot stage, therefore yield reduction and wheat height reduction was greatest in fall and late-spring treatment. These data support those of Klingman (1953). Martin et al. (1989) found that wheat kernels per spike, kernel weight, volume weight, or germination were not affected.

Quimby and Nalewaja (1966) found that the greatest reduction in the height and yield in 'Selkirk' hard red spring wheat treated with dicamba came at the late tiller and boot stages, whereas Schroeder and Banks (1989) found that 'Florida 302' soft red winter wheat was more sensitive to dicamba and dicamba + 2,4-D at midtillering. Nalewaja (1970) also found that 'Selkirk' wheat was the most susceptible to picloram at the late tiller stage. Heering and Peeper (1991) found that hard red winter wheat yield reductions were more severe from picloram applied to the first joint growth stage than the late tillering stage. Both Nalewaja (1970) and Heering and Peeper (1991) found wheat sometimes more severely injured by picloram plus 2,4-D than by picloram alone.

In the greenhouse 'Milam' wheat was noticeably taller when the soil was watered with water containing 0.25, 0.50, 1.0, 10, and 100 ppb of picloram compared to untreated seedlings or seedlings treated with 1000 ppb. There were no differences, however, in the fresh and dry weights of wheat growth in any treatment (Baur et al., 1970).

'Caddo' wheat seedlings were not injured by rates of 2,4,5-T or clopyralid from 0.002 to 0.14 kg/ha. 'Caddo' wheat tolerated triclopyr at rates of 0.002 to 0.56 kg/ha (Bovey et al., 1981).

2. Sulfonylurea herbicides

Triasulfuron at 30 g/ha or chlorsulfuron plus metsulfuron at 26 g/ha (21.7 + 4.3 g/ha) applied preemergence, metribuzin applied early postemergence alone at 280 g/ha or tank-mixed with triasulfuron at 158 + 30 g/ha, or chlorsulfuron plus metsulfuron at 210 + 21 g/ha all decreased total forage production of 'Karl,' '2180' hard red winter wheat (Koscelny et al., 1996). Conversely, all herbicides treatments except triasulfuron applied preemergence increased the wheat grain yield.

3. Paraquat

Paraquat was applied at 0.28 and 0.56 kg/ha to 'Vona' winter wheat at five stages at 0800, 1300, and 1600 hours (Anderson and Nielsen, 1991). Biomass reduction was 84% when winter wheat was in the one- to three- leaf stage, but only 68% when treated at tillering. Paraquat phytotoxicity continued to decrease at later growth stages. The time of day for application did not affect paraquat actively.

4. Tebuthiuron

'Caddo' wheat and 'Tomcot' cotton were used as bioassey plants to detect tebuthiuron in an Axtell fine sandy loam (Udertic Paleustalfs) applied as pellets by aircraft at 2.2 and 4.4 kg/ha (Bovey et al., 1982). Both plants were good indicators for estimating tebuthiuron concentration in the treated soil in field plots and produced similar results. Tebuthiuron concentrations ranged from 0.08 to 0.42 μg/g 2 years after application of 2.2 kg/ha and 0.21 to 0.49 μg/g after application of 4.4 kg/ha. The fresh and drys weights of a second crop of cotton and wheat were similar to the first crop's. The data indicated that the soil residues of tebuthiuron were not leached out of pots by daily watering and that growing plants did not remove all phytotoxic amounts of tebuthiuron. Switchgrass (*Panicum coloratum* L.) tolerated and grew in soils from all treatments and soil depths, but some injury occurred in soils treated with 4.4 kg/ha. Kleingrass (*Panicum coloratum* L.) was injured in soils receiving 4.4 kg/ha of tebuthiuron.

B. Grain Sorghum [*Sorghum bicolor* (L.) Moench]

1. 2,4-D

a. Foliar Application Phillips (1958) sprayed 'Midland' grain sorghum with ester and amine salt formulations of 2,4-D in 1950, 1952, 1953, and 1954 at 0.14, 0.28, 0.56, and 1.12 kg/ha each year with each formulation. Growth stages were five- to seven-leaf, nine- to 11-leaf, or early floral development and pollinating. At the five- to seven-leaf stage, 2,4-D at less than 1.12 kg/ha did not reduce grain sorghum yield but damaged roots. At the nine- to 11-leaf stage 2,4-D at 0.56 and 1.12 kg/ha reduced the yield in every year. At the pollinating stage 2,4-D damaged yields only in 1950, but in 1952 and 1953 only the 1.12 kg/ha rate reduced yields. No differences were noted between the ester and amine forms of 2,4-D.

b. Soil Application 'Combine Kafir-60' grain sorghum could be grown in soils treated with 2,4-D + 2,4,5-T (1:1) at 26.9 kg/ha as early as 3 months after treatment without a reduction in the fresh weight after 58 cm of rainfall in the tropics (Bovey et al., 1968).

2. Dicamba

Dicamba was applied to grain sorghum at 0.14 and 0.28 kg/ha (Fields, 1968). When applied from emergence to 10 days after emergence, dicamba can cause lodging. When applied to plants 10 to 15 cm tall and 10 days old, dicamba had little effect on roots or lodging. A small percentage of head blasting can occurr if dicamba is applied 20 days prior to head emergence from the boot and up to the flowering period.

3. Picloram

a. Foliar Application Foliar sprays of 0.025 and 0.07 kg/ha of picloram increased seedling weights of Pioneer 820, RS-625, and RS-626 grain sorghum (Scifres and Bovey, 1970). 'GA-615' grain sorghum seedling weight was reduced by 0.018 kg/ha of picloram, whereas the dry weights of Tophand and RS-671 were reduced by 0.026 kg/ha to 0.14 kg/ha. Pioneer 820 and RS-625 tolerated picloram rates of 0.28 kg/ha, and were the most tolerant sorghum varieties. Significant increases in the dry weight of Pioneer 820, RS-625, and PA 6-665 occurred when treated preemergence with irrigation water containing 0.5, 1, and 2 ppm of picloram, whereas GA-615, RS671, Topland, and RS-626 were reduced in growth by these concentrations. Dry weights of Pioneer 820 and RS-625 were increased by irrigation water containing 0.001 and 0.005 ppm of picloram applied postemergence.

b. Soil Application Bovey et al. (1968) found that 'Combine Kafir 60' grain sorghum could be grown in Puerto Rico in a Mucara silty clay loam soil as early as 3 months after application of 6.72 kg/ha of the potassium salt of picloram without a reduction in fresh weight. Rainfall was 58 cm during the 3 months.

In Texas, 'Tophand' grain sorghum could be grown in Wilson clay loam soil 12 months after picloram application at 1.12 kg/ha without a reduction in plant numbers, dry weight production, flowering, or germination (Bovey et al., 1975). No picloram was detected in sorghum seed (immature and mature) seeded 6 and 12 months after picloram application.

4. Other Hormonelike Herbicides

Seedling 'MS 398' grain sorghum in the three-leaf stage was not injured by 2,4,5-T and triclopyr rates of 0.002, 0.009, 0.03, and 0.14 kg/ha applied as foliar sprays (Bovey and Meyer, 1981). Slight injury with 2,4,5-T and triclopyr occurred at 0.56 kg/ha. Seedling grain sorghum showed slight injury by clopyralid at rates of 0.14 and 0.56 kg/ha.

5. Paraquat

Scifres and Santlemann (1966) applied paraquat at 0.14 and 0.28 kg/ha to RS-610, RS-613, OK627, OK632, and OK612 hybrid grain sorghum and their parents combine Kafir 60 and 7078, Dwarf Redlan, Wheatland, Redlan, Derk 8-2, OKRY8, and OKRY10. All varieties suffered sheath and leaf margin burn on the lower vegetation. Combine Kifir 60, a experimental hybrid (Redlan X OKRY10), and OKRY8 were the least affected. Treatment of 15-cm-tall sorghum reduced the yield, but plants treated when taller than 15 cm were not injured.

6. Glyphosate and Tebuthiuron

Baur and Bovey (1975) found that the order of decreasing effectiveness of five herbicides on 'Tophand' grain sorghum and 'Southern Blackeye' cowpeas were paraquat, glyphosate, tebuthiuron, 2,4-D, and endothall. Herbicide treatments were applied to seedling plants at 1.4 to 1,434 μg/plant to one unifoliolate leaf of cowpeas and to the partially unfurled true leaf of sorghum. The absorption of all herbicides occurred within the first hour. Glyphosate and tebuthiuron had little inhibitory effect on germination of sorghum, 'Era' wheat, or cowpea seeds.

C. Rice (*Oryza sativa* L.)

1. Phenoxy Herbicides

Kaufman and Crafts (1956) indicated that the most striking deleterous effects of phenoxy herbicides on rice were at young growth stages and when the seed was sown. Pellet formulation was more selective than sprays at equivalent dosages for weed control. MCPA and 2,4-D appeared to stimulate rice growth at low dosages.

2. Triclopyr

Smith (1988) indicated that triclopyr was applied alone at 0.3 or 0.6 kg/ha and tank-mixed at 0.3 or 0.6 kg/ha plus propanil at 3.4 kg/ha at early tillering (40 cm tall), jointing (66 cm tall; internodes 0.6 cm long), early booting (76 cm tall; panicles 14 cm long in the sheath), and late booting (86 cm tall; panicles 15 cm long in the flag leaf sheath). Triclopyr alone and in mixture with propanil reduced rough grain yields of rice by 18% when applied at the late-booting growth stage, but did not affect yields when applied at the early-tillering, jointing, or early-booting growth stages. The whole-grain milling yield and germination of rice seed were not affected by triclopyr alone or in mixture with propanil.

Pantone and Baker (1991) found that the rice cultivar 'Tebonnet' was more tolerant than 'Lemant' to triclopyr, with an average percentage grain yield reduction across treatments of 6% compared to 9% for 'Mars' and 12% for Lemont. Triclopyr rates were 0.4 and 0.8 kg/ha applied at the two- to three-leaf, four- to five-leaf, and panicle initiation stages of growth.

3. Picloram

Seedling 'Bluebell' rice tolerated one application of irrigation water containing up to 100 ppb of picloram applied to the soil as measured by fresh and dry plant weight growth (Baur et al., 1970). Bovey et al. (1968) grew 'Mentana' rice 3 months after application of 6.72 kg/ha of picloram in a tropical environment (Mucara silty clay-loam soil) without a reduction in the fresh weight of the rice.

D. Corn (*Zea Mays* L.)

1. 2,4-D

Lee (1949) summarized the work of several investigators and concluded that the most damage occurred when corn was 25 to 90 cm tall when sprayed with 2,4-D, but was less susceptible at 5 to 10 cm tall. 2,4-D can cause brittleness, lodging, curvature of the stalk, and development of abnormal brace roots. Reduced yields usually resulted from lodging because of brittleness. Inbred lines varied in tolerance. Hybrids appeared more sensitive to 2,4-D than inbreeds. Corn tolerated respraying with 2,4-D as two postemergence or as one preemergence and one postemergence treatment.

2. Picloram

Arnold and Santlemann (1965) found that picloram applied preemergence and postemergence at rates 0, 0.14, 0.28, and 0.56 kg/ha caused some crop injury. Corn stand was not reduced, but height was reduced early in the season. Corn injury increased as the concentration of picloram increased. Corn yield was not affected by any rate, however. Sorghum yield in 1964 was reduced by picloram at rates above 0.14 kg/ha.

In the greenhouse, 'Texas No. 30' hybrid corn tolerated 1000 ppb of picloram in irrigation water without any reduction in fresh and dry weights (Baur et al., 1970). 'USDA-34' corn could be grown within 2 months after treatment of picloram at 6.72 kg/ha in a tropical soil without any reduction in fresh weights (Bovey et al., 1968).

3. Other Hormonelike Herbicides

Seedling corn '5855 X 127C' in the three- to four-leaf stage tolerated rates of 0.002, 0.009, 0.03, 0.14, and 0.56 kg/ha of 2,4,5-T and clopyralid without significant injury. Triclopyr caused injury at 0.14 and 0.56 kg/ha, however (Bovey and Meyer, 1981).

E. Additional Data

During the late 1940s and early 1950s considerable effort was given to the phenoxy herbicides, especially 2,4-D, on weeds and brush control and the response of different crops. This is clearly documented in the research reports and the proceedings of the annual meetings of the North Central Weed Control Conference and other conferences. The Western Weed Control Conference was organized in Denver in 1938; the North Central in 1944; the Northeastern and Eastern and Western Canada in 1947; and finally the Southern in 1948. The national society, the Weed Science Society of America, was organized in Fargo, North Dakota, in 1954, and held its first meeting in New York City in 1956 (Timmons, 1970).

Viehmeyer (1947) summarized the work of several investigators on various crops. Sprays of 0.56, 1.12, and 2.24 kg/ha did not reduce the yield, seed germination, or kernel weight of 'Oderbrurker' barley (*Hordeum vulgare* L.) and 'Vicland' oats (*Avena sativa* L.). The kernel weight of barley was increased when sprayed at 10 and 20 cm tall. Lodging was reduced by 2,4-D treatment. Higher rates of 2,4-D (4.5 kg/ha) depressed yields and germination. The effect of 2,4-D on corn, clover, alfalfa, wheat, grain sorghum, flax, and soybeans have been discussed. Bovey and Meyer (1981) found that seedling 'TAM 0312' oats were not affected by clopyralid rates up to and including 0.56 kg/ha. Oats were slightly injured by 2,4,5-T at 0.56 kg/ha, and triclopyr at 0.14 and 0.56 kg/ha.

F. Sugarcane (*Sacchurum* sp.)

Field studies were conducted in Louisiana to determine the growth and yield of sugarcane from accidental application of the isopropylamine formulation of glyphosate (Richard, 1991). Glyphosate was applied in water diluent at 0.1, 0.2, 0.4, and 0.8 kg/ha in May, June, and August. Sugar yields were reduced 44% when glyphosate was applied over the top at 0.2 kg/ha in June; similar reductions for May and August applications required 0.4 kg/ha. The recoverability of the subsequent ratoon crop was also affected—particularly in May and June applications to the previous crop—at rates of 0.4 kg/ha and 0.8 kg/ha of glyphosate.

G. Perennial Grasses (Pasture and Turf)

Glyphosate was applied foliar at 0, 0.14, 0.28, 0.56, and 1.12 kg/ha to 2- and 4-month-old perennial ryegrass (*Lolium perenne* L.), orchardgrass (*Dactylis glomerata* L.), Kentucky bluegrass (*Poa pratensis* L.), red fescue (*Festuca rubra* L), and highland bentgrass (*Agrostis tenuis* Sibth.) (Bingham et al., 1980). All 2-month-old grasses were killed by glyphosate at 0.28 kg/ha or more. At 0.14 kg/ha red fescue was moderately resistant, and bluegrass, orchardgrass, and perennial ryegrass were moderately susceptible. Bentgrass was very susceptible. When 4-month-old grasses were treated, bluegrass was as tolerant to 0.28 kg/ha of glyphosate as was red fescue. Orchardgrass and perennial ryegrass were moderately susceptible, whereas bentgrass remained the most susceptible. Dosages at 0.14 kg/ha had little effect, whereas dosages higher than 0.28 kg/ha injured all grasses. ^{14}C-glyphosate absorption and transport was related to phytotoxicity since more radioactivity was transported in orchardgrass and perennial ryegrass than red fescue.

H. Introduced and Native Grasses

Lym and Kirby (1991) found that western wheatgrass [*Pascopyrum smithii* (Rydb.) Love] production decreased more than any other following either spring or fall glyphosate application. Glyphosate was applied at 0.2 or 0.4 kg/ha alone or with 2,4-D at 0.35 or 0.7 kg/ha. In seeded plots, intermediate wheatgrass [*Elymus intermedia* (Host) Nevski] was the most, and 'Fairway' crested wheatgrass [*Agropyron cristatum* (L.) Beauv.] the least susceptible to glyphosate plus 2,4-D, with average yields of 57 and 97%, compared with untreated controls of each species, respectively. Glyphosate reduced the yield of 'Nordan' standard crested wheatgrass [*A. desertorum* (Fisch ex. Link) Schult] more than Fairway or other diploid cultivars. Glyphosate plus 2,4-D can be used for pasture and rangeland weed control, provided some yield reduction is acceptable.

I. Pasture and Rangeland Grasses

1. Foliar Application

In the field, picloram, tebuthiuron, and 2,4-D at 0.4, 1.1, and 2.2 kg/ha did not reduce protein concentrations in kleingrass (*Panicum coloratum* L.), buffelgrass (*Cenchrus cilaris* L.), and coastal bermudagrass [(*Cynodon dactylon* (L.) Pers.], but did in a buffel X birdwood hybrid (*Cenchrus setigenis* Vahl.) (Baur et al., 1977). Glyphosate increased the protein content in buffelgrass and kleingrass, but sometimes reduced the production of buffelgrass and buffelgrass X birdwood hybrid if applied in April or June. All grasses tolerated (with regard to production) 2,4-D applied in March, April or June and March application of glyphosate. Common bermudagrass and kleingrass tolerated March and April application of tebuthiuron. Coastal bermudagrass tolerated tebuthiuron applied in March but not in April or June. Picloram reduced common bermudagrass production when applied in March and all application dates for coastal bermudagrass.

In other studies 2,4-D, 2,4,5-T, and dicamba applied in the spring or fall usually did not reduce yields of bermudagrass (Bovey et al., 1974a). Picloram reduced the density and yield of bermudagrass during dry periods. The degree of injury was directly related to the herbicide rate. Common, coastal, and coastcross-1 varieties responded the same to all herbicides.

At early postemergence kleingrass tolerated rates up to and including 0.28, 0.56, and 1.12 of picloram, 2,4-D, dicamba, and propazine, respectively, but did not tolerate tebuthiuron or hexazinone at any rate in the greenhouse (Bovey et al., 1979). At the 5-to-12.5-cm-tall stage or when mature, kleingrass tolerated higher rates of all herbicides. Most herbicides were phytotoxic applied preemergence, but kleingrass tolerated 0.56 kg/ha of 2,4-D.

Common buffelgrass in the greenhouse tolerated 1.1 kg/ha or less of clopyralid applied preemergence. All other treatments were phytotoxic, except picloram and 2,4,5-T at 0.3 kg/ha each (Bovey et al., 1984). Buffelgrass tolerated postemergence sprays of 2,4-D, picloram, and tebuthiuron at 0.3 kg/ha; dicamba and 2,4,5-T at 0.6 kg/ha; and clopyralid at 2.2 kg/ha based on overdry shoot production 1 month after treatment. The regrowth of buffelgrass from stubble 1 month after the original harvest of early postemergence treatments occurred only with clopyralid at all rates and 2,4,5-T at 0.3 kg/ha. When treated postemergence 45 days after planting, buffelgrass tolerated dicamba, 2,4-D, 2,4,5-T, clopyralid, and picloram, all at 2.2 kg/ha, and triclopyr at 0.6 kg/ha, but did not tolerate any rate of tebuthiuron or hexazinone. Mature buffelgrass (90 or 150 days old) responded similarly to herbicides applied at 45 days after planting.

Bovey (1998) applied foliar sprays of 2,4-D amine, clopyralid, metsulfuron, picloram, triclopyr, and tebuthiuron at 0.07 and 0.14 kg/ha to 14 tame, introduced, or native grasses. 2,4-D caused slight injury only on blue panicgrass (*Panicum antidotale* Retz.) and 'TAM Wintergreen' hardinggrass [*Phalaris tuberosa* var. *stenoptera* (Hack.) Hitchc.] at 0.07 and 0.14 kg/ha. Clopyralid caused no apparent injury to any grass. Metsulfuron caused slight injury to King Ranch bluestem [*Bothriochloa ischaemum* (L.) King. var. *songaricus* (Rupr.) Celeries & Harlan], moderate injury to blue panicgrass at both rates, and moderate injury to Indiangrass [(*Sorghastrum nutans* (L.) Nash ex. Small)] at 0.14 kg/ha. Severe injury occurred at both rates to 'TAM Wintergreen' hardinggrass.

Picloram showed no injury to any at the 14 grasses at 0.07 kg/ha. Slight injury occurred to sideoats grama at 0.14 kg/ha. Triclopyr showed slight injury to blue panicgrass at both rates. Tebuthiuron showed slight injury to sideoats grama at 0.07 kg/ha and 0.14 kg/ha, slight injury to blue panicgrass at 0.14 kg/ha, slight injury to 'TAM Wintergreen' hardinggrass at 0.07 kg/ha, and severe injury at 0.14 kg/ha. Other grasses in the study included Kleberg bluestem (*Dichanthium annulatum* Stapf), green sprangletop [*Leptochloa dubia* (H.B.K.) Nees.], sorghum almum (*Sorghum X almum* Parodi), 'Morpa' weeping lovegrass [*Eragrostis curvula* (Schard.) Nees.] 'Selection 75' kleingrass, 'Llano' buffelgrass, 'Alamo' switchgrass, little bluestem [*Schizachyrium scoparium* (Michx.) Nash], and blue grama [*Boutelocia gracilis* (H.B.K.) Lag. ex. Steud.].

2. Soil Application

Seedling of 'Blackwell' switchgrass (*Panicum virgatum* L.), 'Premier' sideoats grama [*Bouteloua curtipendula* (Michx.) Torr.], and native vine mesquite (*Panicum obtusum* H.B.K.) tolerated 0.28 kg/ha of dicamba applied preemergence (Halifax and Scifres, 1972). After emergence, sideoats grama tolerated 0.56 kg/ha dicamba. Preemergence or postemergence applications of 1.12 or 2.24 kg/ha severely retarded shoot production of all three species. Susceptibility of species to dicamba were vine mesquite > switch grass > sideoat grama.

Scifres and Halifax (1972a) found that picloram did not affect grass seed germination, but did affect radicle elongation of sideoats grama, buffalograss [*Buchloe dactyloides* (Nutt. Engelm)], and switchgrass in petri dishes at 1.25 ppb.

Shoot elongation was not affected by 1000 ppb in any species. Picloram did not reduce top growth production of the three grasses growing in soil containing 50 ppb. Top growth production of Arizona cottontop [*Digitaria California* (Benth.) Henr.] and vine mesquite was reduced by 125 to 250 ppb of picloram in soil.

Scifres and Halifax (1972b) found that root production and root: shoot ratios of switchgrass were decreased when 1000 to 2000 ppb of picloram were placed on the soil surface or at 7.5 cm deep. Sideoats grama root production decreased by application of 1000 ppb placed 2.5 cm deep, but production was increased in the soil with 1000 ppb of picloram placed 15 cm deep. Root: shoot ratios in picloram-treated soil were typically no different from those in untreated soil, but root growth pattern was affected.

Tebuthiuron at 2 or 4 ppmw placed 0 to 3, 8 to 11, or 15 to 18 cm deep in soil columns reduced the root and shoot weight of buffelgrass 30 days after emergence (Rasmussen et al., 1985). Plains bristlegrass (*Setaria macrostachya* H.B.K.) seedling shoot weights were not reduced when 2 ppmw of tebuthiuron were placed 8 to 11 cm deep or deeper.

In the field, tebuthiuron pellets (20% ai) applied at 0.6, 1.1, or 2.2 kg/ha in the spring at two locations and in the fall at a third location on the South Texas plains did not reduce buffelgrass standing crop or foliar cover compared to untreated areas (Hamilton and Scifres, 1983).

J. Turf Grasses

The effects of the alkanolamine salt of 2,4-D and the sodium salt of 4-hydroxy-3,5-diiodobenzonitrile (ioxynil) upon seedlings of Chewings fescue (*Festuca rubra* var. *commutata* Gaud.), Park bluegrass (*Poa pratensis* L.), and Highland bentgrass (*Agrostis tenuis* Sibth.) were evaluated in growth-room and field studies (Adamson and Turley, 1970). In the former, injury to Park bluegrass from both 1.12 and 3.4 kg/ha applications of 2,4-D was severe, with pronounced but less marked effects on fescue and bentgrass. The results were similar from applications shortly after emergence and after clipping the grass. Negligible injury was caused by 1.12 kg/ha of ioxynil. In the field, ioxynil gave better weed control than 2,4-D when applied at a seedling height of 2.5 cm. No injury was caused by ioxynil. Although 2,4-D reduced yields from early clippings, there was marked recovery, even at 3.4 kg/ha. No treatments significantly affected plant stand.

IX. HERBICIDE RUNOFF FROM TREATED TO NONTREATED AREAS

A. Movement of Picloram in Surface Water

Trichell et al. (1968) determined the movement of picloram in runoff water from small plots 24 hr after application. The loss of picloram was greater from sod than from fallow. The maximum loss obtained for picloram, dicamba, or 2,4,5-

T was 5.5%, and the average was approximately 3%. The time interval from picloram application to the first rainfall determined the amount of picloram that moved into the soil profile and/or the amount that moved away from the point of application with surface runoff. Four months after application, picloram losses were <1% of that lost during the initial 24 hr after application.

Scifres et al. (1971) indicated that picloram moved in surface runoff when 0.28 kg/ha was applied in the rolling plains of Texas for controlling honey mesquite. Irrigation the first 10 days after application resulted in a concentration of 17 ppb of picloram in surface runoff. Irrigation at 20 or 30 days resulted in <1 ppb of picloram residue in runoff water. No more than 1 or 2 ppb of picloram was detected after dilution of runoff water into ponds.

Baur et al. (1972) studied picloram residues from a 6.1 -ha watershed treated with the potassium salt of picloram at 1.12 kg/ha near Carlos, Texas, on an Axtell fine sandy loam soil. Samples were collected directly below the treated area and in streams below the plots after each heavy rainfall. Within 4 days after treatment, picloram residues in runoff water ranged from 9 to 168 ppb after heavy rainfall. After 3 months, concentrations of 5 ppb or less of picloram were found in runoff water. After 1.5 weeks with initial treatments in April, no picloram was found in streams from 0.8 to 3.2 km from the treated area. After a 6.1-cm rainfall 10 months after treatment, no picloram was detected in runoff water regardless of sampling location.

Research shows that herbicide residues can occur in surface runoff water if heavy rainfall occurs soon after treatment. When pelleted picloram was applied at 2.24 kg/ha to a 1.3-ha rangeland watershed, surface runoff of 1.5 cm from a 2.1-cm rain received 2 days after treatment contained an average of 2.8 ppm of picloram (Bovey et al., 1978a). The picloram content declined rapidly in each successive runoff event, however, and runoff water contained <5 ppb by 2.5 months after application. Loss of the potassium salt of picloram from grassland watersheds in surface runoff water was similar whether the picloram was applied as aqueous sprays or as pellets on a Houston black clay soil. Picloram plus 2,4,5-T at 0.56 kg/ha each were applied May 4, 1970, December 4, 1970, May 14, 1971, October 8, 1971, and May 5, 1972 (Bovey et al., 1974b). No runoff event occurred until July 25, 1971, 72 days after the third herbicide treatment. The concentrations of 2,4,5-T and picloram averaged 7 and 12 ppb, respectively, in runoff water and <5 ppb during subsequent runoff events. The data indicated that picloram or 2,4,5-T content was typically <5 ppb in runoff if major storms occurred 1 month or more after treatment on Houston black clay.

On sandy soils, Scifres et al. (1977) found only trace amounts of picloram or 2,4,5-T, which had been applied at 0.56 kg/ha each, in surface runoff water following storms about 30 days after application.

Mayeux et al. (1984) found that maximum concentrations of picloram were 48 and 250 ppb in initial runoff from an 8-ha area treated with 1.12 kg/ha in

1978 and 1979, respectively. Herbicide concentration decreased with distance from the treated area in proportion to the size of adjacent untreated watershed subunits that contributed runoff water to streamflow. About 6% of the applied picloram was lost from the treated area during active transport.

B. Movement of Phenoxy Herbicides in Surface Water

Using gas chromatographic and bioassay detection techniques. Trichell et al. (1968) investigated the loss of 2,4,5-T, dicamba, and picloram from bermudagrass and fallow plots of 3 and 8% slope. When determined 24 hr after application of 2.24 kg/ha, a maximum of about 2, 3, and 5 ppm picloram, 2,4,5-T, and dicamba, respectively, were found in runoff water after 1.3 cm of simulated rainfall. Losses of dicamba and picloram were greater from sod than from fallow plots, whereas 2,4,5-T losses were approximately equal. Four months after application, picloram, 2,4,5-T, and dicamba concentration in runoff water from sod plots had diminished to 0.03, 0.04, and 0 ppm, respectively. The maximum loss of any herbicide from the treated area was 5.5% and averaged 3%.

Bovey et al. (1974b) sprayed a 1:1 mixture of the triethylamine salts of 2,4,5-T plus picloram at 1.12 kg/ha every 6 months on a native-grass watershed for a total of five treatments. Plant "washoff" was the main source of herbicide detected in runoff water. Concentrations of both herbicides were moderately high (400 to 800 ppb) in runoff water if 3.8 cm of simulated rainfall was applied immediately after herbicide application. If major natural storms occurred 1 month or longer after herbicide treatment, concentration in runoff water was <5 ppb.

Norris and Moore (1970) and Norris (1971) indicated that concentrations of 2,4-D, 2,4,5-T, picloram, and amitrole seldom exceed 0.1 ppm in streams adjacent to carefully controlled forest spray operations in Oregon. Concentrations exceeding 1 ppm have never been observed and are not expected to occur. Chronic entry of these herbicides into streams did not occur for long periods after application.

C. Movement of Dicamba in Surface Runoff Water

Trichell et al. (1968) studied dicamba runoff from sloping sod plots in Texas. They found that as much as 5.5% of the applied dicamba was recovered in runoff water when 1.3 cm of artificial rain were applied 24 hr after herbicide application. No dicamba was found in runoff water from a similar artificial rain application of 1.3 cm 4 months later after 21.6-cm natural rainfall. Approximately 8% of the artificial rain was recovered as runoff.

Norris and Montgomery (1975) found maximum dicamba levels of 37 ppb about 5.2 hr after treatment at 1.3 km from the point where the sample stream entered the treatment unit in Oregon. Dicamba residues detected the first 30 hr after application resulted from drift and direct application to exposed surface water. By 37.5 hr, residue levels had declined to background levels; no dicamba residues were found more than 11 days after application. Dicamba levels found in streams were several orders of magnitude below threshold response levels for fish and mammals.

D. Movement of Triclopyr in Surface Runoff Water

Using a helicopter, Schubert et al. (1980) treated the upper part of a watershed in West Virginia with 11.2 kg/ha triclopyr. Two streams traversed the treated area. The movement of triclopyr residues in soil and water downslope from the treated area was insignificant.

The maximum concentration of triclopyr in stream water was 95 ppb the first 20 hr after application, similar to that observed for other herbicides applied to forest streams (Norris et al., 1987). A reduction in the concentration the first 20 hr after application was attributed to photodecomposition. In September, during the first significant rains after application in May, maximum triclopyr residues of 12 ppb were found in a small pond at the site. A 6-cm rain on November 9, causing a 6500-L stream discharge, increased triclopyr concentrations to 15 ppb, but after November 11 no more triclopyr was detected.

E. Movement of Tebuthiuron in Surface Runoff Water

Pelleted tebuthiuron was applied at 2.24 kg/ha to a 1.3-ha rangeland watershed. A 2.8-cm rain 2 days after application produced 0.94 cm of runoff, which contained an average of 2.2 ppm of tebuthiuron (Bovey et al. 1978b). Tebuthiuron concentration decreased rapidly with each subsequent runoff event. After 3 months, tebuthiuron concentration was <0.05 ppm; none was detected in runoff water 1 year after treatment. Concentration of tebuthiuron, applied as a spray at 1.12 kg/ha, decreased to <0.01 ppm in runoff within 4 months from a small plot receiving simulated rainfall. On 0.6-ha plots, mean tebuthiuron concentration from sprays and pellets was 0.50 ppm or less in water when the first runoff event occurred 2 months after application. Concentrations of tebuthiuron in the soil and grass from pellet applications were <1 ppm, and decreased with time.

Tebuthiuron applied at 1 kg/ha as 20% ai pellets to dry Hathaway gravelly, sandy loam soil in the spring diminished by 5% at the first simulated rainfall event (37 mm) in runoff water and sediment (Morton et al., 1989). The second and third simulated rainfall events (22 and 21 mm, respectively) removed an additional 2% of tebuthiuron. When tebuthiuron was applied to wet soil in the spring, the initial simulated rainfall events, totaling 42 mm, removed 15% of the tebuthiuron. When tebuthiuron was applied to wet soil in the fall, the initial rainfall events, totalling 40 mm, removed 48% of the tebuthiuron in runoff water and sediment. No significant differences were found in the total amount of tebuthiuron within the soil profile after application to dry and wet soils. More than half of the tebuthiuron had moved into the upper 7 cm 1 day after application. Tebuthiuron was not detected below 90 cm after 165 mm of simulated rainfall and 270 mm of natural rainfall.

F. Movement of Hexazinone in Surface Runoff Water

Lavy et al. (1989) found relatively small amounts of hexazinone in runoff water from a spot-gun application to a forest floor in Arkansas. Forest litter was highly effective in absorbing surface applications of hexazinone. In another study the maximum concentration of hexazinone was 14 ppm in the stream that drained a 11.5-ha watershed treated with 2 kg/ha (Bouchard et al., 1985). Hexazinone residues of <3 ppm were detected in stream discharge for 1 year after application. The amount of hexazinone transported from the watershed in stream discharge represented only 2 to 3% of the amount initially applied.

Neary et al. (1986) found only a 0.53% loss of hexazinone in streamflow of the applied herbicide in Georgia. Residues in streamflow peaked at 442 ppb in the first storm, but declined rapidly and disappeared within 7 months. Total sediment yield increased by a factor of 2.5 because of the increased runoff associated with site preparation using herbicide and salvage logging. Sediment loading remained below those produced by mechanical techniques, however, and overall water quality changes were small and short-lived. Leitch and Flinn (1983) applied hexazinone at 2 kg/ha from a helicopter to a 46.4-ha catchment. Only six of 69 samples analyzed contained hexazinone, which was well below the maximum allowable concentration of 600 μg/L for potable water.

X. SUMMARY

An added benefit of woody plant control is usually the control of the herbaceous weeds associated with the woody plants. The herbicide is sometimes selected to manage both herbaceous and woody vegetation. Because of this, herbaceous crop and desirable vegetation may be extremely sensitive to the chemicals used. Desirable plants adjacent to weed and brush areas may be damaged by volatility (movement by vapors in the air) of the herbicide moving to nontarget areas. The amount of fumes or vapors given off by 2,4-D, for example, is related to the vapor pressure of the chemical and high temperature. 2,4-D acid amine salts have very low volatility and cause little or no volatility hazard, whereas methyl, ethyl, and isopropyl esters of 2,4-D are very volatile. Low-volatility esters of 2,4-D and other hormonelike herbicides are made by using high-molecular-weight alcohols. In addition, an ether linkage (-0-) further reduces the volatility.

The relative volatilities of 2,4-D esters have been measured by exposing selected test plants in bags or bell jars to vapors of herbicides for a designated time and air temperature. Leaf epinasty, stem curvature, leaf modification, stem proliferation, and growth inhibition are some manifestations of herbicide effect to demonstrate its presence. Volatility may also be measured by trapping the herbicide and quantifying it by gas chromatography or other chemical analysis. Damage to a number of plants, especially broadleaf plants, has been demonstrated in the laboratory at extremely low rates of 2,4-D and also observed in the field. In the field, however, injury to plants (crops) by volatile herbicides is sometimes difficult to demonstrate because of the dilution effect and dispersion by air currents. In crop-sensitive areas, high-volatility esters of 2,4-D are usually prohibited.

Spray drift or accidental application of herbicides can cause injury and can be readily demonstrated in agronomic and horticulture crops, especially with dicots. Application of 0.01 kg/ha of 2,4-D at the seedling stage of cotton both delayed maturity and reduced yield. Most other herbicides are not as injurious as 2,4-D on cotton. Soybeans tolerated 2,4-D amine in the early stages of growth, but became more sensitive as they matured; 0.28 kg/ha reduced yields at all stages of growth. Picloram and dicamba were more injurious than 2,4-D to soybeans applied at prebloom. At flowering, dicamba at 0.035 kg/ha and picloram at 0.009 kg/ha reduced soybean yield about 50%. Tomatoes and root crops were more sensitive to 2,4-D than the other vegetable crops tested. Lettuce showed epinasty, necrotic, and chlorotic areas within 24 hr after treatment with >50 μg/plant. Imazapyr persisted in soil in the greenhouse for over 1 year, and forage grasses and herbs were highly sensitive to residual carryover damage. Beans and squash were tolerant of imazapyr residues. Hormonelike herbicides were generally more phytotoxic to dicots and tolerated by monocots at reduced rates that simulate spray drift. The effects on soil residues from herbicide application or runoff from adjacent treated areas are discussed.

REFERENCES

Adamson RW, Turley RH. Effects of 2,4-D and ioxynil on seedling fescue, bluegrass and bentgrass. Weed Sci 18:77–80, 1970.

Al-Khatib K, Parker R, Fuerst EP. Rose (*Rose dilecta*) response to simulated herbicide drift. Hort Tech 2:394–398, 1992a.

Al-Khatib K, Parker R, Fuerst EP. Sweet cherry (*Prunus avium*) response to simulated drift from selected herbicides. Weed Tech 6:975–979, 1992b.

Al-Khatib K, Parker R, Fuerst EP. Alfalfa (*Medicago sativa*) response to simulated herbicide spray drift. Weed Tech 6:956–960, 1992c.

Anderson RL, Nielson DC. Winter wheat (*Triticum aestivum*) stage effect on paraquat bioactivity. Weed Tech 5:439–441, 1991.

Arle HF. The sensitivity of Acala 44 cotton to 2,4-D. Proc West Weed Control Conf 14: 20–25, 1954.

Arnold WR, Santlemann PW. The effects of 4-amino-3,5,6-trichlorpicolinic acid on corn and sorghum. Proc South Weed Sci Soc 8:56–62, 1965.

Banks PA, Santelmann PW. Glyphosate as a postemergence treatment for johnson grass control in cotton and soybeans. Agron J 69:579–582, 1977.

Baskins AD, Walker EA. The response of tomato plants to vapor of 2,4-D and/or 2,4,5-T formulations at normal and higher temperature. Weeds 2:280–287, 1953.

Baur JR, Bovey RW, Holt EC. Effect of herbicides on production and protein levels in pasture grasses. Agron J 69:846–851, 1977.

Baur JR, Bovey RW. Herbicidal effects of tebuthiuron and glyphosate. Agron J 67:547–553, 1975.

Baur JR, Bovey RW, Merkle MG. Concentration of picloram in runoff water. Weed Sci 20:309–313, 1972.

Baur JR, Bovey RW, Benedict CR. Effect of picloram on growth and protein levels in herbaceous plants. Agron J 62:627–630, 1970.

Behrens R, Lueschen WE. Dicamba volatility. Weed Sci 27:486–493, 1979.

Behrens R, Hall WC, and Fisher CE. Field response of cotton to four phenoxy-type herbicides. Proc South Weed Conf 8:72–75, 1955.

Bennet RJ. The effects of 2,4-D isooctyl ester/ioxynil herbicide in the liquid and vapour phases on the growth of tomato (*Lycopersicon esculentum* Mill.) plants. So Afr J Plant Soil 6:24–31, 1989.

Betts M, Ashford R. The effect of 2,4-D on rapeseed. Weed Sci 24:356–360, 1976.

Bhatti MA, Al-Khatib K, Parker R. Wine grape (*Vitis vinifera*) response to repeated exposure of selected sulfonlurea herbicides and 2,4-D. Weed Tech 10:951–956, 1996.

Bhatti MA, Al-Khatib K, Parker R. Wine grape (*Vitis vinifera*) response to fall exposure of simulated drift from selected herbicides. Weed Tech 11:532–537, 1997.

Bingham SW, Segura J, Foy CL. Susceptibility of several grasses to glyphosate. Weed Sci 28:579–585, 1980.

Bouchard DC, Lavy TL, Lawson ER. Mobility and persistence of hexazinone in a forest watershed. J Environ Qual 14:229–233, 1985.

Bovey RW, Senseman SA. Response of food and forage crops to soil-applied imazapyr. Weed Sci 46:614–617, 1998.

Bovey RW, Miller FR. Desiccation and defoliation of plants by different herbicides and mixtures. Agron J 60:700–702, 1968.

Bovey RW, Meyer RE. Effects of 2,4,5-T, triclopyr and 3,6-dichloropicolinic acid on crop seedlings. Weed Sci 29:256–261, 1981.
Bovey RW, Senseman SA. Response of food and forage crops to soil-applied imazapyr. Weed Sci 46:614–617, 1998.
Bovey RW, Baur JR, Bashaw EC. Tolerance of kleingrass to herbicides. J Range Mgt 32: 337–339, 1979.

Bovey RW, Hein H Jr, Meyer RE. Effect of herbicides on the production of common buffelgrass (*Cenchrus ciliaris*). Weed Sci 32:8–12, 1984.

Bovey RW, Meyer RE, Hein H Jr. Soil persistence of tebuthiuron in the claypan resource area of Texas. Weed Sci 30:140–144, 1982.

Bovey RW, Meyer RE, Holt EC. Tolerance of bermudagrass to herbicides. J Range Mgt 27:213–296, 1974a.

Bovey RW, Miller R, Diaz-Colon J. Growth of crops in soils after herbicidal treatment for brush control in the tropics. Agron J 60:678–679, 1968.

Bovey RW, Miller FR, Baur JR, Meyer RE. Growth of sorghum and soybeans in picloram-treated soil. Agron J 67:433–436, 1975.

Bovey RW, Burnett E, Meyer RE, Richardson C, Loh A. Persistence of tebuthiuron in surface runoff water, soil and vegetation in the Texas Blacklands Prairie. J Environ Qual 7:233–236, 1978b.

Bovey RW, Burnett E, Richardson C, Baur JR, Merkle MG, Knisel WG. Occurrence of 2,4,5-T and picloram in surface runoff water in the blacklands of Texas. J Environ Qual 3:61–64, 1974b.

Bovey RW, Richardson C, Burnett E, Merkle MG, Meyer RE. Loss of spray and pelleted picloram in surface runoff water. J Environ Qual 7:178–180, 1978a.

Breeze VG. Uptake by tomato plants of the herbicide [^{14}C] 2,4-D butyl in the vapour phase. Pestic Sci 29:9–18, 1990.

Breeze VG. Herbicide vapour phytotoxicity: Laboratory facts and field speculation in 1994. British Crop Protection Council Monograph no. 59: Comparing Glasshouse and Field Pesticide Performance II. International Symposium, Canterbury, England: 1994, pp. 85–95.

Breeze VG, van Rensburg E. Vapour of the free acid of the herbicide 2,4-D is toxic to tomato and lettuce plants. Environ Pollu 72:259–267, 1991.

Breeze VG, van Rensburg E. Uptake of the herbicide [^{14}C] 2,4-D iso-octyl in the vapour phase by tomato and lettuce plants and some effects on growth and phytotoxicity. Ann Appl Bio 120:493–500, 1992.

Breeze VG, West CJ. Effects of 2,4-D butyl vapour on the growth of six crop species. Ann Appl Bio 111:185–191, 1987a.

Breeze VG, West CJ. Long-and short-term effects of vapour of the herbicide 2,4-D butyl on the growth of tomato plants. Weed Res 27:13–21, 1987b.

Breeze VG, Simmons JC, Roberts MO. Evaporation and uptake of the herbicide 2,4-D-butyl applied to barley leaves. Pestic Sci 36:101–107, 1992.

Brown CA, Hodleman QL, Hagood ES. Injuries to cotton by 2,4-D. bull. 426. Baton Rouge: Louisiana Agric Exp Stn., 1948.

Carns HR, Goodman VH. Responses of cotton to 2,4-D. bull. 544. State College, MS: Mississippi Agric Exp Stn, 1956.

Coffman CB, Frank JR, Potts WE. Crop responses to hexazinone, imazapyr, tebuthuron and triclopyr. Weed Tech 7:140–145, 1993.

Dawson JH. Response of alfalfa (*Medicago sativa*) grown for seed production to glyphosate and SC-0224. Weed Tech 6:378–381, 1992.

Derksen DA. Dicamba, chlorsulfuron and clopyralid as sprayer contaminants on sunflower (*Helianthus annus*), mustard (*Brissica juncea*) and lentil (*Lens culinaris*), respectively. Weed Sci 37:616–621, 1989.

Dilbeck RE, Quisenberry JE, Wiese AF, Regier CG. (1987). Comparison of hairy and smooth leaf phenotype to drift and direct application of 2,4-dichlorophenoxy acid. 1987 Proceedings Beltwide Cotton Prod Research Conference, 1987, p. 132.

Dunlap AA. 2,4-D injury to cotton from airplane dusting of rice. Phytopath 38:638–644, 1948.

Ergle DR, Dunlap AA. Responses of cotton to 2,4-D. bull. 713. College Station, TX: Texas Agric Exp Stn, 1949.

Fields RW. Broadleaf weed control in grain sorghum with Banvel (dicamba). Proc South Weed Conf 21:133–135, 1968.

Flint GW, Alexander JJ, Funderburk OP. Vapor pressure of low-volatile esters of 2,4-D. Weed Sci 16:541–544, 1968.

Foy CL, Witt HL. Effects of paraquat on weed control and yield of alfalfa (*Medicago sativa*) in Virginia. Weed Tech 7:495–506, 1993.

Fratesi KT, Wray MW, Paroonagian DL. Cotton and soybeans tolerance to triclopyr amine, 2,4-D amine and bromoxynil. Proc South Weed Sci Soc 41:355, 1988.

Gentner WA. Herbicidal activity of vapor of 4-amino-3,5,6-trichloropicolinic acid. Weeds 12:239–240, 1964.

Gile JD. Relative airborne losses of commercial 2,4-D formulations from a simulated wheat field. Archives Environ Contam Toxicol 12:456–469, 1983.

Goodman VH. The yield and pogeny seedling responses to treatment with 2,4-D, 2,4,5-T and MCP, Proc South Weed Conf 6:47–56, 1953.

Goodman VH, Ennis WB Jr, Palmer RD. Cotton response to 2,4-D, 2,4,5-T, MCP and related growth regulations. Proc South Weed Conf 8:76–81, 1955.

Grover R. A method for determing the volatility of herbicides. Weed Sci 23:529–532, 1975.

Halifax JC, Scifres CJ. Influence of dicamba on development of range grass seedlings. Weed Sci 20:414–416, 1972.

Hamilton WT, Scifres CJ. Buffelgrass (*Cenchrus ciliaris*) responses to tebuthiuron. Weed Sci 31:634–638, 1983.

Hart SE, Glenn S, Kenworthy WW. Tolerance and basis for selectivity to 2,4-D in perennial *Glycine* species. Weed Sci 39:535–539, 1991.

Heering DC, Peeper TF. Winter wheat (*Triticum aestivum*) response to picloram and 2,4-D. Weed Tech 5:317–320, 1991.

Hemphill DD Jr, Montgomery ML. Response of vegetable crops to sublethal applications of 2,4-D. Weed Sci 29:632–635, 1981.

Hitchcock AE, Zimmerman PW, Kirkpatrick H Jr. A simple, rapid biological method for determining the relative volatility of esters of 2,4-D and 2,4,5-T. Contrib Boyce Thomp Inst 17(3):243–263, 1953.

James LF, Evans JO, Ralphs MH, Child RD, eds. Noxious Range Weeds. Boulder, CO: Westview, 1991.

Johnson JP, Stelzer HE. Loblolly pine photosynthesis is enhanced by sublethal hexazinone concentrations. Tree Physiol 8:371–379, 1991.

Jordan TN, Bridge RR. Tolerance of cotton to herbicide glyphosate. Agron J 71:927–928, 1979.

Kaufman PB, Crafts AS. Responses of the rice plant to different formulations and methods of application of 2,4-D, MCP, and 2,4,5-T. Hilgardia 24:411–452, 1956.

Kay BL. Paraquat for selective control of range weeds. Weeds 12:192–194, 1964.

Keeley PE, Thullen RJ, Carter CH, Miller JH. Control of johnsongrass in cotton with glyphosate. Weed Sci 32:306–309, 1984a.

Keeley PE, Thullen RJ, Carter CH, Miller JH. Comparison of ropewick applicators for control of johnsongrass in cotton with glyphosate. Weed Sci 32:431–435, 1984b.

Klingman DL. Effects of varying rates of 2,4-D and 2,4,5-T at different stages of growth on winter wheat. Agron J 45:606–610, 1953.

Klingman DL, Bovey RW, Knake EL, Lange AH, Meade JA, Shroch WA, Stewart RE, Wyse DL. Systemic herbicides for weed control: Phenoxy herbicides, dicamba, picloram, amitrole and glyphosate. A.D. BU-2281. Washington DC: Extension Service, U.S. Department of Agriculture, 1983.

Klingman GC. Weed Control: As a Science. New York: Wiley, 1961.

Klingman GC, Guedez H. Picloram and its effect on field-grown tobacco. Weeds 15:142–146, 1967.

Koscelny JA, Peeper TF, Krenzer EG Jr. Sulfonlylurea herbicides affect hard red winter wheat (*Triticum aestivum*) forage and grain yield. Weed Tech 10:531–534, 1996.

Krausz RF, Kapusta G, Matthews JL. Soybean (*Glycine max*) tolerance to 2,4-D ester applied preplant. Weed Tech 7:906–910, 1993.

Lavy TL, Mattice JD, Kochenderfer JN. Hexazinone persistence and mobility of a setup forested watershed. J Environ Qual 18:504–514, 1989.

Lee OC. Effect of herbicides in growing crops. North Central Weed Control Conf Res Rept 6:55–59, 1949.

Leitch CJ, Flinn DW. Residues of hexazinone in streamwater after aerial application to an experimental catchment planted with radiata pine. Aust For 46:126–131, 1983.

Lym RG, Kirby DR. Effect of glyphosate on introduced and native grasses. Weed Tech 5:421–425, 1991.

Lyon DJ, Wilson RG. Sensitivity of field beans (*Phaseblus vulgaris*) to reduced rates of 2,4-D and dicamba. Weed Sci 34:953–956, 1986.

Marth PC, Mitchell JW. Comparative volatility of various forms of 2,4-D. Bot Gaz 110: 632–636, 1949.

Martin JA, Fletcher JT. The effects of sublethal doses of various herbicides on lettuce. Weed Res 12:268–271, 1972.

Martin DA, Miller SD, Alley HP. Winter wheat (*Triticum aestivum*) response to herbicides applied at three growth stages. Weed Tech 3:90–94, 1989.

Mayeux HS Jr, Richardson CW, Bovey RW, Burnett E, Merkle MG, Meyer RE. Dissipation of picloram in storm runoff. J Environ Qual 13:44–49, 1984.

McIlrath WJ, Ergle DR. Further evidence of persistence of the 2,4-D stimulus in cotton. Plant Physiol 28:693–702, 1953.

McIlrath WJ, Ergle DR, Dunlap AA. Persistence of 2,4-D stimulus in cotton plants with reference to its transmission to the seed. Bot Gaz 112:511–518, 1951.

Meinhardt HR, Vorster LM, Van Dyk LP. Threshold concentrations of 2,4-D acid or 2,4-D dimethylamine for causing morphological symptoms on lettuce seedlings. Applied Plant Sci 5:90–93, 1991.

Miller JH, Kempen HM, Wilkerson JA, Foy CL. Response of cotton to 2,4-D and related phenoxy herbicides. Washington, DC: USDA agric tech bull. no. 1289. 1963.

Morgan PW, Hall WC. Metabolism of 2,4-D by cotton and grain sorghum. Weeds 11: 130–135, 1963.

Morton HL, Johnsen TN, Semanton JR. Movement of tebuthiuron applied to wet and dry rangelands soils. Weed Sci 37:117–122, 1989.

Moyer JR. Sulfonylurea herbicide effect on following crops. Weed Tech 9:373–379, 1995.

Moyer JR, Bergen P, Schaalje GB. Effect of 2,4-D and dicamba residues on following crops in conservation tillage systems. Weed Tech 6:149–155, 1992.

Mullison WR, Hummer RW. Some effects of the vapor of 2,4-dichlorophenoxyacetic acid derivatives on various field crops and vegetables seeds. Bot Gaz 111 (1):77–85, 1949.

Nalewaja JD. Reaction of flax to dicamba applied at several stages. Weed Sci 17:385–387, 1969.

Nalewaja JD. Reaction of wheat to picloram. Weed Sci 18:276–278, 1970.

Neary DG, Bush PB, Grant MA. Water quality of ephemeral forest streams after site preparation with the herbicide hexazinone. For Ecol Mgt 14:23–42, 1986.

Newton M. Phenoxy herbicides in rights-of-way and forestry in the United States. In: Biologic and Economic Assessment of Benefits from Use of Phenoxy Herbicides in the United States. USDA-NAPIAP report no. 1-PA-96. Washington, DC; 1996, pp. 165–178.

Noble A, Hamilton DJ. Relation between volatility rating and composition of phenoxy herbicide ester formulations. Pestic Sci 28:203–214, 1990.

Norris LA. The behavior of chemicals in the forest. Pesticides, Pest Control and Safety on Forest Range Lands: Proceedings, Short Course for Pesticidal Application, Oregon State University, Corvallis, 1971, pp. 90–106.

Norris LA, Montgomery ML. Dicamba residues in streams after forest spraying. Bull Environ Contam Toxicol 13:1–8, 1975.

Norris LA, Moore DG. The entry and fate of forest chemicals in streams. Proceedings, Symposium or Forest Land Uses and Stream Environment, Oregon State University, Corvallis. 1970, pp. 138–158.

Norris LA, Montgomery ML, Warren LE. Triclopyr persistence in western Oregon hill pastures. Bull Environ Contam Toxicol 39:134–141, 1987.

Ogg AG Jr, Ahmedullah MA, Wright GM. Influence of repeated applications of 2,4-D on yield and juice quality of Concord grapes (*Vitis labruscana*). Weed Sci 39:284–295, 1991.

Pantone DJ, Baker JR. Varietal tolerance of rice (*Oryza sativa*) to bromoxynil and triclopyr at different growth stages. Weed Tech 6:968–974, 1991.

Phillips WM. The effects of 2,4-D on the yield of Midland grain sorghum. Weeds 6:271–280, 1958.

Porter WK Jr, Thomas CH, Baker JB. A three-year study of the effect of some phenoxy herbicides on cotton. Weeds 7:341–348, 1959.

Que Hee SS, Sutherland RG. Volatilization of various esters and salts of 2,4-D. Weed Sci 22:313–318, 1974.

Quimby PC Jr, Nalewaja JD. Effect of dicamba on wheat and wild buckwheat at various stages of development. Weeds 14:229–232, 1966.

Rasmussen AG, Smith RP, Scifres CJ. Seedling growth responses of buffelgrass (*Pennisetum ciliare*) to tebuthiuron and honey mesquite (*Prosopis glandulosa*). Weed Sci 34:88–93, 1985.

Regier CG, Dilbeck RE, Undersander DJ, Quisenberry JE. Cotton resistance to 2,4-dichlorophenoxy acid spray drift. Crop Sci 26:376–377, 1986.

Richard EP Jr. Sensitivity of sugarcane (*Saccharum* sp.) to glyphosate. Weed Sci 39:73–77, 1991.

Schroeder GL, Cole DF, Dexter AG. Sugarbeet (*Beta vulgaris* L.) response to simulated herbicide spray drift. Weed Sci 31:831–836, 1983.

Schroeder J, Banks PA. Soft red winter wheat (*Triticum aestivum*) response to dicamba and dicamba plus 2,4-D. Weed Tech 3:67–71, 1989.

Schubert DE, McKellar RL, Stevens LP, Byrd BC. Triclopyr movement on a small watershed. abstract no. 105. Toronto, Canada: Weed Science Society of America, 1980, p. 50.

Schweizer EE. Response of sugarbeets (*Beta vulgaris*) to sublethal rates of 2,4-D. Weed Sci 26:629–631, 1978.

Scifres CJ, Bovey RW. Differential response of sorghum varieties to picloram. Agron J 62:775–777, 1970.

Scifres CJ, Halifax JC. Development of range grass seedlings germinated in picloram. Weed Sci 20:341–344, 1972a.

Scifres CJ, Halifax JC. Root production of seedling grasses in soil containing picloram. J Range Mgt 25:44–46, 1972b.

Scifres CJ, Santlemann PW. Response of cotton and sorghum to post-emergence applications of paraquat. Weeds 14:86–88, 1966.

Scifres CJ, Allen TJ, Leinweber CL, Pearson KH. Dissipation and phytotoxicity of dicamba residues in water. J Environ Qual 2:306–309, 1973.

Scifres CJ, Hahn RR, Diaz-Colon J, Merkle MG. Picloram persistence in semiarid rangeland soils and water. Weed Sci 19:381–384, 1971.

Scifres CJ, McCall HG, Maxey R, Tai H. Residual properties of 2,4,5-T and picloram in sandy rangelands soils. J Environ Qual 6:36–42, 1977.

Sciumbato AS. Volatility of the diglycolamine salt of dicamba. M.S. thesis. College Station, TX: Department of Soils and Crop Science, Office of Graduate Studies, Texas A&M University, 1999.

Simpson SW, Neher LA, Mischke TM, Oshima RJ. Monitoring the air for the presence of 2,4-D in Kern County, Kings County and San Luis Obispo County: A Cooperative California Study. Evironmental Hazard Assessment Program, California Department of Food and Agriculture, 1980.

Slife FW. The effect of 2,4-D and several other herbicides on weeds and soybeans when applied as post-emergence sprays. Weeds 4:61–68, 1956.

Smit C, deBeer PR, Van Dyk LP. Stability and retention of 2,4-D isooctyl ester on low volume air monitoring. Chemosphere 24:261–269, 1992.

Smith AL, ed. Handbook of Weed Management Systems. New York: Marcel Dekker, 1995.

Smith DT, Wiese AF. Cotton response to low rates of 2,4-D and other herbicides. TX: Texas Agric Exp Stn., bull. 1120 Lubbock, 1972.

Smith RJ Jr. Tolerance of rice (*Oryza sativa*) to aciflouorfen and triclopyr applied alone and in mixtures with propanil. Weed Sci 36:378–383, 1988.

Snipes CE, Street JE, Mueller TC. Cotton (*Gossypium hirsutum*) response to simulated triclopyr drift. Weed Tech 3:493–498, 1991.

Taylor SG, Shilling DG, Quesenberry KH, Chaudhry GR. Phytotoxicity of 2,4-D and 2,4-dichlorophenol to red clover (*Trifolium pratense*). Weed Sci 37:825–829, 1989.

Thorsness KB, Messersmith CG. Clopyralid influences rotational crops. Weed Tech 5: 159–164, 1991.

Timmons FH. A history of weed control in the United States and Canada. Weed Sci 18: 294–307, 1970.

Trichell DW, Morton HL, Merkle MG. Loss of herbicides in runoff water. Weed Sci 16: 447–449, 1968.

van Rensburg E, Breeze VG. Uptake and development of phytotoxicity following exposure to vapour of the herbicide ^{14}C 2,4-D butyl by tomato and lettuce plants. Environ Exp Bot 30:405–414, 1990.

Viehmeyer G. Effect of 2,4-D on cereal, flax, grasses, corn and other tolerant crops: Regional summary. Proc North Central Weed Control Conf 4:213–222, 1947.

Vogel A. Methodology and determination of 2,4-D and triclopyr residues employing the GC-ITD in the analysis of lettuce plants cultivated in the Tala Valley, Republic of South Africa. Bull Environ Contam Toxicol 60:371–378, 1998.

Watson AJ. The response of cotton to low rates of 2,4-D, 2,4,5-T, MCP and silvex. Proc South Weed Conf 8:82–86, 1955.

Wax LM, Kruth LA, and Slife FW. Response of soybeans to 2,4-D, dicamba and picloram. Weed Sci 17:388–393, 1969.

Wiese AF, Bovey RW, Eastin EF. Effect of herbicides on growth of cotton and associated crops. In: McWhorter CG, Abernathy JR, eds. Weeds of Cotton: Characterization and Control. Memphis: Cotton Foundation, 1992, pp. 515–547.

Wills GD. Factors affecting toxicity and translocation of glyphosate in cotton. Weed Sci 26:509–513, 1978.

Zimmerman PW, Hitchcock AE, Kirkpatrick H Jr. Methods for determining relative volatility of ester of 2,4-D and other growth regulators based on response of tomato plants. Weeds 2:254–261, 1953.

11

Ecological Impact of Woody Plant Management

I. INTRODUCTION

The primary purpose of using herbicides on range, forest, and noncrop lands is to control undesirable woody and herbaceous weeds, to increase desirable plants, and to maintain or improve plant and animal diversity without harm to the environment. Soil water and watershed yields are sometimes also increased with weed and brush control, and as stated before, one added benefit of woody plant control is the simultaneous control of many undesirable herbaceous weeds. Many forest lands are grazed by livestock and wildlife, and forage and browse plants can be provided with minimum effect on tree growth and production.

It is the purpose of this chapter to explore the ecological changes and shifts caused by herbicides for woody plant management.

II. FORAGE RESPONSE TO HERBICIDES—GRAZING LANDS

A. Mesquite Control

Gibbens (1981) measured grass and forb production for 4 years following the aerial application of 2,4,5-T to 3,634 ha of mesquite duneland on the Jornada Experimental Range in south-central New Mexico. Exclosures were used to exclude cattle from sampling sites on the sprayed and control areas. Perennial grass production was seven-, eight-, and fourfold greater, respectively, on sprayed areas than on control areas the first 3 years following treatment. The maximum perennial grass production was 642 kg/ha in the first season following treatment. In the fourth year the control area received 49 mm more precipitation than the sprayed area, so grass production was nearly equal on the two sites. Mesa dropseed (*Sporobulus flexuosus* [Thurb] Rydb.) was the major grass. The annual forb production varied widely among years but was greatest in the fourth season, when precipitation was best. The authors concluded that mesquite control was effective for improving forage production on arid rangelands, particularly with favorable precipitation following treatment. Herbel et al. (1983) indicated that in dense stands about 30% of the mesquite must be killed before grass yields are significantly increased.

Warren et al. (1996) indicated that honey mesquite canopy levels below 17% had little effect on forage production in New Mexico. In 1974, Martin and Morton (1993) killed velvet mesquite (*Prosapis velutina* Woot.) with diesel oil on watersheds of gully headcuts on the Santa Rita Experimental Range in Arizona. By 1977 Lehmann love grass (*Eragrostis lehmanniana* Nees) was greater where mesquite was killed than in untreated watersheds. Soil loss (mm) and total runoff over a 12-year period was lower where mesquite was killed.

The production of native perennial grasses and seeded Lehmann love grass was measured periodically for 21 years on a semidesert area, in which velvet mesquite was controlled by 2,4,5-T aerial sprays, and on an adjacent unsprayed area to determine how mesquite control would affect grass production and how long the effect would last (Cable, 1976). Grass production on the sprayed area increased dramatically during the first 5 years in a time-dependent relationship in response to the higher levels of available soil moisture. During the last 12 years, changes in love grass production were associated with changes in summer rainfall of the current and previous summers and of the intervening winter (two separate variables). Because of the strong competition from love grass, native grass production during the last 12 years did not show its usual relationship with summer rainfall, but deceased gradually and consistently on both the sprayed and unsprayed areas. At the end of the study period, native grasses provided only 10% of the total perennial grass production on the sprayed area and 20% on the unsprayed. Increased grass production, resulting from the mesquite control treatment and seeding, paid for the treatment within 4 years, and the sprayed area was still producing more grass than the unsprayed area 20 years later.

Martin and Morton (1980) also controlled false mesquite (*Calliandra eriophylla*) in southern Arizona. The greatest vegetation change was an increase in the density of Lehmann love grass on sprayed and unsprayed areas. Perennial forbs were almost completely eliminated, and densities of native perennial grasses were greatly reduced both on treated and untreated plots.

Scifres and Polk (1974) evaluated vegetation changes for 4 years following aerial application of 2,4,5-T + picloram (1:1) at 0.56 kg/ha to a semiarid range in the rolling plains of Texas with a light canopy of honey mesquite (12%) and sand sagebrush (2%). Forage production increased on areas with brush control and protection from grazing only in years of average or above-average rainfall and supported more grasses of fair to good grazing value than did unsprayed areas. Dahl et al. (1978) indicated similar results after aerial sprays of 2,4,5-T in light stands of honey mesquite. With a honey mesquite cover of 30%, however, a plant kill of over 80% during the year of application was required to provide a 605 kg/ha/year grass increase. A 90% kill would provide nearly 840 kg/ha/year extra grass.

Using aerial sprays of 2,4,5-T + picloram and clopralid, respectively, in western Texas, Bedunah and Sosebee (1984) and Jacoby et al. (1982) doubled the forage production in honey mesquite-treated areas compared to unsprayed areas 3 years after application. Forb production at the Crane site (Jacoby et al., 1982) was no different between the treated and untreated plots. McDaniel et al. (1982) found that aerial application of 2,4,5-T and 2,4,5-T + picloram on honey mesquite provided a 7 to 16% increase in grazing capacity over a 4-year period on light and heavy honey mesquite-infested pastures.

Heitschmidt et al. (1986), however, indicated that there were no differences between sprayed and unsprayed plots after 6 and 7 growing seasons relative to species composition, growth dynamics, and the production of herbaceous plants. The data partially validate the view that honey mesquite control may not always be justified based on economic analysis, and that light infestations may present no ecological problems.

Laxson et al. (1997) compared above-ground yields of herbage and wood in undisturbed [7630 stems ± 490 (SE)/kg], cleared, and three levels of thinned (100, 300, and 900 stems/ha) honey mesquite in west Texas. The total removal of the mesquite canopy resulted in a 45% increased standing herbaceous crop compared to the control the first 2 years postclearing. The herbage yield for thinned mesquite was intermediate, except that 900 stem/ha of mesquite produced herbage yield similar to the control. Mesquite growth rates were two- to threefold greater for 100 and 300 stem/ha stands than high-density stands during a wet year, but no different in a drier year. The amount of available forage and the growth rate of mesquite was increased when severely thinned to less than 900 stems/ha.

Meyer and Bovey (1985) applied several herbicides and herbicide combinations to honey mesquite in the post oak savannah region of east-central Texas, including glyphosate, picloram, triclopyr, clopyralid, tebuthiuron, hexazinone, 2,4,5-T, and picloram: 2,4,5-T combinations. Fall treatments reduced the diversity in the herbaceous cover in the following year. Glyphosate, hexazinone, and tebuthiuron reduced the grass cover, and most foliar–active herbicides controlled bitter sneezeweed, western ragweed, and woolly croton. In the fall the year after treatment the grass cover generally increased in most treatments. Bermudagrass and gaping panicum increased in plots treated with picloram, whereas shortspike windmillgrass usually increased in tebuthiuron-treated areas. Broadleaf cover increased in most treatments by the fall of the year following application. Woolly croton was numerous in glyphosate-, picloram-, and in some cases triclopyr-treated areas. The herbaceous cover was similar for all treatments by 2 years after treatment.

B. Huisache Control

Scifres et al. (1982) found that Texas wintergrass (cool season) tended to increase as the huisache canopy cover increased from 9 to 30%, compared to open areas. Grass production decreased as the huisache canopy cover increased beyond 30% on clay soils. Huisache is one of the preferred browse species for white-tailed deer as well as Texas wintergrass. Brush management for the Coastal Prairie should be directed toward thinning heavy stands of huisache to optimize livestock production and wildlife habitat.

C. White Brush Control

Meyer et al. (1969) used spray and granule forms of picloram to control white brush in central Texas. Herbaceous species released by picloram included fringeleaf paspalum (*Paspalum ciliatifolum* Michx var. *ciliatifolium*), tumble love grass (*Eragrostia sessilispica* Buckl.), and Texas wintergrass (*Stipa leucotricha* Trim. and Papr.), but reduced hooded windmillgrass (*Chloris cucullata* Bisch) and buffalo grass [*Buchloe dactyloides* (Nutt.) Engelm.]. Six weeks fescus [*Festuca octoflora* Walt. var. *glauca* (Nutt.) Fern.], mat sandbur (*Cenchrus pauciflorus* Benth.), and fringed signalgrass [*Brachiaria ciliatissima* (Buckl.) Chase] were annual grasses that increased in picloram plots. Picloram reduced stands of perennial forbs and the annual rosering gaillardia (*Gaillardia pulchella* Foug.) in fall, but not in spring treatments.

Fisher et al. (1972) treated pastures primarily with mesquite at nine locations in the Rolling Plains, Trans–Pecos, and Edwards Plateau of Texas of about one section each. In grazing studies calf weights increased an average of 10 kg per head more in aerial-sprayed pastures than in untreated pastures. Grass yields on treated pastures were twofold over untreated areas.

D. Pricklypear Cactus Control

Brownspine pricklypear (*Opuntia phaecantha* Engelm. & Bigel.) was effectively controlled within 2 years following application of a 1:1 mixture of 2,4,5-T and picloram at a rate of 0.6 kg/ha (Price et al., 1985). Brownspine pricklypear canopy cover and dry weight declined from approximately 23% and 3800 kg/ha to 8% and 1600 kg/ha, respectively. No significant difference in the total herbaceous forage dry weight was found between plants growing inside brownspine pricklypear canopy areas and plants growing outside the canopy areas. The differences between areas in species composition were significant in that cool-season grasses dominated the canopy area of the brownspine pricklypear colonies, while warm-season grasses dominated the area outside the canopy. Control of brownspine pricklypear will enhance the livestock-carrying capacity of rangeland in the Rolling Plains of Texas by increasing forage availability but not forage production.

In a range-improvement study, both liquid and pelleted forms of picloram were effective in controlling pricklypear cactus (Johnson et al., 1988). Higher rates of chemical hastened control and gave more complete control. At lower rates, the cactus was recovering in 1985, the fourth year of the study, suggesting that higher rates may be the most cost-effective.

In 1983, noncactus vegetation response was minor. In 1984, "all perennial grass" production increased by more than 50% (350 kg/ha) at higher rates of picloram. Shifts in cool-season and warm-season grass components were not significant. In 1985, cool-season grass was not generally affected by treatments, but warm-season grass increased at several rates of picloram. Cactus continued to decrease.

Grass utilization by cattle in 1984 increased greatly at higher rates of picloram. Desirable forage was increased when cactus was decreased. Forage access improved substantially. Based on livestock-grazing utilization estimates and increased perennial grass production, the potential reduction in acreages required for livestock grazing ranged from 49 to 72%.

E. Redberry Juniper Control

Basal cover, density, biomass, and species richness of the understory were measured in concentric zones from the stem bases of large redberry juniper (*Juniperus pinchotii* Sudw.) trees to 6 meters beyond their canopy edges on a shallow, rocky soil and two deep soils in the northern Edwards Plateau of Texas (Dye et al., 1995). Juniper interference intensified with increasing proximity to the stem bases. Biomass and basal cover of the herbaceous understory responded to a greater extent than did density and species richness 2 years after large redberry junipers were killed with soil injections of picloram. Herbaceous biomass responses after junipers were killed indicated that the sphere of influence of large junipers was more extensive on the shallow soil than on the deep soils. Herbaceous biomass in the presence of interference by large junipers on the Kimbrough, Angelo clay loam, and Tulia loam soils was 1,300, 1,780, and 1,290 kg/ha, respectively, compared to 2,140, 2,140, and 1,560 kg/ha 2 years after the junipers were killed on the three sites, respectively. The projected herbaceous biomass when juniper populations on the sites develop into closed-canopy woodlands was 320, 880, and 270 kg/ha for the Kimbrough, Angelo clay loam, and Tulia loam soils, respectively.

F. Sand Shinnery Oak Control

Jacoby et al. (1983) aerially treated sand shinnery oak (*Quercus havardii* Rydb.) with pelleted tebuthiuron near Andrews and Jayton, Texas. Sand shinnery oak was significantly reduced with pelleted tebuthiuron at 1.1 kg/ha at both locations. The yield of annual and perennial grasses was significantly greater and forbs significantly less on tebuthiuron-treated areas in Andrew. The grass yields in Jayton were greater, but no difference occurred in forb yield.

Sears et al. (1986a) found total sand shinnery oak biomass decreased on sites treated with 0.6 kg/ha tebuthiuron compared to untreated areas. Aboveground herbaceous material increased about sixfold on both treated areas compared to the untreated. Oak-root biomass decreased 12% at 3 years and 37% at 6 years following treatment. Herbaceous root biomass increased threefold at 3 years and was twofold after 6 years. The total nitrogen by weight in the ecosystem showed no change from untreated rangeland at 3 years, but was 14% higher on the 6-year-old treated areas (Sears et al., 1986b).

G. Post Oak, Blackjack Oak, and Associated Woody Species

Darrow and McCully (1959) indicated that thinning the overstory of oaks by basal spray and frilling treatments in a mature oak woodland resulted in a decrease in little bluestem and an increase in basal-area densities of panicums and paspalums. Aerial sprays (2,4,5-T) showed a comparable increase in density and proportion of desirable grasses following treatment with residual plants of little bluestem, Indiangrass, dropseed, panicums, and paspalums, making rapid growth after brush removal. Clearing a dense stand of second-growth oak 2 to 9 meters tall resulted in a 13-fold herbage production increase over a 2-year period. In a mature oak woodland with good forage cover and complete overstory removal a fivefold increase occurred for a 3-year period.

Scifres and Haas (1974) found considerable change after 3 years in herbaceous vegetation in the post oak savannah mixed brush when 2.2 kg/ha picloram was applied. Little bluestem had established 39 months after treatment, and basal cover of brownseed paspalum increased from 5 to 16.5%. The grass production in plots treated with picloram, 2,4,5-T, or 2,4,5-T + picloram was similar to that in plots manually thinned. Plots receiving 2.2 kg/ha of picloram, however, produced more forage (>1900 kg/ha oven-dry weight) than other treatments. Southern dewberry, a vigorous invader, also increased when woody overstory was controlled, regardless of treatment.

H. Live Oak, Yaupon, and Whitebrush Control

Bovey et al. (1972) used picloram, bromacil, 2,4,5-T, and dicamba to control brush in central, south, and east Texas. Bromacil and picloram applied in spring, summer, or fall controlled live oak (*Quercus virginiana* Mill) 1 to 10 meters tall, and significantly increased little bluestem, brownseed paspalum, Indiangrass, threeawn (*Aristida* spp.), love grass, and knotroot bristlegrass (*Sefaria grisebachii* Fourn.) yields compared to the untreated area 7 years after treatment. Forb yields were not reduced. In yaupon (*Ilex vomitoria* Ait), picloram increased the herbage yield (little bluestem, bristlegrass, Indiangrass, brownseed paspalum, Lindheimer croton, and bitter sneezeweed) 6 years after May, June, and October treatments. Whitebrush [*Aloysia gratissima* (Gillies & Hook.) Trohcoso] control by picloram, picloram + 2,4,5-T, and picloram + 2,4-D applied in May or picloram applied in September increased the grass yield of sideoats grama [*Bouteloua curtipendula* (Michx.) Torr.], curly mesquite [*Hilaria belangeri* (Steud.) Nash], vine mesquite (*Panicum obtusum* H.B.K.), and buffalo grass [*Buchloe dactyloides* (Nutt.) Engelm.] by 1 and 4 years following treatment compared to untreated areas.

I. Shrub Live Oak

The basal cover and the production of weeping love grass [*Eragrostis curvula* (Schrad.) Nees.] tended to be inversely proportional to shrub live oak (*Quercus turbinella* Greene) near Glode, Arizona (Pond, 1961). With less than 50% reduction of oak cover, basal cover remained the same for 3 years; with more than 50% oak cover reduction, grass cover continued to increase during the second and third year.

J. Mixed Brush Control

The annual production of native grasses on the Texas coastal prairie was decreased by 90, 66, and 67 kg/ha in 1978, 1979, and 1981, respectively, for each 1% increase in mixed-brush canopy cover in the range, from 12 to 57% (Scifres et al., 1983a). The differences in regression coefficients among the years were not significant ($P < 0.05$), although the rainfall varied from 84% of the 26-year average (91.5 cm) in 1978 to 150% of the annual average in 1981. Grass production during the growing season (May through October) only decreased by 58, 113, and 78 kg/ha in 1978, 1979, and 1981, respectively, for each unit percentage of increase in brush canopy cover. A greater proportion of the annual rainfall occurred during the growing season of 1979 than during the drier spring and summer of 1978. Native grasses typical of the brush-free coastal prairie occupy the interstitial areas among the mixed-brush mottes, but bunch cutgrass (*Leersia monandra* Swartz) is the only grass that persisted from the canopy edge to 1.5

meters inward from the brush driplines. Little or no herbaceous vegetation grows in the center of the mottes.

Production of native grass following aerial applications of 1.12 kg/ha of 2,4,5-T, 2,4,5-T + dicamba, or 2,4,5-T + picloram (1:1) to a south Texas mixed-brush (*Prosopis-Acacia*) community was significantly increased by all herbicide treatments the year of application and by the herbicide combinations during the second year, but only by 2,4,5-T + picloram the third year after treatment (Scifres et al., 1977). Moisture-use efficiency based on kg/ha native grass produced/cm precipitation was the greatest where the herbicide combinations were applied. Defoliation of woody plants in years of above-average rainfall resulted in favorable grass production responses regardless of herbicide(s). Range improvement over the 3-year study was dependent on the maintenance of herbicide effectiveness, however, especially the control of the underbrush, which resulted only where 2,4,5-T + picloram was applied. The consumption of native grass was a direct function of availability in response to brush control as augmented by rainfall. Forb production was reduced by all herbicides the year of treatment and by 2,4,5-T + picloram the year following application, but was not reduced by any treatment during the third growing season.

Aerial application of 2.24 kg/ha (ai) of tebuthiuron pellets to mixed brush and whitebrush-dominated stands in south Texas significantly increased grass standing crops at 1, 2, and 3 years after treatment (Scifres and Mutz, 1978). The higher rates did not significantly increase the grass standing crop over that resulting from 2.24 kg/ha; the lower rates did not increase the grass standing crop compared to that on untreated plots at two of three locations. The genus *Chloris* appeared to be particularly tolerant of the herbicide, and by 2 to 3 years after application the overall grazing value of the grass stand improved where at least 2.24 kg/ha of tebuthiuron were applied. Forb production and diversity decreased where 1 kg/ha or more of the herbicide was applied, and the detrimental effect on forbs increased with the increasing application rate. Forb production was nearly eliminated for 2 years following the application of 4.48 kg/ha of tebuthiuron, but the recovery of the population was evident after 3 years regardless of the rate of application.

The mortality of mesquite, catclaw mimosa, and creosotebush treated with 1.7 kg ai/ha of tebuthiuron in April of 1984 and the resulting grass production were determined on both a draw site and a gravelly site in south Brewster County, Texas (Nelson and Vick, 1988). Brush mortality 3 years posttreatment on the draw site was 10% for mesquite, 55% for creosotebush (*Larrea tridentata* DC.), and 75% for catclaw mimosa (*Mimosa biuncifera* Benth.). Mortality on the gravelly site was 0% for mesquite, 94% for creosotebush, and 100% for catclaw mimosa. Grass production was increased more on the gravelly site (1075 kg/ha on the treated area vs. 180 kg/ha on the untreated) than on the draw site (570 kg/ha on the treated area vs. 149 kg/ha on the untreated control). Forb production was decreased on both treated sites. The lesser degree of grass response on the draw site was attributed to the lack of mesquite control.

III. FORAGE RESPONSE TO INTEGRATED BRUSH MANAGEMENT SYSTEMS

A. Running Mesquite and Whitebrush

Integrated brush management systems (IBMS) use two or more brush management methods in an appropriate sequence to achieve specific resource management goals within an economic framework, with cognizance of critical needs for enhancement of environmental quality and improvement of wildlife habitat. This concept reduces dependency on single methods of brush control by developing a logical series of treatments for application over a defined planning horizon. Integrated brush management systems minimize the weaknesses of each treatment while amplifying the treatment's unique strengths. This can help the resource manager capitalize on ecological and economic synergisms not possible with single-treatment approaches. Research on IBMS reported herein describes herbicide-fire-based systems applied with decision-deferment grazing for improving south Texas rangeland supporting excessive covers of whitebrush or the running mesquite complex (a shrub type composed of a mixture of decumbent honey mesquite and screwbean) (Scifres et al., 1983b).

Prescribed burning in the winter approximately 31 months after aerial application of 1.1 kg/ha of 2,4,5-T + picloram in the spring and burning again at about 57 months after spraying maintained the canopy reduction of running mesquite more effectively than spraying or prescribed burning alone (Scifres et al., 1983b). Grass standing crops were greater, and a higher proportion of the grass stands were composed of species of good to excellent grazing value on plots sprayed and burned than on untreated areas (periodic grazing deferment only) or those treated with herbicide or burning alone.

Aerial application of tebuthiuron pellets at 2 kg/ha in the fall of 1975 effectively controlled heavy stands of whitebrush (Scifres et al., 1983b). Although the subsequent prescribed burn could not improve brush control, forage production was increased, and the botanical composition of the stands was improved compared to areas treated with herbicide only.

Herbicide–fire combinations appeared promising for improving rangeland supporting excessive cover of running mesquite or whitebrush (Scifres et al., 1983b). Herbicide application initially reduced the brush cover and released herbaceous vegetation to serve as fine fuel as well as increased the livestock-carrying capacity. Prescribed burning expedites forage production, improves botanical composition of herbage stands, and suppresses brush regrowth and reinvasion by woody seedlings. Sites may be burned at 3- to 5-year intervals during periods of average rainfall.

B. Macartney Rose

Scifres (1975) found that in Macartney rose-infested pastures on the Gulf Coast Prairie of Texas, sprayed, burned, shredded, sprayed + burned, and sprayed + shredded treatments all significantly increased herbaceous plant production and improved botanical composition for good grazing value (common bermudagrass, little bluestem, and switchgrass). Knotroot bristlegrass, paspalum spp., and panicum spp. were also increased. The burning and spraying treatments had little effect on species of poor grazing value, especially rattail smutgrass [*Sporobolus poiretii* (Roem. & Schult)] or forbs such as Texas croton [*Croton texensis* (Klotzch) Muell. Arg.] and snow-on-the-prairie (*Euphorbia bicolor* Engelm. and Gray). These forbs, however, are important for wildlife.

C. Post Oak, Blackjack Oak, and Associated Woody Species

Excessive cover of post oak, blackjack oak, and associated woody species limit cattle production on about 4.5 million ha in the eastern third of Texas (Scifres et al., 1987). Aerial applications of tebuthiuron pellets at 2.2 or 4.4 kg ai/ha control the oaks and several species of understorey shrubs, and grass production may increase significantly within a growing season after tebuthiuron application (Scifres et al., 1987). Tebuthiuron-tolerant species, however, especially vines such as saw greenbrier, southern dewberry, and peppervine, increase in abundance following the control of the overstory trees and their associated shrubs with tebuthiuron. In addition, the understory shrub American beautyberry may increase dramatically on sites after aerial application of tebuthiuron. The increase in tebuthiuron-tolerant vines, the spread of American beautyberry, and the invasion of other woody species may negate grass production within 3 years after release by herbicide treatment. Prescribed burning as headfires in late winter within two growing seasons after tebuthiuron application and at regular intervals thereafter suppresses the development of secondary woody plant stands. Prescribed burning has little effect on infiltration rates, sediment production, or nutrient loss in runoff on near level sandy loam sites. Forage production and cattle weight gains may not be increased until the growing season following tebuthiuron applications, especially after treatment of heavy brush stands. Follow-up burns should extend animal performance into the fall, when available nutrients are typically less than what is required for maintenance of growing and lactating cattle. Apparently control of the forbs and browse in heavy brush canopies without increased production of grasses accounts for reduced cattle performance during the first growing season after tebuthiuron application. Tebuthiuron can be applied in strips or other suitable patterns to prevent negative alterations in habitat for white-tailed deer and sustain the nutrient intake of cattle. Treated strips probably should be no wider than 325 meters and alternate with untreated strips at least 80 meters wide. Populations of small rodents, such as cotton rats and wood rats, are cyclic following tebuthiuron application in response to shifts in vegetal cover.

Understory vegetation was studied following treatment by herbicides and fire on grazed pastures within the Cross Timbers vegetation type in Oklahoma (Engle et al., 1991). Tebuthiuron and triclopyr at 2.2 kg/ha were applied in March and June of 1983, respectively (Engle et al., 1991). The pastures were burned in late spring of 1985, 1986, and 1987. The frequency of horseweed, rosette panicgrass, and little bluestem increased with both herbicides, but the increase was usually larger with tebuthiuron than with triclopyr. Burning as a follow-up to either herbicide had little effect on the frequency of rosette panicgrass and little bluestem. Horseweed increased following burning in 1985 and 1987. The standing crop of grasses and forbs increased dramatically following the herbicide treatments. Grass production was greater with tebuthiuron, whereas the production of forbs and browse was generally greater with triclopyr. McCollum et al. (1987) found that annual spring burning followed by triclopyr or tebuthiuron application increased steer gain compared to herbicides without burning. Brush management increased grazing capacity compared to control areas. Combined steer gains and greater carrying capacity following brush control has resulted in large increases in per-ha livestock production.

D. Sagebrush Ecosystems

Lancaster et al. (1987) describes weed control-seeding systems (involving the control of brush and herbaceous weeds plus seeding of forage and browse species) for the successful conversion of degraded communities in the sagebrush ecosystem to stable, high-producing rangelands. Asay (1987) indicated that crested wheatgrass [*Agropyron cristatum* (L.) Gaertn. and *A. desertorum* Fisch. ex Link], a cool-season species-complex indigenous to Eurasia, has been the most widely used grass for reseeding the depleted ranges of the western United States and Canada. The cultivars 'Nordan' and 'Fairway' were the first improved cultivars of crested wheatgrass used in North America. Other recent cultivars include 'Ephraim,' 'Ruff,' and 'Hycrest.' Russian wild rye [*Psathyrostachys juncea* (Fisch.) Nevski] from the former USSR is productive and easier to establish than the cultivar 'Vinall.' Other wild rye, wheatgrasses, and hybrid derivatives are also being evaluated.

Monsen and Shaw (1986) evaluated the vegetation response of an alkali sagebrush (*Artemisia longiloba*) site to spraying, burning, chaining, disking, and protection from grazing. All treatments reduced both the shrub density and the vigor of understory species, but some recovery of the shrubs and understory plants occurred. Perennial bunchgrass and broadleaf forb cover increased slowly after treatment due to the low initial density.

E. Prickly Burnet

Henkin et al. (1998) conducted experiments on a Mediterranean hill range dominated by spring dwarf shrubs [mainly prickly burnet—*Sarcopoterium spinosum* (L.) Spach] in order to determine the effects of P fertilizer application, shrub control, and fire on the herbaceous and shrub communities. The application of 2,4-D eliminated most shrubs during the first year. After 6 years shrub regrowth was less than one-third of the original cover. After a fire, the original shrub cover was restored within 6 years. A single application of phosphorus fertilizer (4.5-9.0 g P m^{-2}) significantly retarded shrub regeneration, produced a three- to fivefold increase in herbaceous biomass production, and was still significant after 7 years. In this experiment, prickly burnet dwarf shrub communities with a low forage value were converted to a highly productive herbaceous community by combined control with 2,4-D and P fertilizer application, either with or without previous burning.

IV. FORAGE RESPONSE TO HERBICIDES AT SEVERAL U.S. LOCATIONS

A. Post Oak, Blackjack Oak, and Associated Woody Species

Elwell et al. (1974) indicated that the grass yield from blackjack- and post oak-infested areas in Oklahoma varied from about 110 to 1000 kg/ha, depending upon soil moisture, brush density, management, and location. In all cases more forage was produced on the treated areas. The highest grass yield 2 or 3 years after spraying was 4500 kg/ha in two areas in central Oklahoma, and represents a fourfold increase in grass production over the untreated brush areas. In eastern Oklahoma a 10-fold increase in grass production was obtained in some sprayed plots, but the total grass production in southern and eastern Oklahoma was not as good as in central Oklahoma. This is attributed to the greater number of desirable grass plants in central Oklahoma.

In May 1957, Crawford (1960) sprayed several hardwood tracts in the Ozark Mountains near Paris, Arkansas, with 2,4,5-T in order to control such trees as post oak, blackjack oak, and hickories (*Carya* spp.). The production of grass, forbs, and browse under 1.5 meters tall was measured in September 1957 and 1958 on adjacent sprayed and unsprayed areas. Major grasses were little bluestem

(*Schizachyrium scoparium*), broomsedge (*Andropogon virginicus*), panicums (*Panicum* spp.), and poverty oatgrass (*Danthonia spicata*). The most important forbs were asters (*Aster* spp.), various legumes, horseweed (*Erigeron canadensis*), fireweed (*Erechtites* spp.), and pokeweed (*Phytolacca americana*). Various species of oaks, hickories, blueberries (*Vaccinium* spp.), grapes (*Vitis* spp.), and sumacs (*Rhus* spp.) composed the bulk of the woody growth.

At the end of the 1957 growing season, air-dry grass production was 630 kg/ha on the sprayed area and 485 kg/ha on the unsprayed area. During 1958, grass yields increased to 980 kg/ha on the sprayed area but declined to 360 kg/ha on the unsprayed area. Forb yield was sharply reduced the first year. It increased greatly the second year, but the new growth consisted largely of horseweed, fireweed, and other undesirable species, therefore increased forb yield did little to enhance grazing values for either cattle or deer, but probably had value in reducing the runoff during this period of land conversion.

The large decrease in browse plants in 1957 was caused mainly by defoliation and killing of the oaks—species of low browsing value. In 1958, browse yields increased to 312 kg/ha, an amount equal in weight to that on the unsprayed areas. The sprayed areas, however, produced more browse plants of the species preferred by deer, such as blueberries and grapes.

Halls and Crawford (1965) concluded that aerial spraying of an Ozark woodland with 2,4,5-T temporarily increased the yields of the grasses preferred by cattle. The reinvasion of woody plants and heavy grazing by cattle contributed to a subsequent decline in the yields of grass. The invading shrubs included many species preferred by deer.

Sprayed areas produced more quality herbage than either spring burned or control areas in the Missouri Ozarks (Ehrenreich and Crosby, 1960). Spring burning did not significantly increase the total yield. Five years after treatment with 2,4,5-T, the total herbage yield on the sprayed areas was about 4.5 times greater than on burned areas and about 5.5 times greater than on control areas. Little bluestem and other perennial grasses accounted for practically all of the increased yield on sprayed areas. There was no significant difference in the yield of forbs on sprayed, burned, or control areas. Treated woody plants were mainly post oak and blackjack oak, with some black oak (Q. *velutina*), hickory (*Carya*), sassafras (*Sassafras albidum*), and other hardwood species.

The average yield of little bluestem was about 900 kg/ha on sprayed areas, and only 80 and 56 kg/ha on burned and control areas, respectively. This species increased so greatly on sprayed areas because of the great reduction in hardwood competition. Little bluestem occurred in all the sprayed plots and only about 80% of the burned and control plots. It contributed about 65% of the total herbage yield on sprayed areas, and only about 25% of the total yield on burned and control areas. Indiangrass (*Sorghastrum nutans*) and switchgrass (*Panicum virga-

tum) yielded about 112 kg/ha on the sprayed areas and nothing on the burned or control areas.

Goatsrue, an unpalatable legume, yielded more than all the other forbs combined. There was no significant difference in the yield of goatsrue among sprayed, burned, and control areas. This legume contributed only about 10% of the total herbage yield from the sprayed areas, however it made up more than half the total yield from the burned and control areas. Only about 12% of the total yield on sprayed areas was made up of unpalatable plants, whereas about 65% of the total yield from the burned and control areas consisted of unpalatable plants. The total yield and forage quality was therefore better on the sprayed areas.

B. Blue Oak

Johnson et al. (1959) found that controlling blue oak trees (*Quercus douglasii* H.&A.) in California with undiluted 2,4-D amine in a cut-surface application technique resulted in a fivefold forage yield increase. The quality of the forage with respect to crude protein and phosphorus content was not affected by the treatment. Conversely, Bartolome et al. (1994) removed all oaks from six plots in the central coast of California and found no significant change in understory biomass over 3 years. Understory herb cover averaged 33% on cleared plots compared to 24% on uncut plots, with little change in botanical composition. Site has a great influence on understory growth by blue oak, and controlling it on sites with 50 cm or less of annual precipitation is not recommended for increasing forage production.

C. Gambel Oak

In southwestern Colorado, Jefferies (1965) found that herbage production and regrowth following grazing were greater in openings than under gambel oak (*Quercus gambelli*). Soil moisture was greater under the oaks throughout the season. Haile (1984) found that sprays of 2,4,5-T or silvex applied to gambel oak increased the total herbage yield by about 500 kg/ha compared to untreated areas.

D. Chaparral

Spraying an old stand of chaparral [purple sage (*Salvia leucophylla*), chamise (*Adenostoma fasciculatium*), and California lilac (*Ceanothus* spp.)] with herbicides increased the range-carrying capacity by 37% by the end of the third year (Perry et al., 1967). The control of woody plants was 89 to 100%. Forage production on sprayed areas consisted mostly of annual grasses and forbs and was nearly double that on untreated areas. Purple sage was treated with 2,4-D ester, and the others with a 1:1 mixture of 2,4-D plus 2,4,5-T ester.

Pond (1964) sprayed chaparral in central Arizona, including Wright eriogonum (*Eriogonum wrightii*), rough menodora (*Menodora scabra*), shrubby deervetch (*Lotus rigidus*), and Wright deervetch (*L. wrightii*). The crown cover of forbs increased in all the treated areas between 1954 and 1959, and the understory vegetation increased on soils derived from quartzite, but soils derived from diabase had no grass. Here half shrubs increased greatly in cover following the control of the overstory shrubs, and the forbs increased more than on the quartzite soils.

E. Creosotebush

Gibbens et al. (1987) treated creosotebush [*Larrea tridentata* (Sesse & Moc. ex DC.) Coville] with pelleted tebuthiuron of 0.4 kg ai/ha in southern New Mexico. Total shrub density was reduced from 4440 plants/ha to 570 plants/ha. Total herbaceous above-ground biomass production in the fourth season after treatment was 860 kg/ha, compared to 140 kg/ha on untreated areas. Perennial grass production was increased nearly 11-fold. Morton et al. (1990) applied three tebuthiuron treatments and four mechanical treatments in 1981 at three locations in Chihuahua, Mexico, and one location in Arizona. Forage production averaged across locations and years was 529, 524, 606, 303, 344, 290, 330, and 302 kg/ha for the 0.5, 1.0, and 1.5 kg ai/ha tebuthiuron, land imprinting, two-way railing, disk plowing, disk plowing with furrowing, and untreated check treatment, respectively. When treatments reduced shrub density and remnants of nature forage grasses were present, forage production increased in both wet and dry years.

Morton and Melgoza (1991) monitored further changes in the herbaceous plant density and canopy cover of creosote bush and associated shrubs following brush control treatments in Sonoran and Chihuahuan Desert communities. The treatments were applied in 2 successive years at the Santa Rita Experimental Range in Arizona, and at three locations in Chihuahua, Mexico. Across all locations and years 1.5 kg/ha tebuthiuron $>$ 1.0 kg/ha tebuthiuron $=$ disking $=$ disking with furrowing $>$ 2-way railing $>$ 0.5 kg/ha tebuthiuron $>$ land imprinting in reducing canopy cover of creosotebush and associated shrubs. At the Santa Rita Experimental Range the annual precipitation was above the long-term mean in 1982, 1983, 1984, and 1985, and the grass density increased on both the treated and untreated plots. The annual precipitation was below the long-term mean during 1986 and 1987, and the grass density decreased on both the treated and untreated plots but did not decrease to pretreatment densities. Forb densities were less than 3 plants/m^2 throughout the study, except in 1987, when Russian thistle (*Salsola iberica* Sennen & Pau) increased on all plots. At the Chihuahuan locations, grass densities usually increased during the first year of the study, but very low precipitation throughout the study caused subsequent reductions in grass and forb densities. In dry years brush control treatments did not increase herbaceous plant density.

F. Juniper and Pinyon Juniper

Young et al. (1985) found that the control of western juniper (*Juniperus occidentalis* Hook.) trees released shrub and herbaceous vegetation from dominance of the tree overstory in western Lassen County in California. The revegetation of the woodland with desirable browse and forage species required weed control of both shrub and herbaceous species. The method of tree control (herbicidal, mechanical, or wood harvesting) influenced subsequent weed control–revegetation options in the understory. The physical restraints imposed by standing dead trees resulting from picloram treatment limited the use of weed control and seeding equipment. The accumulation of litter interfered with the use of atrazine for annual grass control on both the picloram and wood harvest plots. The establishment of forage species was most successful on the mechanically cleared plots. Attempts to revegetate with browse species were largely foiled by native herbivores, and selective predation occurred on herbaceous species.

Felling of a 13% cover of alligator juniper (*Juniperus deppeana*) in north-central Arizona resulted in a 38% increase in total herbage production and a 45% increase in forage plant production (Clary, 1974). These increases were highly variable relative to the control. There was little or no apparent response in 3 of the 7 postfelling years.

Herbage production was evaluated after overstory removal from different sites within the pinyon-juniper type (Clary and Jameson, 1981). The average annual production varied from 43 to 643 kg/ha before treatment and 715 to 3703 kg/ha after treatment. Production variation among the sites was related to the annual precipitation, pretreatment tree canopy, pretreatment nitrate-nitrogen, and presence or absence of limestone soils. Grasses increased in the composition from 46 to 73% on the average, while forbs decreased from 21 to 19% and half shrubs and shrubs decreased from 33 to 8%.

The pinyon-juniper vegetation types covers a substantial portion of the western and southwestern United States (17.5 to 32.5 million ha) (Clary and Jameson, 1981). The most consistent tree species is pinyon (*Pinus edulis*); however, one of several juniper species usually dominates the stand, including one-seeded juniper (*J. monosperma*) with blue grama (*Bouteloua gracillis*) in the dry winter areas north of the Mogollon Rim; Utah juniper (*J. osteosperma*) with big sagebrush (*Artemisia tridentata*) in areas with cold, moist winters (Northern Arizona); and alligator juniper at high elevations in Utah juniper at lower elevations with shrub oaks and manzanita (*Arctostaphylos* spp.) south of the Mogollon Rim where winters are cool and moist. Eleven locations across Arizona were used. Overstory pinyon-juniper trees were killed mostly by girdling stumps of alligator juniper, shrub live oak, and herbaceous forbs, and half shrubs were treated with herbicides. Plots were excluded from livestock, but not wildlife.

G. Sagebrush—Short Term

Blaisdell and Mueggler (1956) evaluated 12 large sprayed areas in eastern Idaho for the effects of 2,4-D on forbs, shrubs, and trees commonly associated with big sagebrush within 3 years after treatment. Thirteen of the 38 forbs were moderately or severely damaged, including important forage species such as arrowleaf balsamroot (*Balsamor-hiza sagittata*), milkvetch (*Astragalus stenophyllus*), oneflower sunflower (*Helianthella uniflora*), lupine (*Lupinus caudatus* and *L. leucophyllus*), and bluebell (*Mertensia oblongifolia*). Hawksbeard (*Crepis acuminata*), geraniums (*Geranium viscosissimum*), matroot penstemon (*Penstemon radicosus*), and groundsel (*Senecio integerrimus*) are also important forage plants that showed little or no injury.

Of the 15 shrubs and trees present, only serviceberry (*Amelanchier alnifolia*), three-tip sagebrush (*A. tripartita*), silver sagebrush (*A. cano*), and big sagebrush (*A. tridentata*) suffered moderate or heavy mortality. Aerial portions of snowbrush (*Ceanothus velutinus*), downy rabbitbrush (*Chrysothamnus puberulus*), aspen (*Populus tremuloides*), chokeberrie (*Prunus virginiana*), willow (*Salix* spp.), and snowberrie (*Symphoricarpos oreophilus*) were mostly killed, but a high proportion of the species sprouted profusely. Bitterbrush (*Purshia tridentata*), a particularly valuable forage species, was unharmed.

The sagebrush (*Artemisia*) grasslands of western North America are a tremendous grazing resource that are currently producing forage at only about one-half of its potential (Young et al., 1981). The history of degradation of this environment is traced from the late nineteenth century, when livestock were introduced to the sagebrush rangelands. The dominant species of sagebrush that characterizes the landscape is not preferred by domestic livestock. Continuous excessive grazing greatly reduced the perennial grass cover and allowed the shrubs to increase. Phenoxy herbicides proved to be very valuable and economically feasible tools for improving grass and forage production on shrub-dominated rangelands. Second-generation herbicides were used in controlling alien annual weeds to permit the seeding of degraded sagebrush rangelands. Techniques for herbaceous weed control were vertically integrated with brush control and seeding techniques for total range improvement. Environmental and economic constraints brought sagebrush range improvement to a virtual standstill during the 1970s.

The vegetation canopy cover on upland and lowland sites both inside and outside a 22-year-old exclosure in a big sagebrush range was evaluated by sampling for canopy cover (Holechek and Stephenson, 1983). The area outside the exclosure had received moderate use as grazable forage by cattle in the late winter and spring for 22 years. The two sites did not show a consistent response to grazing. The big sagebrush canopy cover was higher inside the exclosure on the upland site and higher outside the exclosure on the lowland site. Big sagebrush dominated the canopy cover both inside and outside the exclosure on both sites, and relatively little understory was present. Forbs were nearly absent from the area, which is attributed to a past history of heavy sheep grazing. The elimination of grazing had little effect on vegetation composition on either site.

In Paradise Valley, Nevada, Robertson (1969) indicated that the rate of increase in yield of crested wheatgrass following the use of herbicide on associated sagebrush was measured over 4 years, including the year of treatment. Significant increases in yield, which were probably worthwhile economically, did not begin until the third year after spraying.

At Ely, Nevada, Tueller and Evans (1969) sprayed a mixed stand of green rabbitbrush [*Chrysothamnus viscidiflorus* (Hook.) Nutt. var. *puberulus* Jepson] and big sagebrush at different dates for 3 years with the potassium salt of picloram, the propylene glycol butyl ether esters of 2,4-D, and a mixture of the triisopropanolamine salts of picloram plus 2,4-D. Picloram at 0.56 and 1.12 kg/ha controlled rabbitbrush but not sagebrush. Poor rabbitbrush control and fair to good sagebrush control resulted from 2,4-D. The mixture of picloram and 2,4-D controlled rabbitbrush well but only partially controlled sagebrush. Forage release from brush control was negligible the first 2 years after spraying. During the third and subsequent years, the production of crested wheatgrass was greatly increased.

In Reno, Nevada, Evans and Young (1975) found that an aerially applied mixture of 2,4-D and picloram was more effective than 2,4-D alone in controlling green rabbitbrush and big sagebrush in a very droughty big sagebrush/desert needlegrass community. Herbage production had more than doubled by 5 years after the brush stand was reduced. Needle-and-thread was the only species to increase in density because of brush control, although considerable seed was produced from other perennial grasses. Modification of the seedbed environment is apparently necessary to establish additional desert needlegrass or Indian ricegrass plants.

Three habitats dominated by mountain big sagebrush were sprayed with 2,4-D butyl ester in eastern Oregon (Miller et al., 1980). The shrub cover in the sprayed mountain big sagebrush communities was significantly less than in the unsprayed. Perennial grass production and density were twice as high across sprayed stands as compared with nonsprayed stands. Across the three habitat types bluebunch wheatgrass productivity was more responsive on the sprayed stands than Idaho fescue. The production and density of tailcup lupine, the most abundant forb across the three sites, was significantly less in the sprayed communities, causing total forb production to be lower. Other forbs either showed little or no difference in production and density between the sprayed and nonsprayed communities.

The control of low sagebrush in northern Nevada increased the productivity of understory grass species (Eckert et al., 1972). On fair condition sites, climax dominant species such as Idaho fescue, bluebunch wheatgrass, and Thurber needlegrass gave a yield response, but the response was not obtained for 2 years after treatment. The rapid response of Sandberg bluegrass resulted in a significant total perennial grass response the year after treatment, however. On poor condition sites, squirreltail gave a smaller response, and the response was not obtained for 2 to 4 years after treatment. Dense stands of annual species gave a large-yield response the year after treatment and suppressed the response of squirreltail. Scattered stands of annuals did not respond until 4 years after treatment. The soil moisture relations explained the differences in total yield. Differences in early spring growth were attributed to soil nitrogen, however.

Field experiment conditions between 1982 and 1988 compared 2,4-D and metsulfuron for controlling Wyoming big sagebrush (*A. tridentata* ssp. *wyomingensis* Beetle and Young) in northern New Mexico (McDaniel et al., 1991). Broadcast sprays of 2,4-D at 2.2 kg/ha were the most efficacious during rapid shoot elongation, but the mortality averaged less than 38% from treatments applied over 4 separate years. Wyoming big sagebrush shoot growth was greatest in April and May compared to other months, but the growth was highly variable among shrubs, and probably reduced the effectiveness of 2,4-D sprays. The optimum application timing for metsulfuron was during the late-flower growth and fruiting stages. Fall-applied metsulfuron at 0.035 kg/ha provided 65% Wyoming big sagebrush mortality, compared to 27% when spring-applied. When metsulfuron was fall-applied at 0.07 kg/ha or higher, control averaged 88% following three annual applications. Combining metsulfuron at .0175 kg/ha plus 2,4-D at 1.1 kg/ha was comparable to or more effective than either herbicide applied alone in the spring or fall. The total standing crop of grasses increased by nearly 300% after 1 or 2 growing seasons when the Wyoming big sagebrush canopy cover was reduced by at least 75% following herbicide treatments.

Blue grama [*Boutilvua gracilis* (H.B.K.) Lag. ex Steud] was the most important associated species at both sites (McDaniel et al., 1991). No herbaceous species damage was noted after the first growing season. Broadleaf species were primarily woolly plantain (*Plantago patagonica* Jacq.) and scarlet globemallow [*Sphaeralcea coccinea* (Nutt.) Rydb.].

H. Sagebrush—Long Term

In 1960 and 1961, big sagebrush on four cattle ranges in the Bighorn Mountains of Wyoming was sprayed with 2,4-D (Thilenius and Brown, 1974). By 1971, the canopy cover of big sagebrush was 8 to 42% of the pretreatment levels, seedling density ranged from 5.7 to 11.3 plants/3.4 m^2, and herbage production was below the pretreatment levels, with the proportion of graminoids about equal to that prior to spraying. The effects of grazing deferment for as long as 3 years after spraying could not be detected.

Four Wyoming big sagebrush control treatments—burning, spraying with 2,4-D, rotocutting, and plowing, along with no control (rest)—were compared in southwestern Montana (Wambolt and Payne, 1986). Production data (excluding sagebrush) were collected for 10 years, and data on the sagebrush canopy cover and understory basal cover were collected for 8 years during the period from 1963 to 1981. The sagebrush canopy was the most effectively reduced by burning, while plowing with seeding was the least effective. Rest alone resulted in a 29% reduction in the sagebrush canopy during the study period. By 1981, burning provided the most production from the dominant forage species [bluebunch wheatgrass—*Agropyron spicatum* (Pursh) Scribn.] and important vegetal classes, although burning and spraying were equally successful when production was totaled for all the years sampled. Understory basal cover did not prove useful to evaluate treatment effectiveness.

Secondary plant succession was determined in stands of the *Artemisia tridentata/Festuca idahoensis* habitat type treated with 2,4-D in the Madison Ranger District in Beaverhead National Forest in southwestern Montana (Stewart, 1981). Nine stands representing 4-, 6-, 7-, 8-, 10-, 12-, 14-, 18-, and 19-year stages of succession following treatment and nine adjacent untreated stands were sampled for comparison with near-climax vegetation. The coverage and frequency of all shrub and herbaceous species was recorded.

The general trend of species dominance is characterized by (1) a herbaceous phase followed by (2) a shrub phase. Four years following spraying, *Geranium viscosissimum, Festuca idahoensis*, and *Stipa occidentalis* dominated the vegetation. *Festuca idahoensis* and *Stipa occidentalis* were dominant from the sixth through the tenth years. By the twelfth year, *Festuca idahoensis* and *Agropyron smithii* dominated the vegetation, and *Artemisia tridentata* began to increase.

Total shrub cover remained low up to the twelfth growing season. The greatest increase occurred in the fourteenth growing season. The shrub phase was characterized by the dominance of *Artemisia tridentata* and *Festuca idahoensis*. This phase persisted from the fourteenth through the nineteenth year.

I. Sagebrush—Soil-Applied Tebuthiuron

Tebuthiuron pellets containing 20 and 40% ai were applied at 0.6 and 1.1 kg/ha during May 1979 to range sites dominated by mountain big sagebrush, three-tip sagebrush (*A. tripartita* Rydb.), and gray horsebrush (*Tetradymia canescens* DC.) to determine their effectiveness for shrub control and the response of associated grasses and forbs (Murray, 1988). Sagebrush densities were significantly less ($P \leq 0.5$) in 1984 on all herbicide-treated sites compared to the untreated sites. The 1.1 kg/ha (20P) treatment reduced the densities of all shrubs more than other treatments at all sites. Grass production was significantly ($P \leq .05$) greater on plots treated at 1.1 kg/ha (40P) at the mountain big sagebrush and three-tip sagebrush sites. Forb production did not respond to tebuthiuron treatments. Grass production was not increased or decreased significantly by any treatment at the gray horsebrush site. For sites with similar soil and environmental conditions, the 1.1 kg/ha (40P) treatment should give sufficient control of sagebrush to allow for significant increases in total grass and forb production.

Tebuthiuron was evaluated on three Wyoming soil types for the control of both big sagebrush and silver sagebrush (*Artemisia cana* Pursh.), and for phytotoxic effects on associated vegetation (Whitson and Alley, 1984). Big sagebrush was controlled with tebuthiuron at application rates of 0.6 to 1.1 kg ai/ha, whereas silver sagebrush was not completely controlled. Prairie junegrass [*Koeleria pyramidata* (Cam.) Beauv.], western wheatgrass (*Agropyron smithii* Rydb.), needle-and-thread (*Stipa comata* Trin.), and green needlegrass (*Stipa viritdula* Trin.) were not significantly reduced with tebuthiuron up to 1.1 kg ai/ha. Blue grama (*Bouteloua gracilis* H.B.K.), downy brome (*Bromus tectorum* L.), plains prickly pear (*Opuntia polyacantha* Haw.), and broom snakeweed were tolerant to tebuthiuron at rates up to 1.1 kg ai/ha.

Tebuthiuron was evaluated at two Wyoming locations for its effects on big sagebrush control and associated vegetation. Big sagebrush was controlled with tebuthiuron at 0.6 to 1.1 kg ai/ha (Whitson et al., 1988). Green needlegrass, prairie junegrass, blue grama, and Douglas rabbitbrush were unaffected by tebuthiuron treatments. Western wheatgrass and downy brome increased at different study locations and were the dominant grass species at the end of 7 years.

V. FORAGE RESPONSE TO HERBICIDES OUTSIDE THE UNITED STATES

A. Fringed Sagebrush—Canada

Fringed sagebrush (*A. frigida* Willd.) is a native, drought-resistant, increaser species in Saskatchewan pastures and rangelands. Conventional control by cultivating and reseeding is neither highly effective nor sustainable. Two field experiments were conducted in 1988 and 1989 through 1991 to develop a management plan for fringed sagebrush (Peat and Bowes, 1994). A comparison of fringed sagebrush control using various applications of clopyralid, dicamba, dichlorprop, fluroxypyr, picloram, triclopyr, and 2,4-D, showed that control was sufficient using the low-cost herbicide 2,4-D isooctyl ester at 1.5 kg ai/ha. The total yield of crested wheatgrass plus smooth brome was compared, and the plots subjected to two applications of 2,4-D yielded the greatest accumulation of grass over the length of the experiments. The economic threshold of fringed sagebrush is the density at which the yield is reduced by approximately 290 kg/ha; above this density it is economically viable to use 2,4-D at 1.5 kg/ha to control fringed sagebrush. Grazing practices can aid greatly in the management of fringed sagebrush by allowing the forage to outcompete fringed sagebrush.

B. Shrubby Cinquefoil—Canada

Picloram, tebuthiuron, hexazinone, 2,4-D (isooctyl-ester), 2,4-D plus dichlorprop, and clopyralid were applied to native rangeland at two southern Alberta sites with heavy infestations of shrubby cinquefoil (*Potentilla fruticosa* L.) (Moyer and Smoliak, 1987). The yields of shrubby cinquefoil shoots, forbs, and grasses were measured in ungrazed areas for 4 years after herbicide application. Hexazinone (grid balls), 2,4-D, and 2,4-D plus dichlorprop reduced shrubby cinquefoil yields to almost zero. 2,4-D and 2,4-D plus dichlorprop increased the grass yield and reduced the forb yield, compared with the untreated plots. Hexazinone initially decreased the grass yield, but after 4 years, grass yields were higher on the treated than on the untreated plots. Hexazinone and tebuthiuron treatments killed all vegetation within a 15-cm radius of the applied grid balls, and there was little recovery of the grasses on the tebuthiuron-treated areas 4 years later. Picloram and tebuthiuron reduced shrub yields, but many live shrubs remained after treatment. Clopyralid did not reduce shrub yields. In a further experiment, 2,4-D gave good shrubby cinquefoil control and increased grass yield. When this area was grazed the percentage of utilization of the total vegetation was similar for herbicide-treated and untreated areas, but the amount of vegetation utilized increased by about 40% when the shrubs were controlled.

C. Aspen Poplar and Associated Species—Canada

Hilton and Bailey (1974) indicated that an aspen poplar (*Populus tremuloides* Michx.) forest in Alberta, Canada, receiving as little as 3.4 kg/ha of 2,4-D in a single application achieved an annual herbage production of 980 kg/ha 2 years after treatment compared to only 210 kg/ha in the control. Sprayed forest border areas (small aspen) showed a fourfold increase in herbage production. Grasses, sedges, and forbs increased in herbage production in sprayed forests. Only two forbs were detrimentally affected by the herbicides. In the sprayed forest area, even though there was a considerable amount of obstruction, cattle were able to consume 48% of the total herbage and as much as 58% of the green herbage. Even with the inclusion of 2,4,5-T in the second herbicide application, however, there were many woody species in the treatment areas with as high or higher densities than in the control.

Using regression, Woods et al. (1982) found a coefficient of determination of 0.61 between herbage production and overstory basal area of aspen poplar in Colorado. For overstory basal areas <10 m^2/ha, herbaceous understory production varied considerably and was often double that found at higher densities of overstory basal areas. Herbage production at higher densities (10 to 19 m^2/ha) showed less variation, with an average production of 1100 kg/ha. The authors concluded that for creating optimum herbaceous understory production in unmanaged, pure aspen poplar stands, stands of overstory basal areas of <10 m^2/ha are best.

Yields of forage were measured 3 to 5 years after the use of 2,4-D, 2,4-D + 2,4,5-T, and picloram + 2,4-D applied at brush control rates (Bowes, 1982). The yield of bromegrass and alfalfa on the herbicide-treated areas was either reduced or remained unchanged. This occurred because alfalfa cannot tolerate the high herbicide rates required for brush control. On an area that had never received an herbicide treatment, alfalfa was unable to compete with invading aspen poplar and prickly rose. Ranchers should use herbicides to prevent secondary succession from grasslands to the aspen poplar vegetation type, which has a low amount of forage available for cattle. Arguments are presented for including alfalfa in a seeding program when a mixture of 2,4-D + picloram is used for brush control, but not when a mixture of 2,4-D + 2,4,5-T is used.

Seven- and 9-year studies were conducted in northeastern Saskatchewan to compare the long-term control of aspen poplars and western snowberries (*Symphoricarpos occidentalis* Hook.) following single and multiple applications of 2,4-D ester (2,4-DE) or amine (2,4-DA) mixed with dicamba (Bowes, 1991). The area was bulldozed free of trees and was subsequently invaded by aspen poplar and western snowberry regrowth, native forbs, and grasses. Herbicides were applied to 2- and 4-year-old regrowth of aspen poplars growing with western snowberries. A single application of dicamba with either 2,4-DE or 2,4-DA applied at 1.5 + 2.2 kg/ha controlled aspen poplars. In one experiment, two consecutive applications of dicamba + 2,4-D controlled western snowberries, whereas in the other experiment a single application was adequate. The highest forage production of grasses and forbs followed dicamba + 2,4-DE treatments. Following a single dicamba + 2,4-DE application, grass production, averaged over 7 and 9 years, increased from 38 to 98 g m^{-2} and from 31 to 83 g m^{-2}, whereas forb production decreased from 48 to 25 g m^{-2} and from 58 to 48 g m^{-2}, respectively.

Aspen poplar, balsam poplar, prickly rose (*Rosa acicularis* Lindl.), and western snowberry studies were conducted in east-central Saskatchewan to evaluate metsulfuron, 2,4-D, and a mixture of metsulfuron + 2,4-D for long-term control of these woody species (Bowes and Spurr, 1996). Metsulfuron applied at 60 g ai/ha effectively controlled the four woody species for 7 years. Control was not improved when 2,4-D at 2 kg ai/ha was added to the 60 g/ha rate of metsulfuron. Aspen poplar control was improved when sucker regrowth was 2 years old rather than 3 years old. Grass and forb production was increased by 50% from 1023 kg/ha in the untreated areas to 1619 kg/ha following application of 60 g/ha metsulfuron in two experiments averaged over 5 years. Forb production decreased from 44% on the untreated area to 8% on the areas treated with metsulfuron.

D. Oak and Sagebrush—Korea

In Korea, Kim et al. (1990) treated mixed brush dominated by *Quercus dentata* Thunb. with a mixture of glyphosate plus dicamba. Brush control significantly increased the total herbage production of orchard grass (*Dactylis glomerata* L.), timothy (*Phleum pratense* L.), Kentucky bluegrass (*Poa pratensis* L.), and red clover (*Trifolurm pratense* L.). The best treatment and highest dry matter yield resulted from herbicides + burning + cutting + oversowing + raking. Kim et al. (1992) found that dicamba controlled *Artenisia japonica* and *A. montana* and other broadleaved weeds without damage to orchard grass and other pasture grasses. Dry matter yield after reseeding increase production by 181% over the untreated areas.

E. Prickly Burnet—Israel

In Israel, Henkin et al. (1998) found that 2,4-D eliminated most spring dwarf shrubs (mainly prickly burnet) for several years following treatment. A single application of phosphorus fertilizers with 2,4-D significantly retarded shrub regeneration. A three- to fivefold increase in herbaceous biomass was still significantly higher than the untreated areas 7 years after treatment.

F. Jarilla—Argentina

In Argentina, Passera et al. (1992) found that shrub removal of jarilla (*Larrea cuneifolia* Cav.) on grazing lands increased forage cover by 156% by the third growing season and increased the carrying capacity (22%) compared to the untreated areas. Sowing with native grasses produced a response in only one species. The authors concluded that shrub control with or without additional sowing of native grasses can improve the carrying capacity of shrub communities of a typical temperate arid zone in midwest Argentina.

G. Mixed Brush in Conifers—Sweden

In Sweden, Ingelög (1978) found that where phenoxy acid herbicides were used for chemical brush control grasses generally increased temporarily, often to a great degree, while herbs and dwarf shrubs were reduced in biomass. Ingelög

(1978) concluded that vegetation changes after mechanical cleaning were generally smaller than those after chemical control.

VI. HERBICIDE EFFECT ON FORAGE QUALITY

The treatment of a forb-dominated subalpine cattle range in the Bighorn Mountains of Wyoming with 2,4-D did not change the in vitro dry matter digestibility coefficients of the grasses or surviving forbs (Thilenius and Brown, 1976). Forbs and grasses were equally digestible throughout the growing season. The production of total and digestible dry matter was not influenced by 2,4-D, but the proportion of both supplied by grasses was increased.

Various combinations of atrazine, 2,4-D, and NPK fertilizer were applied initially in 1975 on a tallgrass prairie in good condition and again as a retreatment in 1976 on one-half of each treated area (Baker et al., 1980). An application of atrazine with fertilizer increased the crude protein content, crude protein yield, and range condition during the summer. Differences in protein levels in dormant forage were not evident, therefore the application of atrazine and fertilizer appears to be more feasible for summer grazing or the production of high-quality prairie hay than for yearlong or winter grazing.

Meyer et al. (1982) indicated that several herbicides applied at sublethal rates may increase amino acid content and total crude protein and cause other biochemical changes that may improve the growth and yield of plants. Penner and Ashton (1966) discussed the biochemical and metabolic changes in plants as a result of phenoxy herbicides. Phenoxy herbicides may increase phosphorus content, vitamin content, protein content, amino acid content, and proteolytic enzyme activity in plant stems and roots, whereas lowered amounts or activities are apparent in the leaves.

Meyer et al. (1982) found that 2,4-D had little affect on amino acid in tolerant roundseed dicanthelium [*Dicanthelium sphaerocarpon* (Ell.) Gould], a grass. It reduced the concentrations of most amino acids in western ragweed (*Ambrosia psilostachya* DC.) and Macartney rose. 2,4-D may inhibit protein synthesis in these susceptible species at field rates, but at sublethal rates phenoxy herbicides have increased nucleic acid and protein content by enhancing DNA and RNA synthesis in several species (Penner and Ashton, 1966).

Clopyralid was relatively specific for honey mesquite (Meyer et al., 1982). It reduced all amino acid concentrations except aspartic acid and proline in honey mesquite. It did not affect the amino acid concentration of the tolerant yaupon (*Ilex vomitoria* Ait.), but it did increase the amino acid concentration at 2 days and decrease it at 8 days after treatment in Macartney rose. Dicamba slightly increased six amino acids in the resistant roundseed dicanthelium, but reduced the concentrations of most amino acids in the susceptible western ragweed.

Being an effective herbicide for controlling grasses, glyphosate reduced the concentrations of most amino acids in roundseed dicanthelium, but increased the glutamic acid concentration (Meyer et al., 1982). Glyphosate also reduced the concentrations of 10 amino acids in western ragweed foliage and all of the amino acids except proline in honey mesquite. Glyphosate has been postulated either to inhibit aromatic amino acid biosynthesis in plants (Jaworski, 1972) or to reduce aromatic amino acid levels by increasing phenylalanine ammonia-lyase activity (Hoagland et al., 1978; Duke et al., 1980). Our results support the postulate that both tyrosine and phenylalanine concentrations were reduced, at least in some species. Glyphosate has little effect on Macartney roses, possibly because the water-soluble herbicide cannot penetrate the waxy Macartney rose leaves to any extent.

Hexazinone, an effective herbicide for most grasses and some broadleaf species, increased aspartic acid, glycine, phenylalanine, and threonine in the grass, roundseed dicanthelium, and ammonia and aspartic acid in western ragweed through 14 days (Meyer et al., 1982), but it reduced the concentrations of six amino acids in western ragweed 27 days after treatment.

Picloram increased the concentration of six amino acids in the resistant roundseed dicanthelium but reduced the concentration of most amino acids in honey mesquite and Macartney rose (Meyer et al., 1982). The 2,4,5-T increased 12 amino acid concentrations in roundseed dicanthelium 27 days after treatment. It reduced the concentrations of most amino acids in honey mesquite and yaupon. Tebuthiuron increased the concentration of 12 amino acids in roundseed dicanthelium and aspartic acid in western ragweed; the other amino acids were not affected. Triclopyr increased the concentration of 14 amino acids in roundseed dicanthelium, but reduced the concentration of most amino acids in yaupon, honey mesquite, and Macartney rose. Triclopyr caused the most common reduction in amino acids of any herbicide studied, but it was followed closely by picloram.

Baur et al. (1977) applied picloram, tebuthiuron, 2,4-D, and glyphosate on five grasses. The herbicides had little effect on the protein concentration in coastal bermudagrass. Picloram, tebuthiuron, and 2,4-D reduced protein concentrations in the buffel × birdwood hybrid. Glyphosate caused increased protein levels in buffelgrass and kleingrass.

More grass was consumed in all grazing periods on tebuthiuron-treated pastures and in fall- and summer-grazing periods on mechanically treated pastures than on untreated pastures (Kirby and Stuth, 1982a). Cow diets were dominated by grasses, mainly brownseed paspalum and little bluestem, regardless of treatment and season. Similar amounts of forbs were selected from all treatments during all seasons. More woody vegetation was selected from mechanically treated and untreated pastures than from tebuthiuron-treated pastures. Forbs decreased and woody vegetation increased in the diets from the spring through the fall. Grasses and leaves decreased, whereas woody vegetation and stems increased in the diets from the beginning to the end of the grazing periods. Within the grazing periods forb consumption decreased in the fall but increased in the summer and spring with time spent in pastures. Small amounts of dead forage were consumed at irregular intervals.

The nutritive content of seasonal diets following mechanical and chemical brush management on east-central Texas rangeland was determined using esophageally fistulated cows (Kirby and Stuth, 1982b). Brush-managed pastures had a greater herbaceous standing crop, except during the grazing period in the fall of the year of treatment, and generally yielded diets higher in crude protein (CP) and in vitro digestible organic matter (IVDOM) than untreated pastures. The mean crude protein content of cattle diets was higher ($P \leq 0.05$) in all seasons from tebuthiuron-treated pastures and in the summer and fall from pastures that were bulldozed compared to untreated pastures. The mean IVDOM content of cattle diets was higher ($P \leq 0.05$) in the spring and summer from pastures receiving brush management compared to untreated pastures. Crude protein and the IVDOM content of diets decreased in all pastures and seasons between the beginning and end of the grazing periods, with the exception of a higher IVDOM in the spring, suggesting that the cows benefitted by grazing regrowth of herbaceous plants.

Cows allowed free access to randomly placed plots of Bell rhodesgrass, kleingrass, and weeping love grass appeared to prefer to graze plots treated with 2.2 or 4.4 kg/ha (ai) of 20% tebuthiuron pellets compared to untreated plots, regardless of the grass species (Scifres et al., 1983c). The apparent preference was observed during the summer and fall following herbicide application in the spring, but was not detected during the growing season 1 year after herbicide application. The cows also appeared to prefer herbicide-treated [2,4-D or picloram sprays at 1 kg/ha, tebuthiuron pellets at 0.5, 1 or 2 kg/ha (ai)] little bluestem–brownseed paspalum native stands to untreated plots. Moreover, cows usually grazed on plots treated with 1 or 2 kg/ha of tebuthiuron more than on those plots treated with 2,4-D or picloram sprays. Since all plots were mowed prior to the grazing trials, apparent grazing preferences were not attributable to differences in the stage of grass maturity or to the control of broadleaved plants by the herbicides.

Tebuthiuron pellets (20% ai) applied at 0.6, 1.1, or 2.2 kg/ha (ai) to native stands of little bluestem and at 1.1 or 2.2 kg/ha to 1-year-old seeded stands of Bahiagrass, Bell rhodesgrass, green sprangletop, and little bluestem did not significantly alter the IVDOM concentrations of grass leaves (Masters and Scifres, 1984). Leaf water concentrations of Bahiagrass, green sprangletop, and little bluestem were not consistently altered by the application of tebuthiuron, however, application of 1.1 or 2.2 kg/ha of tebuthiuron pellets to seeded stands or to native little bluestem increased the foliar crude protein concentrations. Application of

0.125, 0.188, or 0.25 ppm of tebuthiuron in aqueous solutions to pots containing grasses in the greenhouse significantly increased the foliar crude protein concentrations compared to that of untreated plants. Crude protein concentrations were increased only during the growing season of application in the native stand of little bluestem. These results suggest that the application of tebuthiuron for brush control may enhance rangeland forage crude protein concentrations while not affecting IVDOM.

Controlling natural infestations of dog fennel [*Eupatorium capillifolium* (Lam.)] and woolly croton (*Croton capitatus* Michx.) in a bermudagrass [*Cynodon dactylon* (L.) Pers.] sward ranged from 20 to 100% with surface-applied tebuthiuron in Mississippi (Crowder et al., 1983). Dry matter yields of desirable forage ranged from 60 to 290 kg/ha more with tebuthiuron treatment than in the untreated check. There were no consistent differences in forage quality among treatments, which indicates that tebuthiuron had no detrimental effects upon the bermudagrass or dallisgrass.

The effects of burning, mowing, and 2,4-D on the nutrient contents of antelope bitterbrush (*Purshia tridentata* Pursh.) were evaluated in south-central Wyoming (Kituku et al., 1992). During the first growing season following treatments, spraying 2,4-D increased bitterbrush nitrogen (N) contents from 1.5 to 1.9%, phosphorus (P) from 0.12 to 0.15%, and in vitro digestible dry matter (IVDDM) from 44.1 to 48.4%. Mowing increased N from 1.5 to 1.7%, P from 0.12 to 0.16%, and IVDDM from 44.1 to 46.1%. Burning increased N from 1.4 to 1.9%, P from 0.11 to 0.17%, ash from 3.7 to 5.3%, and IVDDM from 47.4 to 51.0%, and decreased gross energy from 4640 to 4380 kcal/g. There were no differences in N and IVDDM contents among treatments at the end of the second growing season, but the P content was still greater in mowed bitterbrush regrowth than in untreated bitterbrush. Ash contents were not affected by treatments, but were higher in the summer (3.9%) than in the winter (2.4%). Gross energy contents varied only 5 to 10% among all treatments and seasons. Correlation coefficients among N, P, ash, and IVDDM contents varied from +0.54 to +0.76, and all of these nutrients were negatively correlated with gross energy. Bitterbrush nutrient contents can be increased by shrub-management practices, but short-term responses require that small portions of the total area be treated annually in a rotational shrub-management program.

VII. REINVASION OF WOODY PLANTS AFTER CONTROL

Woody plants are usually vigorous competitors and perennial, and tend to regenerate if damaged and disturbed. Control by cutting, fire, herbicides, or biological means may not completely kill the plant and it may regenerate from basal stem buds, root sprouts, and buds from rhizomes or stems. The rapidity of regrowth depends upon the species, environment, competing vegetation, and effectiveness of the original treatment. Follow-up treatments are usually required to maintain weed and woody plant control every few years and to maintain forage production or whatever crop is grown. A common practice is to treat once with an herbicide or nonchemical treatment and expect the problem to be solved. As previously discussed, however, woody plants may rapidly regenerate from parent plants or from seeds, so continued surveillance and action is required. If thinning the brush is desired for wildlife habitat then follow-up treatments may be less rigid and timely.

A. Mesquite and Associated Woody Plants

Hamilton et al. (1981) suggested shredding and/or prescribed burning for low-cost methods to suppress woody plant reinvasion in the South Texas Plains. Knowledge of the rate of top-growth replacement indicates how often these maintenance practices must be repeated. Honey mesquite and twisted acacia attain their pretreatment heights within the first year after shredding or burning. Blackbrush acacia regrowth attains pretreatment height by midway through the second growing season following burning. Whitebrush, lotebrush, spiny hackberries, and Carolina wolfberries attain half their pretreatment height in 10 months or less after shredding. Blackbrush acacia, catclaw acacia, guayacon, and most other species attain pretreatment heights within two growing seasons after shredding. Based on these data, burning and shredding on the south Texas plains should be scheduled every 3 to 5 years.

Root-plowed rangeland in southern Texas is often dominated by fabaceous shrubs. Stewart et al. (1997) tested the hypothesis that the shrub community present 40 years after root plowing does not exhibit successional trends toward the original mixed-brush species community. Twenty shrub clusters, each organized around a central honey mesquite individual, were selected within both a control site and a root-plowed (35 to 40 years ago) site at each of three locations. The number of all woody plants species, including cacti *Opuntia* spp. and *Yucca* spp., beneath the nuclear honey mesquite was determined. Shrub species richness within clusters increased with increasing central honey mesquite basal diameter on control and root-plowed sites. Species richness/honey mesquite in root-plowed (2 $\pm$ 0.5 species, $\pm$ SE) sites was lower than species richness/honey mesquite $>$200 mm in diameter on control sites (7 $\pm$ 0.4 species/honey mesquite). Honey mesquite seedlings (1 to 60 mm basal stem diameter) composed 39 $\pm$ 14% of the shrubs beneath honey mesquite canopies on root-plowed sites compared to $\leq$3% of the woody plants present on untreated sites. Honey mesquite may continue to dominate root-plowed sites for some time, since honey mesquite was the major subordinate shrub species on root-plowed sites.

Rootplowing and front-end stacking are two mechanical manipulation practices that were used to reduce brush density (Gonzalez, 1990). The rate of brush reinfestation including mesquite was faster for front-end stacking than rootplowing; however, pricklypear density was not reduced by rootplowing. Grass seeding following mechanical brush manipulation had no major long-term effect on brush reinfestation. Brush reinfestation was negatively correlated to rainfall, suggesting a faster rate of brush reinfestation in drier years.

Honey mesquite regrowth on the Texas Experimental Ranch in the Rolling Plains of northwest Texas was evaluated 8 years after aerial spraying with 2,4,5-T (Scifres et al., 1974). Canopy cover, density, and top-growth production of honey mesquite regrowth were significantly greater under moderate, deferred-rotation grazing than under heavy, continuous grazing of the deep upland site. Honey mesquite density, canopy cover, and top-growth production did not differ between grazing systems on the rocky hill site. Regrowth on the rolling hill site was usually intermediate between the rocky hill and deep upland sites. Honey mesquite plant density, top-growth production, canopy cover, and rate of new stem initiation were greater under moderately stocked, deferred rotation than heavily stocked, continuous grazing. Averaged across grazing systems, regardless of the variable evaluated, regrowth was greatest on the deep upland site.

Brown and Archer (1989) indicated that openings in the herbaceous layer created by moderate defoliation of grasses on a long-term continuous grazing site increased germination and/or survival of honey mesquite seven- to eightfold. Honey mesquite survival was comparably high in dung and bare seed placements after one growing season in east-central Texas.

Honey mesquite populations continue to expand and become more dense, even on areas once ''successfully'' treated either with herbicides or by bulldozing in southern New Mexico (Gibbens et al., 1992). Areas treated from 1958 to 1964 for mesquite control on the USDA-ARS Jornada Experimental Range and the New Mexico State University College Ranch were sampled to determine mesquite-density changes. On herbicide-treated areas sampled in 1976 and again in 1988, mesquite densities increased 10 to 128% and had densities from 67 to 494 plants/ha. Two areas treated either by bulldozing or fenuron in 1959 to 1960 and with original kills near 100%, had an average density of 377 plants/ha by 1988, with an establishment rate of 13.5 plants/ha/year. On the college ranch, mesquite densities increased 11%, from 130 (1982) to 147 (1988) plants/ha. Only 19% of a cohort of mesquite seedlings that germinated in 1989 were still alive in May 1990. Even though only a small percentage of the mesquite that germinated survived into the second year, this is enough to change former grasslands into mesquite-dominated rangelands.

B. Big Sagebrush

Five chemically sprayed and 15 plowed and seeded areas in southwestern Montana were examined to determine the influence of several environmental factors on big sagebrush reinvasion (Johnson and Payne, 1968). Sagebrush surviving the treatments was found to be the most important factor related to reinvasion. Plowing near or after sagebrush seed maturation resulted in heavy reinfestation of seeded stands. Sagebrush adjacent to treated areas was of no practical importance as a seed source for reinvasion. Non-sagebrush vegetation, slope, erosion, soil texture, and precipitation were seldom related to sagebrush reinvasion. Northwest exposures favored reinvasion.

Six stands of big sagebrush, which had been plowed or sprayed earlier to remove brush and enhance understory vegetation, were sampled in southeastern Oregon to determine the age structure of the shrubs and to evaluate the rates of reinvasion (Bartolome and Heady, 1978). Five of the six stands contained big sagebrush older than the treatment. In three project areas, plants established the first year following treatment formed the largest age class, 12 to 24% of the stand, indicating that reinvasion begins immediately after treatment. Most reestablishment occurred in the first several years after treatment for all locations. Establishment occurred either from seeds present in the soil at the time of treatment or from seeds produced as the plants became established and seed-bearing. Treated sagebrush/grass ranges should remain highly productive under proper grazing use despite the reinvasion of big sagebrush.

Four Wyoming big sagebrush (*Artemisia tridentata* ssp. *wyomingensis* Beetle and Young) control treatments: burning, spraying with 2,4-D, rotocutting, and plowing, along with no control (rest) were compared in southwestern Montana (Wambolt and Payne, 1986). Data on sagebrush canopy cover were collected for 8 years during the period from 1963 to 1981. Sagebrush canopy was the most effectively reduced by burning, while plowing with seeding was the least effective. Rest alone resulted in a 29% reduction in the sagebrush canopy during the study period. Further evaluations of the same treatments by Watts and Wambolt (1996) indicated that burning had the longest-term effect on sagebrush cover, but growth eventually returned to levels not significantly different ($P \leq 0.025$) from untreated areas. The other three treatments exceeded untreated levels over the long term. Sagebrush that was sprayed, plowed, and rotocut equalled the untreated areas after 18.1, 10.4, and 18.1 years, respectively. The exclusion of grazing for 30 years had no effect on the sagebrush canopy in the untreated plots. These findings clarify successional trends following disturbances in Wyoming big sagebrush habitat types. This information should enhance management opportunities of this important vegetative type for a variety of resources.

C. Baccharis

A remeasurement of transects established in 1952 and a comparison of early vegetation maps with maps prepared in 1961 indicate that a brush species, *Baccharis pilularis*, had invaded grassland areas of the East Bay Regional Parks near

Oakland, California (McBride and Heady, 1968). The common movement of the species has advanced on exposed soil occurring at baccharis–grassland boundaries. Baccharis seedlings and young plants are very susceptible to damage with controlled burning and grazing animals. These experiments supported the hypothesis that baccharis has increased due to the reduction of wildfires and the elimination of grazing.

D. Aspen Poplar

Changes in aspen reproduction and undergrowth production and composition were recorded over a 3-year period following clear-cutting (Bartos and Mueggler, 1982). Aspen suckers increased from 2300 per ha prior to cutting to a maximum of 44,000 per ha by the third year. Undergrowth production on the cut units increased from 1013 kg/ha prior to cutting to 3000 kg/ha after three growing seasons; production on the uncut control areas increased from 1199 kg/ha to 1539 kg/ha during this period. The significant increase in undergrowth is attributed to the reduction in competition from the aspen overstory removal. Clear-cutting appeared to increase the proportion of shrubs in the undergrowth and decrease the proportion of forbs. The greatest change in species composition occurred the first year after cutting, with a gradual return toward the precut conditions.

E. Lantana

Gentle and Duggin (1997) indicated that successful *Lantana camara* invasions occur whenever canopy disturbances create patches of greatly decreased competition and/or increased resource availability. Shading plays a greater role as a limiting factor than any other, while surface soil macronutrient levels are also important, particularly when combined with canopy disturbances that increase light availability. The effects of biomass reduction and soil disturbance associated with fire and cattle grazing are significant in the successful invasion of *L. camara*. Management strategies to reduce weed encroachment and community degradation must identify and maintain ecological barriers to *L. camara* invasion in order to promote rainforest conservation and biodiversity.

F. Ashe Juniper

The reinvasion of mechanically disturbed juniper communities is possible through contributions from the soil seedbank, seed rain, and the juvenile seedling bank (Owens and Schliesing, 1995). The treatment associated with single trees caused the litter layer to be removed, resulting in the removal of that portion of the seedbank. Most seeds (>80%) were consequently found under the canopy of mature, seed-producing trees. Soil disturbance was less severe in small and large motte arrangements, so only 65% of the soil seed bank was under mature trees. In undisturbed communities, the seed population was distributed evenly under tree canopies and in interspaces. Viability and germinability within the seedbank were low (4% and 0%, respectively). The viability of new seed was 47%, and germinability was approximately 5%. The juvenile seedling bank contained a sufficient number of seedlings (408 seedlings/ha) for ashe juniper to regain dominance on the site through growth. There was no advantage to any spatial pattern of tree distribution in terms of invasive potential when fewer than 10 trees/ha were left on a site. When 20 to 50 trees/ha are left on a site, however, tree spatial arrangement has a significant effect on reinvasion rates.

VIII. COMPETITION BETWEEN HERBACEOUS AND WOODY VEGETATION

Once woody vegetation is suppressed, herbaceous vegetation usually increases dramatically. A vigorous forb and grass stand should discourage brush invasion or reinvasion. During periods of drought or soil disturbance, however, woody plants sometimes gain establishment opportunity and may reinvade. There are many questions concerning the ability of herbaceous vegetation to compete with trees and brush on grazing and forest land. It is obvious from our earlier discussion that woody plants affect herbaceous plant production. Now we will explore the effect of herbaceous vegetation on woody plant growth and production.

A. Honey Mesquite

Van Auken and Bush (1990) conducted greenhouse experiments with honey mesquite seedlings and bermudagrass [*Cynodon dactylon* (L.) Pers.]. When one honey mesquite seedling was grown with two or more bermudagrass plants, the above ground, below ground, and total dry weight of honey mesquite was decreased by 84, 71, and 79%, respectively. Conversely, honey mesquite caused a significant reduction in the dry weight of bermudagrass, whether planted at the same time or 2 months after the bermudagrass was planted. Van Auken and Bush (1990) concluded that both plants compete about equally with each other, and honey mesquite establishment in the field may require reduced herbaceous plant density or vegetation gaps. Van Auken and Bush (1990) indicated that buffalo grass, little bluestem, and sideoats grama are good competitors of honey mesquite, and that seedling growth of honey mesquite in established grassland would be very limited.

Scifres et al. (1971) studied honey mesquite succession in an exclosure protected from domestic livestock since 1941. Only 14% of the honey mesquite stand recorded in 1941 remained in 1968. Age estimation indicated that no honey mesquite established after 1959. The average height of the mesquite was 0.5 meters. The herbaceous vegetation was primarily tobosa (*Hilaria mutica* Buckl.), buffalo grass [*Buchloe dactyloides* (Nutt.) Englem.] and vine mesquite (*Panicum obtusum* H.B.K.). An adjacent grazed area from which mesquite had been removed by hand periodically during the 27 years was dominated by annual herbs and tobosa.

Meyer and Bovey (1982) planted honey mesquite and huisache seed on a native ungrazed pasture (but plowed several years earlier) to study plant establishment as influenced by no treatment and mechanical and chemical treatments. No treatment entirely prevented the establishment of either species during a 3- to 5-year period. After 5 years no more than 1 and 2% of the original honey mesquite and huisache seeds established plants. The most effective treatment was 1.12 kg/ha of picloram spray-applied the fall following spring seeding.

Brown and Archer (1990) concluded that the rapid development of honey mesquite roots of seedlings during their first year of growth enhanced survival by enabling them to access soil moisture beyond the zone effectively utilized by grasses. On grazed sites, competition for water from herbaceous vegetation may not limit the establishment of honey mesquite seedlings or the growth of mature plants in most years.

This author has observed no honey mesquite or huisache on undisturbed true native prairie in east-central Texas, while disturbed adjacent grazed areas may contain honey mesquite or huisache, depending upon the location. Remnants of the true nature prairie are very rare in central Texas.

B. Big Sagebrush

Robertson (1947) grew 17 perennial grasses and five other species for 3 years in competition with big sagebrush in northern Nevada. Root trenching to 25 cm increased herbage yield the second and third year but not the first. Exotic species excelled natives in stands and yield. Crested wheatgrass was the best.

Blaisdell (1949) planted a grass mixture 1 and 2 years before sagebrush, simultaneously with sagebrush, and 1, 2, and 3 years after sagebrush. In addition, crested wheatgrass was drilled into 2-, 3-, 4-, and 5-year-old stands of sagebrush. Grass yields were much larger than sagebrush on plots seeded to grass prior to or as late as 1 year after sagebrush, but sagebrush yields were larger than grass when grass was seeded 2 or 3 years after sagebrush. Plant suppression was chiefly responsible for yield differences rather than the failure of plants to become established. Reseeding depleted ranges during the same year of destruction by fire or mechanical means gives advantage to grass over sagebrush. Reseeding areas established in sagebrush was not recommended.

Cook and Lewis (1963) found that the zone of root concentration for sagebrush and wheatgrasses were about the same, so direct competition for soil moisture is indicated. Spraying sagebrush with 2,4-D increased the yields of crested, intermediate, and tall wheatgrass by as much as 770, 360, and 960 kg/ha, respectively, by removing the sagebrush and increasing the soil moisture.

Frischknecht (1963) found that the depressed yield of grass around big sagebrush is associated with highly developed lateral roots in the grass-root zone. Also, the most active growth periods of crested wheatgrass and big sagebrush coincide. Rubber rabbitbrush in the same area makes the most of its growth after crested wheatgrass and does not significantly decrease its yield. Rabbitbrush presence improves fall grazing because understory grass remains more succulent and abundant than in open areas or near sagebrush.

Robertson (1972) found crested wheatgrass more tolerant of drought than big sagebrush in an experiment in which water was withheld on a shallow claypan soil. Water was withheld until the leaves were air-dry. Crested wheatgrass was the only survivor in all replications.

C. Juniper

Jameson (1970) found that one-seed juniper reduced the basal area of blue grama, but the effect was small.

D. Turfgrass

Fales and Wakefield (1981) found that turfgrass significantly reduced the growth of flowering dogwood (*Cornus florida* L.) and forsythia (*Forsythia intermedia* Spaeth). Competition for N and not water was indicated by the results of a leaf tissue analysis as being the factor responsible. Also, leachates from the roots of ryegrass (*Lolium perenne* L.), red fescue (*Festuca rubra* L.), and Kentucky bluegrass (*Poa pratensis* L.) inhibited the top growth of the forsythia.

E. Loblolly and Shortleaf Pine

Cain (1995) studied herbaceous weed effects on loblolly and shortleaf pine seedlings and hardwood resprouts. Crop hardwoods, including oaks, hickories (*Carya* spp.), sweetgum (*Liquidambar styraciflua* L.), and blackgum (*Nyssa sylvatica* Marsh.), were 1.5 meters taller and 1.3 cm longer in diameter at breast height (DBH) where herbaceous vegetation was controlled with herbicides than on check plots 9 years after treatment. On check plots and herbaceous control plots, crop pines averaged 1 to 2 meters taller and 2.5 to 5 cm larger in DBH than did crop hardwoods.

IX. FOREST LAND

McMahon et al. (1994) suggested that there are six roles that herbicides play in forest management: 1) they create and maintain desired habitats, 2) they create mixed and uneven aged stands, 3) they restore damaged landscapes, 4) they control exotic, noxious, and poisonous plants, 5) they maintain recreational areas, trails, and scenic vistas, and 6) they manage rights-of-way for multiple use.

Newton (1996) stated that to avoid losses in plantation management of forests it has become necessary to control shrubs and hardwood species. Traditional mechanical methods of weeding forests are dangerous to workers, costly, and labor-intensive. Herbicides have become the vegetation management tool most widely applicable for silvicultural use because of their efficacy, safety, and low cost. Newton (1996) indicated that 2,4-D and 2,4,5-T controlled brush in conifers and were the only herbicides of major use in forestry before 1970. Other herbicides, such as atrazine, dichlorprop, glyphosate, hexazinone, imayazpyr, metsulfuron, oxyfluorfen, sulfometuron, and triclopyr, are sometimes highly effective, but none is as economical to use as 2,4-D.

A. Woody Plant Control

Gratkowski (1959) found that 2,4-D controlled hairy, hoary, greenleaf, and Howell manzanita, scrub tanoak, saskatoon serviceberry, and cayon live oak, whereas 2,4,5-T controlled snowbrush, varnishleaf, mountain whitethorn ceanothus, golden chinkapin, and golden evergreen chimkspin in southwestern Oregon.

Gratkowski (1961) indicated that by 1961 herbicides (phenoxys) had proven to be effective silvicultural tools for preparing sites for reforestation and releasing conifers from brush competition in southwestern Oregon.

Stewart (1974) showed that red alder, vine maple, and California hazel could be adequately controlled in Douglas firs by budbreak with sprays of 2,4,5-T applied in diesel oil. Aerial sprays of dicamba plus 2,4,5-T were useful for site preparation and preburn desiccation on areas dominated by red alder, salmonberry, and vine maple.

In eastern Maine in cutover forest sites, Sprague and McCormack (1982) obtained suppression of striped maple (*Acer pensylvanicum* L.), red maple (*A. rubrum* L.), aspen poplar (*Populus tremuloides* Michx.), paper brush (*Betula papyrifera* Marsh.), beech (*Fagus grandifolia* Ehrh.), pin cherry (*Prunus pensylvanica* L.F.), raspberry [*Rubus idaeus* L. var. *strigosus* (Michx.) Maxim.], eastern hemlock [*Tsuga canadensis* (L.) Carr.], and balsam fir [*Abies balsamea* (L.) Mill] with fosamine or hexazinone. These hardwoods typically interfere with spruce, fir, and pine crop tree development. Reynolds et al. (1986) applied hexazinone in New Brunswick as aerial sprays of 2, 3, and 4 kg/ha. Similar to the main studies (Sprague and McCormack, 1982), the weeds were about 1 meter tall. Hexazinone was generally unsuccessful except for raspberry, pin cherry, and striped maple.

Spotgun application of hexazinone in northern Ontario at 480 mg ai/spot (1.6 kg ai/ha) resulted in 88% aspen poplar stem dieback and variable suppression of white birch and pin cherry (Prasad and Feng, 1990). The height and basal diameter of treated red pine increased 131 and 150% of control, respectively, after 3 years.

Hexazinone applied at rates of 1.0, 1.5, and 1.75 mL per spot controlled competing hardwoods around 7-year-old underplanted white pine (Wendel and Kochenderfer, 1988). The growth response was more pronounced when hexazinone was applied to release individual pines than when the entire area was treated on a 1.8-by-1.8-meter grid. In most cases, hexazinone killed white oak, chestnut oak, American beech, sourwood, red maple, and black gum. Yellow poplar, red oak, cucumber tree, black cherry, sugar maple, sweet birch, and sassafras showed variable sensitivity to hexazinone. Three years after treatment, the resprouting of killed stems and the mortality of white pine remained low.

In an upland northern hardwood site in Albany, New York, picloram plus triclopyr mixtures injected by Hypohatchet into tall trees after leaf emergence (June) controlled white ash, beech, sugar maple, hemlock, and white pine (Jackson, 1988). In northern California, ponderosa pine (*Pinus ponderosa* Dougl. ex Laws. var. *ponderosa*) was released by a hydro-ax or hydro-ax plus 2,4-D of several shrubs (Fiddler et al., 1992). After 6 years, pine volume, diameter, and height increments were significantly larger for areas treated with the hydro-ax plus 2,4-D. The hydro-ax alone did not increase pine height or volume growth. The herbicide treatment increased pine diameter growth by 79%, height growth by 7%, and volume growth by 260%.

In the Arkansas Ozarks, Yeiser and Cobb (1988) found that the height growth of shortleaf and loblolly pine seedlings underplanted in hardwood stands was greatest in plots with the greatest overstory hardwood control. Glyphosate plus 2,4-D, picloram plus 2,4-D (Tordon 101R), and hexazinone (Velpar L, tree-injected) and hexazinone (Pronone 10G pellets) and Velpar L (soil-applied) provided greater than 80% hardwood overstory control.

Preharvest hardwood control by stem injection, soil application of herbicide, and rotary mowing with an herbicide spray facilitated the natural regeneration of loblolly and shortleaf pines in a 75- to 80-year-old stand in southern Arkansas (Cain, 1988). After 3 years, the preharvest hardwood control provided optimum pine seedlings numbers that were significantly taller and more competitive than those on untreated areas.

Haywood (1986) indicated that brush was defined as hardwood trees, shrubs, blackberries, vines, and volunteer pines in northern Louisiana loblolly pines. When above-ground brush was cut annually, or both above-ground stems and roots were controlled, both treatments produced 10-year-old trees that averaged 0.9 meters taller with a 2.5-cm greater stem diameter at breast height (1.4 meters) than those untreated.

Haywood (1995) also suggested that hexazinone can be used to accelerate a change in the stand composition of a 7-year-old stand to favor pine where overstocked hardwood-pine mixtures exist in the southeastern United States. Two dormant-season prescribed burns did not significantly influence the natural shift in species composition.

Zutter et al. (1988) found hexazinone effective in the piedmont of Alabama in releasing newly planted and 1-year-old loblolly pine in a dense stand of hardwoods. Hexazinone pellets (1 cc and 0.5 cc 10%), ai pellets, and 20% ai granules were effective.

Quicke et al. (1994) applied herbicide on 2-year-old loblolly pine in the Virginia piedmont and 3-year-old loblolly pine on the coastal plain of Arkansas. The predominant hardwoods at the Virginia site were black tupelo, red maple, red oak, and chestnut oak. Pine increase in basal area was 27 to 83%, and the increase in the mean pine height ranged from 0.8 to 1.7 meters over the control area. Imazapyr outperformed glyphosate and at 1.12 kg/ha it was very effective in reducing hardwoods. In Arkansas hardwood species were winged elm, hickory, red oak, and white oak. Imazapyr decreased the hardwood basal area by 28 to 76%, and increased pine basal area by 7 to 27% over the controls. Measurements were taken 7 years after herbicide application at both locations.

Watson and Zedaker (1996) studied early pine release treatments at varying intensity levels on competing vegetation 12 years after treatment. Treatments included total, two-thirds, one-third, or no woody stem control in combination with either total or no herbaceous vegetation control. Woody stems received basal sprays of 4% triclopyr ester in diesel-oil carrier. Herbaceous plants were treated with sulfometuron (Oust). Treating all the woody vegetation resulted in the highest percentage of loblolly pine increase (19%) over the check at DBH. As treatment intensity decreased, pine mean DBH decreased; the 1/3 woody without herbaceous weed control was included. Treatments that combine moderate to high levels of woody stem control with herbaceous control are effective at increasing the growth of loblolly pine without affecting overstory species richness 12 years after treatment.

Harty et al. (1995) studied a site with vigorous competition between loblolly pine and sweetgum (*Liguidambar styraciflua* L.) in a mature, 40-year-old loblolly pine stand. Herbicides (imazapyr, 0.56 kg/ha plus metsulfuron, .042 kg/ha) significantly reduced the hardwood basal area (91%) and density (89%). Burning and herbicide by burning interaction did not reduce hardwoods, however. The burn-herbicide tended to increase pine growth (11%) but was not significantly different from the control.

Minogue et al. (1997) compared imazapyr, imazapyr plus glyphosate, imazapyr plus triclopyr, and imazapyr plus nitrogen (9 kg/ha) at six locations in the southeastern United States for hardwood control and other vegetation for site preparation. After 1 year imazapyr alone was highly effective on hardwoods. The effect of imazapyr and tank mixes on *Rubus*, vines, broadleaf, and grass cover were discussed.

Heinze et al. (1997) treated stumps of felled mixed hardwoods in the northern piedmont region of Virginia after whole-tree removal. Loblolly pine seedlings were planted after tree falling. Another treatment consisted of basal bark spray and soil-applied herbicide 1 year after planting pine. The loblolly pine basal area was 11% greater on lots with hardwood stumps treatment compared to planted plots with no hardwood control. The loblolly pine basal area was 20% greater on plots with both stump and release treatments compared to stump treatments alone, and was 31% greater than planted pine with no competition control 13 years after treatment. Loblolly pine heights and diameters had similar responses. Overstory richness was 7% less on released plots than on natural regeneration plots.

B. Woody Plant Control Effects on Herbaceous Vegetation

Elwell (1967) chemically controlled oaks, hickories, and assorted species in southeastern Oklahoma for shortleaf pine (*Pinus echinata* Mill.) release. Native grass growth was increased twice. Where dense shortleaf pine seedlings developed, however, the native grasses were smothered and the volume of herbage decreased.

McConnell and Smith (1969) found in north-central Washington that pine thinning caused highly significant increases in understory vegetation. After 8 years, the understory yield was 84 kg/ha on unthinned ponderosa pine to 467 kg/ha under pines spaced at 8 meters. Composition was 51% grass, 37% forbs, and 12% shrubs. When the pine canopy exceeded 45%, the forbs produced more than grasses and vice versa below 45% pine canopy. Wolters et al. (1982) found similar results to McConnell and Smith (1969) in Louisiana where, after selectively cutting uneven-aged stands of loblolly pine to various densities, herbage and browse production was generally related to the residual pine basal area and site quality. Uniolas were the principal forage species under stands having high-residual pine basal area. Bluestems were the major component in clearings. Browse made up one-fourth of the forage on the high-residual pine basal area.

Wolters (1982) found that slash pine decreased herbage production as early as plantation age 17 for longleaf pine and plantation age 10 for slash pine in Louisiana. Pinehill and slender bluestem were the principal species on nonforested plots after 15 years with a mixture of forbs and pinehill bluestem, and other bluestem grasses were the most common on forested plots.

Pyke and Zamora (1982) also found canopy coverage most significantly correlated to total understory production in a grand fir/myrtle boxwood habitat in north-central Idaho.

C. Herbaceous Weed Control Effects on Conifers and Tree Crops

In the late 1960s Christensen et al. (1974) used 1.12 kg/ha of atrazine to control medusa-head and downy brome for the establishment of perennial wheatgrass in ponderosa pine woodlands previously burned by wildfire. A fall application of atrazine improved the survival of pine or bitterbrush and resulted in the greater growth of ponderosa pine seedlings. The wheatgrass did not establish because of the reinvasion of annual grasses at the northeastern California site.

In Georgia, hexazinone effectively controlled herbaceous weeds in a loblolly pine plantation. Pine-seedling height increased as a result of weed control, but phytotoxic symptoms were evident (Fitzgerald and Fortson, 1979). Surfactants added to hexazinone increased phytotoxicity. Activated charcoal root coatings reduced the phytotoxicity of hexazinone to the pine.

In Pennsylvania, Horsley (1981) found that glyphosate was the most economical herbicide for controlling several herbaceous weeds for at least 3 years in Allegheny Plateau hardwood sites. An application of 1.1 kg/ha between August 1 and September 1 was the best.

Weed control had a positive effect on the fourth-year height of longleaf pine (*P. palustris*), along with groundline diameter and the percentage of pine seedlings out of the grass stages, while survival was unaffected (Nelson et al., 1985). Trees on plots receiving 2 years of weed control were 1 meter taller and 1.3 cm greater in groundline diameter than the untreated in southwest Georgia. From a 12-year study on three locations in Mississippi and Arkansas, herbaceous weed control demonstrated consistent and substantial volume gain in loblolly pine growth (Glover et al., 1989).

In Arkansas, Yeiser et al. (1987) found that herbicides controlled herbaceous competitors in short leaf and loblolly pine sites better than mechanical treatments. Glyphosate and hexazinone (Velpar L) mixed with sulfometuron (Oust) provided the best overall control of herbaceous competition. The greatest growth response of pines was in hexazinone plus sulfometuron-treated areas.

In Georgia, Pehl and Shelnutt (1990) also found that hexazinone plus glyphosate controlled herbaceous weeds, and after three growing seasons pines had greater heights and diameters than nontreated plots.

White et al. (1990) found that reducing herbaceous competition with hexazinone increased noble fir survival by about 100%, and Douglas fir and noble fir diameters by 38 and 25%, respectively, on a wet coastal site and two successively drier interior sites in Oregon.

At 13 loblolly pine plantations in southeast Oklahoma and southwest Arkansas, Yeiser and Williams (1994) treated pine seedlings with spot, band, or broadcast applications of hexazinone plus sulfometuron. Seedlings treated for herbaceous weed control exhibited significant increases in first-year survival at 11 sites and in height and groundline diameter of 12 of 13 sites compared to untreated areas. At the end of five growing seasons, the advantages from herbaceous weed control were maintained with significant differences in survival, height, and DBH at most of the study sites.

Cain (1991) found that after treatment of herbaceous weeds for 4 consecutive years, pines were significantly taller and larger in groundline diameter (GLD) and volume per tree than pines on untreated areas. Control of woody vegetation alone did not improve pine growth compared to untreated checks. After 5 years, the pines on total vegetation control had significantly larger GLDs and more volume per tree than any other treatment.

After one growing season Williams et al. (1996) found that herbaceous weed control using combinations of sulfometuron (Oust) plus hexazinone (Velpar L) or imazapyr (Asenal AC) applied over the top of loblolly pine seedlings resulted in significantly greater height and diameter growth at all sites in southeast Arkansas.

Despite its marked increase in use, the current and predicted area receiving herbaceous weed control in the South is still small relative to its potential (Teeter et al., 1993). The most important factors cited by survey respondents in this study are the lack of available funds, the existence of attractive alternative investments, the fact that it is not required on many sites, and the fact that site-preparation treatments provide sufficient vegetation control. Economic and silicultural constraints will likely continue to diminish as companies gain experience with the technology, but environmental and public concerns will probably increase concerning herbaceous weed control with herbicides.

D. Herbaceous and Woody Plant Effects on Tree Crops

At 14 locations southside, Miller et al. (1991) found that during the first 5 years the general pine response by treatment has been a greater growth in height, diameter, and volume index as follows: total herbaceous weed and woody plant control > herbaceous weed control > woody control > no control after 5 years.

Miller et al. (1991) found similar results to Cain (1991) in that the control of herbaceous vegetation resulted in significant increases in pine height, GLD, and volume per tree. The control of woody competition did not improve pine growth compared to untreated checks in naturally regenerated loblolly and shortleaf pines in southern Arkansas.

Further investigation of 13 locations across the South to examine loblolly pine plantations established using four vegetation control treatments after mechanical site preparation indicated that pine heights were greater with herbaceous control (Miller et al., 1995). The development of a pine canopy cover was similar with woody and herbaceous control after 8 years, however. There were 101 prevalent genera of herbaceous plants and 76 species/genera of woody plants present.

After 12 years, Watson and Zedaker (1995) found in Virginia piedmont plantations that treatments combining total woody stem control and herbaceous control resulted in the highest percentage increase (19%) of loblolly pine DBH over untreated pine. Species richness was independent of herbaceous control.

In north Florida and south Georgia, Lauer and Glover (1996) indicated that loblolly pines averaged 6 meters in height and 8 cm in DBH across four sites 5 years after planting when both life forms were controlled. Responses due to shrub control were about twice that of herbaceous weed control with height responses above the check of 0.7, 1.5, and 2.2 meters due the first year to herbaceous control, shrub control, and both herbaceous and shrub control, respectively.

Minogue and Burns (1994) found that the greatest pine seedling diameter resulted from 1.12 kg/ha of imazapyr where herbaceous and woody plants were controlled in Alabama and South Carolina.

E. Effect of Herbicides on Plant Diversity in Forests

The maintenance of biodiversity is a goal of forest management. Zutter and Zedaker (1988) found that woody species diversity decreased significantly with increasing hexazinone rates in young loblolly pine plantations. The most susceptible woody plants were sumacs (*Rhus* spp.) and oaks (*Quercus* spp.); the least susceptible were loblolly and other pines and blueberries (*Vaccinium* spp.). Using hexazinone rates of 1.2 to 1.5 kg ai/ha, the differences between treated and untreated were small at four Virginia sites after 2 years.

In Georgia, Brooks et al. (1992) found that plant species diversity varied with the herbicide used. For example, picloram plus triclopyr treatment after 1 year had the highest woody plant diversity. Imazapyr and hexazinone reduced woody plant diversity, but the highest herbaceous diversity occurred in the imazapyr treatment. Herbaceous diversity was similar for hexazinone and picloram plus triclopyr treatments, which favored grass species. Imazapyr plots had the highest number of forb species.

In Mississippi, Wilson et al. (1993) indicated that plant species richness and diversity were similar for all treatments after 1 and 2 years with few exceptions after using roll-chop-burn, imazapyr with and without burning, and burn-felling.

In central Georgia, Boyd et al. (1995) found no effect of treatment on total species richness or species richness divided into growth form categories 7 years after treatment in an area planted with loblolly pine. The herbicides used were imazapyr, glyphosate, and liquid and pellet forms of hexazinone.

In South Carolina on three common soil types, Bourgeois (1996) found that the lowest number of species found in the untreated plots was 27 the first growing season, but was highest at one location in the next growing season. By the second growing season all plots had 97 to 100% cover, including sulfometuron/hexazinone treatments.

In Louisiana, Reed and Noble (1994) found that in the clear-cut plots, herbicides reduced first-year vegetative composition and density. Species richness improved the second year, however, because forb and grass components increased. Burned plots in previous years had less woody plant components than those without a burn history.

In California, Ditomaso et al. (1997) concluded that in coniferous crops, despite initial reduction in plant diversity and species richness in herbicide treated areas, the native plant diversity was not statistically different from that of the unburned forest sites 8 years after treatment. In contrast, untreated burned areas demonstrated a long-term reduction in plant diversity and species richness compared to the unburned site. The untreated site was dominated by only a few shrubby species, particularly *Ceanothus* spp. and *Arctostaphylos patula*, even though vegetative cover was similar to the unburned or herbicide-treated site.

Miller (1995) named the 16 most prevalent exotic trees, shrubs, vines, and grasses that infest southern forests. Miller (1995) describes their origin, biology, range, uses, and herbicidal control. It is important that these species be controlled or they will cause extensive ecological damage and economic loss. Such plants include kudzu, Japanese honeysuckle, chinaberry, tallowtree, multiflora rose, Japanese privet, and cogongrass.

F. Herbicides and Wildlife Effects in the Forest Environment

1. Managing Food Plants for Wildlife

Gill and Healy (1974) discussed management of 97 native and three naturalized shrubs or woody vines important to wildlife in the northeastern states and Canada. Herbicides are important to manage these plants because they may become overabundant and cause management problems or become too tall for wildlife use.

Patton (1977) indicated that aspen poplar is an important wildlife food in the southwest United States. Unless stands are regenerated by cutting, burning, or herbicides, many ha will gradually decline and have to be managed for browse production. Clear-cutting and fire can produce a stand with thousands of vigorous sprouts per ha.

Witt et al. (1990) used herbicides for site preparation in loblolly pine plantations in Georgia. At age 2, imazapyr treatments produced quail food plants four times as high as the control and other chemical treatments. Means of soft mast and winter browse at age 2 were higher in hexazinone treatments. At age 3, control treatment means for soft mast and winter browse were highest, with triclopyr only slightly lower. At age 4 triclopyr produced the most soft mast vegetation. Overall, hexazinone and triclopyr promoted the most soft mast and winter vegetation browse.

Hurst (1997) indicated that weeds are wildlife food and cover plants. The impacts of forest herbicides for site preparation on habitats, while significant, are short-lived, usually one growing season. The fast growth of pine seedlings and their dominance over the plant community has greater effect than herbicides. Herbicides to release pines can improve some habitat conditions, such as deer forage. Herbicides can be used by habitat managers to increase deer, gopher, and tortoise forage, improve northern bobwhite and redcockaded woodpecker habitats, improve roadside vegetation, control exotic plants, and restore wetlands and native prairies.

Felix et al. (1986) indicated that conversion from second-growth hardwood and pine-hardwood forest to loblolly pine plantations in the central piedmont of Virginia resulted in very adequate deer and turkey habitat. Upon conversion there was a rapid increase in the available food plants, but a loss of mast. Cover and principal food plants for bobwhite quail and wild turkey were the highest in 1-year-old plantations; by 3 years, plantations were deficient in seed-producing plants. Deer browse was adequate in plantations up to 9 years, but declined following crown closure at 10 to 15 years. Between crown closure and the first commercial thinning (about 20 years), plantations were of negligible food value to the three wildlife species. The data suggest that the optimum conditions for bobwhite are in young plantations, while those of deer and turkey are supplied by a mix of second-growth forests and various-aged pine plantations.

2. Herbicide Effects on Wildlife Food and Habitat

On a Mississippi site, Blake and Hurst (1987) found that total herbaceous plant biomass in a loblolly pine plantation was greater on control plots than on plots receiving broadcast or banded applications of granular hexazinone the first growing season. At the end of the second growing season the number of plant species on all treatments was similar, including white-tailed deer forage after site preparation.

Brooks et al. (1993) indicated that for site preparation in loblolly pine in the Georgia sandhills hexazinone-treated areas produced more quail and deer food than picloram plus triclopyr or imazapyr 1 year posttreatment. The abundance of gopher tortoise foods was similar among treatments, although plant species were different. Herbaceous plant species diversity was highest on the imazapyr treatments and lowest on the hexazinone treatments, whereas woody species diversity was lowest on hexazinone treatments and highest on picloram plus triclopyr treatments.

Work by Miller et al. (1997) in the Georgia sandhills for chemical site preparation for loblolly pine indicated that total herbaceous vegetation occurred more frequently on hexazinone and imazapyr treatments 4 and 5 years posttreatment. At 1 year posttreatment imazapyr produced lower resprouting woody vegetation than hexazinone. At 6 years posttreatment herbaceous plant species richness was greatest on imazapyr and picloram plus triclopyr-treated areas. Vines were the most abundant on imazapyr plots at 4 to 6 years posttreatment. Northern bobwhites preferred vines and legume areas on imazapyr-treated sites. Hexazinone and imazapyr treatments had the greatest amounts of highly preferred deer food plants, but the overall differences in wildlife habitat, including small mammals, were minor among treatments and diminished over time.

Leopold et al. (1994) examined imazapyr as a selective herbicide for hardwood tree species at a Mississippi pine site. The treatments compared were imazapyr with and without burning, roller chopping and burning, and fell and burning. The total biomass (grass, woody, forb, vine) for all growth forms was 50% less by the second year, but did not reduce deer forage. No differences were detected for forage and the total biomass between mechanical and chemical treatments, indicating that imazapyr is effective for site preparation without reducing deer forage abundance.

Trichet et al. (1987) indicated that glyphosate was generally satisfactory for bramble control but it was not entirely destroyed. Roe deer depend upon bramble (*Rubus frutcosis*) for winter food. The treatments did not change use intensity by roe deer, and when the control of bramble exceeded 80%, roe deer fed on shrubs until brambles recovered (in 4 to 5 years).

Campbell et al. (1981) indicated that black-tailed deer readily browsed dormant Douglas fir seedlings and salal treated with 2,4-D or 2,4,5-T, but reduced acceptance occurred with glyphosate. The investigators indicated that physiological changes in Douglas fir caused the change. Deer accepted Douglas fir treated with atrazine, fosamine, and dalapon.

In Canada, Sullivan and Sullivan (1981) investigated forest application of glyphosate on the deer mouse, Oregon vole, Townsend chipmunk, and shrews. Treatment of the 20-year-old Douglas fir plantation did not have any vegetative effects on the distribution and abundance of the small mammal populations during the first year after this habitat alternation.

Sullivan et al. (1995) found no difference in the number of plant species between herbicide-treated plots and the control (5-year-old successional stage), but shrub species richness declined. Species diversity of small mammal communities was not affected. The herbicide was applied for conifer release in a subboreal spruce forest.

Borrecco et al. (1979) treated three clear-cuttings in western Oregon with atrazine and 2,4-D to control herbaceous vegetation. The herbicides reduced the number of plant species, chiefly grasses, forbs, and ground cover, but Douglas fir and most shrubs increased. Eleven species of small mammals were collected. Effective herbaceous vegetation control altered species composition. Species of small mammals preferring grassy habitats were less abundant on treated than untreated plots.

Kirkland (1978) found that the total number of small mammals tended to increase after aerial application of 2,4,5-T in a West Virginia forest. Shrews and mice increased significantly; voles declined slightly. The decrease in the abundance of microtine rodents resulted from a differential decline in the abundance of diurnal species.

Miller and Chapman (1995) compared mechanical (V-blade and windrow) and hexazinone, imazapyr, and picloram plus triclopyr for site preparation in the upper coastal plain physiographic province of the southeastern United States. Although differences occurred in plant and vertebrate communities, the differences were short-lived and were no longer evident 5 years posttreatment.

Webber et al. (1992) showed that few differences occurred among streams that were treated with hexazinone and those that were not treated in six microinvertebrate biocriteria, and concluded that hexazinone did not adversely impact macroinvertebrates. Fish populations were low in diversity and abundance, but there was no evidence that hexazinone adversely impacted these communities.

Cone et al. (1993) studied the impact of imazapyr, hexazinone, and picloram + 2,4-D + triclopyr on songbird use of clear-cuts in Georgia. Vegetative composition, abundance, and diversity were estimated prior to and 1 year after herbicide application. The hexazinone treatments had more wintering birds than the other treatments. There was no significant difference among treatments for species diversity or breeding-bird abundance the first year after treatment.

Miller (1997) indicated that numerous publications report enhanced wildlife habitat values following herbicide use for forestry-site preparation or release, range management, or rights-of-way management. When used according to label guidelines, herbicides can create specific habitat conditions for targeted wildlife species. Herbicides can be used with prescribed fire, mechanical disturbance, and planting for the manipulation of wildlife habitat.

G. Herbicide Effects on Pine Seedlings

Seedling growth and mycorrhizal frequency of *Pinus contorta* Dougl. Var. *latifolia* Engelm (lodgepole pine) and *Picea glauca* (Moench) Voss (white spruce) were inhibited by high concentrations (2 and 4 kg/ha) of hexazinone (Chakravarty and Sidhu, 1987). With time, however, seedling growth and mycorrhizal infection significantly increased. At 1 kg/ha, seedlings recovered from hexazinone, and mycorrhizal frequency was about the same as the control 4 to 6 months after treatment.

Chakravarty and Chatarpaul (1988) investigated hexazinone on ectomycorrhizal (*Laccaria laccata*) and nonmycorrhizal *Pinus resinosa*. Hexazinone reduced seedling growth at 1 to 4 kg/ha, but significant recovery occurred later.

X. EFFECT ON SOIL MICROORGANISMS

A. Phenoxy Herbicides

As early as 1947 it was recognized from the research by several investigators that the disappearance of 2,4-D and 2,4,5-T from the soil was due largely to microbial action (Newman, 1947). Certain groups of soil microorganisms, as determined by carbon dioxide evolution, nitrification plate counts, and the growth of fungi, were injured more readily than others. The workers concluded, however, that the quantity of phenoxy herbicides reaching the soil from weed control would probably not have a serious effect on most soil microorganisms.

Fletcher (1956) investigated the effect of the sodium salts of 2,4-D, MCPA, 2,4,5-T, 2,4-DB, and MCPB on the growth of *Rhizobium trifolii*. Since growth was not affected at concentrations of 25 ppm by any phenoxy studied, it was assumed that concentrations used in agriculture of 1.12 kg/ha (2 to 2.5 ppm) in soil would have no adverse effect on the growth of *Rhizobium trifolii*, a nodule-forming organism of clover.

Gil et al. (1970) found that the butyl ester of 2,4-D at 4 and 40 ppm had no effect on the growth of *Azotobacter chroococcum* in vitro.

Venkataramay and Rajyalakshmi (1971) found that several blue-green algae strains, both nitrogen- and non-nitrogen-fixing, were highly tolerant to several pesticides, including 2,4-D, at rates up to 200 kg/ha. Bakalivanov (1972) found 2,4-D stimulated the growth of the mycelium of five species of fungi (*Paecilomyces varioti*, *Aspergillus niger*, *A. flavus*, *A. tamarii*, and *Penicillium funiculosum*). Unfortunately the dosage or formulation of 2,4-D was not given.

Large doses (25 to 250 ppm) of 2,4-D, 2,4,5-T, and silvex were required to cause inhibition of growth and inhibition of oxygen evolution by 50% (EC_{50}) in four species of unicellular algae (Walsh, 1992). Silvex was more inhibitory to growth than 2,4-D or 2,4,5-T. The acid formulation of 2,4-D was more inhibitory than the butoxyethanol ester of 2,4-D.

Poorman (1973) indicated that 50 and 100 ppm of 2,4-D and 2,4,5-T, respectively, were required to inhibit growth of *Englena gracilis* cultures. Cells were morphologically altered by the herbicides, but recovered rapidly and completely when transferred to an herbicide-free medium.

2,4-D stimulated growth of the soil ameba *Acanthamoeba castellanii* at 0.1 to 1 ppm, but stimulation was less pronounced at 10 to 100 ppm (Prescott and Olson, 1972). The investigators indicated that *Acanthamoeba* may be able to degrade 2,4-D and use it as a carbon and/or energy source.

Under field conditions, some workers have found that phenoxy herbicides have little or no effect on microbial populations (McCurdy and Molberg, 1974; Chulakov and Zharasov, 1973), whereas others have shown both the depression and stimulation of numbers and the growth of some soil organisms (Audus, 1964).

Microbial studies by Stark et al. (1975) have shown that the application of 2,4-D and 2,4,5-T at massive rates (5000 to 40,000 ppm) did not sterilize the soil, but stimulated the growth of certain microflora. These bacteria, actinomycetes, and fungi proliferated to such an extent that they used the herbicides as a carbon source and contributed to their degradation.

See Bovey and Young (1980) and others for details on the effects of phenoxy herbicides on soil microorganisms.

B. Other Herbicides with and Without 2,4-D

Triclopyr and 2,4-D were applied together to study 2,4-D mineralization in Palouse soils (Lobaugh et al., 1993). No effect on either the 2,4-D mineralization rates or the most probable numbers (MPN) of 2,4-D degraders was observed in soil previously exposed to both 2,4-D and triclopyr. In soil that had not already been exposed to either herbicide, there was significant inhibition of the 2,4-D mineralization rates, and the MPN of 2,4-D degraders were <0.1% of those observed when 2,4-D was applied alone. The inhibition mechanism has not been identified.

Camargo (1981) showed that picloram + 2,4-D, dicamba + 2,4-D, and 2,4-D all inhibited the fungi *Metarhizium anisopliae* in vitro, a fungus common in Brazilian pastures.

Picloram + 2,4-D and glyphosate inhibited the decomposition of pasture plants (*Trifolium repens* L.) and (*Lolium perennel*) shoot and (*Carduus nutans* L.) root tissue, but stimulated that of *C. nutans* shoot tissue (Wardle et al., 1994). The rapid decomposition of the plant tissue, however, suggested that herbicides had little long-term effect on plant litter persistence. Microbial basal respiration and substrate-induced respiration initially were strongly enhanced by the herbicides; however, this effect was highly transitory in all tissue types except one.

Estok et al. (1989) studied the effect of 2,4-D, glyphosate, hexazinone, and triclopyr on the growth of three species of ectomycorrhizal fungi. *Cenococum geophilum* growth was usually not affected adversely until concentrations of 1000 ppm of herbicide were used. Glyphosate inhibited growth on agar medium at 10 ppm. *Hebeloma longicaudum* was more sensitive, and all herbicides inhibited growth at 100 ppm. *Pisolithus tinctorius* was inhibited at 1 ppm by all herbicides. These mycorrhizal fungi are important in symbiotic association with gymnosperm tree growth.

Three strains of dicamba-degrading microorganisms (DI-6 and DI-8 *Pseudomonas* sp. and one strain of DI-7 *Moraxella* sp.) were able to reduce the herbicidal activity of dicamba in the rhizosphere quickly enough to allow dicamba-susceptible crop species to grow (Krueger et al., 1991). Pea seedlings planted immediately after inoculation had higher weights over the uninoculated controls at the 0.56 and 4.5 kg/ha rates in growth chamber studies. Pea seedlings planted 2 or 5 days after inoculation had a higher mass over the uninoculated controls at all treatment rates. The concentration of dicamba in the soil was reduced dramatically at all treatment rates, compared to uninoculated controls. Dicamba-degrading bacteria also showed activity in field-test plots, where soybeans were protected from dicamba damage even at the 9 kg/ha application rate.

Low concentrations of glyphosate had little effect on the total populations of soil actinomycetes, bacteria, and fungi during a 214-day experiment, while high concentrations initially increased actinomycete and bacterial numbers by 2 and 1½ logs, respectively (Roslycky, 1982a). The stimulation was followed by a decline and fluctuation showing a gradual increase in numbers. The respiration rates of the soil microbiota in soil suspensions showed some irregular stimulation and retardation with up to 10 μg glyphosate/mL. In contrast, high doses suppressed O_2 uptake by the microbiota. Fungi were the least affected. Pronounced inhibition of actinomycete and bacterial respiration was in agreement with the results from isolate replication. The results indicated both stimulation and inhibition of O_2 uptake by some organisms within these groups. In contrast to some reports of limited, short-term inquiries, these results showed considerable effects of glyphosate on soil microorganisms.

Various concentrations of paraquat, atrazine, simazine, linuron, diuron, and paraquat in combinations with each, including simazine + diuron and terbacil alone, did not inhibit lytic activity of four bacteriophages of *Agrobacterium radiobacter*, three bacteriophages of *Rhizobium meliloti*, three bacteriophages of *R. trifolii*, or two bacteriophages of *Streptomyees chrysomallus* (Roslycky, 1982b). Generally the herbicides had no effect on the neutralization of radiobacteriophage PR-1001 with its homologous antiserum, the length of the latent period, the percentage of adsorption, or the average burst size. In contrast, paraquat concentrations from 20 to 400 μg/mL gradually reduced the adsorption from 38 to 21% and the average burst size from 67 to 9 in the PR-1001 : R-1001 phage: host system. The same concentrations, however, showed no effect on the particle attachment or the length of the latent period.

XI. AQUATIC ORGANISMS

A. Phenoxy Herbicides and Others

Hooper (1953) applied 2,4-D attaclay pellets as the acid (22.4 kg/ha) and ester (44.8 kg/ha) to the bottom of a lake. Population changes noted that benthos such as midges, mayflies, odonata, caddis flies, and tubificids (bottom invertebrates) appeared similar to those in untreated areas. The investigator found there were no data obtained to indicate that 2,4-D was deleterious to bottom fauna. Haven (1963), however, found benthic (molluscs, worms, or amphipods) organism numbers were reduced after control of the water milfoil (*Myriophyllum spicatum*) with granular 2,4-D (2-ethylhexyl ester). The authors suggested that the reduced numbers of organisms living in or on the mud were not due to the toxicity of the herbicide, but to the collapse and decay of the water milfoil on the bottom of the river, limiting oxygen to bottom feeders. Changes in the ecology of the weed (water milfoil) layer also make the amphipods more subject to predation by fish and crabs.

Several investigators (Mackenthun and Kemp, 1972; Marshall and Rutschky, 1974; Mullison, 1970; Sears and Meehan 1971; Whitney et al., 1973;

Wojtalik et al., 1971; Smith and Isom, 1967) have studied the effect of various formulations of 2,4-D under actual field conditions at rates used for aquatic weed control and have found no adverse effects on aquatic fauna or water quality.

Smith and Isom (1967) applied the butoxyethanol ester of granular 2,4-D (20%) at rates from 44.8 to 112 kg/ha on 3240 ha in a Tennessee Valley Authority (TVA) reservoir to control the Eurasian water milfoil. The 2,4-D had no measurable toxic effect on benthic fauna or changes in burrowing mayfly (*Hexagenia*) numbers. The loss of some invertebrates resulted from the loss of the water milfoil used as food for the bottom fauna. Little uptake of 2,4-D occurred by fish, but some occurred by mussels. The concentration of 2,4-D was <1 ppb in water, although some 2,4-D persisted in isolated mud sediment samples up to 10 months after treatment.

Wojtalik et al. (1971) studied the effect of the dimethylamine salt of 2,4-D (liquid) at rates of 22.4 and 44.8 kg/ha to control the Eurasian water milfoil in TVA reservoirs. No harmful response was observed in zooplankton, phytoplankton, benthic macroinvertebrates, or fish. Plankton sorbed large amounts of 2,4-D and retained it for long periods (up to 6 months), whereas the water with its small content of widely dispersed planktons lost its 2,4-D load in approximately 2 weeks. Finished drinking water from municipal treatment plants on occasion contained 2,4-D.

In other studies in the southeast for controlling the Eurasian water milfoil, Whitney et al. (1973) applied the butoxyethanol ester of 2,4-D (20% granules) at 112 kg/ha by air to 81 ha. No adverse effects occurred on fish, macroinvertebrates, or water quality.

In Pennsylvania, 2,4-D was applied to control the Eurasian water milfoil and other aquatic weeds (Marshall and Rutschky, 1974). No significant changes in benthic and epiphytic macroinvertebrate diversity occurred, but there was an absolute decrease in the total numbers of organisms and damselfly naiads. The numbers of midge larvae and aquatic worms increased, however. The investigators indicated reasons other than the toxicity of 2,4-D as the probable cause for the drop in macroinvertebrate fauna numbers.

In Alaska, Sears and Meehan (1971) applied 2,4-D at 2.24 kg/ha by helicopter at a 162-ha watershed on the Nakwasina River for weed control. Test animals were placed live in boxes in the spray area. No immediate mortality to salmonoid fish or aquatic invertebrates was attributable to the spray. Samples of water and fish had concentrations of 2,4-D well below the level generally considered to be lethal to aquatic organisms. Further research to assess the immediate and long-term effects was recommended.

Hansen et al. (1972) indicated that when given a choice, mosquito fish avoided higher concentrations of 2,4-D (butoxyethanol ester at 10 ppm) in preference to low concentrations. Work with the grass shrimp also indicated similar results (Hansen et al., 1973), in that grass shrimp selected the lower concentration of 2,4-D when given a choice. The shrimp did not attempt to avoid DDT, endrin, dursban, malathion, and carbaryl. The phenomenon may function under field conditions.

Johnson (1978) found fenoprop, sodium arsenite, and picloram, respectively, were the most toxic to mosquito fish, while amitrole T and 2,2-DPA were the least toxic. Dicamba, 2,4-D amine, and paraquat were intermediate in effect. McBride et al. (1981), however, indicated that the minimal concentration of the butoxyethyl ester of 2,4-D required to induce a stress response is only marginally (1–5 mg/L) below the lethal level in sockeye smolts.

The toxic effects of 2,4-D on *Phaedactylum tricornutum* (Bohlin) and *Dunaliella tertiolecta* (Butcher), two species of phytoplankton well suited to bioassay studies and responsive to pollutants, were studied by monitoring changes in growth in terms of cell populations, chlorophyll fluorescence, and the rate of $^{14}CO_2$ assimilation (Okay and Gaines, 1996). Short-term bioassays and batch and continuous cultures were studied. Pure 2,4-D acid appeared more toxic than the commercial amine form of the herbicide, but this may have been due to the small quantities of acetone present in the solvent. Concentrations of amine herbicide in excess of 100 mg/L extended the duration of the lag phase and inhibited growth, but smaller concentrations stimulated growth, the amine being consumed by phytoplankton in preference to nitrate. Continuous culture confirmed the ability of phytoplankton to adapt slowly to herbicide concentrations even as high as 500 mg/L. It is suggested that green algae adapt more rapidly to environmental change than do diatoms.

Herbicide 2,4-D at a dosage of 500 mg/L stimulated the proliferation of the heterotrophic bacterial community present in the water of three fishponds over a period of 1 year (Jana and De, 1982). Due to its toxic action, 2,4-D might kill the planktonic algal population occurring in the pond water, which on degradation generated more nutrients conducive to bacterial proliferation. The microbial density in different hours was closely correlated ($p < 0.01 < 0.05$) with the variations of each of NH_4-N, NO_2-N, and the specific conductivity of water, while the relationship between the former and PO4-P was reciprocal ($P < 0.001$).

When one considers the safety of high rates of 2,4-D applied directly to water sources relative to water quality and the effect on aquatic organisms, the safety margin of 2,4-D or other phenoxys for aquatic organisms by movement from agricultural lands is readily apparent. Since most phenoxys are applied on land at relatively low rates and since a very small portion of the water sources in the United States are treated for aquatic weed control, the chances for adverse effects on aquatic biota are highly remote. See Bovey and Young (1980) for further details.

B. Glyphosate

Hildebrand et al. (1982) reported on the effects of glyphosate on rainbow trout viability and behavior in several field experiments at the University of British

Columbia Research Forest. Laboratory and field 96-hr LC_{50} values were similar: 54.8 and 52.0 mg/L. Avoidance-preference data indicated that the fish would avoid lethal levels of glyphosate. Operational application of glyphosate at the recommended field dose (of 2.2 kg ae/ha), as well as at 10× and 100× the field dose, resulted in no mortality to rainbow trout in field streams. The results indicated that operational spraying with this herbicide for weed control should not be detrimental to rainbow trout populations. Improper use or accidental spills of glyphosate could be avoided by rainbow trout and should not be lethal if diluted in a moderately flowing stream.

C. Hexazinone

Experiments were conducted to measure the acute lethal response of aquatic insects to hexazinone (Velpar L) and triclopyr ester (Garlon 4) in flow-through laboratory bioassays, and to determine the lethal and behavioral effects of these herbicides on insects in outdoor stream channels (Kreutzweiser et al., 1992). No significant mortality (x^2 $P > 0.05$) occurred in 13 test species exposed to hexazinone in laboratory flow-through bioassays (1-hr exposure, 48-hr observation) at the maximum test concentration of 80 mg/L. The survival of insects exposed to 80 mg/L of hexazinone in outdoor stream channels was likewise unaffected. Significant drift (x^2 $P < 0.001$) of *Isonychia* sp. occurred during a hexazinone treatment of the stream channels, but only at the maximum concentration of 80 mg/L, and survival of the displaced *Isonychia* sp. was not affected. In flow-through bioassays with triclopyr ester, 10 of 12 test species showed no significant mortality at concentrations greater than 80 mg/L. Survival of *Isogenoides* sp. and *Dolphilodes distinctus* was significantly affected at less than 80 mg/L. Lethal concentrations were estimated by probit analysis of concentration-response data (1-hr exposure, 48-hr observation) for *Simulium* sp. (LC_{50} = 303 mg/L), *Isogenoides* sp. (LC_{50} = 61.7 mg/L), and *D. distinctus* (LC_{50} = 0.6 mg/L). Triclopyr ester applications to the stream channels resulted in significant drift and mortality of *D. distinctus* at 3.2 mg/L (no effects at 0.32 mg/L), *Isogenoides* sp. at 32 mg/L, and *Hydropsyche* sp. and *Espeorus vitrea* at 320 mg/L.

The 24-hr LC_{50} for chorus frog tadpoles exposed to 2,4-D was 100 ppm (Sanders, 1970). Cooke (1972) found that 2,4-D (acid) had no visible effect and no tissue residues on frog tadpoles (*Rana tenporaia*) up to 50 ppm for 48-hr exposure. Nishiuchi and Yosida (1974) obtained TL_m values of >40 and 16 ppm, respectively, using the potassium salt and butyl ester of MCPA on tadpoles (*Bufo bufo japonius*).

Rodriguez-Lopez (1961) found that 2,4-D acid or its sodium salt had no effect on the mobility of frog spermatozoon. A 0.5% solution of the sodium salt of 2,4-D was found to inhibit the development of frog (*Runa temporaria*) eggs (Lhoste and Roth, 1946).

XII. INSECTS AND NEMATODES

A. Bees

Glynne-Jones and Connell (1954) found that the salts of (4-chloryl-2-methylphenoxy)acetic acid (MCPA) and 2,4-D had no appreciable contact effect on adult worker honeybees and a very low toxicity in stomach poison tests. Beran and Neururer (1955) indicated that the butyl ester of 2,4-D and 2,4,5-T, 2,4,5-T acid, and the sodium salt of 2,4-D were also of very low toxicity to bees.

King (1964), however, indicated that the ester formulations of 2,4-D were toxic to bees by direct contact (immersion and spray solutions) and that the amine salt should be considered in areas of high bee activity. The investigator indicated that bees ceased foraging after 2 days on most plants sprayed with 2,4-D and related herbicides, but that contamination with 2,4-D up to 250 ppm within the colony did not present a threat to colony survival.

Palmer-Jones (1964) found that aerial application of sodium 2,4-D superphosphate (dusted) caused adult bee mortality and a reduced honey crop, but no adverse effect was observed in the apiary in which the bees lived. In laboratory tests, however, no effect occurred when bees had been directly dusted with 2,4-D or when they were made to crawl through 2,4-D dust to enter the hive. Since the bees were not harmed by direct dusting with 2,4-D, the investigators concluded that poisoning may have occurred via the nectar. The reason for the poisoning was not determined. Johansen (1969) and King (1964) indicated that the most damaging effect of phenoxy herbicides on beekeeping is the widescale elimination of bee pastures. Conversely, Berliman (1950) reported that 2,4-D was used to restore bee pastures along the side of the road by destroying brush and other weeds, thus encouraging sweet clover to regain its dominance.

Moffett et al. (1972) found that herbicides such as monosodium salt of Maa (MSMA), paraquat, and cacodylic acid were highly toxic to honeybees in small cages when applied as water sprays, whereas various formulations (amine salts and ester) of 2,4-D, 2,4,5-T, silvex, and picloram were not toxic when applied in water carrier. Diesel oil–water and diesel oil–water–dimethylsulfoxide (DMSO) combination carriers were less toxic than straight diesel oil, but more toxic than water alone. It was concluded that the substituted phenoxy and picolinic acid herbicides have relatively low toxicity to honeybees, but that the oil carriers are toxic to bees.

Morton et al. (1972) fed herbicides to newly emerged worker honeybees in 60% sucrose syrup at concentrations of 0, 10, 100, and 1000 ppm. Herbicides that were relatively nontoxic at all concentrations included 2,4-D, 2,4,5-T, silvex, 2,4-DB, dicamba, 2,3,6-TBA, picloram, and others. Those that were extremely toxic at 100 and 1000 ppm concentrations included paraquat and several organic arsenicals. At 1000 ppm 2,4-D, 2,4,5-T, silvex, and 2,4-DB severely reduced or eliminated brood production in both ester and salt formulations (Morton and Moffett, 1972). At 10 ppm the phenoxy herbicides caused no adverse effect on brood development, but reduced the amount of brood when fed at concentrations of 100 ppm. The adverse effects of the phenoxy herbicides were temporary since once the herbicide was removed, normal brood development was resumed.

In other studies Morton et al. (1974) placed apiaries where the bees' only source of water contained paraquat or 2,4,5-T (triethylamine salt) at 1000 ppm. Large numbers of bees died immediately when exposed to paraquat. When colonies were exposed to 2,4,5-T, large numbers of bees drowned because of the lower surface tension of the water. Production of the brood was reduced below that of check colonies during the period the treated water was used and for 3 months thereafter; however, in the subsequent 9 months, production returned to normal. Concentrations of 2,4,5-T in bees using water containing 2,4,5-T were as high as 149 ppm, but dropped to approximately 5 ppm as soon as the bees used untreated water. Honey contained 2,4,5-T as high as 50 ppm, but dropped to 5 ppm within 1 week after the bees began using untreated water. Trace amounts of 2,4,5-T could be detected in bees and honey after more than a year from the time of exposure. The occurrence of 2,4,5-T or other phenoxy herbicides at this high dosage (1000 ppm) after normal use is very unlikely, however.

Way (1969) indicated that the hazard to bees and possibly other nectar-feeding insects from applications of phenoxy herbicides occurs to plants in flower (apparently from the toxicity of the nectar or the loss of nectar from herbicide treatment), otherwise there appears to be little hazard to insects from direct toxicity of the compound at normal agricultural rates of application.

Foraging honeybees were offered various sucrose–herbicide solutions. Despite the visual attractiveness of the feeding dishes to foragers, six of the seven herbicides significantly reduced the incidence of feeding and were judged to be olfactory and gustatory repellents (Elliott et al., 1979). The most repellent herbicide was 2,4,5-T, which totally inhibited feeding at concentrations as low as 1000 ppm. The next most repellent was 2,4-DB, followed by linuron, picloram, 2,4-D, and monuron. Paraquat was the only herbicide that did not exhibit a marked repellency at concentrations up to 4000 ppm.

B. Other Insects

Adams (1960) reported that the amine salt of 2,4-D at the rate of 0.56 kg/ha increased the mortality of coccinellid larva (a predator of aphids in grain fields throughout Canada) and increased the mean time to pupation in all age groups except the 1-day-old larvae.

Gall and Dogger (1967) found that the larvae mortality of the wheat stem sawfly (*Cephus cinctus* Norton) was highest when 2,4-D was applied immediately before or during oviposition and decreased as the time between application and oviposition (before or after) was increased. Adult sawflies and eggs, however, appeared to be unaffected.

Müller (1971) found 2,4-D more toxic than sodium chlorate, dalapon, simazine, dinoseb, simazine + amitrole, pyrazon, and diquat in the laboratory on *Carobidae* (horned beetle).

Maxwell and Harwood (1958) found that 2,4-D increased the rate of reproduction of pea aphids on broad beans. This was related to an increase in total nitrogen, amino nitrogen, and certain free amino acids in broad beans. Robinson (1961), however, found only three of 30 herbicides that affected the mortality or fecundity of the aphids, including amitrole, DMPA, and maleic hydrazide. Phenoxy herbicides having no effect on pea aphids caged for 5 days on treated broad bean plants included 2,4-D, 2,4-DB, erbon, MCPA, 4-(4-chloryl-2-methylphenoxy)butanoic acid (MCPB), sesone, silvex, and 2,4,5-T at 1000, 100, 10,000, 10, 1000, 8000, 100, and 0.1 ppm, respectively.

Fox (1964) applied atrazine, dalapon, monuron, trichlorylacetic acid (TCA), and 2,4-D to grassland to evaluate their effects for at least two summers after application on the numbers of wireworms (*Agriotes* spp.), springtails (*Collembola* suborder *Arhropleona*), mites (*Acarina*), earthworms (*Lumbricidae*), and millipedes (*Diplopoda*) in the soil. Treatment with 2,4-D did not affect the numbers of wireworms, springtails, or mites.

The effects of 2,4-D upon the chrysomelid beetle *Gastrophysa polygoni* were investigated. Larvae feeding upon either host plant material or their own egg cases sprayed with herbicide suffered significantly higher mortalities than those larvae feeding upon unsprayed material (Sotherton, 1982). Application of 2,4-D did not affect hatching success, nor did contact with the chemical cause significant losses among larvae.

The use of 2,4-D in a mixture with MCPP in a field of spring barley significantly reduced host plant densities and the numbers/m^2 of egg batches in sprayed areas of the field compared with unsprayed areas. The use of specific ‘‘anti-*Polygonum*’’ herbicides eradicated the host plants and thus populations of *G. polygoni* on a field scale.

The population dynamics of litter and foliage arthropods in undisturbed and intensively managed big sagebrush (*Artemisia tridentata*) and bitterbrush (*Purshia tridentata*) habitats in southeastern Wyoming were assessed by the measurement of density and the determination of indices of diversity, richness, and evenness (Christiansen et al., 1989). Brush management consisted of either mowing to a 20-cm stubble or applying the herbicide 2,4-D butyl ester. A total of 63 arthropod species were found in foliage and 150 species in litter. Mowing and herbicide applications resulted in significant changes in the density of 16 of the 46 major arthropod foliage species and 56 of the 70 major litter species. Diversity increased, except in Hymenoptera and Coleoptera, in both mowed and herbicide-treated foliage. In foliage, richness generally increased in all orders following mowing, and evenness tracked diversity. In litter, the diversity of Coleoptera decreased following mowing and herbicide application in Collembola, Homoptera, and Diptera. Evenness followed diversity in trends in both treatments. Soil arthropods were absent in this habitat before and after treatments.

C. Nematodes

Webster (1967) found more nematodes (*Ditylenchus dipsaci*) per plant after spraying 2,4-D (140 mg/100/mL/m^2) on both resistant and susceptible cultivars. The spray decreased the resistance of oats but not barley to nematodes as a result of cell hypertrophy and proliferation, a condition favorable for nematode development.

In 1956 Chappell and Miller (1956) found 2,4-D or silvex at 2.24 kg/ha did not affect most fungi studied and did not appear to inhibit sting nematodes in laboratory studies unless exceptionally high rates of the herbicides were used (1000 or 10,000 ppm). Data on the toxicity of all the herbicides discussed in this book are given in Chapter 7.

XIII. WILDLIFE—GRAZING LANDS

A. Bird Eggs

Dunachie and Fletcher (1970) injected 2,4-D, 2,4-DB, MCPA, MCPB, mecoprop, and dichlorprop into fertile chicken eggs. The herbicides had little or no effect up to 100 ppm. Except for feather blanching, caused by some of the phenoxy herbicides, no teratogenic effects were found. Conversely, DeLage and Alnot (1974) found high fetal death, cervical lordosis, and muscular dystrophy of the feet with total or partial paralysis in pheasants when eggs were incubated in 2,4-D.

Whitehead and Pettigrew (1972), however, could find no adverse effects on the rate of egg production, egg or yolk weight, eggshell thickness, hatchability, or growth rate of the progeny when laying hens were fed 2,4-D at a rate equivalent to 50 and 150 mg/kg diet from 28 to 48 weeks of age. Likewise, most workers could not find any affect on egg embryos, hatchability, and progeny when the eggs of chickens or game birds were sprayed directly with 2,4-D at recommended or higher than recommended field rates (Dobson, 1954; Kopischke, 1972; Somers et al., 1974; Seutin et al., 1981).

Lutz-Ostertag and Lutz (1970) documented the severe consequences of spraying pheasants and red-legged and grey partridge eggs with recommended field rates of 2,4-D amine. Dead embryos, paralyzed chicks, and structural malformations resulted. Damaging effects were also noted when unsprayed eggs were placed in sprayed nests.

Seutin et al. (1981) found no harmful effects on hatching in domestic hen or pheasant eggs if the usual dosage rates were used with herbicides MCPA,

2,4-D, 2,4,5-T, mecoprop, and ioxynil. Somers et al. (1974) found no adverse effects of aqueous solutions of 2,4-D : picloram and 2,4-D : 2,4,5-T mixtures when applied to fertile pheasant eggs preceding incubation at 10 times the normal field rate. No treatments caused any adverse effects on hatching success, incidence of malformed embryos, or subsequent chick mortality.

The hatchability and time to hatch of chicken eggs were found to be unaffected by the application of glyphosate herbicide at three different concentrations and at four different embryo ages (Batt et al., 1980). They concluded that the use of the chemical as a weed control agent in zero tillage farming should not negatively affect the hatchability of the eggs of upland nesting birds.

B. Birds

Best (1972) found strip spraying of sagebrush (*Artemisia tridentata*; 50% reduction) resulted in no notable damage in breeding pairs of Brewer's (*Spizella breweri*) or vesper (*Pooecetes gramineus*) sparrow in Montana. Total kill of sagebrush did significantly reduce nesting pairs of the Brewer's sparrow, but not the vesper sparrow. The decline of the Brewer's sparrow was attributed to the reduced nest cover or the diversity of flora and associated insect fauna.

Schroder and Sturges (1975) sprayed Brewer's sparrow nests with eggs in sagebrush (*A. Tridentata*) with 2,4-D. 2,4-D did not reduce nest success, but sparrows used the sprayed sagebrush stand 1 and 2 years after spraying 67 and 94% less, respectively, than unsprayed brush, and no evidence of nesting occurred in the sprayed stand.

Martin (1970) observed that sage grouse (*Centrocercus urophasianus*) were more abundant in unsprayed big sagebrush than in the sprayed areas because their favored food plants were in the unsprayed areas. Although 2,4-D residues were detected in sage grouse muscles and brains, its effects were not apparent (Carr and Glover, 1970). By 1½ years after spraying, sage grouse strutting activities or grounds were not affected. Nesting density or success were also not affected, although the nests on sprayed areas tended to be near unsprayed areas. Brood production, survival, and movement were also apparently not affected.

Martin (1965), however, found that areas of post oak and blackjack oak forest sprayed with 2,4,5-T provided a significantly more suitable habitat for pairs of eastern bluebirds, eastern meadowlarks, mockingbirds, mourning doves, and bobwhites. The eastern woody pewee, blue gray gnatcatcher, and brown-headed cowbird had higher populations in the treated area than in the control. The investigator concluded that there was no marked adverse effect upon any nesting species of birds and that 2,4,5-T actually improved the habitat for a few species.

Schulz et al. (1992) censused breeding nongame birds on replicated 5- and 6-year post-herbicide-treated (tebuthiuron and triclopyr) and untreated cross-timbers rangeland in central Oklahoma. Twenty species of breeding birds were observed. No treatment effects were detected for total bird density, species diversity, or richness; however, species composition varied considerably among the treatments. The control sites supported species associated with closed canopy woodlands, and the treated sites supported species associated with brushy and prairie habitat. Generally, the control sites had greater foliar cover, fewer snags, and less slash and herbaceous cover than treated sites. Densities of six of the seven most abundant bird species were correlated variously with habitat variables. They concluded that changes in habitat structure resulted in differences in the bird species' composition among treatments.

Rio Grande turkeys used brush-controlled and untreated rangeland equally when suitable roosting and other cover was available, but were absent in areas having an adequate food supply with little available cover (Quinton et al., 1980). Food selection of turkeys was based upon the availability of their preferred foods at different seasons. The two most prevalent foods in each of grass, forb, mast, and cactus classes were the same for both treated and untreated areas 83% of the time. Similarity, indices of diets between brush-controlled versus untreated areas were 60 and 73% for summer and fall, respectively.

Gruver and Guthery (1986) observed the responses of nongame birds to brush suppression and habitat management for game birds in the rolling plains of Texas during the period from 1981 to 1983. The data from line transects were used to describe density, species diversity, species richness, and equitability. They observed no difference in these variables between untreated sites and sites late sprayed with herbicides in 1969. The density of northern mockingbirds (*Mimus polyglottos*) was lower on treated than untreated areas, but no other species were affected. Habitat management to favor mourning doves (*Zenaida macroura*) and bobwhites (*Colinus virginianus*) was associated with a 54% increase in the combined density of nongame birds and a reduction in equitability. Species diversity and species richness were similar on managed and unmanaged sites. In the study area, the past herbicide treatment of mesquite and habitat management for game birds was compatible with nongame birds.

Twenty-one years after chaparral was converted to grassland and riparian habitat in central Arizona, bird population density ranged from a high of 321 pairs per 10 ha in the riparian stringer to a low of 24 pairs per 10 ha on the grassland, compared to 103 pairs per 10 ha in the chaparral (Szaro, 1981). The conversion of 10 ha of chaparral resulted in 38 ha of grassland and 2 ha of riparian habitat, however. Bird density on both areas together was 37 pairs per 40 ha, thus although population density, species richness, and species diversity were higher in the riparian area when contrasted with the chaparral, there was an overall net difference of 66 fewer pairs per 40 ha in the converted area.

C. Small Animals

1. 2,4-D

Several investigators (Hull, 1971; Keith et al., 1959; Tietjen et al., 1967) indicated that pocket gopher (*Thomsmys talpoides*) populations are reduced by more than 80% where rangelands are sprayed with 2,4-D. Reduced gopher numbers on sprayed areas were not caused by the toxicity of the herbicide, but by the reduced food supply (forbs), which had been eliminated by 2,4-D.

In other studies, Johnson and Hansen (1969) also found that the least chipmunk (*Eutamias minimus*) and the northern pocket gopher were reduced in density by 2,4-D, whereas the deer mouse (*Peromyscus maniculatus*) was little affected and montane voles (*Microtus montanus*) increased in abundance in 2,4-D-treated areas. With the reestablishment of forb dominance, pocket gopher and vole populations returned to pretreatment levels. The density changes of pocket gophers on treated ranges were due to changes in food (forbs), those of chipmunks to both food and cover, and those of voles to cover changes.

At two established prairie dog (*Cynomys ludovicianus*) colonies in Montana, Fagerstone et al. (1977) found that the prairie dog diet changed significantly from forbs to grass after 2,4-D sprays. Before spraying, prairie dogs ate 73% forbs and 5% grass. Afterward, they ate 9% forbs and 82% grass. The availability of these forbs appeared to be responsible for the diet change. Despite the diet change, the 2,4-D treatment appeared to have little detrimental effect on prairie dogs.

2. Other Herbicides

After 3 years, rodent abundance was 71% higher on creosotebush areas treated with tebuthiuron than control plots in southeastern Arizona (Standley and Smith, 1988). Arizona cotton rats (*Sigmodon arizonae*) and Western harvest mice (*Reithrodontomys megalotis*) were more abundant, while Merriam's kangaroo rats (*Dipodomys merrami*) was similar. The authors concluded that tebuthiuron is safe to use in semidesert grasslands unless rare or endangered species would preclude any alterations to the community. In studies with cottontail rabbits (*Sylvilagus floridanus*) in the cross-timbers vegetation type in Oklahoma, Lochmiller et al. (1991) concluded that tebuthiuron and triclopyr effectively decreased hardwood overstory and increased the preferred habitats for cottontail rabbits. In further studies using tebuthiuron or triclopyr, fire, or a combination of both, Lockmiller et al. (1995) investigated the parameters of the cottontail rabbits' physical condition among habitat types. Brush management using herbicides influenced rabbit condition, but the type of habitat disturbance was not important. The body condition of rabbits from burning were not apparent until later seral stages, in which the production of herbaceous dicots declined and the vegetative composition resembled undisturbed areas.

Nolte and Fullbright (1997) treated honey mesquite with a 1:1 mixture of triclopyr and picloram. Their results indicated that in the Texas Coastal Bend this treatment did not appear to reduce plant (herbaceous) and vertebrate species richness and diversity within the first 2 years after treatment.

D. Large Animals

In Colorado, Anderson (1960) found that year-long distribution and abundance patterns of cattle and deer during the study period (1959) were generally similar to those prior to spraying sagebrush areas with 2,4-D. Cattle use shifted within the sagebrush type to the oak–snowberry–sagebrush type, however. Both animals were the most abundant on sprayed areas during fall–winter–spring periods. Year-round cattle use, in order of decreasing importance, was sagebrush, oak–snowberry–sagebrush, and conifer–sagebrush–oak. Deer use on the oak–snowberry–sagebrush type and sagebrush–oak type was similar, followed by lesser use in conifer–sagebrush–oak and sagebrush types. Rabbit use on the areas as a whole increased, and grouse use was most prominent in the sagebrush type. The sprayed sagebrush area yielded the least number of rodents. No significant difference was noted between forage yield estimates on sprayed and control areas.

Newton and Norris (1968) studied black-tail deer on the Oregon Coast Range after treatment with 2,4,5-T and atrazine, and concluded that deer did not leave the treated area or accumulate 2,4,5-T or atrazine, that detectable levels of herbicide in the deer was rare, and that the ruminant was able to degrade the herbicides almost completely soon after ingestion.

In a 19-year study in Indiana, Bramble and Byrnes (1972) found that a diversity of food plants useful to wildlife developed on the rights-of-way following spraying. These plants included common herbs of the forest, along with such invaders as the common goldenrods and sheep sorrel. These herbs were highly nutritious and provided food for wildlife both in the summer and winter. Woody plants were interspersed throughout this plant community and included the common blueberries, huckleberries, and teaberries as low shrubs, along with sweetferns and bear oaks as tall shrubs or small trees. The taller woody plants supply food throughout the year and are particularly valuable as an emergency food when deep snow covers the ground in the winter. The rights-of-way were heavily used by common wildlife species, such as white-tailed deer, rabbits, grouse, and wild turkeys. A special study made of the white-tailed deer on the rights-of-way showed a consistent and heavy use in all seasons, indicating that attractive food and cover had been developed.

Data from Germany indicated that herbicides (including phenoxys) had no harmful effects on deer, wild pigs, hares, rabbits, pheasants, and wood pigeons (Madel, 1970). A decline in the population of partridges has been ascertained due to the removal of weed seed and protective hedges, and also because of pheasant competition.

Giban (1972) concluded that phenoxy herbicides and other herbicides used in forestry posed no appreciable risk to game animals since only a small fraction of the land was treated and only at extended intervals.

Ward (1973) found that elk did not change their calving behavior or feeding habits on a site in which 97% of the big sagebrush (*Artemisia tridentata*) cover had been killed with 2.24 kg/ha 2,4-D.

Wilbert (1963) found an increase in elk activity of about 40% 1 year after treatment inside a sagebrush area treated with 2,4-D. By 2 years after treatment an increase of 55% activity was indicated.

In the Rolling Plains of Texas, Darr and Klebenow (1975) indicated that herbicides had little detrimental effect and in some situations may have been beneficial. Grazing by sheep was negatively related to deer densities except in the bottomland habitat. In mesquite–juniper and mimosa–erioneuron uplands, replacing sheep with cattle should increase deer populations.

In South Texas, Beasom and Scifres (1977) found that aerial spraying 80% of mature honey mesquite brushland in alternating strips with 2,4,5-T plus picloram did not adversely affect populations of white-tailed deer, nilgai antelopes, wild turkeys, or feral hogs. Complete control of the brush was detrimental to all species except nilgai antelopes. With the restoration of the forbs 27 months after treatment the deer population recovered. Javelina populations were reduced when pricklypear was reduced by the herbicide.

In another south Texas location, Beasom et al. (1982) found that discriminate treatment of brush with 2,4,5-T plus picloram did not cause consistent differences in white-tailed deer use of that habitat. Any differences in deer use between sprayed and unsprayed habitats were attributed to minor impacts of sprays on forb populations.

Quinton et al. (1979) studied the botanical composition of white-tailed deer fecal pellets from untreated and brush-controlled areas of the Texas rolling plains by microscope. Deer showed a marked preference for 11 of 54 species from a total of 254 identified. On nontreated areas, mistletoe was used the most, with pricklypear used the most on treated areas. Brush control affected dietary selection but did not appear to affect overall deer usage of the habitats studied.

E. Multiple Animal Species

Selected wildlife species were observed for 2 years in velvet mesquite (*Prosopis juliflora* var. *velutina*) near Tucson, Arizona (Germano et al., 1983). More black-tailed jackrabbits (*Lepus californicus*), antelope jackrabbits (*Lepus alleni*), and Gambel's quail (*Lophortyx gambelii*) were seen, and more bird calls were heard in undisturbed mesquite and mesquite with clearings than on the mesquite-free range. The differences in bird and mammal populations were not significant between undisturbed stands and partially cleared stands.

Two aerial applications of 2,4,5-T were applied to 3,634 ha of mesquite dunelands in southern New Mexico (Gibbens et al., 1986). Herbaceous plant production was greater on the sprayed than on the unsprayed area for the first 3 years, and was about the same on both areas for the next 2 years. Microbial populations were not numerically different between treatments, but dehydrogenase activity and CO_2 evolution were greater in dunal than interdunal soils. The numbers of tenebrionid beetles (*Coleoptera: Tenebrionidae*) did not differ between treatments. More mesquite leaf tiers (*Tetralopha euphemella*) were found on the sprayed area than on the untreated area. The population statistics for small mammals were similar on both treatments. More bird species were found on untreated than on sprayed areas. Cattle weights, travel, and time budgets did not differ between treatments, and there were only minor differences between treatments in cattle diet quality. The sprayed area supported over twice as many AUMs of grazing as the untreated area in the first 3 posttreatment years. In the second posttreatment year, the cattle liveweight produced was 2.9 and 1.5 kg/ha on the sprayed and untreated areas, respectively. Overall, the 2,4,5-T treatment caused relatively minor perturbations in measured ecosystem components.

Terrel and Spillet (1975) stated that pinyon–juniper conversion to grasslands has been positive for mule deer and small animals such as rodents and cottontail rabbits in the western states, but indicated that they were in near total ignorance of the impact on hundreds of vertebrate species.

Quimby (1966) stated that browse (big sagebrush) and forbs are important food for wildlife in Montana. Any program that removes or drastically reduces these forage classes will be detrimental to antelope, sage grouse, mule deer, elk, white-tailed deer, and moose under some conditions. The removal of sagebrush on winter ranges would apparently be most detrimental to sage grouse, antelope, and mule deer. The data also suggest that white-tailed deer, elk, and bighorn sheep may also be impacted. The differences in spray and unsprayed strips in brush were caused by food plants, since sage grouse numbers were sometimes lower in the sprayed areas.

Olson et al. (1994) used low rates of tebuthiuron (0.3 to 0.6 kg/ha) to thin big sagebrush and other weeds and enhance wildlife habitat in Wyoming. The treatment increased plant and animal diversity, and increased grass-cover dominance. Increased ground squirrel, antelope, and elk have been reported.

Johnson and Fitzhugh (1990) found that grazing by deer and cattle reduced the development of the brush stand after brush control. The most desirable plant community for deer and cattle forage, fire hazard reduction, and soil protection resulted when both deer and cattle grazed on a rangeland site in the Sierra Nevada foothills of California.

XIV. WATERSHED RESPONSE TO WOODY PLANT CONTROL

Hydrologic and vegetative changes induced by controlling big sagebrush with 2,4-D were evaluated over a 23-year period on paired watersheds in south-central Wyoming (Sturges, 1994). The annual water yield increased by 20% for 11 years after brush control, but then returned to the pretreatment level. Two-thirds of the increase came as snowmelt discharge and one-third from groundwater discharge. The total sediment transport averaged nearly seven times less on the 2,4-D-treated watershed. Grass production more than doubled in the 5 years after spraying, and was 1.5 times greater in the ninth and tenth years. After 11 years sagebrush returned to its pretreatment density on the 2,4-D-treated watershed.

Rowe (1963) showed that removing the tree–brush cover shading a stream in southern California resulted in an increase in streamflow. Hill and Rice (1963) indicated that streamflow gains were highest during the dry season when water was needed and lowest during the rainy season. They demonstrated at San Dimas, California, that water yields can be improved by converting canyon-bottom vegetation from woodland brush to a grass cover. Grassed areas also provided better wildlife habitat and better livestock grazing than dense brush.

Wood et al. (1991) reported that on a range site in New Mexico dominated by creosotebush, plots that had been root plowed, seeded, and treated with tebuthiuron and untreated plots showed no difference in runoff the first year, although treated plots reduced sediment loss. In the second year, root plowing and seeding increased sediment yield, but herbicide-treated plots decreased sediment yield. Runoff and sediment yield the third and fourth year were greater from the untreated areas. The untreated plots had 13% shrub cover and large bare ground areas that contributed to higher runoff and erosion rates.

It has long been recognized that mesquite is a heavy user of soil water (Cable, 1977; Knight, 1991). Cable (1977) found that the competitive effect of velvet mesquite on perennial grasses was the most severe in the upper 37.5 cm of soil under or near the mesquite crowns with the highest extraction rates in the early spring and early summer in Arizona. In the Blackland prairie of Texas, two small watersheds infested with honey mesquite were investigated by comparing mesquite removal from one watershed by chemicals. Killing the mesquite reduced evapotranspiration about 8 cm per year and increased surface runoff about 10% (Richardson et al., 1979). In the Edward Plateau, two watersheds infested with brush were used to determine the hydrologic effect of mechanical brush control. Root plowing one watershed to remove brush reduced the surface runoff about 20% (Richardson et al., 1979). Knight (1991) concluded that removing honey mesquite, blackbrush, and huisache from herbicide-treated watersheds in south Texas had less runoff and greater forage production than untreated watersheds. He concluded that it was possible that aquifer recharge during years of good precipitation may be increased through deep drainage on the treated areas.

In the Rolling Plains, Carlson et al. (1990) indicated that when using nonweighable lysimeters there was no net change in deep drainage, evapotranspiration, or runoff on sites in which the herbaceous component increases in response to honey mesquite removal. Likewise, on the eastern edge of the rolling plains of Texas, Heitschmidt and Dowhower (1991) found in intact tree lysimeters that honey mesquite removal resulted in significantly greater amounts of herbaceous leaf area and standing crop, but that honey mesquite control would not enhance water yields dramatically in this region in the absence of livestock grazing. Working in the same general area as Heitschmidt and Dowhower (1991), Dugas and Mayeux (1991) reached similar conclusions—that while honey mesquite used substantial amounts of water, evaporation from the rangeland from which it was removed was just slightly lower due to the increased herbaceous evaporation associated with increases in standing crops.

Terminal infiltration rates were similar in soils with dense whitebrush on herbicide-treated and/or burned areas compared to untreated sites (Knight et al., 1983). Sediment production was greater on herbicide versus herbicide-fire system or untreated. Infiltration rates on running mesquite (*Prosopis reptans*) sites were similar for herbicide-fire treatments versus untreated. Sediment production, however, was greater on sprayed areas than on untreated areas. The difference in sediment production is attributed to the slightly reduced mulch loads in areas in which herbicides were used. Prescribed burning compensated for the loss of brush leaf mulch by promoting grass cover on herbicide-treated areas.

Bedunah and Sosebee (1985) found that removal of honey mesquite by herbicides, mechanical grubbing, or mechanical grubbing and planting kleingrass did not increase infiltration. Infiltration was increased by vibratilling and shredding.

Tremendous effort and man hours have been expended in evaluating the prospect of brush management and water yields in Texas (McCarl et al., 1987; Texas State Soil & Water Conservation Board, 1991). The practical aspects (Walker, 1998) and feasibility of increasing streamflow by different degrees of honey mesquite and juniper control are being investigated (Conner and Bach, 1998).

XV. OTHER CONSIDERATIONS

One added benefit to woody plant management with herbicides is the control of noxious and poisonous plants toxic to livestock. Some of these plants are woody and cause considerable economic loss to producers. Woody plants may also harbor destructive insects that attack crops and livestock and predators that reduce wildlife populations and destroy livestock. Herbicides have also been used to control weeds to preserve endangered species in some areas. Herbicides are usually required to control weeds after brush control to establish native or introduced forage and food plants on noncrop land, pasture, and rangeland; or otherwise the forage stand may fail. Herbicides also kill plants that cause allergies and hay fever, thus relieving human suffering. State and federal laws exist that require the control of noxious weeds growing on private and public lands.

Herbicides are used on rights-of-way and roadsides for safety and improved visibility and to maintain utility rights-of-way. They are used around industrial buildings, airports, parking lots, and many other areas to maintain the environment and prevent fire hazards. Wildlife benefits from using utility rights-of-way and pipelines. Herbicides are used on recreational sites for the maintenance of pathways and other high-use areas. The control of brush and weeds on watersheds may significantly increase off-site water yield for domestic and urban use. Controlling brush on depleted land with herbicides increases standing cover herbaceous crop, stabilizes the soil, and reduces wind and water erosion. Numerous studies have shown that herbicide residues are not a problem on forest and grazing lands and that soil fertility is not affected at recommended rates. (See Chap. 6.) Certain herbicides can also selectively control some woody problem plants, leaving desirable woody plants. For example, Whisenant (1987) used clopyralid to control mountain big sagebrush without any injury to desirable shrubs, antelope bitterbrush, and/or Saskatoon serviceberry in the western United States. Clopyralid can also be used for honey mesquite and huisache control in Texas without any injury to desirable oaks.

XVI. SUMMARY

Numerous studies show that grass and forb production can be significantly increased by adequate woody plant control with herbicides in many different climatic regions with differing woody plants. Increased production was usually only measured 2 to 5 years following treatment, therefore the treatment benefit may not always be long-term. Resulting woody plant control on grazing lands increased livestock production and many times improved wildlife habitat and numbers. Herbicide treatment may reduce forb production during the first growing season, but usually the forbs recover by the second growing season. Herbicide treatment may not be justified in light or scattered brush stands or stands on poor sites. Herbicides are useful in IBMSs in combination with fire or mechanical methods. For example, low rates of picloram can be used on prickly pear cactus after fire with excellent results in west Texas. Soil-applied herbicides such as tebuthiuron can be used at reduced rates (from 0.3 to 0.6 kg/ha) to thin big sagebrush and increase plant and animal diversity. Spraying strips of brush and leaving alternate untreated areas may significantly increase livestock forage as well as maintain wildlife habitat and food plants. Forage quality is not adversely affected by herbicide use and is sometimes improved (crude protein). Cattle may prefer forage in herbicide-treated plots.

Woody plants are usually vigorous competitors and tend to regenerate rapidly if disturbed and left untreated, therefore follow-up treatments with some control method are desired to maintain herbaceous vegetation and maintain productivity.

In forest land, herbicides are used to create and maintain desired wildlife habitats, create mixed- and uneven-aged stands of plants, restore damaged landscapes, control exotic, noxious, and poisonous plants, maintain recreational areas, and manage rights-of-way for multiple use. To avoid economic loss in forests, controlling shrubs and hardwood species is necessary. Several herbicides frequently mentioned in this book, such as 2,4-D, triclopyr, imazapyr, and hexazinone, are used in site preparation for conifers and for maintenance. The maintenance of wildlife habitat is important, and like rangelands, herbicides can be used to improve wildlife habitat with only temporary reduction in forbs and other food plants. Conifers are usually tolerant of the herbicides used. The beneficial effects of weed control and conifer release are usually long-term (10 years or more). Provisions to control exotic competing weeds and brush are needed in forests. In rangeland, forest, and noncrop lands, herbicides may temporarily reduce wildlife numbers not because of toxicity but because of reduced food plants. The effects of herbicides on microorganisms, insects and nematodes, aquatic organisms, birds, small mammals, and large mammals are discussed.

REFERENCES

Adams JB. Effects of spraying 2,4-D amine on coccinellid larvae. Can J Zool 38:285–288, 1960.

Anderson AE. Effects of sagebrush eradication by chemical means on deer and related wildlife. Federal Aid in Wildlife Restoration Project W-38-R-13. Deer-Elk Invest., work plan 7, job no. 1. Colorado Game and Fish Department, 1960.

Asay KH. Revegetation in the sagebrush ecosystem. In: Onsager JA, ed. Integrated Pest Management on Rangeland: State of the Art in the Sagebrush Ecosystem. USDA-ARS, ARS 50, 1987, pp. 19–27.

Audus LJ. Herbicide behavior in the soil. II. Interactions with soil microorganisms. In: Audus LJ, ed. The Physiology and Biochemistry of Herbicides. London: Academic, 1964, pp. 163–206.

Bakalivanov D. Biological activity of certain herbicides on microscopic soil fungi. Symp Bio Hung 11:373–377, 1972.

Baker RL, Powell J, Morrison RD, Stritzke JF. Effects of atrazine, 2,4-D and fertilizer on crude protein content of Oklahoma tall-grass prairie. J Range Mgt 33:404–407, 1980.

Bartolome JW, Allen-Diaz BH, Tietje WD. The effect of *Quercus douglasii* removal on understory yield and composition. J Range Mgt 47:151–154, 1994.

Bartolome JW, Heady HF. Ages of big sagebrush following brush control. J Range Mgt 31:403–406, 1978.

Bartos DL, Mueggler WF. Early succession following clearcutting of aspen communities in northern Utah. J Range Mgt 35:764–768, 1982.

Batt BD, Black JA, Cowan WF. The effects of glyphosate herbicide on chicken eggs hatchability. Can J Zool 58:1940–1942, 1980.

Baur JR, Bovey RW, Holt EC. Effect of herbicides on production and protein levels in pasture grasses. Argon J 69:846–851, 1977.

Beasom SL, Scifres CJ. Population reactions of selected game species to aerial herbicide applications in South Texas. J Range Mgt 30:138–142, 1977.

Beasom SL, Inglis JM, Scifres, CJ. Vegetation and white-tailed deer responses to herbicide treatment of a mesquite drainage habitat type. J Range Mgt 35:790–794, 1982.

Bedunah DJ, Sosebee RE. Forage response of a mesquite–buffalograss community following range rehabilitation. J Range Mgt 37:483–487, 1984.

Bedunah DJ, Sosebee RE. Influences of site manipulation on infiltration rates of a depleted West Texas range site. J Range Mgt 38:200–205, 1985.

Beran F, Neururer J. Effect of protection materials on the honeybee. I. Toxicity to bees. Pflanzenschutzberichte 15:97–147, 1955.

Berliman AP. Weed killers and bee pasture. Amer Bee J 90:542–543, 1950.

Best LB. First-year effects of sagebrush control on two sparrows. J Wildl Mgt 36:534–544, 1972.

Blaisdell JP. Competition between sagebrush seedlings and reseeded grasses. Ecology 30: 512–519, 1949.

Blaisdell JP, Mueggler WF. Effect of 2,4-D on forbs and shrubs associated with big sagebrush. J Range Mgt 9:38–40, 1956.

Blake PM, Hurst GA. Responses of vegetation and deer forage following applications of hexazinone. South J Appl For 11:176–180, 1987.

Borrecco JE, Black HC, Hooven EF. Response of small mammals to herbicide-induced habitat changes. Northwest Sci 53:97–106, 1979.

Bourgeois DM. Documentation of species richness and plant succession in intensively managed loblolly pine plantations in southern woodlands. Proc South Weed Sci Soc 49:82–83, 1996.

Bovey RW, Young AL. The Science of 2,4,5-T and Associated Phenoxy Herbicides. New York: Wiley, 1980.

Bovey RW, Meyer RE, Morton HL. Herbage production following brush control with herbicides in Texas. J Range Mgt 25:136–142, 1972.

Bowes GG. Changes in the yield of forage following the use of herbicides to control aspen poplar. J Range Mgt 35:246–248, 1982.

Bowes GG. Long-term control of aspen poplar and western snowberry with dicamba and 2,4-D. Can J Plant Sci 71:1121–1131, 1991.

Bowes GG, Spurr DT. Control of aspen poplar, balsam poplar, prickly rose and western snowberry with metsulfuron-methyl and 2,4-D. Can J Plant Sci 76:885–889, 1996.

Boyd RS, Freeman JD, Miller JH, Edwards MB. Forest herbicide influences on floristic diversity seven years after broadcast pine release treatments in central Georgia, USA. New For 10:17–37, 1995.

Bramble WC, Byrnes WR. A Long-Term Ecological Study of Game Food and Cover on a Sprayed Utility Right-of-Way. Agric. Exp. Stn. Res. bull. no. 885, Lafayette, IN: Purdue University, 1972.

Brooks JJ, Johnson AS, Miller KV. Effects of chemical site preparation on wildlife habitat and plant species diversity in the Georgia sandhills. Presented at 7th Biennial South Silvicultural Res. Conf. Mobile, AL. New Orleans: USDA South. For. Exp. Stn. gen. tech. rept. 50, 1993, pp. 605–612.

Brooks JJ, Johnson AS, Miller KV, Bush PB, Taylor JW. Effects of chemical site preparation on plant diversity in the Georgia Coastal Plain. Proc South Weed Sci Soc 45: 229, 1992.

Brown JR, Archer S. Woody plant invasion of grasslands: Establishment of honey mesquite (*Prosopis glandulosa* var. *glandulosa*) on sites differing in herbaceous biomass and grazing history. Oecolgia 80:19–26, 1989.

Brown JR, Archer S. Water relations of a perennial grass and seedling vs. adult woody plants in a subtropical savanna, Texas Oikos 57:366–374, 1990.

Cable DR. Twenty years of changes in grass production following mesquite control and reseeding. J Range Mgt 29:286–289, 1976.

Cable DR. Seasonal use of soil water by mature velvet mesquite. J Range Mgt 30:4–11, 1977.

Cain MD. Hardwood control before harvest improves natural pine regeneration. Res. paper 50-249. New Orleans: USDA-Forest Serv., South. For. Exp. Stn., 1988.

Cain MD. The influence of woody and herbaceous competition on early growth of naturally regenerated loblolly and short leaf pines. South J Appl For 15:179–187, 1991.

Cain MD. A 9-year evaluation of the effects of herbaceous competition on upland hardwoods that developed from sprouts on cutover sites. Res. paper 50-284. New Orleans: USDA-Forest Serv., South. For. Exp. Stn., 1995.

Camargo LMPC de A. Efeito dos herbicidas utilizados nas pastagens sorbre o fungo entomopatogênico *Metarhizium anisopliae* (Metsch.) Sorokin. Biologico Sao Paulo 47(4):99–102, 1981.

Campbell DL, Evans J, Lindsey GD, Dusenberry WE. Acceptance by black-tailed deer of foliage treated with herbicide. Res. paper PNW-290. Portland, OR: USDA-Forest Serv., Pacific NW For. & Range Exp. Stn., 1981.

Carlson DH, Thurow TL, Knight RW, Heitschmidt RK. Effect of honey mesquite on the water balance of Texas Rolling Plains rangeland. J Range Mgt 43:491–496, 1990.

Carr HD, Glover FA. Effects of sagebrush control on sage grouse. 35th North Amer Wildl Conf 35:205–215, 1970.

Chakravarty P, Chatarpaul L. The effects of Velpar L (hexazinone) on seedling growth and ectomycorrhizal symbiosis of *Pinus resinosa*. Can J For Res 18:917–921, 1988.

Chakravarty P, Sidhu SS. Effect of hexazinone (Pronone ™5G) on the seedling growth and mycorrhizal incidence of *Pinus cortorts* var. *latifolia* and *Picea glauca*) Eur J For Path 17:282–291, 1987.

Chappell WE, Miller LI. The effects of certain herbicides on plant pathogens. Plant Dis Reporter 40:52–56, 1956.

Christensen MD, Young JA, Evans RA. Control of annual grasses and revegetation in ponderosa pine woodlands. J Range Mgt 27:143–145, 1974.

Christiansen TA, Lockwood JA, Powell J. Arthropod community dynamics in undisturbed and intensively managed mountain brush habitats. Great Basin Nat 49:570–586, 1989.

Chulakov SA, Zharasov SU. The biological activity of southern soils of Kazakhstan with the use of herbicides. Izestiya Akademii Nauk Kayakhghoi SSR, Seriyn Biologicheskaya 11:7–13, 1973.

Clary WP. Response of herbaceous vegetation to felling of alligator juniper. J Range Mgt 27:387–389, 1974.

Clary WP, Jameson DA. Herbage production following tree and shrub removal in the pinyon-juniper type of Arizona. J Range Mgt 34:109–113, 1981.

Cone MA, Brooks JJ, Chapman BR, Miller KV. Effects of chemical site preparation on songbird use of clearcuts in Georgia. Proc South Weed Sci Soc 46:175, 1993.

Conner JR, Bach JP. Economics of brush control for water yield. Texas Plant Protect Assoc Conf 10:18, 1998.

Cook CW, Lewis CE. Competition between big sagebrush and seeded grasses on foothill ranges in Utah. J Range Mgt 16:245–250, 1963.

Cooke AS. The effect of DDT, dieldrin and 2,4-D on amphibian spawn and tadpoles. Environ Pollu 3:51–68, 1972.

Crawford HS. Effect of aerial 2,4,5-T sprays on forage production in west-central Arkansas. J Range Mgt 13:44, 1960.

Crowder SH, Cole AW, Watson VH. Weed control and forage quality in tebuthiuron treated pastures. Weed Sci 31:585–587, 1983.

Dahl BE, Sosebee RE, Goen JP, Brumley CS. Will mesquite control with 2,4,5-T enhance grass production? J Range Mgt 31:129–131, 1978.

Darr GW, Klebenow DA. Deer, brush control, and livestock on the Texas rolling plains. J Range Mgt 28:115–119, 1975.

Darrow RA, McCully WG. Brush control and range improvement in the post oak–blackjack oak area of Texas. Bull. 942. Texas Agric. Exp. Stn., College Station, TX. 1959.

Delage C, Alnot MO. Teratogenic effect of certain pesticides on avian and mammal embryos. Econ Med Anim 14:141–150, 1974.

Ditomaso JM, Marcum DB, Rasmussen MS, Healy EA, Kyser GB. Post-fire herbicide sprays enhance native plant diversity. Calif Agr 51:6–11, 1997.

Dobson N. Chemical sprays and poultry. Agriculture 61:415–418, 1954.

Dugas WA, Mayeux HS Jr. Evaporation from rangelands with and without honey mesquite. J Range Mgt 44:161–170, 1991.

Duke SO, Hoagland RE, Elmore CD. Effects of glyphosate on metabolism of phenolic compounds. V. L-α-aminooxy-β-phenyl-propionic acid and glyphosate effects on phenylalanine ammonia-lyase in soybean seedlings. Plant Physiol 65:17–21, 1980.

Dunachie JF, Fletcher WW. The toxicity of certain herbicides to hen's eggs assessed by the egg-injection technique. Ann Appl Bio 66:515–520, 1970.

Dye KL, Ueckert DN, Whisenant SG. Redberry juniper–herbaceous understory interactions. J Range Mgt 48:100–107, 1995.

Eckert RE Jr, Bruner AD, Klomp GJ. Response of understory species following herbicidal control of low sagebrush. J Range Mgt 25:280–285, 1972.

Ehrenreich JH, Crosby JS. Forage production on sprayed and buried areas in the Missouri Ozarks. J Range Mgt 13:68–70, 1960.

Elliott RH, Cmiralova D, Wellington WG. Olfactory repellency of herbicides to foraging honeybees (*Hymenoptera apidae*). Can Entomol 111:1131–1135, 1979.

Elwell HM. Herbicides for release of short-leaf pine and native grasses. Weeds 15:104–107, 1967.

Elwell HM, Santelmann PW, Stritzke JE, Greer H. Brush control research in Oklahoma. Bull. B-712. Stillwater, OK: Oklahoma Agric. Exp. Stn., 1974.

Engle DM, Stritzke JF, McCollum FT. Vegetation management in the Cross Timbers: Response of understory vegetation to herbicides and burning. Weed Tech 5:406–410, 1991.

Estok D, Freedman B, Boyle D. Effects of the herbicides 2,4-D, glyphosate, hexazinone and triclopyr on the growth of three species of ectomycorrhizal fungi. Bull Environ Contam Toxicol 42:835–839, 1989.

Evans RA, Young JA. Aerial application of 2,4-D plus picloram for green rabbitbrush control. J Range Mgt 28:315–318, 1975.

Fagerstone KA, Tietjen HP, LaVoie K. Effect of range treatment with 2,4-D on prairie dog diet. J Range Mgt 30:57–60, 1977.

Fales SL, Wakefield RC. Effects of turfgrass on the establishment of woody plants. Agron J 73:605–610, 1981.

Felix AC III, Sharik TL, McGinnes BS. Effects of pine conversion on food plants of northern bobwhite quail, eastern wild turkey, and white-tailed deer in the Virginia Predmont. South J Appl For 10:47–52, 1986.

Fiddler GO, McDonald PM, Chapman CR. Releasing young conifers with herbicides: Fast and effective. Proc South Weed Sci Soc 45:209–218, 1992.

Fisher CE, Wiedemann HT, Walter JP, Meadors CH, Broch JH, Cross BT. Brush Control Research on Rangeland. MP-1043 Texas Agric. Exp. Stn., College Station, TX. 1972.

Fitzgerald CH, Fortson JC. Herbaceous weed control with hexazinone in loblolly pine (*Pinus taeda*) plantations. Weed Sci 26:583–588, 1979.

Fletcher WW. Effect of hormone herbicides on the growth of *Rhizobium trifolii*. Nature 177:1244, 1956.

Fox CJS. The effects of five herbicides on the number of certain invertebrate animals in grassland soil. Can J Plant Sci 44:405–409, 1964.

Frischknecht NC. Contrasting effects of big sagebrush and rubber rabbitbrush on production of crested wheatgrass. J Range Mgt 16:70–74, 1963.

Gall A, Dogger JR. Effect of 2,4-D on the wheat stem sawfly. J Econ Entomol 60:75–77, 1967.

Gentle CB, Duggin JA. *Lantuna camara* L. invasions in dry rainforest–open forest ecotones: The role of disturbance associated with fire and cattle grazing. Austr J Ecol 22:298–306, 1997.

Germano DJ, Hungerford R, Martin SC. Responses of selected wildlife species to the removal of mesquite from desert grassland. J Range Mgt 36:309–311, 1983.

Giban J. Does the use of weedkillers in forestry pose a threat to game? Rev For Franc 24:421–428, 1972.

Gibbens RP. Grass and forb production on sprayed and nonsprayed mesquite (*Prosopis glandulosa* Torr.) dunelands in south-central New Mexico. Proceedings XIV International Grasslands Congress, 1981, pp. 437–440.

Gibbens RP, Herbel CH, Lenz JM. Field-scale tebuthiuron application on brush-infested rangeland. Weed Tech 1:323–327, 1987.

Gibbens RP, Beck RF, McNeely RP, Herbel CH. Recent rates of mesquite establishment in the northern Chihuahuan desert. J Range Mgt 45:585–588, 1992.

Gibbens RP, Herbel CH, Morton HL, Lindemann WC, Ryder-White JA, Richman DB, Huddleston EW, Conley WH, Davis CA, Reitzel JA, Anderson DM, Guiao A. Some impacts of 2,4,5-T on a mesquite duneland ecosystem in southern New Mexico: A synthesis. J Range Mgt 39:320–326, 1986.

Gil J, Morales J, Martin A, Ruano C, Aragones F. Influence of some pesticides on *Azotobacter*. Microbiol Espan 23:271–277, 1970.

Gill JD, Healy WM. Shrubs and Vines for Northeastern Wildlife. Gen. tech. rept. NE-9. Upper Darby, PA: USDA For. Serv., USDA-Northeastern For. Exp. Stn., 1974.

Glover GR, Creighton JL, Gjerstad DH. Herbaceous weed control increases loblolly pine growth. J For 87:47–50, 1989.

Glynne-Jones, GD, Connel JU. Studies on the toxicity to worker honeybees (*Apis mellifera*) of certain chemicals used in plant protection. Ann Appl Bio 41:271–279, 1954.

Gonzalez CL. Brush reinfestation following mechanical manipulation. J Arid Environ 18: 109–117, 1990.

Gratkowski H. Effects of herbicides on some important brush species in southwestern Oregon. Res. paper 31. Portland, OR: USDA-Forest Serv., Pacific Northwest For. and Range Exp. Stn., 1959.

Gratkowski H. Use of herbicides on forest lands in southwestern Oregon. Res. note No. 217. Portland, OR: USDA-Forest Serv., Pacific Northwest For. and Range Exp. Stn., 1961.

Gruver BJ, Guthery FS. Effects of brush control and game-bird management on nongame birds. J Range Mgt 39:251–253, 1986.

Haile A. A study of the effects of oak defoliation on understory plants. J Agric Sci UK 102:247–249, 1984.

Halls LK, Crawford HS. Vegetation response to an Ozark woodland spraying. J Range Mgt 18:338–340, 1965.

Hamilton WT, Kitchen LM, Scifres CJ. Height Replacement of Selected Woody Plants Following Burning or Shredding. B-1361. Texas Agric. Exp. Stn., College Station, TX. 1981.

Hansen DJ, Schimmel SC, Keltner JM Jr. Avoidance of pesticides by grass shrimp (*Palaemonities pugio*). Bull Environ Contam Toxicol 9:129–133, 1973.

Hansen DJ, Matthews E, Nall SL, Dumas DP. Avoidance of pesticides by untrained mosquito fish, *Gambusia affinis*. Bull Environ Contam Toxicol 8:46–51, 1972.

Harty RL, Friend AL, Watkins RM. Response of a mature loblolly pine (*Pinus taeda* L.) stand to hardwood reduction by imazapyr and fire. Proc South Weed Sci Soc 48: 150, 1995.

Haven D. Mass treatment with 2,4-D of milfoil in tidal creeks in Virginia. Proc South Weed Conf 16:345–350, 1963.

Haywood JD. Response of planted *Pinus taeda* L. to brush control in northern Louisiana. For Ecol Mgt 15:139–134, 1986.

Haywood JD. Prescribed burning and hexazinone herbicide as release treatments in a sapling hardwood-loblolly pine stand. New For 10:39–53, 1995.

Heinze JA, Zedaker SM, Smith DW. Effects of low-input vegetation management on pine-hardwood mixed stands. Proc South Weed Sci Soc 50:114, 1997.

Heitschmidt RK, Dowhower SL. Herbage response following control of honey mesquite within single tree lysimeters. J Range Mgt 44:144–149, 1991.

Heitschmidt RK, Schultz RD, Scifres CJ. Herbaceous biomass dynamics and net primary production following chemical control of honey mesquite. J Range Mgt 39:67–71, 1986.

Henkin Z, Seligman NG, Noy-Mein I, Kafkafi W, Gutman M. Rehabilitation of Mediterranean drarf-shrub rangeland with herbicides, fertilizers, and fire. J Range Mgt 51: 193–199, 1998.

Herbel CH, Gould WL, Leifeste WF, Gibbens RP. Herbicide treatment and vegetation responses to treatment of mesquites in southern New Mexico. J Range Mgt 36: 149–151, 1983.

Hildebrand LD, Sullivan DS, Sullivan TP. Experimental studies of rainbow trout populations exposed to field applications of Roundup herbicide. Arch Environ Contam Toxicol 11:93–98, 1982.

Hill LW, Rice RM. Converting from brush to grass increases water yield in southern California. J Range Mgt 16:300–305, 1963.

Hilton JE, Bailey AW. Forage production and utilization in a sprayed aspen forest in Alberta. J Range Mgt 27:375–380, 1974.

Hoagland RE, Duke SO, Elmore D. Effects of glyphosate on metabolism of phenolic compounds. II. Influence on soluble hydroxyphenolic compound, free amino acid and soluble protein levels in dark-grown maize roots. Plant Sci Lett 13:291–299, 1978.

Holechek JL, Stephenson T. Comparison of big sagebrush vegetation in north-central New Mexico under moderately grazed and grazing excluded conditions. J Range Mgt 36:455–456, 1983.

Hooper FN. The effect of application of pelleted 2,4-D upon the bottom fauna of Kent Lake, Oakland County, Michigan. Proc North Central Weed Conf 15:41, 1953.

Horsley SB. Control of herbaceous weeds in Allegheny hardwood forests with herbicides. Weed Sci 29:655–662, 1981.

Hull AC Jr. Effect of spraying with 2,4-D upon abundance of pocket gophers in Franklin Basin, Idaho. J Range Mgt 24:230–232, 1971.

Hurst GA. Plants equal habitat: Symp.–Wildlife Habitat Management. Proc South Weed Sci Soc 50:126–127, 1997.

Ingelög T. Effects of the silvicultural use of phenoxy acid herbicides on forest vegetation in Sweden. Ecol Bull 27:240–254, 1978.

Jackson LW. Picloram and triclopyr injection-results and followup. Proc Northeast Weed Sci Soc 41:158–162, 1988.

Jacoby PW, Slosser JE, Meadors CH. Vegetational responses following control of sand shinnery oak with tebuthiuron. J Range Mgt 36:510–512, 1983.

Jacoby PW, Meadors CH, Foster MA, Hartmann FS. Honey mesquite control and forage responses in Crane County, Texas. J Range Mgt 35:424–426, 1982.

Jameson DA. Juniper root competition reduces basal area of blue grama. J Range Mgt 23:217–218, 1970.

Jana BB, De UK. Stimulatory effect of herbicide 2,4-D on the heterotrophic microbial community in the water of three fish ponds. Acta Microbial Acad Sci Hung 29: 77–82, 1982.

Jaworski FG. Mode of action of *N*-(phosphonomethyl)-glycine: Inhibition of aromatic amino acid biosynthesis. J Agr Food Chem 20:1195–1198, 1972.

Jefferies NW. Herbage production on a Gambel oak range in southwestern Colorado. J Range Mgt 18:212–213, 1965.

Johansen C. The Bee Poisoning Hazard from Pesticides. Bull no. 709. Pullman, WA: Washington Agric. Exp. Stn., 1969.

Johnson CR. Herbicide toxicities in the mosquito fish, *Gambusea affinis*. Proc R Soc Queens 89:25–27, 1978.

Johnson DR, Hansen RM. Effects of range treatment with 2,4-D on rodent populations. J Wildl Mgt 33:125–132, 1969.

Johnson JR, Payne GE. Sagebrush reinvasion as affected by environmental influences. J Range Mgt 21:209–213, 1968.

Johnson JR, Tucker WL, Stymiest CE, Bowker EJ. Pricklypear cactus control in western South Dakota. South Dakota Beef Rept. Brookings, SD: South Dakota Agric. Exp. Stn. and South Dakota Coop. Ext. Serv., 1988, pp. 128–134.

Johnson W, McKell CM, Evans RA, Berry LJ. Yield and quality of annual range forage following 2,4-D application on blue oak trees. J Range Mgt 12:18–20, 1959.

Johnson WH, Fitzhugh EL. Grazing helps maintain brush growth on cleared land. Calif Agr 44:31–32, 1990.

Keith JO, Hansen RM, Ward AL. Effect of 2,4-D on abundance and foods of pocket gophers. J. Wildl Mgt 23:137–145, 1959.

Kim HS, Chung YH, Lee JY. Studies on the renovation of inferior grassland dominated by mixed brush. Res Rept Rural Dev Adm Suweon Korea 32:24–29, 1990.

Kim JG, Han HJ, Han MS. Improving of botanical composition and yield performance of *Artemisia* spp. dominated pasture mixtures by dicamba herbicide. Res Rept Rual Dev Adm Suweon Korea 34:86–93, 1992.

King CC. Effects of herbicides on honeybees. Glean Bee Cult 92:230–233, 251, 1964.

Kirby DR, Stuth JW. Botanical composition of cattle diets grazing brush managed pastures in east-central Texas. J Range Mgt 35:434–436, 1982a.

Kirby DR, Stuth JW. Brush management influences the nutritive content of cattle diets in east-central Texas. J Range Mgt 35:431–433, 1982b.

Kirkland GL Jr. Population and Community Responses of Small Mammals to 2,4,5-T. Res. note PNW-314. Portland, OR: USDA, For. Serv. Pacific NW For. Range. Exp. Stn., 1978.

Kituku VM, Powell J, Smith MA, Olson RA. Increasing bitterbrush nutrient quality with 2,4-D, mowing, and burning in southcentral Wyoming. J Range Mgt 45:488–493, 1992.

Knight RW. Water use by mesquite. In: Risks in Ranging. PR-4872. College Station, TX: Texas Agric. Exp. Stn., 1991, pp. 8–11.

Knight RW, Blackburn WH, Scifres CJ. Infiltration rates and sediment production following herbicide/fire brush treatments. J Range Mgt 36:154–157, 1983.

Kopischke ED. The effect of 2,4-D and diesel fuel on egg hatchability. J Wildl Mgt 36: 1353–1356, 1972.

Kreutzweiser DP, Holmes SB, Behmer DJ. Effects of the herbicides hexazinone and triclopyr ester on aquatic insects. Ecotoxicol Environ Safety 23:364–374, 1992.

Krueger JP, Butz RG, Cork DJ. Use of dicamba-degrading microorganisms to protect dicamba susceptible plant species. J Agr Food Chem 39:1000–1003, 1991.

Lancaster DL, Young JA, Evans RA. Weed and brush control tactics in the sagebrush ecosystem. In: Onsager JA, ed. Integrated Pest Management on Rangeland: State of the Art in the Sagebrush Ecosystem. USDA-ARS, ARS 50. Reno, NA. 1987, pp. 11–18.

Lauer DK, Glover GR. Pine response to herbicidal control of flatwoods vegetation. Proc South Weed Sci Soc 49:86, 1996.

Laxson JD, Schacht WH, Owens MK. Above-ground biomass yield at different densities of honey mesquite. J Range Mgt 50:550–554, 1997.

Leopold BD, Hurst GA, Watkins RM. White-tailed deer forage following chemical or mechanical site preparation for pine regeneration. Proc South Weed Sci Soc. 47: 114, 1994.

Lhoste J, Roth P. Sur l'action des solutions agueuses de 2,4-dichlorophenoxyacetate de sodium sur l'evolution des oeues de *Rana temporaria* L. CR Soc Biol 140:272–273, 1946.

Lobaugh S, Farrow F, Feng X, Orgam A. Inhibition of 2,4-D degrading bacteria and 2,4-D mineralization by triclopyr in Palouse soil. 93rd Meeting of American Society of Microbiology, Atlanta, 1993.

Lochmiller RL, Boggs JF, McMurry ST, Leslie DM, Engle DM. Response of cottontail rabbit populations to herbicide and fire applications on cross timbers rangeland. J Range Mgt 44:150–155, 1991.

Lochmiller RL, Pietz DG, McMurray ST, Leslie DM, Engle DM. Alteration in condition of cottontail rabbits (*Sylvilagus floridanus*) on rangelands following brush management. J Range Mgt 48:232–239, 1995.

Lutz-Ostertag Y, Lutz H. Deleterious effect of the herbicide 2,4-D on the embryonic development and fecundity of winged game. CR Acid Sci (Paris) 271 (Series):2418–2421, 1970.

Mackenthun KM, Kemp LE. Water pollution freshwater macroinvertebrates. J Water Pollu Contr Fed 44:1137–1150, 1972.

Madel N. Herbicides and conservation in the Federal Republic of Germany. Proc British Weed Contr Conf 10:1078–1088, 1970.

Marshall CD, Rutschky CW III. Single herbicide treatment: Effect on the diversity of aquatic insects in Stone Valley Lake, Huntington County, PA. Proc Pa Acad Sci 48:127–131, 1974.

Martin NS. Sagebrush control related to habitat and sage grouse occurrence. J Wildl Mgt 34:313–320, 1970.

Martin RP. Effects of the herbicide 2,4,5-T on breeding bird populations. Proceedings of the Oklahoma Academy of Science, 1965, pp. 235–237.

Martin SC, Morton HL. Response of false mesquite, nature grasses and forbs and Lehmann lovegrass after spraying with picloram. J Range Mgt 33:104–106, 1980.

Martin SC, Morton HL. Mesquite control increases grass density and reduces soil loss in southern Arizona. J Range Mgt 46:170–175, 1993.

Masters RA, Scifres CJ. Forage quality responses of selected grasses to tebuthiuron. J Range Mgt 37:83–87, 1984.

Maxwell RC, Harwood RF. Increased reproduction of aphids by the herbicide 2,4-dichlorophenoxyacetic acid. Paper no. 225. Bull Entomol Soc Am 4:100, 1958.

McBride J, Heady HF. Invasion of grassland by *Baccharis pilularis*. J Range Mgt 21: 106–108, 1968.

McBride JR, Dye HM, Donaldson EM. Stress response of juvenile sockeye salmon (*Oncorhynchus nerka*) to the butoxyethanol ester of 2,4-dichlorophenoxy acetic acid. Bull Environ Contam Toxicol 27:877–884, 1981.

McCarl BA, Griffin RC, Kaiser RA, Freeman LS, Blackburn WH, Jordan WR. Brushland Management for Water Yield: Prospects for Texas. B-1569. College Station, TX: Committee for Water Policy, Texas Agric. Exp. Stn., 1987.

McCollum FT, Engle DM, Stritzke JF. Brush Management on the Cross Timbers Experimental Range III: Carrying Capacity and Steer Performance. Animal sci. res. rept. 113. OK: Oklahoma Agric. Exp. Stn., Stillwater, OK. 1987, pp. 110–112.

McConnell BR, Smith JG. Response of understory vegetation to ponderosa pine thinning in eastern Washington. J Range Mgt 22:208–212, 1969.

McCurdy EW, Molberg ES. Effects of the continuous use of 2,4-D and MCPA on spring wheat production and weed populations. Can J Plant Sci 54:241–245, 1974.

McDaniel KC, Anderson DL, Balliette JF. Wyoming big sagebrush control with metsulfuron and 2,4-D in northern New Mexico. J Range Mgt 44:623–627, 1991.

McDaniel KC, Brock JH, Haas RH. Changes in vegetation and grazing capacity following honey mesquite control. J Range Mgt 35:551–557, 1982.

McMahon CK, Miller JH, Thomas DF. Ecosystem management and our natural forests—Is there a role for forest herbicides? Proc South Weed Sci Soc 47:131–134, 1994.

Meyer RE, Bovey RW. Establishment of honey mesquite and huisache on a native pasture. J Range Mgt 35:548–550, 1982.

Meyer RE, Bovey RW. Response of honey mesquite (*Prosopis glandulosa*) and understory vegetation to herbicides. Weed Sci 33:537–543, 1985.

Meyer RE, Riley TE, Morton HL, Merkle MG. Control of Whitebrush and Associated Species with Herbicides in Texas. MP-930. Texas. Agric. Exp. Stn., College Station, TX. 1969.

Meyer RE, Smalley HE, Cooper JF, Farr FM. Herbicide influence on foliar amino acid content in five representative southwestern range plant species. ARS-S-13. USDA ARS, College Station, TX. 1982.

Miller JH. Exotic plants in southern forests: Their nature and control. Proc South Weed Sci Soc 48:120–126, 1995.

Miller JH, Zutter BR, Zedaker SM, Edwards MB, Newbold RA. Early plant succession in loblolly pine plantations as affected by vegetation management. South J Appl For 19:109–126, 1995.

Miller JH, Zutter BR, Zedaker SM, Edwards MD, Haywood JD, Newbold RA. A regional study on the influence of woody and herbaceous competition on early loblolly pine growth. South J Appl For 15:169–179, 1991.

Miller KV. Use of herbicides in wildlife management: Forestry and waterfowl. Proc South Weed Sci Soc 50:132–133, 1997.

Miller KV, Chapman BR. Responses of vegetation, birds and small mammals to chemical and mechanical site preparation: Popular Summaries from 2nd International Conference Forest and Veg. Management. FRI Bulletin no. 192. Rotorua, New Zealand, 1995, pp. 146–148.

Miller KV, Chapman BR, Moore WF. Wildlife habitat conditions following chemical site preparation in the Georgia sandhills: A 6-year study. Proc South Weed Sci Soc 50: 114–115, 1997.

Miller RF, Findley RR, Alderfer-Findley J. Changes in mountain big sagebrush habitat types following spray release. J Range Mgt 33:278–281, 1980.

Minogue PJ, Burns AJ. Herbaceous and woody plant responses following forest site preparation with imazapyr tank mixes. Proc South Weed Sci Soc 47:114–115, 1994.

Minogue PJ, Robins W, Quicke HE, Reynaud LE, Harrison JL. Forest vegetation development following herbicide site preparation with glyphosate, imazapyr, and triclopyr tank mixtures in oil emulsion carrier. Proc South Weed Sci Soc 50:113, 1997.

Moffett JO, Morton HL, Macdonald RH. Toxicity of some herbicidal sprays to honeybees. J Econ Entomol 65:32–36, 1972.

Monsen SB, Shaw N. Response of an alkali sagebrush/fescue site to restoration treatments. In: McArthers ED, Welch BL, compilers Proc., Symposium on Biology of *Artemisia* and *Chrysothamnus*, 1984 July 9–13, Provo, UT. Gen. tech. rept. INT-200. Ogden, UT: USDA, Forest Serv., Intermountain, Res. Stn., 1986, pp. 125–133.

Morton HL, Melgoza A. Vegetation changes following brush control in creosotebrush communities. J Range Mgt 44:133–139, 1991.

Morton HL, Moffett JO. Ovicidal and larvicidal effects of certain herbicides on honeybees. Environ Entomol 5:611–614, 1972.

Morton HL, Moffett JO, Macdonald RH. Toxicity of some newly emerged honeybees. Environ Entomol 1:102–104, 1972.

Morton HL, Moffett JO, Martin RD. Influence of water treated artificially with herbicides on honeybee colonies. Environ Entomol 3:808–812, 1974.

Morton HL, Ibarra-F FA, Martin-R MH, and Cox JR. Creosotebrush control and forage production in the Chihuahuan and Sonoran deserts. J Range Mgt 43:43–48, 1990.

Moyer JR, Smoliak S. Shrubby cinquefoil control changes range forage production. Can J Plant Sci 67:727–734, 1987.

Müller G. Laboratory studies on the effect of herbicides on *Carabidae*. Arch Pflanzenschutz 7:351–364, 1971.

Mullison WR. Effects of herbicides on water and its inhabitants. Weed Sci 18:738–750, 1970.

Murray RB. Responses of three shrub communities in southeastern Idaho to spring-applied tebuthiuron. J Range Mgt 4:16–22, 1988.

Nelson JT, Vick C. Chemical control of mesquite, creosotebrush, and catclaw mimosa with tebuthiuron and subsequent grass production, Tex J Agr Nat Resources 2:30–31, 1988.

Nelson LR, Zutter BR, Gjerstad DH. Planted longleaf pine seedlings respond to herbaceous weed control using herbicides. South J Appl For 9:236–240, 1985.

Newman AS. The effects of certain plant growth-regulators on soil microorganisms and microbial processes. Soil Sci Soc Am 12:217–221, 1947.

Newton M. (1996). Phenoxy herbicides in rights-of-way and forestry in the United States. In: Burnside OC, ed. Biologic and Economic Assessment of Benefits from use of

Phenoxy Herbicides in the United States. USDA-NAPIAP rept. no. 1-PA-96. Washington, DC. 1996, pp. 165–178.

Newton M, Norris LA. Herbicide residues in blacktail deer from forests treated with 2,4,5-T and atrazine. Proceedings Western Society Weed Science, 1968, pp. 32–34.

Nishiuchi Y, Yosida K. Effects of pesticides on tadpoles. Part 3: Nayaku Kensasha Hokoku. Bull Agr Chem Inspect Stn 14:66–68, 1974.

Nolte KR, Fullbright TE. Plant, small mammal and avian diversity following control of honey mesquite. J Range Mgt 50:205–212, 1997.

Okay OS, Gaines A. Toxicity of 2,4-D acid to phytoplankton. Wat Res 30:688–696, 1996.

Olson R, Hansen J, Whitson T, Johnson K. Tebuthiuron to enhance rangeland diversity. Rangelands 16:197–201, 1994.

Owens MK, Schliesing TG. Invasive potential of ashe juniper after mechanical disturbance. J Range Mgt 48:503–507, 1995.

Palmer-Jones T. Effect on honeybees of 2,4-D. New Zealand. J Agr Res 7:339–342, 1964.

Passera CB, Borsetto O, Candia RJ, Stasi CR. Shrub control and seeding influence on grazing capacity in Argentina. J Range Mgt 45:480–482, 1992.

Patton DR. Managing aspen for wildlife in the southwest. General tech. rept. RM-37. USDA For. Serv., Fort Collins, CO. 1977.

Peat HC, Bowes GG. Management of fringed sagebrush (*Artemisia frigida*) in Saskatchewan. Weed Tech 8:553–558, 1994.

Pehl CE, Shelnutt HF. Hexazinone influences on *Pinus taeda* seedlings. For Ecol Mgt 35: 271–276, 1990.

Penner D, Ashton FM. Biochemical and metabolic changes in plants induced by chlorophenoxy herbicides. In: Gunthen FA, ed. Residue Reviews, vol 14. Berlin: Springer-Verlag, 1966, pp. 39–113.

Perry CA, McKell CM, Goodin JE, Little TM. Chemical control of an old stand of chaparral to increase range productivity. J Range Mgt 20:166–169, 1967.

Pond FW. Basal cover and production of weeping lovegrass under varying amounts of shrub live oak crown cover. J Range Mgt 14:335–337, 1961.

Pond FW. Response of grasses, forbs and half shrubs to chemical control of chaparral in central Arizona. J Range Mgt 17:200–203, 1964.

Poorman AE. Effects of pesticides on *Euglena gracilis*. I. Growth studies. Bull Environ Contam Toxicol 10:25–28, 1973.

Prasad R, Feng JC. Spotgun-applied hexazinone: Release of red pine (*Pinus resinosa*) from quaking aspen (*Populus tremuloides*) competition and residue persistence in soil. Weed Tech 4:371–375, 1990.

Prescott LM, Olson DL. The effect of pesticides on the soil ameba *Acanthamoeba castellanii* (Neff.) Proceed S Dak Acad Sci 51:136–141, 1972.

Price DL, Heitschmidt RK, Dowhower SA, Frasure JR. Rangeland vegetation response following control of brownspine pricklypear (*Opuntia phaecantha*) with herbicides. Weed Sci 33:640–643, 1985.

Pyke DA, Zamora BA. Relationship between overstory structure and understory production in the grand fir/myrthe boxwood habitat type of northern Idaho. J Range Mgt 35:767–773, 1982.

Quicke HE, Lauer DK, Glover GR. Hardwood control and loblolly pine response seven years after chemical release treatments. Proc South Weed Sci Soc 47:126–127, 1994.

Quimby DC. A review of literature relating to the effects and possible effects of sagebrush control on certain game species in Montana. Proc West Assoc State Game Fish Comm 46:142–149, 1966.

Quinton DA, Horejsi RG, Flinders JT. Influence of brush control on white-tailed deer diets in north-central Texas. J Range Mgt 32:93–97, 1979.

Quinton DA, Montel AK, Flinders JT. Brush control and Rio Grande turkeys in north-central Texas. J Range Mgt 33:95–99, 1980.

Reed DP, Noble RE. Effects of timber management activities on understory plant succession in loblolly pine plantations. Proc South Weed Sci Soc 47:109–113, 1994.

Reynolds PE, MacKay TS, McCormack ML Jr. One year results for a hexazinone conifer release trial. Proc Northeast Weed Sci Soc 40:218–222, 1986.

Richardson CW, Burnett E, Bovey RW. Hydrologic effects of brush control on Texas rangelands. Trans ASAE 22:315–319, 1979.

Robertson JH. Responses of range grasses to different intensities of competition with sagebrush (*Artemisia tridentata* Nutt.). Ecology 28:1–16, 1947.

Robertson JH. Yield of crested wheatgrass following release from competition by 2,4-D. J Range Mgt 22:287–288, 1969.

Robertson JH. Competition between big sagebrush and crested wheatgrass. J Range Mgt 25:156–157, 1972.

Robinson AG. Effects of amitrole, zytron and other herbicides on plant growth regulators on the pea aphid, *Acyrthosiphon pisum* (Harris) caged on broad bean, *Vicia faba* L. Can J Plant Sci 41:413–417, 1961.

Rodriquez-Lopez M. Actum of 2,4-D acid and its sodium salt in the frog spermatozoan. Bol R Soc Esp Hist Natl (B) 59:219–226, 1961.

Roslycky EB. Glyphosate and the response of the soil microbiota. Soil Bio Biochem 14: 87–92, 1982a.

Roslycky EB. Influence of selected herbicides in phages of some soil bacteria. Can J Soil Sci 62:217–220, 1982b.

Rowe PB. Streamflow increases after removing woodland-riparian vegetation from a southern California watershed. J For 61:365–370, 1963.

Sanders HO. Pesticide toxicities to tadpoles of the Western chorus frog, *Pseudacris triseriata* and Fowler's toad, *Bufo woodhousii fowleri*. Coperia 2:246–251, 1970.

Schroeder MH, Sturges DL. The effect on the Brewer's sparrow of spraying big sagebrush. J Range Mgt 28:294–297, 1975.

Schulz CA, Leslie DM Jr, Lochmiller RL, Engle DM. Herbicide effects on cross timbers breeding birds. J Range Mgt 45:407–411, 1992.

Scifres CJ. Systems for Improving Macartney Rose Infested Coastal Prairie Rangeland. MP-1225. Texas Agric. Exp. Stn., College Station, TX. 1975.

Scifres CJ, Haas RH. Vegetation Changes in a Post Oak Savannah Following Woody Plant Control. MP-1136. Texas Agric. Exp. Stn., College Station, TX. 1974.

Scifres CJ, Polk DB Jr. Vegetation response following spraying a light infestation of honey mesquite. J Range Mgt 27:462–465, 1974.

Scifres CJ, Mutz JL. Herbaceous vegetation changes following applications of tebuthiuron for brush control. J Range Mgt 31:375–378, 1978.

Scifres CJ, Brock JH, Hahn RR. Influence of secondary succession on honey mesquite invasion in north Texas. J Range Mgt 24:206–210, 1971.

Scifres CJ, Durham GP, Mutz JL. Range forage production and consumption following aerial spraying of mixed brush. Weed Sci 25:48–54, 1977.

Scifres CJ, Kothmann MM, Mathis GW. Range site and grazing system influence regrowth after spraying honey mesquite. J Range Mgt 27:97–100, 1974.

Scifres CJ, Scifres JR, Kothmann MM. Differential grazing use of herbicide-treated areas by cattle. J Range Mgt 36:65–69, 1983c.

Scifres CJ, Stuth JW, and Koerth BH. Improvement of Oak-Dominated Rangeland with Tebuthiuron and Prescribed Burning. B-1567. Texas Agric. Exp. Stn., College Station, TX. 1987.

Scifres CJ, Mutz JL, Rasmussen GA, Smith RP. Integrated Brush Management Systems (IBMS): Concepts and Potential Technologies for Running Mesquite and Whitebrush. B-1450. Texas Agric. Exp. Stn., College Station, TX. 1983b.

Scifres CJ, Mutz JL, Whitson RE, Drawe DL. Interrelationships of huisache canopy cover with range forage on the Coastal Prairie. J Range Mgt 35:558–562, 1982.

Scifres CJ, Mutz JL, Whitson RE, Drawe DL. Mixed-brush canopy cover-rainfall interrelationships with native grass production. Weed Sci 31:1–4, 1983a.

Sears HS, Meehan WR. Short-term effects of 2,4-D on aquatic organisms in the Nakwasina River Watershed, Southeastern Alaska. Pestic Mon J 5:213–217, 1971.

Sears WE, Britton CM, Wester DB, Pettit RD. Herbicide conversion of a sand shinnery oak (*Quercus havardii*) community: Effects on biomass. J Range Mgt 39:399–403, 1986a.

Sears WE, Britton CM, Wester DB, Pettit RD. Herbicide conversion of a sand shinnery oak (*Quercus havardii*) community: Effects on nitrogen. J Range Mgt 39:403–407, 1986b.

Seutin E, Baurant R, Salembier JF, Henriet J, Detroux L. Influence of some herbicides on the hatching of domestic hen (*Gallus gallus*) and pheasant (*Phasianus colchicus* eggs). Bull Rech Agron Gembloux 16:137–162, 1981.

Smith GE, Isom BG. Investigation of effects of large scale applications of 2,4-D on aquatic fauna and water quality. Pestic Mon J 1:16–21, 1967.

Somers J, Moran ET, Reinhart BS. Effect of external application of pesticides to the fertile egg and hatching success and early chick performance 2. Commercial-herbicide mixtures of 2,4-D with picloram or 2,4,5-T using the pheasant. Bull Environ Contam Toxicol 11:339–342, 1974.

Sotherton NW. The effects of herbicides on the chrysomelid beetle *Gastrophysa polygoni* (L.) In the laboratory and field. Zeitshrift Fur Angewandte Entolologie 94:446–451, 1982.

Sprague E, McCormack M Jr. Fosamine hexazinone, and oxyfluorfen effects on forest vegetation. Proc Northeast Weed Sci Soc 36:235–236, 1982.

Standley WG, Smith NS. Effects of treating creosotebush with tebuthiuron on rodents, Symposium Management of Amphibians, Reptiles and Small Mammals in North America, Flagstaff, AZ, 1988, pp. 422–424.

Stark HE, McBride JK, Orr GF. Soil Incorporation/Biodegradation of Herbicide Orange. Volume I. Microbial and Baseline Ecological Study of the U.S. Air Force Logistic Command Test Range. Doc. no. DPG-FR-C615F. Dugway, UT: U.S. Army Dugway Proving Ground, 1975.

Stewart KM, Bonner JP, Palmer GR, Patten SF, Fulbright TE. Shrub species richness beneath honey mesquite on root-plowed rangeland. J Range Mgt 50:213–216, 1997.
Stewart RE. Effects of Herbicide on Specific Species. Res. paper. PNW-176. Portland, OR: USDA-Forest Serv., Pacific Northwest For. and Range Exp. Stn., 1974.
Stewart WL. Succession following 2,4-D application on the *Artemisia tridentata/Festuca idahoensis* habitat type in southwestern Montana. Disser Abstr Internal B 42(5): 1706, 1981.
Sturges DL. High-Elevation Watershed Response to Sagebrush Control in South Central Wyoming. Res. paper RM-318. Fort Collins, CO: USDA-For. Sur., Rocky Mountain Forest and Range Exp. Stn., 1994.
Sullivan TP, Sullivan DS. Responses of small-mammal populations to a forest herbicide application in a 20 year conifer plantation. J Appl Ecol 19:95–106, 1981.
Sullivan TP, Lautenschlages RA, Wagner RG. Changes in diversity of plant and small mammal communities after herbicide application in sub-boreal spruce forest. Popular Summaries from 2nd International Conference on For. Veg. Management, FRI bull. no. 192. Rotorua, New Zealand, 1995, pp. 143–145.
Szaro RC. Bird population responses to converting chaparral to grassland and riparian habitats. Southwest Natural 26:251–256, 1981.
Teeter L, Bliss JC, Henry WA. Adoption of herbaceous weed control by southern forest industry. Can J For Res 23:2312–2316, 1993.
Terrel TL, Spillet JJ. Pinyon-juniper: Its impact on mule deer and other wildlife, Pinyon-Juniper Ecosystem: A Symposium, Logan, UT, 1975, pp. 105–119.
Texas State Soil & Water Conservation Board. A Comprehensive Study of Texas Watersheds and Their Impact on Water Quality and Water Quantity. Temple, TX: Texas State Soil & Water Conservation Board, 1991.
Thilenius JF, Brown GR. Long-term effects of chemical control of big sagebrush. J Range Mgt 27:223–224, 1974.
Thilenius JF, Brown GR. Effect of 2,4-D on digestibility and production of subalpine herbage. J Range Mgt 29:63–65, 1976.
Tietjen HP, Halvorson CH, Hegdal PL, Johnson AM. 2,4-D herbicide, vegetation and pocket gopher relationships, Black Mesa, Colorado. Ecology 48:634–643, 1967.
Trichet P, Boisaubert B, Frochot H, Picard JF. Impact of herbicide treatments against bramble (*Rubus fruticosus* L. agg) on roe deer (*Capreolus capreolus* L.). Gibier Faune Sauvage 4:165–188, 1987.
Tueller PT, Evans RA. Control of green rabbitbrush and big sagebrush with 2,4-D and picloram. Weed Sci 17:233–235, 1969.
Van Auken OW, Bush JK. Importance of grass density and time of planting on *Prosopis glandulosa* seedling growth. Southwest Natural 35:411–415, 1990.
Venkataramay GS, Rajyalakshmi B. Tolerance of blue-green algae to pesticides. Current Sci 40:143–144, 1971.
Walker JW. The north Concho river brush control project. Texas Plant Protect Assoc Conf 10:17, 1998.
Walsh GE. Effects of herbicides on photosyntheses and growth of massive unicellular algae. Hyacinth Contr J 10:45–48, 1992.
Wambolt CL, Payne GF. An 18-year comparison of control methods for Wyoming big sagebrush in southwestern Montana. J Range Mgt 39:314–419, 1986.

Ward LA. Sagebrush control with herbicide had little effect on elk calving behavior. Forest Serv. Res. note RM-240. USDA, 1973.

Wardle DA, Nicholson KS, Rahman A. Influence of herbicide applications on the decomposition, microbial biomass and microbial activity of pasture shoot and root-litter. New Zeal J Agr Res 37:29–39, 1994.

Warren A, Holechek J, Cardenas M. Honey mesquite influence on Chihuahuan desert vegetation. J Range Mgt 49:46–52, 1996.

Watson LE, Zedaker SM. Twelfth-year response of loblolly pine to eight levels of release. Proc South Weed Sci Soc 49:85, 1995.

Watson LE, Zedaker SM. Twelfth-year response of loblolly pine to eight levels of release. Proc South Weed Sci Soc 49:85, 1996.

Watts MJ, Wambolt CL. Long-term recovery of Wyoming big sagebrush after four treatments. J Environ Mgt 46:95–102, 1996.

Way JM. Toxicity and hazards to man, domestic animals and wildlife from some commonly used auxin herbicides. In: Gunther FA, ed. Residue Reviews. New York: Springer-Verlag, 1969, pp. 37–62.

Webber EC, Seesock WC, Bayne DR, Michael JL. Impacts of hexazinone (Velpar) on aquatic communities in small streams. Proc South Weed Sci Soc 45:227, 1992.

Webster JM. Some effects of 2,4-dichorophenoxyacetic acid herbicide on nematode-infested cereals. Plant Pathol 16:23–26, 1967.

Wendel GW, Kochendufer JN. Release of 7-year-old underplanted white pine using hexazinone applied with a spot gun. Res. paper NE-614. Broomall, PA: USDA-Forest Serv., Northeastern Forest Exp. Stn., 1988.

Whisenant SG. Selective control of mountain big sagebrush (*Artemisia tridentata* ssp. *vaseyama*) with clopyralid. Weed Sci 35:120–123, 1987.

White DE, Witherspoon-Joos L, Newton M. Herbaceous weed control in conifer plantations with hexazinone and nitrogen formulations. New For 4:97–105, 1990.

Whitehead CC, Pettigrew RJ. The subacute toxicity of 2,4-dichlorophenoxy acetic acid and 2,4,5-T richlorophenoxy acetic acid to chicks. Toxicol Appl Pharmacol 21: 348–354, 1972.

Whitney EW, Montogomery AB, Martin EC, Gangstad EO. The effects of a 2,4-D application on the biota and water quality in Currituck Sound, North Carolina. Hyacinth Contr J 11:13–17, 1973.

Whitson TD, Alley HP. Tebuthiuron effects on *Artemisia* spp. and associated grasses. Weed Sci 32:180–184, 1984.

Whitson TD, Ferrell MA, Alley HP. Changes in rangeland canopy cover seven years after tebuthiuron application. Weed Techn 2:486–489, 1988.

Wilbert DE. Some effects of chemical sagebrush control on elk distribution. J Range Mgt 16:74–78, 1963.

Williams RA, Yeiser JL, Earl JA. Loblolly pine seedling performance following herbaceous weed control with hexazinone or imazapyr mixed with sulfometuron on fine-textured soils in S.E. Arkansas. Proc South Weed Sci Soc 49:122–126, 1996.

Wilson LM, Leopold BD, Hurst GA, Watkins B. Plant species richness following mechanical or chemical site preparation in eastern Mississippi. Proc South Weed Sci Soc 46:177, 1993.

Witt JS, Johnson AS, Miller PM. Responses of wildlife food plants to herbicide site preparation in the Georgia piedmont. Proc South Weed Sci Soc 48:201, 1990.

Wojtalik TA, Hall TF, Hill LO. Monitoring ecological conditions associated with wide-scale applications of DMA and 2,4-D to aquatic environments. Pestic Mon J 4: 184–203, 1971.

Wolters GL. Longleaf and slash pine decreases herbage production and alters herbage composition. J Range Mgt 35:761–763, 1982.

Wolters GL, Martin A, Pearson HA. Forage response to overstory reduction on loblolly-shortleaf pine-hardwood forest range. J Range Mgt 34:443–446, 1982.

Wood KL, Garcia EL, Tromble JM. Runoff and erosion following mechanical and chemical control of creosotebrush (*Larrea tridentata*). Weed Tech 5:48–53, 1991.

Woods RF, Betters DR, Mogren EW. Understory herbage production as a function of rocky mountain aspen stand density. J Range Mgt 35:380–381, 1982.

Yeiser JL, Cobb SW. Herbicides for Releasing Pine Seedlings in the Arkansas Ozarks. Arkansas Agric. Exp. Stn., AR: Fayetteville, 1988.

Yeiser JL, Williams RA. Performance of loblolly pine seedlings with and without herbaceous weed control—5th year results. Proc South Weed Sci Soc 47:115, 1994.

Yeiser JL, Sundell E, Boyd JW. Preplant and postplant treatments for newly planted pine. bull. 902. Fayetteville, AR: Arkansas Agric. Exp. Stn., 1987.

Young JA, Evans RA, Echert RE Jr. Environmental quality and the use of herbicides on (*Artemisia*)/grasslands of the U.S. Intermountain Area. Agr Environ 6:53–61, 1981.

Young JA, Evans RA, Rimbey C. Weed control and revegetation following western juniper (*Juniperus accidentalis*) control. Weed Sci 33:513–517, 1985.

Zutter BR, Zedaker SM. Short-term effects of hexazinone application on woody species diversity in young loblolly pine (*Pinus taeda*) plantations. For Ecol Mgt 24:183–189, 1988.

Zutter BR, Gjerstad DH, Webb AL, Glover GR. Response of a young loblolly pine (*Pinus taeda*) plantation to herbicide release treatments using hexazinone. For Ecol Mgt 25:91–104, 1988.

12

The Response of Woody Plants to Herbicides

I. INTRODUCTION

When woody plants become so persistent and aggressive that they cause economic losses in forage or tree crops, interfere with off-site water yield, safety, and rights-of-way, or cause problems around structures, buildings and industrial sites, wildlife habitats, and recreation areas, then some management technique is necessary.

Selected herbicides can be used to control woody plants by the methods described in Chapter 5. Herbicides can also be applied to manage weeds and brush to fit ecological requirements, as described in Chapter 11. Woody plants may be difficult to manage, but many are susceptible to small amounts of designated herbicides, making them a modern scientific wonder and a great success story (Chap. 3). The fate and mode of action of herbicides in plants (Chap. 9) and their persistence and degradation (Chap. 6) in plants, soil, and water sources indicate that they are not a problem on treated areas or the environment when applied according to label instructions.

The information in this chapter will indicate the specific herbicides required to control specific woody plant problems and the methods used to evaluate their effectiveness.

II. WOODY PLANT CONTROL EVALUATION

Each woody plants species reacts differently to a specific herbicide. There are different methods to evaluate the effectiveness of herbicides on woody plants.

The method most commonly used in research is to make a visual rating by estimating the percentage of defoliation or desiccation of the leaves of individual plants by species within the first growing season to get some idea of herbicide efficacy. After 1 year or longer, visual canopy (leaves) reduction estimates are made on each tree. Ratings should also be made on larger trees at least 2 and 3 years after initial treatment for both the percentage of canopy reduction and/or the percentage of mortality. If the shrubs or trees have no live sprouts or tissue after 1 or 2 years (or longer), they are considered dead.

Where possible, Tsichirley (1967), indicated that in the tropics a reliable estimate of whole-plot defoliation was obtained with a sample size of 50 plants per treatment plot. Fifty plants provided an estimate within 7% of whole-plant defoliation. A range of 7% is within the limit of accuracy of ocular defoliation estimates for individual trees; therefore, a sample size of 50 trees appeared adequate.

III. WHAT TO EXPECT

Controlling the various types of woody plants with herbicides can sometimes be a difficult and challenging task. The herbicide applied in small amounts must kill or suppress woody plants ranging from small shrubs to large trees. The response of woody plants to herbicides is usually relatively rapid with contact herbicides such a paraquat and glyphosate and some of the growth-regulator-type materials. Fast-acting herbicides, however, may rapidly brown out or defoliate the plants without sufficient transport to kill woody tissue. Paraquat is a good example of a contact herbicide with limited transport into the cambium layers of the stem or roots. Within a few weeks, regrowth on the stems or base of the plant is evident and the plant usually recovers. The same process can occur with hormone-type herbicides, whereby rapid brownout occurs. This may be because the rates are too high, or because of phytotoxic diluents or herbicide-resistant species. Depending upon the species, the regrowth time after treatment may vary from 1 month to several years and is influenced by plant size, previous treatment, environment, and time of treatment. For the best results, some woody plants will require retreatment with either the same herbicide or other herbicides or control by other means 1 or more years after the original application. Some woody plants may be so vigorous or resistant to treatment that despite repeated treatment, only temporary canopy reduction can be obtained with little plant mortality.

Even though excellent woody plant control may be obtained, a small percentage of the plants may slowly (1 or more years after treatment) regrow from stems or rootstocks, and follow-up treatment is recommended to keep them from spreading and dominating the landscape, regardless of the methods(s) used.

Controlling woody plants before they become a big problem is desired, and knowing the species' name and knowledge of its life history will help determine the action to take.

Although some soil- or foliar-applied herbicides are slow to act on certain woody species, the land manager must be patient and not disturb the treatment, or regrowth may occur. A good example was the use of 2,4,5-T on honey mesquite. Mechanical disturbance of the trees several years after treatment sometimes caused resprouting from the basal buds on the trunks, and regrowth occurred unless the tree was removed from the soil.

Soil-applied herbicides are notoriously slow-acting, and patience is required not to abandon the treatment as a failure and retreat the brush too soon. Bromacil, hexazinone, picloram, and tebuthiuron are examples of effective soil-applied herbicides that kill or suppress brush through exposure to the roots. Rainfall is needed to leach the material in the root zone of the woody plant to be effective. They are usually applied to or near the base of the plant in liquid or pellet form. Imazapyr, hexazinone, picloram, dicamba, clopyralid, metsulfuron, and triclopyr have both soil and foliar activity sometimes increasing their effectiveness. Materials such as bromacil, hexazinone, and tebuthiuron have similar modes of action, and woody plants may partially refoliate several times for a year (or more) before the herbicide actually kills the plant.

IV. HOW TO PROCEED

A. Know the Weed Problem

If you are certain that you have a woody plant problem and can identify all the woody species involved then you are on your way to solving the problem. If you are uncertain of the species to control you may need help from local authorities, such as the county agent, a weed extension specialist, or scientists from industry, a university, or the U.S. Department of Agriculture. Chances are your state university has a published document on suggestions for weed and brush control. In this document woody species will be listed under pastures, rangeland, forestry, or noncrop areas, and the labeled herbicides will be listed for each problem species. In many cases there may be one or more alternatives that fit your budget and situation. Control recommendations for herbaceous weeds are usually also given for each crop and/or grazing land situation. If a published source is not available from your local area or neighboring states a great deal of information can be found on the herbicide label.

B. Know the Herbicide Products

Information on woody plant control can also be obtained from the local agricultural chemical dealer and representatives from chemical companies. As was dis-

TABLE 1 Selected List of Herbicides, Formulations, Cost, and Manufacturers

Trade name	Common name	Formulation	Approximate cost ($/gal/unit)	Manufacturer
Ally 60DF	Metsulfuron-methyl	60%	23/oz	DuPont
Arsenal 2S	Imazapyr	2 lb/gal	209	American Cyanamid
Amitrol T	Amitrole	2 lb/A		Rhône-Poulenc
Banvel	Dicamba	4 lb/gal	80	BASF
Banvel 720	Dicamba + 2,4-D amine	1 + 1.9 lb/gal		BASF
Brushkiller 875	2,4-D + 2,4-DP + dicamba	2 + 1.9 + 0.34 lb/gal	43	PBI Gordon
Brushkiller BK 800	2,4-D + 2,4-DP + dicamba	2 + 2 + 0.5 lb/gal	57	PBI Gordon
Brushmaster	2,4-D + 2,4-DP + dicamba	1.1 + 1.1 + 0.25 lb/gal	42	PBI Gordon
Chopper RTU	Imazapyr	3.6%		American Cyanamid
Crossbow 3E	2,4-D + triclopyr	2 + 1 lb/gal	49	Dow AgroSciences
Contain 1S	Imazapyr	1 lb/gal	125	American Cyanamid
Diquat	Diquat	2 lb/gal	40	Zeneca
Escort 60DF	Metsulfuron-methyl	60%	32/oz	DuPont
Garlon 3A	Triclopyr	3 lb/gal	81	Dow AgroSciences
Garlon 4 (Remedy)	Triclopyr	4 lb/gal	82	Dow AgroSciences
Gramoxone Extra 2.5L	Paraquat	2.5 lb/gal	35	Zeneca
Grazon P + D	2,4-D + picloram	2 + 0.54 lb/gal	24	Dow AgroSciences
Hi-Dep	2,4-D	4 lb/gal	22	PBI Gordon
Hyvar X-L	Bromacil	2 lb/gal	61	DuPont
Hyvar X	Bromacil	80%	15	DuPont
Krenite S, Krenite UT	Fosamine	4 lb/gal	52	DuPont

Pathway	Picloram + 2,4-D amine	5.4% + 20.9%	28	Dow AgroSciences
Remedy 4E	Triclopyr	4 lb/gal	82	Dow AgroSciences
Roundup Ultra 3AS	Glyphosate	3 lb/gal	44	Monsanto
Roundup PRO	Glyphosate	3 lb/gal	65	Monsanto
Accord	Glyphosate	3 lb/gal	65	Monsanto
Spike 20P	Tebuthiuron	20%	12/lb	Dow AgroSciences
Spike 40P	Tebuthiuron	40%		Dow AgroSciences
Spike 80W	Tebuthiuron	80%	25/lb	Dow AgroSciences
Super Brushkiller	2,4-D, + 2,4-DP + dicamba	2 + 2 + 0.5 lb/gal	58	PBI Gordon
Stalker 2S	Imazapyr	2 lb/gal	209	American Cyanamid
Stinger 3E, Reclaim	Clopyralid	3 lb/gal	480	Dow AgroSciences
Tordon 22K	Picloram	2 lb/gal	83	Dow AgroSciences
Tordon 101 Mixture	Picloram + 2,4-D amine	0.54 + 2 lb/gal	37	Dow AgroSciences
Tordon RTU	Picloram + 2,4-D amine	5.4% + 20.9%	29	Dow AgroSciences
Touchdown 6AQ	Glyphosate trimesium	6 lb/gal	66	Zeneca
Velpar 2L	Hexazinone	2 lb/gal	77	DuPont
Velpar 75DF	Hexazinone	75%		DuPont
Velpar	Hexazinone	90%		DuPont
2,4-D amine	2,4-D amine	Several	13	Several
2,4-D ester	2,4-D ester	Several	18	Several
Weedmaster	Dicamba + 2,4-D amine	1 + 2.8 lb/gal	25	BASF
Weedone 170	2,4-D ester + 2,4-DP	1.8 + 1.85 lb/gal		Rhône-Poulenc
Weedone CB	2,4-D ester + 2,4-DP	0.66 + 0.66 lb/gal		Rhône-Poulenc
Weedone	Dichloprop (2,4-DP)	3.7 lb/gal	35	Rhône-Poulenc

Source: Modified from Johnson and Kendig, 1997; Baldwin and Boyd, 1998; Regehr et al., 1998.

cussed in the proceeding chapters and especially in Chapter 7, the herbicide label is essential reading. It not only gives the woody species controlled but also the best application methods to use and the proper herbicide rates for the best control. The proper application timing and placement of the herbicide is also indicated for optimum results and safety to the crops and the environment.

Selecting the most inexpensive, selective, effective, safe, and environmentally compatible material is always sought if available for a given weed problem, but it is not always easy. Many factors need to be considered in the selection process. First, various herbicides may have several names and may cause confusion. Herbicides have commercial names, and each company has its own commercial name. If several companies sell the product, there will be several different names. There is usually only one common name, however, and familiarity with the common and commercial name is recommended. Table 1 gives the most commonly used commercial and common names of woody plant herbicides as well as the active ingredient, manufacturer, and cost. Table 1 should be referred to often, both to prevent confusion and for the proper identification of the product(s) of choice.

V. REVIEW OF HERBICIDES USED

The cost of herbicides by the gallon or pound can be readily obtained from your local agriculture chemical dealer. These costs vary slightly, based on the amounts purchased, from year to year and between dealers. Prices are given in Table 1 but may become outdated.

A. 2,4-D

2,4-D is used as a foliar spray in amine or ester formulations. A large number of woody species are susceptible or moderately susceptible to it (Table 2). It is desirable to use since it is inexpensive (Table 1) and has a short residue in the environment. The ester formulation is also useful for stump treatment and basal sprays when mixed with diesel fuel oil. The amine form can be used undiluted for cut-surface applications and is very effective on some plants using this method. 2,4-D is very important in mixtures with other herbicides, such as dicamba (Weedmaster), picloram (Grazon P+D), triclopyr (Crossbow), dichlorprop (Weedone 170), and other herbicides used in brush and weed control.

Dichlorprop (2,4-D) is the propionic form of 2,4-D and is useful on many woody plants. It is also used in three-way mixtures with 2,4-D and dicamba (Brushmaster). Dairy cattle should not graze treated areas until 14 days after treatment (Boyd, 1995). Meat animals should be removed from treated areas 7 days before slaughter, and forage cut for hay should not occur until 30 days after treatment.

B. Dicamba (Banvel)

Dicamba is used as a foliar spray at full leaf in the spring or as a spot treatment. Dicamba is effective on a large number of woody species listed on the herbicide label. Table 2 also lists some woody species not labeled but provided for informational purposes only. Plants' response to dicamba is similar to the phenoxys (2,4-D), so drift control is essential from treated areas to nontarget areas. Dicamba may be applied as a cut-surface treatment for controlling unwanted trees and preventing sprouts of cut stumps. The freshly cut surface of stumps should be thoroughly wet. 2,4-D and other herbicides may be mixed with dicamba. Basal bark or spot soil treatments of dicamba can also be used on some woody plants. Wiper application can be used. Some grazing restrictions apply.

C. Picloram (Tordan 22K)

Picloram is an excellent weed and brush herbicide and is used alone or in mixtures with other compounds. The commercial mixture with 2,4-D (Grazon P+D) is popular on grazing lands (Table 1). Tank mixtures with dicamba, clopyralid, or triclopyr are highly useful in woody plant control. The water mobility and solubility of picloram coupled with its persistence makes it a potential problem on sensitive crops irrigated with contaminated water. It is a restricted-use pesticide.

D. Clopyralid (Reclaim or Stinger)

Clopyralid is discussed next since it is related to picloram. It differs from picloram in that it is effective on only a few woody plants (Leguminosae; some Rubus spp.). Clopyralid is also effective on fewer herbaceous weeds. It is highly effective on honey mesquite. It degrades more rapidly in soil than picloram, but is more resistant to photodegradation. Clopyralid readily injures legume crop plants, so applications should be avoided where alfalfa, cover, or lespedeza or vegetable crops may be damaged from direct application or spray drift. There are no grazing restrictions at label-use rates.

E. Triclopyr (Remedy)

Remedy is the ester form of triclopyr used as a foliar spray in the spring and sometimes in the fall when most woody plants have full, mature leaf cover. It contains 4 lb/gal of acid-equivalent triclopyr. It is readily used as a basal spray with diesel fuel oil or kerosene on freshly cut stumps or as a basal bark application (several methods). Lactating dairy animals are not allowed to graze green forage for 14 days after treatment if application is 2 qt/A or less. If rate exceeds 2 to 6 qt/A do not graze until the next growing season. Other livestock have no restrictions on green forage treated at 2 qt/A or less, but cannot be grazed on treated areas exceeding 2 to 6 qt/A for 14 days after treatment. Lactating cows cannot

TABLE 2 The Response of Woody Species to 19 Commonly Used Herbicides*

Herbicide‡,§	2,4-D†			Triclopyr			Clopyralic		Bromacil			Tri+2,4D(1:2)	Dicamba	
	Y			Y			Y		N			Y	Y	
	N			N			N		Y			N	N	
	N			Y			Y		Y			Y	Y	
Species#	BS	FS	I/CS	BS	FS	I/CS	BS	FS	BS	FS	ST	FS	BS	FS
Acacia (*Acacia* spp.)														
blackbrush (*A. rigidula*)[a]		R		S	S									R
catclaw (*A. greggii*)[b]		R		S					S	S	S			I-R
guajillo (*A. berlandieri*)				S										
huisache (*A. farnesiana*)[c]	R	R	S	S				S	S	S	S			I
twisted (*A. tortuosa*)		R		S	S									R
white-thorn (*A. constricta*)								S						
Agarito (*Berberis trifoliolata*)		R											I-S	R
Albizia (*Albizia lebbeck*)				S										
Alder (*Alnus* spp.)[d]	S-I	S-I			S				S	S	S		S	S
common (*A. serrulata*)														
red (*A. rubra*)[e]	S-I	S-I	S						S		R		S	S
speckled (*A. rugosa*)	S-I	S-I							S		S			
Apple (*Malus* spp.)														
common (*M. sylvestris*)	S-I	S-I							S	S	S		S	S
crab (*M. ioensis*)	S-I	S-I							S-I	S	I		S	S
Apple-of-sodom (*Solanum sodomeum*)														
Ash (*Fraxinus* spp.)	I	R	R	I	I	I				I	I	I		R
blue (*F. quadrangulata*)	I-R	I-R	S						S		S			
green (*F. pennsylvanica*)[f]	R	R	S						S		S			R
Oregon (*F. oregona*)		S												
white (*F. americana*)[g]	R	R	I						S	S-R	S		S-I	S-R
Aspen, quaking (*Populus tremuloides*)[h]	S-I	S-I		S	S				S	S			S	I
bigtooth (*P. grandidentata*)														
Autumn olive (*Elaeagnus umbellata*)														

* S, susceptible; S-I, susceptible to intermediate; I, intermediate; I-R, intermediate to resistant; R, resistant.

† Selective; Nonselective; Soil Effective for top, middle, and bottom row for each herbicide, respectively (Y = yes; N = no).

‡ Application Methods: BS, basal spray; FS, foliar spray; I/CS, injection/cut surface treatment; ST, soil treatment; IPT, individual plant treatment.

§ Always check and double check precautions in this publication and on container label before using any herbicide.

Amitrole T is labeled for poison ivy, poison oak, honeysuckle, kudzu, salmonberry, blackberry (western dewberry), big leaf maple, wild cherry, ash, locust, and sumac.

[a] Spray triclopyr anytime to lower stem. S to hormone-like herbicide mixtures as IPT.

[b] Same as above.

[c] BS effective with diesel fuel oil or triclopyr in oil. Apply anytime. Apply foliar sprays spring or fall. S to triclopyr + clopyralid.

[d] S to glyphosate cut stump treatment.

[e] 2,4-D amine during growing season only. I/CS effective August–October. R to amitrole by FS.

[f] IPT with tebuthiuron.

[g] IPT with tebuthiuron. I/CS effective August–October. Wide range of responses to bromacil, dicamba, and picloram by FS according to location; consult local authorities before use.

[h] Wide range of response to picloram by FS according to location; consult local authorities before such use.

Dicamba		Hexazinone		Imazapyr		Dichlorprop		Pic+2,4D(1:4)	Dic+2,4D(1:3)	Glyphosate	Tebuthiuron			Metsulfuron	Picloram				Fosamine		Pic+Tri(1:1)	Pic+dic(1:1)	Pic+clop(1:1)
Y		N		N		Y		Y	Y	N	N			Y	Y				Y		Y	Y	Y
N		Y		Y		N		N	N	Y	Y			N	N				N		N	N	N
Y		Y		Y		N		Y	Y	N	Y			Y	Y				N		Y	Y	Y
I/CS	ST	BS	ST	BS	ST	BS	FS	FS	FS	FS	BS	FS	ST	FS	BS	FS	I/CS	ST	BS	FS	FS	FS	FS
													S			S		S			S	S	S
	R	S	S				R			I			S			S		S			S	S	S
													S			S							
		I	I			R	R						I			S		S			S	S	S
													S			S					S	S	S
		S	S										S								S		S
	R															R		S					
		S	S	S	S	S	S			S					S	S							
S	S			S	S	S-I	S						I		S	S	S	S		S			
													I			S							
						S-I	S			S					S	S							
						S-I	S			S					S	S							
													S										
		I	I	S	S	R	R	I	R	I			I	I							I		
						R	R			S	S		S										
S						R	R			S-I	S		S		S-I	S-R		I		I-R	I-R		
S		S	S	S	S	S-I	S-I			S	S		I		S	S-R				S			
		S	S										I										
				S	S																		

TABLE 2 (Continued)*

Herbicide[‡,§]	2,4-D[†]			Triclopyr			Clopyralid		Bromacil			Tri+2,4D(1:2)	Dicamba	
	Y			Y			Y		N			Y	Y	
	N			N			N		Y			N	N	
	N			Y			Y		Y			Y	Y	
Species[#]	BS	FS	I/CS	BS	FS	I/CS	BS	FS	BS	FS	ST	FS	BS	FS
Azalea (*Rhododendron* spp.)														
piedmont (*R. canescens*)[a]	S-I	I							S-R	S	S		R	S
western (*R. occidentale*)		I-R							S		S			
Baccharis (*Baccharis* spp.)														
coyote brush (*B. consanguinea*)[b]														
kidneywort (*B. pilularis*)[c]	S	S							S		S			
Roosevelt (*B. neglecta*)		S		S										
sea myrtle (*B. halimifolia*)														
seepwillow (*B. glutinosa*)		S		S										
willow (B. salicina)	S	S		S									I	I-R
Barberry (*Berberis* spp.)	S	I							S	S	S			S
Basswood, American linden (*Tilia americana*)	R	R							S	S	S			S
Bayberry (*Myrica* spp.)														
northern (*M. pensylvanica*)	S	S							S		S			
southern (*M. cerifera*)		I-R							I		I			R
Bearberry (*Arctostaphylos uva-ursi*)[d]		I-R							S		S			
Bearmat (*Chamaebatia foliolosa*)		S-I												S
Beautyberry, American (*Callicarpa americana*)		R									I-R			
Beech, American (*Fagus grandifolia*)[e]	S-I	R	I-R		S				S	S	S			S-R
Birch (*Betula* spp.)	I	I	I	I	I	I				I	I	I		S
gray (*B. populifolia*)														
paper (*B. papyrifera*)		S												
yellow (*B. alleghaniensis*)														
Blackberry (*Rubus* spp.)	R	R	R	S	S	R		I	S	I	I	S		I
Blackgum (*Nyssa sylvatica*)	R	R	S						S	S	S	S		I
Blueberry (*Vaccinium* spp.)[f]	S-I	I-R							S	S	I		S	S-R
Bluewood (Brazil) (*Condalia obovata*)														
Boxelder (*Acer negundo*)[g]	S	S-I							S	S	S		S	
Broom, Scotch (*Cytisus scoparius*)	S-I	I							S		S			S-I
French (*C. monspessulanus*)														
Buckbrush (*Symphoricarpos orbiculatus*)[h]	R	S	R	R	R	R				I	I	I		I-R

* S = Susceptible; S-I = Susceptible to Intermediate; I = Intermediate; I-R = Intermediate to Resistant; R = Resistant.

† Selective; Nonselective; Soil Effective for top, middle, and bottom row for each herbicide, respectively (Y = yes; N = no).

‡ Application Methods: BS = Basal Spray; FS = Foliar Spray; I/CS = Injection/Cut Surface Treatment; ST = Soil Treatment; IPT = Individual Plant Treatment.

§ Always check and double check precautions in this publication and on container label before using any herbicide.

Amitrole T is labeled for poison ivy, poison oak, honeysuckle, kudzu, salmonberry, blackberry (western dewberry), big leaf maple, wild cherry, ash, locust, and sumac.

a Wide range of responses to bromacil by BS according to location; consult local authorities before such use.

b S to glyphosate cut stump treatment.

c IPT with tebuthiuron.

d Regrowth following a fire is easily killed with the herbicides indicated. Old growth is much more difficult to kill.

e Wide range of responses to dicamba by FS according to location; suggest consulting local authorities before such use.

f Wide range of responses to dicamba by FS according to location; suggest consulting local authorities before such use.

g Wide range of responses to dicamba by FS according to location; suggest consulting local authorities before such use.

h Must use 2,4-Dester formulation.

Dicamba		Hexazinone		Imazapyr		Dichlorprop		Pic+2,4D(1:4)	Dic+2,4D(1:3)	Glyphosate	Tebuthiuron			Metsulfuron	Picloram				Fosamine		Pic+Tri(1:1)	Pic+dic(1:1)	Pic+clop(1:1)
Y		N		N		Y		Y	Y	N	N			Y	Y				Y		Y	Y	Y
N		Y		Y		N		N	N	Y	Y			N	N				N		N	N	N
Y		Y		Y		N		Y	Y	N	Y			Y	Y				N		Y	Y	Y
I/CS	ST	BS	ST	BS	ST	BS	FS	FS	FS	FS	BS	FS	ST	FS	BS	FS	I/CS	ST	BS	FS	FS	FS	FS
						S	S			S					R	S							
										S													
							S						S										
		S	S																				
										S													
		S	S																				
		S	S										S		S	S							
							I			S						S							
S							R			S						S					I		
																S							
						I-R			S							S-I							
															S-I	S-I							
									I														
																S		S					
S				S	S		I-R			I			I			S-I	S-I						
		I	I	R	R	I	I	I	I	I			I	R		S				I			
													S										
							S																
							S-I			S						S				S			
			S	R	R	I	I	R	I	I			I	S	S	S		S		S			
S		S	S			I-R	I-R			S			S		S	S				S			
				S	S	S	I-R			S			S		S	S-I				I			
										S													
				S	S	S	I-R			S			I		S	S-I							
				S	S		I-R			S						S-I							
										S													
			I	R	R	I	I	R	S	I			I	I		R				I			

Table 2 (Continued)*

Herbicide[‡,§]	2,4-D[†]			Triclopyr			Clopyralid		Bromacil			Tri+2,4D(1:2)	Dicamba	
	Y			Y			Y		N			Y	Y	
	N			N			N		Y			N	N	
	N			Y			Y		Y			Y	Y	
Species[#]	BS	FS	I/CS	BS	FS	I/CS	BS	FS	BS	FS	ST	FS	BS	FS
Buckeye (*Aesculus* spp.)[a]	S-R	R							S	S				S-I
California (*A. californica*)[b]			S											
Buckthorn (*Rhammus* spp.)														
California (*R. californica* var. *ursina*)[c]	I	I-R			S									
cascara (*R. purshiana*)														
common (*R. cathartica*)	I	I							S		S			
hollyleaf (*R. crocea* var. *ilicifolia*)[d]	S	I							S		S			
Bumelia, gum (*Bumelia lanuginosa*)														
Bunchberry (*Cornus canadensis*)		R												
Burroweed (*Haplopappus tenuisectus*)														
Bushhoneysuckle, dwarf (*Diervilla lonicera*)		S												
Buttonbush, common (*Cephalanthus occidentalis*)	I	I-R							I	I				
Cactus (*Opuntia* spp.)														
cholla (*Opuntia* spp.)[e]														
pricklypear (*Opuntia* spp.)[f]		R							R	R	R			
tasajillo (*O. leptocaulis*)		R							R	R	R			
California-holly (see Christmasberry)														
Camelthorn (*Alhagi camelorum*)[g]		S-I												
Catalpa (*Catalpa* spp.)	I	I-R							S	S	S			S
Catclaw mimosa (*mimosa pigra*)[h]				S				S						
Catsclaw (*Caesalpinia decapetala*)				S										
Ceanothus (*Ceanothus* spp.)[i]		S-I			S									
blueblossom (*C. thyrsiflorus*)	S	S-I							S		S			
chaparral whitethorn (*C. leucodermis*)[j]	S	S-I							S		S			
deerbrush (*C. integerrimus*)[k]	S	S							S					
desert (*C. greggii*)											S			

* S, susceptible; S-I, susceptible to intermediate; I, intermediate; I-R, intermediate to resistant; R, resistant.
† Selective; Nonselective; Soil Effective.
‡ Application Methods: BS, basal spray; FS, foliar spray; I/CS, injection/cut surface treatment; ST, soil treatment; IPT, individual plant treatment.
§ Always check and double check precautions in this publication and on container label before using any herbicide.
[#] Amitrole T is labeled for poison ivy, poison oak, honeysuckle, kudzu, salmonberry, blackberry (western dewberry), big leaf maple, wild cherry, ash, locust, and sumac.
[a] Wide range of responses to 2,4-D by BS according to location; consult local authorities before such use.
[b] Picloram most effective by I/CS.
[c] Spray sprouts after burning. Usually killed by two repeat treatments with 2,4-D.
[d] Spray sprouts after burning.
[e] S to picloram IPT and broadcast spray, S to hexazinone IPT.
[f] Wet plants with IPT. S to picloram + paraquat. S to prescribed burn + picloram.
[g] Repeat treatment is necessary.
[h] S to triclopyr + clopyralid.
[i] Spray when soil moisture is plentiful and plant actively growing.
[j] IPT with tebuthiuron.
[k] S-I to amitrole and I to 2,4-DB amine by FS.

Dicamba		Hexazinone		Imazapyr		Dichlorprop		Pic+2,4D(1:4)	Dic+2,4D(1:3)	Glyphosate	Tebuthiuron			Metsulfuron	Picloram				Fosamine		Pic+Tri(1:1)	Pic+dic(1:1)	Pic+clop(1:1)
Y		N		N		Y		Y	Y	N	N			Y	Y				Y		Y	Y	Y
N		Y		Y		N		N	N	Y	Y			N	N				N		N	N	N
Y		Y		Y		N		Y	Y	N	Y			Y	Y				N		Y	Y	Y
I/CS	ST	BS	ST	BS	ST	BS	FS	FS	FS	FS	BS	FS	ST	FS	BS	FS	I/CS	ST	BS	FS	FS	FS	FS
						I	R									S-I							
																	S						
																I							
										I													
																		S					
																S		S					
													S										
			S													S		S					
	I-R							S							S	S		S			S	S	S
															S	S		S			S	S	S
							I-R			S						S							
													S								S		
				S	S					I	S		S										
													S										
		S	S													S							
																		S					

Table 2 (Continued)*

Herbicide‡,§	2,4-D†			Triclopyr			Clopyralid		Bromacil			Tri+2,4D(1:2)	Dicamba	
	Y			Y			Y		N			Y	Y	
	N			N			N		Y			N	N	
	N			Y			Y		Y			Y	Y	
Species#	BS	FS	I/CS	BS	FS	I/CS	BS	FS	BS	FS	ST	FS	BS	FS
[Ceanothus (*Ceanothus* spp.)]														
mountain whitethorn (*C. cordulatus*)[a]	S	S-I												
redstem (*C. sanguineus*)		S												
snowbrush (*C. velutinus*)	S-I	I												
varnishleaf (*C. velutinus* var. *laevigatus*)[b]	S-I	S-I												
wedgeleaf (*C. cuneatus*)[c]														
Cedar, incense (*Libocedrus decurrens*)		I-R												
Cedar, northern-white (*Thuja occidentalis*)	R	I							S	S	S			S
Cenzio (*Leucophyllum frutescens*)														
Chamise (*Adenostoma fasciculatum*)[d]	S-I	S-I								I				S-R
Cherry (*Prunus* spp.)	I-R	I-R		S					S-I	S-I	S-I		S	S
bitter (*P. emarginata*)[e]			S											
black (*P. serotina*)											S			S
pin (*P. pensylvanica*)														
Chestnut, American (*Castanea dentata*)	I-R	S-I							S-I	S	S-I		S	S
Chinaberry, (*Melia azedarach*)														
Chinese tallowtree (*Sapium sebiferum*)[f]				S										
Chinkapin														
Allegheny (*Castanea pumila*)		R							S	S	S			S
golden (*Castanopsis chrysophylla*)[g]		I-R												
golden evergreen (*Castanopsis chrysophylla* var. minor)[h]		I												
Sierra evergreen (*Castanopsis sempervirens*)		R												
Chokeberry, black (*Pyrus melanocarpa*)		S-I												
Chokeberry, common (*Prunus virginiana*)[i]	S-I	I-R							S	S	S			S

* S, susceptible; S-I, susceptible to intermediate; I, intermediate; I-R, intermediate to resistant; R, resistant.

† Selective; Nonselective; Soil Effective.

‡ Application Methods: BS, basal spray; FS, foliar spray; I/CS, injection/cut surface treatment; ST, soil treatment; IPT, individual plant treatment.

§ Always check and double check precautions in this publication and on container label before using any herbicide.

Amitrole T is labeled for poison ivy, poison oak, honeysuckle, kudzu, salmonberry, blackberry (western dewberry), big leaf maple, wild cherry, ash, locust, and sumac.

[a] I to amitrole by FS.

[b] Retreatment may be needed. I to 2,4-DB amine by FS. R to amitrole by FS.

[c] IPT with tebuthiuron.

[d] Wide range of responses to dicamba by FS according to location; consult local authorities before such use. Foliage treatment: early spring. Spray sprouts after burning or mechanical top removal. Old growth readily removed by picloram or other soil treatments.

[e] Effective August–October.

[f] S to picloram + 2,4-DB amine (1:1) IPT with tebuthiuron.

[g] I to amitrole and 2,4-DB amine by FS.

[h] I to 2,4-DB amine, and amitrole by FS.

[i] Spray early summer after leaves fully expand. Repeat treatment in 1 or 2 years.

Dicamba		Hexazinone		Imazapyr		Dichlorprop		Pic+2,4D(1:4)	Dic+2,4D(1:3)	Glyphosate	Tebuthiuron			Metsulfuron	Picloram				Fosamine		Pic+Tri(1:1)	Pic+dic(1:1)	Pic+clop(1:1)
Y		N		N		Y		Y	Y	N	N			Y	Y				Y		Y	Y	Y
N		Y		Y		N		N	N	Y	Y			N	N				N		N	N	N
Y		Y		Y		N		Y	Y	N	Y			Y	Y				N		Y	Y	Y
I/CS	ST	BS	ST	BS	ST	BS	FS	FS	FS	FS	BS	FS	ST	FS	BS	FS	I/CS	ST	BS	FS	FS	FS	FS
		S	S																				
													S										
																S							
						R	S-I			S						S							
										S													
							I-R		S-I	I			I			S		S-I			S-I		
		S	S	S	S	S	I-R			S			S		S	S		S					
										S				S			S						
S		S	S							S			I					S-I			S		
										S				S									
						S-I	S-I			S					S	S		S					
		I	I	S	S											S		S					
		S	S	S	S			S		S			S			S				I	S		
																S							
				S	S																		
																I							
							I			S			I			S				I			

TABLE 2 (Continued)*

	2,4-D†			Triclopyr			Clopyralid		Bromaci			Tri+2,4-D(1:2)	Dicamba	
	Y			Y			Y		N			Y	Y	
	N			N			N		Y			N	N	
Herbicide‡,§	N			Y			Y		Y			Y	Y	
Species#	BS	FS	I/CS	BS	FS	I/CS	BS	FS	BS	FS	ST	FS	BS	FS
Christmasberry (*Photinia arbutifolia*)[a]	S-I	I							S		S			
Coffeetree, Kentucky (*Gymnocladus dioicus*)	I-R	I-R							S		S			
Condalia (*Condalia* spp.)														
lotebush (*C. obtusifolia*)	S-I	R							S				S-I	I-R
southwestern (*C. lycioides*)[b]														S
Cottonwood (*Populus* spp.)[c]		S-R	S		S						S	S		I-R
eastern (*P. deltoides*)			S-I			S-I								
plains (*P. sargentii*)	S	S-I							S		S			S
Rio Grande (*P. fremontii* var. *wislizenii*)	S	I							S		S			S
Coyotillo (*Karwinskia humboldtiana*)[d]														
Creosotebush (*Larrea tridentata*)[e]		R									S			I
Currant (*Ribes* spp.)														
gooseberry (*R. montigenum*)		S							S		S			
nutmeg (*R. glutinosum*)	S								S		S			
prickly (*R. lacustre*)									S		S			
Sierra (*R. nevadense*)	S	S							S		S			
sticky (*R. viscosissimun*)	S								S		S			
stink (*R. bracteosum*)	S	S							S		S			
trailing black (*R. laxiflorum*)	S								S		S			
wax (*R. cereum*)	S	S							S		S			
western black (*R. petiolare*)	S	S							S		S			

* S, susceptible; S-I, susceptible to intermediate; I, intermediate; I-R, intermediate to resistant; R, resistant.

† Selective; Nonselective; Soil Effective.

‡ Application Methods: BS, basal spray; FS, foliar spray; I/CS, injection/cut surface treatment; ST, soil treatment; IPT, individual plant treatment.

§ Always check and double check precautions in this publication and on container label before using any herbicide.

Amitrole T is labeled for poison ivy, poison oak, honeysuckle, kudzu, salmonberry, blackberry (western dewberry), big leaf maple, wild cherry, ash, locust, and sumac.

[a] Treat sprouts after burning. Plants killed by retreatment.

[b] Spray at same time as creosotebush or tarbush. Aerial spray with 1½ to 2 pounds per acre in successive years; or 3 to 4 pounds per acre in a single application.

[c] Wide range of responses to 2,4-D by FS according to location; consult local authorities before such use.

[d] Spray base during period of rapid growth.

[e] S to tebuthiuron pellets by ST.

Dicamba		Hexazinone		Imazapyr		Dichlorprop		Pic+2,4D(1:4)	Dic+2,4D(1:3)	Glyphosate	Tebuthiuron			Metsulfuron	Picloram				Fosamine		Pic+Tri(1:1)	Pic+dic(1:1)	Pic+clop(1:1)
Y		N		N		Y		Y	Y	N	N			Y	Y				Y		Y	Y	Y
N		Y		Y		N		N	N	Y	Y			N	N				N		N	N	N
Y		Y		Y		N		Y	Y	N	Y			Y	Y				N		Y	Y	Y
I/CS	ST	BS	ST	BS	ST	BS	FS	FS	FS	FS	BS	FS	ST	FS	BS	FS	I/CS	ST	BS	FS	FS	FS	FS
																I							
	I-R	S	S				R		S-I				I		S	I-R		S					
			S	S	S		R						S							S			
								S					S	S			S			I			
						S	S									S							
						S	S									S							
	R																	S					
	S		S				R						S			S-I		S					
																S							

TABLE 2 (Continued)*

Herbicide[‡,§]	2,4-D[†]			Triclopyr			Clopyralid		Bromacil			Tri+2,4D(1:2)	Dicamba	
	Y			Y			Y		N			Y	Y	
	N			N			N		Y			N	N	
	N			Y			Y		Y			Y	Y	
Species[#]	BS	FS	I/CS	BS	FS	I/CS	BS	FS	BS	FS	ST	FS	BS	FS
[Currant (*Ribes* spp.)]														
winter (*R. sanguineum*)	S	S							S		S			
Cypress (*Taxodium distichum*)									S					
Dangleberry (*Gaylussacia frondosa*)	I	I							S		S			
Deervetch, broom (*Lotus scoparius*)		S							S		S			
Devil's-walking-stick (*Aralia spinosa*	S	R							S		S			S
Dewberry (*Rubus* spp.)		I-R		S	S-I			I	S	I	S	I		I
grapeleaf (*R. unsinus*)	I-R										R		I	I
Dogwood (*Cornus* spp.)[a]	R	R	I	S	I	I				I	I	R		S-I
bunchberry (*C. canadensis*)		R												
flowering (*C. florida*)			S-I								S			
roughleaf (*C. drummondii*)[b]			S-I											
Douglas-fir (*Pseudotsuga menziesii*)		I-R	R											S
Downy rosemyrtle (*Rhodomyrtus tomentosa*)[c]				S										
Elder (*Sambucus* spp.)	S-I	S-I							S	S	S		S	S
American (*S. canadensis*)									S					
Pacific red (*S. callicarpa*)[d]											R			
Elm (*Ulmus* spp.)														
American (*U. americana*)	S-I	I	S-I	S	I	S	S-I			S	S	S-I	S	I
Chinese (*U. parvifolia*)[e]		S										S		
Siberian (*U. pumila*)[f]		S										S		
slippery (*U. rubra*)[g]	I	I									S			I
winged (*U. alata*)	S-I	R	S								I		S	I-R
Eucalyptus (*Eucalyptus* spp.)[h]			I											
Fern-bush (*Chamaebatiaria millefolium*)														
Fir (*Abies* spp.)														
balsam (*A. balsamea*)	I	R							S		S			

* S, susceptible; S-I, susceptible to intermediate; I, intermediate; I-R, intermediate to resistant; R, resistant.
† Selective; Nonselective; Soil Effective.
‡ Application Methods: BS, basal spray; FS, foliar spray; I/CS, injection/cut surface treatment; ST, soil treatment; IPT, individual plant treatment.
§ Always check and double check precautions in this publication and on container label before using any herbicide.
\# Amitrole T is labeled for poison ivy, poison oak, honeysuckle, kudzu, salmonberry, blackberry (western dewberry), big leaf maple, wild cherry, ash, locust, and sumac.
[a] Controlled by triclopry + 2,4-D in Nebraska. S to glyphosphate cut stump treatment.
[b] Most S during growing season, least S in spring (pre-bud-burst). S to picloram anytime.
[c] Spray entire plant. S to paraquat by I/CS.
[d] I to amitrole-T by FS.
[e] IPT with tebuthiuron.
[f] Foliage application of phenoxy herbicides generally ineffective because of root sprouting.
[g] IPT with tebuthiuron.
[h] Very S to picloram by I/CS. S to glyphosate cut stump treatment.

Dicamba		Hexazinone		Imazapyr		Dichlorprop		Pic+2,4D(1:4)	Dic+2,4D(1:3)	Glyphosate	Tebuthiuron			Metsulfuron	Picloram				Fosamine		Pic+Tri(1:1)	Pic+dic(1:1)	Pic+clop(1:1)
Y		N		N		Y		Y	Y	N	N			Y	Y				Y		Y	Y	Y
N		Y		Y		N		N	N	Y	Y			N	N				N		N	N	N
Y		Y		Y		N		Y	Y	N	Y			Y	Y				N		Y	Y	Y
I/CS	ST	BS	ST	BS	ST	BS	FS	FS	FS	FS	BS	FS	ST	FS	BS	FS	I/CS	ST	BS	FS	FS	FS	FS
				S	S																		
							S									S							
				I	I	S	I		I	I				S		S-I		S					
	R					I-R									S-I	S-I		S-I					
S		I	I	S	S	R	R	I	R	R		S	S	I		S-I				I			
		S	S										I				S	S					
								S					S	S			S						
	S	R	R	R	R								I			S	S	S					
						R	R							S				S					
						S	I			S					S	S							
																				I			
	I																	S					
	S-I	S	S			S-I	I		S	I			S		S	S-I	S	S		I			
		S	S										S							S			
		S	S										S							S			
							S-I						S			S-I		S		I			
	S-I	S	S			S-I	R	S	S		S		S	S	S	S	S		I				
				S	S					S							S						
															S-I								
											S		I			S-I							

Table 2 (Continued)*

	2,4-D†			Triclopyr			Clopyralic		Bromacil			Tri+2,4D(1:2)	Dicamba	
	Y			Y			Y		N			Y	Y	
	N			N			N		Y			N	N	
Herbicide‡,§	N			Y			Y		Y			Y	Y	
Species#	BS	FS	I/CS	BS	FS	I/CS	BS	FS	BS	FS	ST	FS	BS	FS
[Fir (*Abies* spp.)]														
Douglas (see Douglas-fir)														
white (*A. concolor*)		S-I												
Gallberry, tall (*Ilex coriace*)														
Goldenweed, fleece (*Haplopappus arborescens*)		S												
Gooseberry (*Ribes* spp.)														
California (*R. californicum*)					S									
Canada (*R. oxyacanthoides*)		S												
desert (*R. velutinum*)					S									
Hupa (*R. marshallii*)					S									
Lobbs (*R. lobbii*)		S			S									
Menzies (*R. menziesii*)					S									
Sierra (*R. roezlii*)		S			I									
Siskiyou (*R. binominatum*)		I			S									
Tulare (*R. tularense*)					S									
Gorse, common (*Ulex europaeus*)	S	I		S										
Grape (*Vitis* spp.)			S			S		R					S	S-I
riverbank (*V. vulpina*)		S-I												I-R
Greasewood, black (*Sarcobatus vermiculatus*)[a]		S												S
Greenbriar (*Smilax* spp.)[b]	R	R	R	S	I-R				R	R	R	R	R	R
Guayacan (*Porlieria angustifolia*)		R												R
Guava (*Psidium cattleianum*)	IR			S										
Guava (*Psidium guajava*)				S										
Hackberry (*Celtis* spp.)	S	I-R	S	S	S-I	S			S		S			I
netleaf (*C. reticulata*)			S								S	S		
spiney (granjeno) (*C. pallida*)		R			S									R
western (*C. occidentalis*)														
Hasardia (*Haplopappus squamosus*)														
Hawthorn (*Crataegus* spp.)	I	I-R	I	I	I	I			S	I	I	R	S-R	I
black (*C. douglasii*)		S-I												
cockspur (*C. crus-calli*)[c]														

* S, susceptible; S-I, susceptible to intermediate; I, intermediate; I-R, intermediate to resistant; R, resistant.

† Selective; Nonselective; Soil Effective.

‡ Application Methods: BS, basal spray; FS, foliar spray; I/CS, injection/cut surface treatment; ST, soil treatment; IPT, individual plant treatment.

§ Always check and double check precautions in this publication and on container label before using any herbicide.

Amitrole T is labeled for poison ivy, poison oak, honeysuckle, kudzu, salmonberry, blackberry (western dewberry), big leaf maple, wild cherry, ash, locust, and sumac.

[a] Spray during rapid growth.

[b] S to dicamba + 2,4-D LV ester (1:2). IPT with tebuthiuron.

[c] IPT with tebuthiuron.

Dicamba		Hexazinone		Imazapyr		Dichlorprop		Pic+2,4D(1:4)	Dic+2,4D(1:3)	Glyphosate	Tebuthiuron			Metsulfuron	Picloram				Fosamine		Pic+Tri(1:1)	Pic+dic(1:1)	Pic+clop(1:1)
Y		N		N		Y		Y	Y	N	N			Y	Y				Y		Y	Y	Y
N		Y		Y		N		N	N	Y	Y			N	N				N		N	N	N
Y		Y		Y		N		Y	Y	N	Y			Y	Y				N		Y	Y	Y
I/CS	ST	BS	ST	BS	ST	BS	FS	FS	FS	FS	BS	FS	ST	FS	BS	FS	I/CS	ST	BS	FS	FS	FS	FS
				S	S																		
				R						I						S							
				S	S				S	S				S						I			
							S-I									S-I							
							R									S				S			
	R		R	S	S	R	R	I-R	R	R	S		I	R	I-R	I-R		R		R			
													I			R							
													S										
		S	S			S	I-R			S						S-I		S		S-I	I		S
	S-I																S	S					
													S			S-I		S			S	S	S
													S										
										I													
	R	I-R	I-R	S	S	S-R	R	I	I	I			S-I	S	S	I	S-I	S		S			
													S										

TABLE 2 (Continued)*

Herbicide‡,§	2,4-D†			Triclopyr			Clopyralic		Bromacil			Tri+2,4D (1:2)	Dicamba	
	Y			Y			Y		N			Y	Y	
	N			N			N		Y			N	N	
	N			Y			Y		Y			Y	Y	
Species#	BS	FS	I/CS	BS	FS	I/CS	BS	FS	BS	FS	ST	FS	BS	FS
[Hawthorn (*Crataegus* spp.)]														
little-hip (*C. spathulata*)											S			
parsley (*C. marshalli*)[a]			S-I											
Hazel (*Corylus* spp.)														
American (*C. americana*)[b]		S												
beaked (*C. cornuta*)	S-I	S							S		S			
California (*C. cornuta* var. *californica*)[c]	S-I	S-I							S		S		I	I-R
Hemlock (*Tsuga* spp.)[d]	S-R	I-R							S		S		S	S
eastern (*T. canadensis*)			R											
western (*T. heterophylla*)		I-R	R											
Hickory (*Carya* spp.)[e]	I-R	I-R	I	S		I			S	I	I	I	S-I	I-R
bitternut (*C. cordiformis*)														
black (*C. texana*)			S-I			S-I						I-R		
mockernut (*C. tomentosa*)			S-I								S			
pignut (*C. glabra*)														
shagbark (*C. ovata*)														
Hogplum (*Colubrina texensis*)														
Holly, American (*Ilex opaca*)														
Honeylocust, common (*Gleditsia triacanthos*)	R	R	I	I	I	I		S	S	I	I	I	S	I
Honeysuckle (*Lonicera* spp.)	I	R	I	R	R	R		R	I-R	I	R	R	S	I
Hophornbean, eastern (*Ostrya virginiana*)	S-I	S-I							S		S	S		S
Hornbean, American (*Carpinus caroliniana*)	S	S	S-I						S		I			S
Horsebrush (*Tetradymia* spp.)		S												
littleleaf (*T. glabrata*)[f]		I-R							S		S			
Huckleberry (*Galylussacia* spp.)										I-R			S	S
Hydrangea, smooth (*Hydrangea arborescens*)	S	S							S		S			
Inkberry, holly (*Ilex glabra*)[g]	I	I								S				I
Javelinabrush (*Microrhamnus ericoides*)		R												R
Juniper (*Juniperus* spp.)	R	R							S		S			I
Ashe (*J. ashei*)														
Eastern red cedar (*J. virginiana*)[h]														S

* S, susceptible; S-I, susceptible to intermediate; I, intermediate; I-R, intermediate to resistant; R, resistant.

† Selective; Nonselective; Soil Effective.

‡ Application Methods: BS, basal spray; FS, foliar spray; I/CS, injection/cut surface treatment; ST, soil treatment; IPT, individual plant treatment.

§ Always check and double check precautions in this publication and on container label before using any herbicide.

\# Amitrole T is labeled for poison ivy, poison oak, honeysuckle, kudzu, salmonberry, blackberry (western dewberry), big leaf maple, wild cherry, ash, locust, and sumac.

[a] S to I/CS treatment of 2,4-D + picloram (4:1).

[b] Spray in May and June.

[c] Spray in May and June.

[d] Spray entire plant. R to phenoxy herbicides when dormant. Wide range of S to 2,4-D by BS; consult local authorities before such use.

[e] S to glyphosate cut stump treatment.

[f] Treat individual plants in spring.

[g] Treat in August.

[h] IPT, picloram; hexazinone soil-applied. Picloram also FS on Ashe juniper. IPT on Eastern red cedar with tebuthiuron.

Dicamba		Hexazinone		Imazapyr		Dichlorprop		Pic+2,4D(1:4)	Dic+2,4D(1:3)	Glyphosate	Tebuthiuron			Metsulfuron	Picloram				Fosamine		Pic+Tri(1:1)	Pic+dic(1:1)	Pic+clop(1:1)
Y		N		N		Y		Y	Y	N	N			Y	Y				Y		Y	Y	Y
N		Y		Y		N		N	N	Y	Y			N	N				N		N	N	N
Y		Y		Y		N		Y	Y	N	Y			Y	Y				N		Y	Y	Y
I/CS	ST	BS	ST	BS	ST	BS	FS	FS	FS	FS	BS	FS	ST	FS	BS	FS	I/CS	ST	BS	FS	FS	FS	FS
																		S					
		S	S							S													
										S													
						I	S-I								S-I	S							
						S	S		S	S	S				S	S							
	S																S	S					
I		I-R	I-R	S	S	I	I	I	R	R			I	R	S	R	S-I			R			
													S										
													I	I			S						
																		S					
				I	I								I										
													I										
													S					S					
				I	I																		
			S	S	S	R	R	I	R	R			S	R		I				I			
		I-R	I-R	S	S	S	S-I	I	I	I	S		S	R	S	S-I				I			
							S-I			S						S							
										I						S		S-I					
				S	S					S			S					S					
						S-I	S-I									S							
																R							
		S	S			I-R	I-R			S	S		S		S	S-I		S					
			S													S		S					
		S	S										S					S					

TABLE 2 (Continued)*

	2,4-D†			Triclopyr			Clopyralid		Bromacil			Tri+2,4-D(1:2)	Dicamba	
[Selective]	Y			Y			Y		N			Y	Y	
[Nonselective]	N			N			N		Y			N	N	
Herbicide‡,§ [Soil effective]	N			Y			Y		Y			Y	Y	
Species#	BS	FS	I/CS	BS	FS	I/CS	BS	FS	BS	FS	ST	FS	BS	FS
[Juniper (*Juniperus* spp.)]														
oneseeded (*J. monsperma*)														
redberry (*J. pinchotii*)[a]		R												R
Kalmia (*Kalmia* spp.)														
lambkill (see sheeplaurel)														
mountainlaurel (*K. latifolia*)	R	R							S	S	S			R
sheeplaurel (*K. angustifolia*)	R	I-R							S		S			
Karakanut (*Corynocarpus laevigatus*)				R										
Kidneywood, Texas (*Eysenhardia Texana*)														
Koahaole (*Leucaena leucocephalla*)														
Kudzu (*Pueraria lobata*)	R	R		R	R	R		S		R	R	R		S
Lantana (*Lantana camara*)[b]				S										
Larch (*Larix* spp.)	S	R							S		S			
Laurel, California (*Umbellularia californica*)			S											
Leatherstem (*Jatropha dioica*)		R			R									R
Leatherwood, Atlantic (*Dirca palustris*)		S							S		S			
Lilac, common (*Syringa vulgaris*)	I	I-R							S		S			
Locust, black (*Robinia pseudoacacia*)	I	S-I	S				R		S	S	S	S	S-I	S
water (*Gleditsia aquatica*)			S											
Lotebush (see Condalia)														
Lyonia (*Lyonia* spp.)														
fetterbush (*L. lucida*)														
staggerbush (*L. mariana*)														
Madrone, Pacific (*Arbutus menziesii*)[c]	S	S-I	S											
Magnolia (*Magnolia* spp.)														
cucumbertree (*M. acuminata*)	I	I							S		S			S
sweetbay (*M. virginiana*)[d]	SR	R	S						I		I			R
Manzanita (*Arctostaphylos* spp.)[e]	S	S												
Del Norte (*A. cinerea*)		S												
greenleaf (*A. patula*)[f]		I			I									
hairy (*A. columbiana*)		S												
hoary (*A. canescens*)		S												
Howell (*A. hispidula*)		S												

* S, susceptible; S-I, susceptible to intermediate; I, intermediate; I-R, intermediate to resistant; R, resistant.

† Selective; Nonselective; Soil Effective.

‡ Application Methods: BS, basal spray; FS, foliar spray; I/CS, injection/cut surface treatment; ST, soil treatment; IPT, individual plant treatment.

§ Always check and double check precautions in this publication and on container label before using any herbicide.

Amitrole T is labeled for poison ivy, poison oak, honeysuckle, kudzu, salmonberry, blackberry (western dewberry), big leaf maple, wild cherry, ash, locust, and sumac.

[a] IPT, picloram spray; hexazinone soil-applied.

[b] IPT with tebuthiuron.

[c] S to glyphyosate cut stump treatment.

[d] Wide range of responses to 2,4-D by BS according to location; consult local authorities before such use.

[e] Spray during rapid growth. Repeat treatment when necessary to kill sprouting species.

[f] IPT with tebuthiuron. Resprouts from burls. Killed by retreatment.

Dicamba		Hexazinone		Imazapyr		Dichlorprop		Pic+2,4D(1:4)	Dic+2,4D(1:3)	Glyphosate	Tebuthiuron			Metsulfuron	Picloram				Fosamine		Pic+Tri(1:1)	Pic+dic(1:1)	Pic+clop(1:1)
Y		N		N		Y		Y	Y	N	N			Y	Y				Y		Y	Y	Y
N		Y		Y		N		N	N	Y	Y			N	N				N		N	N	N
Y		Y		Y		N		Y	Y	N	Y			Y	Y				N		Y	Y	Y
I/CS	ST	BS	ST	BS	ST	BS	FS	FS	FS	FS	BS	FS	ST	FS	BS	FS	I/CS	ST	BS	FS	FS	FS	FS
	I-R		S													R S		S-I					
							R									I							
			R	S	S	R	R	R	I	I S			I S R S	I		S				S			
													I			R							
		S	S			I-R	S-I	I	S	I	S	S	S	I	S-I	S	S	S		S			
				S S S	S S S					I-R							S						
				I	I	S	R									I							
		S	S							I			S										
		S	S										S										

TABLE 2 (Continued)*

	2,4-D†			Triclopyr			Clopyralid		Bromacil			Tri+2,4-D(1:2)	Dicamba	
	Y			Y			Y		N			Y	Y	
	N			N			N		Y			N	N	
Herbicide‡,§	N			Y			Y		Y			Y	Y	
Species#	BS	FS	I/CS	BS	FS	I/CS	BS	FS	BS	FS	ST	FS	BS	FS
[Manzanita (*Arctostaphylos* spp.)]														
pine (*A. parryana* var. *pinetorum*)		S												
pointleaf (*A. pungens*)											S			
Maple (*Acer* spp.)											S			S
bigleaf (*A. macrophyllum*)[a]		I-R	R		R							I-R		
mountain (*A. spicatum*)		I-R			S									
Norway (*A. platanoides*)[b]														
red (*A. rubrum*)	R	R	I-R						S	S	S	S-I	S	I-R
silver (*A. saccharinum*)[c]	I	I-R		S	S				S	S	S			
sugar (*A. saccharum*)		I-R	R	S	S				S			S-I		
vine (*A. circinatum*)[d]					R						R		I-R	R
Melaleuca (*Melaleuca quinquenervia*)														
Melastoma (*Melastoma condidum*)				I										
Mesquite (*Prosopis juliflora*)[e]	I	I-R		S	S				R		R		S-I	S
honey mesquite (*P. juliflora* var. *glandulosa*)[f]	I	I-R	S-I	S	I		S-I	S			R			I
Mountainmahogany (*Cercocarpus montanus*)		I			I									
birchleaf (*C. betuloides*)					I-R						I			
Mulberry, red (*Morus rubra*)	I-R	I-R	I	I	I	I			I	I	I	R	S	S-I
Oak (*Quercus* spp.)[g]	I	I	I	I	S	S				I	I	I		I
bigelow (bigleaf shin) (*Q. durandii* var. *breviloba*)														
black (*Q. velutina*)	S-I	I-R	R						S	S	S	S	S	S-I
blackjack (*Q. marilandica*)[h]	I	I-R	S		S-I				S	S	I		S	I-R

* S, susceptible; S-I, susceptible to intermediate; I, intermediate; I-R, intermediate to resistant; R, resistant.

† Selective; Nonselective; Soil Effective.

‡ Application Methods: BS, basal spray; FS, foliar spray; I/CS, injection/cut surface treatment; ST, soil treatment; IPT, individual plant treatment.

§ Always check and double check precautions in this publication and on container label before using any herbicide.

Amitrole T is labeled for poison ivy, poison oak, honeysuckle, kudzu, salmonberry, blackberry (western dewberry), big leaf maple, wild cherry, ash, locust, and sumac.

[a] S to glyphosate cut stump treatment.

[b] IPT with tebuthiuron.

[c] IPT with tebuthiuron.

[d] IPT with tebuthiuron.

[e] S to hexazinone by ST. Apply before expected rainfall.

[f] R to most soil-applied herbicides. Killed by basal treatment with diesel oil or kerosene. S to triclopyr + clopyralid; I to triclopyr + dicamba or fall clopyralid treatment.

[g] Regrowth highly S to treatment. Oaks S to glyphosate by cut stump treatment.

[h] S-I to paraquat by I/CS.

Dicamba		Hexazinone		Imazapyr		Dichlorprop		Pic+2,4D(1:4)	Dic+2,4D(1:3)	Glyphosate	Tebuthiuron			Metsulfuron	Picloram				Fosamine		Pic+Tri(1:1)	Pic+dic(1:1)	Pic+clop(1:1)
Y		N		N		Y		Y	Y	N	N			Y	Y				Y		Y	Y	Y
N		Y		Y		N		N	N	Y	Y			N	N				N		N	N	N
Y		Y		Y		N		Y	Y	N	Y			Y	Y				N		Y	Y	Y
I/CS	ST	BS	ST	BS	ST	BS	FS	FS	FS	FS	BS	FS	ST	FS	BS	FS	I/CS	ST	BS	FS	FS	FS	FS
																		S					
				S	S					S							SI						
				S	S								I							S			
S-I		S	S	S	S	R	R			S					S	S-I	S-I	S-I		S			
													S										
													S			S		I					
S										S			S				S-I						
	R					I-R	R			I			S		S	S		I					
				S	S								I										
	R	S	S			R	I		I		I		I	R	R	S		R			S	S	S
S		S-I	S-I				I-R	I-R	I-R				I-R			S	S	R			S	S	S
													I					S					
		I	I	S	S	I-R	I-R	I	R	R	S	S	S	R	S	S-I				I			
		S	S	S	S	I	I	I	R	S			S	I						I			
							S						S			S							
				S	S	S-I	I-R			S			S	I	S	S-I	S	S-I					
S	R	S	S			S-I	I			S			S	I	S	I	S-I	I					

TABLE 2 (Continued)*

Herbicide[‡,§]	2,4-D[†]			Triclopyr			Clopyralid		Bromacil			Tri+2,4D(1:2)	Dicamba	
	Y			Y			Y		N			Y	Y	
	N			N			N		Y			N	N	
	N			Y			Y		Y			Y	Y	
Species[#]	BS	FS	I/CS	BS	FS	I/CS	BS	FS	BS	FS	ST	FS	BS	FS
[Oak (*Quercus* spp.)]														
blue (*Q. douglasii*)[a]	S	R	S						I					S
bur (*Q. macrocarpa*)	I	I							S		S			
California black (*Q. kelloggii*)[b]			S											
California live (*Q. agrifolia*)			S											
California scrub (*Q. dumosa*)[c]	I	I							S		S			
canyon live (*Q. chrysolepis*)[d]		I	S-I											
chestnut (*Q. prinus*)	S-I	I-R	I-R						S	S-I	S		S	S-I
Emory (*Q. emory*)	I-R										S			
Gambel (*Q. gambelii*)		I-R							S					I
interior live (*Q. wislizenii*)[e]		I-R	S								R			R
leather (*Q. durata*)[f]		I-R									I			R
live (*Q. virginiana*)[g]	I	R							S	S-I	S		R	R
mohr (*Q. mohriana*)														
myrtle (*Q. myrtifolia*)														
northern red (*Q. rubra*)[h]	I	I-R	R						S	S	S-I		S	I
Oregon white (*Q. garryana*)[i]			S											
overcup (*Q. lyrata*)			S											
Palmer (*Q. Chrysolepis* var. *palmeri*)		I-R									I			
Pin (*Q. palustris*)	I	I							S		S			I
post (*Q. stellata*)	S-I	I-R	S		S-I			R	S	S	S		S	S-I
southern red (*Q. falcata*)														

* S, susceptible; S-I, susceptible to intermediate; I, intermediate; I-R, intermediate to resistant; R, resistant.

† Selective; Nonselective; Soil Effective.

‡ Application Methods: BS, basal spray; FS, foliar spray; I/CS, injection/cut surface treatment; ST, soil treatment; IPT, individual plant treatment.

§ Always check and double check precautions in this publication and on container label before using any herbicide.

[#] Amitrole T is labeled for poison ivy, poison oak, honeysuckle, kudzu, salmonberry, blackberry (western dewberry), big leaf maple, wild cherry, ash, locust, and sumac.

[a] Wide range of response to picloram by BS and ST according to location; consult local authorities before such use. Amines better than phenoxy esters. Spray when growth most active and repeat in 2 years.

[b] Use 2,4-D amine.

[c] Burn, then spray sprouts annually. Usually killed by two retreatments.

[d] Killed by retreatment.

[e] Readily killed with retreatment.

[f] Readily killed with retreatment.

[g] Most S to tebuthiuron.

[h] Badly damaged in northwest by 2,4-D and cannot be used silviculturally (in forest management).

[i] Use 2,4-D amine. Stump spray.

Dicamba		Hexazinone		Imazapyr		Dichlorprop		Pic+2,4D(1:4)	Dic+2,4D(1:3)	Glyphosate	Tebuthiuron			Metsulfuron	Picloram				Fosamine		Pic+Tri(1:1)	Pic+dic(1:1)	Pic+clop(1:1)
Y		N		N		Y		Y	Y	N	N			Y	Y				Y		Y	Y	Y
N		Y		Y		N		N	N	Y	Y			N	N				N		N	N	N
Y		Y		Y		N		Y	Y	N	Y			Y	Y				N		Y	Y	Y
I/CS	ST	BS	ST	BS	ST	BS	FS	FS	FS	FS	BS	FS	ST	FS	BS	FS	I/CS	ST	BS	FS	FS	FS	FS
						S	S-I						S		S-R		S						
													S										
																	S						
													I										
																I							
S						S-I	I-R			S				S-I	S	I	S	S-I					
																		I					
		S	S								S		S		S			S-I					
							I-R										S	R					
							I-R											R					
							R			S			S			S-I		S					
S						I	I-R			S					S	I	S	S-I		S			
													S										
			S	S																			
																		I					
						R	I-R			S	S		I		S	S-I							
S	R	S	S			S-I	S-I	I-R	I-R	S			S		S	I	S-I	S-I					
										S			S							S			

Table 2 (Continued)*

	2,4-D†			Triclopyr			Clopyralic		Bromacil			Tri+2,4D(1:2)	Dicamba	
	Y			Y			Y		N			Y	Y	
	N			N			N		Y			N	N	
Herbicide‡,§	N			Y			Y		Y			Y	Y	
Species#	BS	FS	I/CS	BS	FS	I/CS	BS	FS	BS	FS	ST	FS	BS	FS
[Oak (*Quercus* spp.)]														
sand shinnery (*Q. havardii*)[a]	I	I							S		S			I
scarlet (*Q. coccinea*)	S-I	I-R	S						S		S		S	I
shrub live (*Q. turbinella*)[b]		I-R			I						S			
Spanish (*Q. shumardii* var. *texana*)[c]	I-R	I-R							S		S			I
swamp white (*Q. bicolor*)	I	I		S-I	S-I				S		S			
turkey (*Q. laevis*)		I							S		S			
valley (*Q. lobata*)		S												
Vasey shin (*Q. pungens* var. *vaseyana*)[d]										S				
water (*Q. nigra*)		R	S								S			
white (*Q. alba*)	S-I	I-R	I						S	S	S		S	I
Osageorange (*Maclura pomifera*)[e]	R	I-R	S-I	I	I	I				I	I	I	S	I-R
Paloverde (*Cercidium* spp.)														
Pecan (*Carya illinoensis*)	I	I-R							I		I			
Pecan, bitter (*Carya X lecontei*)			S-I											
Peppertree, Brazilian (*Schinus terebinthifolius*)[f]														
Peppervine (*Ampelopsis arborea*)				S										
Persimmon (*Diospyros* spp.)														
common (eastern) (*D. virginiana*)[g]	I-R	R	I	I	I	I				R	R	I	S	S
texas (*D. texana*)[h]		R		S									S-I	R
Pinchot Juniper (see Redberry Juniper)														
Pine, Australian (*Casuarina* spp.)														
Pine (*Pinus* spp.)[i]	I	R	I	S	S	I				I	I	I	S	S
digger (*P. sabiniana*)[j]			S											
eastern white (*P. strobus*)			S											
loblolly (*P. taeda*)[k]		R			R									
lodgepole (*P. contorta*)														

* S, susceptible; S-I, susceptible to intermediate; I, intermediate; I-R, intermediate to resistant; R, resistant.

† Selective; Nonselective; Soil Effective.

‡ Application Methods: BS, basal spray; FS, foliar spray; I/CS, injection/cut surface treatment; ST, soil treatment; IPT, individual plant treatment.

§ Always check and double check precautions in this publication and on container label before using any herbicide.

Amitrole T is labeled for poison ivy, poison oak, honeysuckle, kudzu, salmonberry, blackberry (western dewberry), big leaf maple, wild cherry, ash, locust, and sumac.

[a] Most S to tebuthiuron.

[b] Spray during rapid growth. Repeat treatment following year.

[c] Regrowth highly S to treatment.

[d] Regrowth highly S to treatment.

[e] S-R to triclopyr.

[f] IPT with tebuthiuron.

[g] S to paraquat by I/CS.

[h] S to stump treatment.

[i] S-I to paraquat. Some defoliation but will recover from phenoxy herbicides.

[j] 2,4-D and picloram equivalent in effectiveness.

[k] Some defoliation but will recover from phenoxy herbicides.

Dicamba		Hexazinone		Imazapyr		Dichlorprop		Pic+2,4D(1:4)	Dic+2,4D(1:3)	Glyphosate	Tebuthiuron			Metsulfuron	Picloram				Fosamine		Pic+Tri(1:1)	Pic+dic(1:1)	Pic+clop(1:1)
Y		N		N		Y		Y	Y	N	N			Y	Y				Y		Y	Y	Y
N		Y		Y		N		N	N	Y	Y			N	N				N		N	N	N
Y		Y		Y		N		Y	Y	N	Y			Y	Y				N		Y	Y	Y
I/CS	ST	BS	ST	BS	ST	BS	FS	FS	FS	FS	BS	FS	ST	FS	BS	FS	I/CS	ST	BS	FS	FS	FS	FS
	R		S				I-R						S			S-I		S-R					
S						S-I	I-R								S	S-I	S						
	I																	I					
							I-R									I							
																S							
							S																
																		S-I		S			
S						S-I	I			I	S		I		S	S-I	S	S-I					
		I	I	R	R	R	R	I	R	R			R	R	S		S-I				I		
													S										
							I-R																
				S	S					I			S	S									
S	S	I-R	I-R	I	I	R	I-R	I	I	I			R	R	S	S-I	S	S		I			
	R														S	I		S-I					
														I									
	S	R	R	R	R	I	R	S	I	R			I	R	S	S		S		S			
																	S						
				R	R												S			I			
		R	R	R	R					S						S-I				S			
		R	R	R	R									S			S						

TABLE 2 (Continued)*

Herbicide[‡,§]	2,4-D[†]			Triclopyr			Clopyralid		Bromacil			Tri+2,4D(1:2)	Dicamba	
	Y			Y			Y		N			Y	Y	
	N			N			N		Y			N	N	
	N			Y			Y		Y			Y	Y	
Species[#]	BS	FS	I/CS	BS	FS	I/CS	BS	FS	BS	FS	ST	FS	BS	FS
[Pine (*Pinus* spp.)]														
longleaf (*P. palustris*)											S			
pinyon (*P. edulis*)		R												
ponderosa (*P. ponderosa*)[a]		I-R	S											
shortleaf (*P. echinata*)[b]		I									S			
sugar (*P. lambertiana*)[c]		S												
western white (*P. monticola*)														
Plum, wild (*Prunus* spp.)	S-I	S-I							S		S		S	S
Poison ivy, common (*Rhus radicans*)[d]	R	R		R	I	S		R		I	I	I		I
Poison oak (*Rhus toxicodendron*)[e]	I	I		R	I			R	S	S-I	S-I	I		I
Pacific poison oak (*R. diversiloba*)														
Popular, balsam (*Populus balsamifera*)[f]	S-I	I		I	I	I				I	I	I		S
Prickly-ash, common (*Zanthoxylum americanum*)	I-R	R							S		S			R
Hercules club (*Z. clava-herculis*)				S										
lime (*Z. fagara*)		R												R
Privet, swamp (*Forestiera acuminata*)[g]			S-I											
Rabbitbrush (*Chrysothamnus* spp.)[h]		S-I			I			I	S		S			I
big (rubber) (*C. nauseosus*)		S												
little (small) (*C. viscidiflorus* var. *stenophyllus*)		S												
Raspberry, red (*Rubus idaeus*)	I-R	I-R							S		S			
black (*R. occidentalis*)														
Ratama (*Parkinsonia aculeata*)														
Rayless goldenrod (*Haplopappus heterophyllus*)[i]														
Redbud, eastern (*Cercis canadensis*)	I-R	R							S	S	S		S	I

* S, susceptible; S-I, susceptible to intermediate; I, intermediate; I-R, intermediate to resistant; R, resistant.

† Selective; Nonselective; Soil Effective.

‡ Application Methods: BS, basal spray; FS, foliar spray; I/CS, injection/cut surface treatment; ST, soil treatment; IPT, individual plant treatment.

§ Always check and double check precautions in this publication and on container label before using any herbicide.

[#] Amitrole T is labeled for poison ivy, poison oak, honeysuckle, kudzu, salmonberry, blackberry (western dewberry), big leaf maple, wild cherry, ash, locust, and sumac.

[a] More sensitive than many pines but effectiveness minimized if treated after shoot growth, or by using lower dosages.

[b] Some defoliation but will recover from phenoxy herbicides.

[c] S in spring and summer.

[d] S to amitrole by FS.

[e] S-I to amitrole by FS. Amitrole and glyphosate are most effective when spraying is delayed until shoot elongation is slowed. Wide range of S to dichlorprop by FS; consult local authorities before such use.

[f] S to glyphosate by cut stump treatment.

[g] IPT with tebuthiuron.

[h] S to 2,4D + picloram (6:1) 2 lb + 1/3 lb per acre. Treat in spring when new growth is 3 inches long with 2,4-D.

[i] S to tebuthiuron-IPT or broadcast.

Dicamba		Hexazinone		Imazapyr		Dichlorprop		Pic+2,4D(1:4)	Dic+2,4D(1:3)	Glyphosate	Tebuthiuron			Metsulfuron	Picloram				Fosamine		Pic+Tri(1:1)	Pic+dic(1:1)	Pic+clop(1:1)
Y		N		N		Y		Y	Y	N	N			Y	Y				Y		Y	Y	Y
N		Y		Y		N		N	N	Y	Y			N	N				N		N	N	N
Y		Y		Y		N		Y	Y	N	Y			Y	Y				N		Y	Y	Y
I/CS	ST	BS	ST	BS	ST	BS	FS	FS	FS	FS	BS	FS	ST	FS	BS	FS	I/CS	ST	BS	FS	FS	FS	FS
		R	R	R	R					S						S-I		S					
		R	R	R	R											S-I	S						
S		R	R	R	R					S						S-I		S					
S																							
		S	S			S-I	I			S				I	S	S				I			
			I	S	S	I	I	S	I	I			R	R		S				R			
				S	S		S-R		I	I				I	S								
															S	S							
		I	I	I	I	I	I	R	I	I			I	R						I			
							R									R							
																R		R					
			S										S										
		S	S	S	S								S										
			S-I				R	I								S							
										S													
														I									
																S					S	S	S
						R	R			S			S		S	I							
													S	S		S							

TABLE 2 (Continued)*

	2,4-D†			Triclopyr			Clopyralid		Bromacil			Tri+2,4D(1:2)	Dicamba	
	Y			Y			Y		N			Y	Y	
	N			N			N		Y			N	N	
Herbicide‡,§	N			Y			Y		Y			Y	Y	
Species#	BS	FS	I/CS	BS	FS	I/CS	BS	FS	BS	FS	ST	FS	BS	FS
Redvine (*Brunnichia cirrhosa*)														
Rhododendron (*Rhododendron* spp.)														
rosebay (great laurel) (*R. maximum*)		R							S	S	S			I-R
Rhodora (*R. canadense*)		R							S		S			
Rose (*Rosa* spp.)		R												
Arkansas (*R. arkansana*)	R	R							R		R			
California (*R. californica*)									R		R			
Cherokee (*R. laevigata*)		I							R		R			
Macartney (*R. bracteata*)[a]	I	I							R		R			I
multiflora (*R. multiflora*)[b]	I	R	I	I	I	I		R		I	I	I		I
prickly (*R. acicularis*)														
sunshine (*R. arkansan* var. *suffulta*)									R		R			
wild (*R. pisocarpa*)		I												
woods (*R. woodsii*)	R	R		I					R		R			
Russian-olive (*Elaeagnus angustifolia*)[c]		I												S
Sage, black (*salvia melifera*)														
Sagebrush (*Artemisia* spp.)														
big (*A. tridentata*)[d]		S			S-I			R						I
California (*A. californica*)[e]		S												
fringed (*A. frigida*)[f]		S						S						S
green sagewort (*A. glauca*)		S												
low (*A. arbuscula*)[g]		S												
sand (*A. filifolia*)[h]	S	S							S		S			
silver (*A. cana*)[i]		S												
threetip (*A. tripartita*)[j]		S												
Salal (*Gaultheria shallon*)[k]		R								I				R

* S, susceptible; S-I, susceptible to intermediate; I, intermediate; I-R, intermediate to resistant; R, resistant.

† Selective; Nonselective; Soil Effective.

‡ Application Methods: BS, basal spray; FS, foliar spray; I/CS, injection/cut surface treatment; ST, soil treatment; IPT, individual plant treatment.

§ Always check and double check precautions in this publication and on container label before using any herbicide.

Amitrole T is labeled for poison ivy, poison oak, honeysuckle, kudzu, salmonberry, blackberry (western dewberry), big leaf maple, wild cherry, ash, locust, and sumac.

[a] S to 2,4-D + picloram (4:1) by FS. Spray from full leaf to bud stage with 2,4-D. Picloram sprays effective in spring and fall, granules most months.

[b] Spray from full leaf to bud stage. Repeat treatment when necessary.

[c] Repeat treatment following year for complete control. IPT with tebuthiuron.

[d] S to glyphosate by BS.

[e] Spray during rapid growth.

[f] Soil moisture and timing critical with phenoxy herbicides. Spray when new growth is 2 inches or more tall. S to chopyralid + 2,4-D.

[g] Spray during rapid growth.

[h] Spray during rapid growth and when 6–8 inches new twig growth. Repeat if necessary.

[i] Spray after 3–4 inches new growth and when growing rapidly.

[j] Soil moisture and timing critical with phenoxy herbicides.

[k] IPT with tebuthiuron.

Dicamba		Hexazinone		Imazapyr		Dichlorprop		Pic+2,4D(1:4)	Dic+2,4D(1:3)	Glyphosate	Tebuthiuron			Metsulfuron	Picloram				Fosamine		Pic+Tri(1:1)	Pic+dic(1:1)	Pic+clop(1:1)
Y		N		N		Y		Y	Y	N	N			Y	Y				Y		Y	Y	Y
N		Y		Y		N		N	N	Y	Y			N	N				N		N	N	N
Y		Y		Y		N		Y	Y	N	Y			Y	Y				N		Y	Y	Y
I/CS	ST	BS	ST	BS	ST	BS	FS	FS	FS	FS	BS	FS	ST	FS	BS	FS	I/CS	ST	BS	FS	FS	FS	FS
				S	S																		
							R			S						I-R							
							R			S													
											S				S-I	I							
				S	S			S			S		I			S		S					
		S	S	S	S	S	S	S	R	I			S	I		S				I	S		
			I							S				S-I									
																		S					
									I	I			I										
										S			S										
			S-I				R	I		S			S	R		I							
										S													
								S								S							
							S						S			S							
																R							
						R							S			I							

TABLE 2 (Continued)*

Herbicide[‡,§]	2,4-D[†]			Triclopyr			Clopyralid		Bromacil			Tri+2,4D(1:2)	Dicamba	
	Y			Y			Y		N			Y	Y	
	N			N			N		Y			N	N	
	N			Y			Y		Y			Y	Y	

Species[#]	BS	FS	I/CS	BS	FS	I/CS	BS	FS	BS	FS	ST	FS	BS	FS
Salmonberry (*Rubus spectabilis*)[a]	S	I-R									R			R
Saltcedar (five-stamen tamarisk) (*Tamarix pentandra*)[b]	S	S-I							S		S		R	S
Sassafras (*Sassafras albidum*)	I	R	S	I	I	I				R	R	I	S	I
Serviceberry (*Amelanchier* spp.)														S
saskatoon (*A. alnifolia*)[c]		S-I									S			
shadblow (*A. canadensis*)	S	S-I								I	S			S
Silktassel (*Garrya* spp.)														
yellowleaf (*G. flavescens*)											S-I			
Silverberry (*Elaeagnus commutata*)[d]		S						R						S
Snakeweed, broom (*Gutierrezia sarothrae*)[e]		S						I						I
Snowberry, western (*Symphoricarpos occidentalis*)[f]	S-I	S							S		S			R
Soapweed, small (*Yucca glauca*)[g]		R												R
Sourbrush (*Pluchea symphytifolia*)				S										
Sourwood (*Oxydendrum arboreum*)	S-I	I-R	I						S	S	S		S	I-R
Spicebush (*Lindera benzoin*)[h]	S-R	I							S		S			S
Spirea (*Spiraea* spp.)														
hardhack (*S. tomentosa*)	I	I							S		S			
narrowleaf meadowsweet (*S. alba*)	S-I	S-I							S		S			

* S, susceptible; S-I, susceptible to intermediate; I, intermediate; I-R, intermediate to resistant; R, resistant.

† Selective; Nonselective; Soil Effective.

‡ Application Methods: BS, basal spray; FS, foliar spray; I/CS, injection/cut surface treatment; ST, soil treatment; IPT, individual plant treatment.

§ Always check and double check precautions in this publication and on container label before using any herbicide.

[#] Amitrole T is labeled for poison ivy, poison oak, honeysuckle, kudzu, salmonberry, blackberry (western dewberry), big leaf maple, wild cherry, ash, locust, and sumac.

[a] S-I to amitrole by FS. Treat in midsummer.

[b] S to glyphosate by cut stump treatment. More effective on young than old plants. Mechanical removal followed by treatment of regrowth more effective than herbicide alone. S to phenoxy herbicides when sprayed with 1/8 to 1/4 pounds per acre twice a year for 2 years.

[c] S-I to 2,4-DB amine by FS. R to amitrole by FS.

[d] Treat in early summer and repeat the following year.

[e] Treat early spring. Timing of application critical with phenoxy herbicides. May treat in late summer or fall with picloram.

[f] Retreatment following year sometimes necessary.

[g] Spray soon after full bloom.

[h] Wide range of S to 2,4-D by BS according to location; consult local authorities before such use.

Dicamba		Hexazinone		Imazapyr		Dichlorprop		Pic+2,4D(1:4)	Dic+2,4D(1:3)	Glyphosate	Tebuthiuron			Metsulfuron	Picloram				Fosamine		Pic+Tri(1:1)	Pic+dic(1:1)	Pic+clop(1:1)
Y		N		N		Y		Y	Y	N	N			Y	Y				Y		Y	Y	Y
N		Y		Y		N		N	N	Y	Y			N	N				N		N	N	N
Y		Y		Y		N		Y	Y	N	Y			Y	Y				N		Y	Y	Y
I/CS	ST	BS	ST	BS	ST	BS	FS	FS	FS	FS	BS	FS	ST	FS	BS	FS	I/CS	ST	BS	FS	FS	FS	FS
	R						R			S						S		S-I		S			
	S			S	S	R	S-I			S	S		S		S			S					
		I-R	I-R	S	S	I	I	I	R	R			R	R		S		S-I		I-R	S		
							S-I					S				S							
																		S					
			I					S	S				I	S		S		S					
							S									R							
		S	S				R									R							
		S	S	S	S	S-I	I-R			I					S	S		I		I-R			
							S-I									S							
													I			S							

TABLE 2 (Continued)*

Herbicide‡,§	2,4-D†			Triclopyr			Clopyralid		Bromacil			Tri+2,4-D(1:2)	Dicamba	
	Y			Y			Y		N			Y	Y	
	N			N			N		Y			N	N	
	N			Y			Y		Y			Y	Y	
Species#	BS	FS	I/CS	BS	FS	I/CS	BS	FS	BS	FS	ST	FS	BS	FS
Spruce (*Picea* spp.)[a]	R	I-R							S		S			I-R
Sitka (*P. sitchensis*)		I-R												
White (*P. glauca*)														
Sumac (*Rhus* spp.)[b]	R	I		S	S	S		I		I	I	S		I
laurel (*R. laurina*)[c]	S-I	S-I		S-I	S-I				S		S			
littleleaf (*R. microphylla*)		S												S
shining (*R. copallina*)[d]					S				S-I		S			
skunkbush (*R. aromatica*)[e]		S-I									S			S-I
smooth (*R. glabra*)														
staghorn (*R. typhina*)														
sugar (*R. ovata*)											S			
Sweet-fern (*Comptonia peregrina*)	I	I							S		S			
Sweetgum (*Liquidambar styraciflua*)[f]	I-R	I-R	S	S	S	S				I	I	I		I-R
Sycamore, American (*Platanus occidentalis*)	S-I	I		I	I	I			I	I	I	I	S	S
Tamarack (*Larix laricina*)														
Tanoak (*Lithocarpus densiflorus*)[g]	S	I-R	S-I											
scrub (*L. densiflorus* var. *echinoides*)		I	S											
Tarbush (*Flourensia cerna*)[h]		S-I									S			I
Thimbleberry, western (*Rubus parviflorus*)[i]		I-R							S		R		I-R	I-R
Thornapple desert (*Datura discolor*)														
Titi (*Cyrilla racemiflora*)														
Toyon (see Christmasberry)														
Tree-of-heaven (*Ailanthus altissima*)[j]	S-R	S-I							S	S	S		S	S-I
Tree tobacco (*Nicotiana glauca*)		S							S		S			
Tropical soda apple (*Salanum viarum*)					S									
Trumpetcreeper (*Campsis radicans*)[k]	R	R	S	R	R			R	R		R	S		I
Tuliptree (see Yellow-popular)														
Viburnum (*Viburnum* spp.)														
nannyberry (*V. lentago*)	R	R							S-I		S			
rusty blackhaw (*V. rufidulum*)	R	I-R							S-I		S			
southern arrowwood (*V. dentatum*)	R	I							S		S			S

* S, susceptible; S-I, susceptible to intermediate; I, intermediate; I-R, intermediate to resistant; R, resistant

† Selective; Nonselective; Soil Effective.

‡ Application Methods: BS, basal spray; FS, foliar spray; I/CS, injection/cut surface treatment; ST, soil treatment; IPT, individual plant treatment.

§ Always check and double check precautions in this publication and on container label before using any herbicide.

Amitrole T is labeled for poison ivy, poison oak, honeysuckle, kudzu, salmonberry, blackberry (western dewberry), big leaf maple, wild cherry, ash, locust, and sumac.

[a] Most R when dormant.

[b] Top kill only; spray annually.

[c] IPT with tebuthiuron.

[d] S to picloram + 2,4-D amine (1:2) FS.

[e] Spray during rapid growth. Repeat as necessary.

[f] Apply triclopyr + 2,4-D (1:2) in 4% diesel oil to freshly cut surface or stump for sweetgum control. IPT with tebuthiuron. S to glyphosate by cut stump treatment on sweetgum, sycamore, and tanoak.

[g] More than one treatment needed.

[h] S-1 to 2,4-D + dicamba (4:1) by FS. Spray in pre-bud or bud stage. ST before expected rainfall with tebuthiuron.

[i] I-R to amitrole by FS. Treat in mid-summer.

[j] Wide range of S to 2,4-D by BS according to location; consult local authorities before such use on tree-of-heaven.

[k] S to glyphosphate by I/CS.

Dicamba		Hexazinone		Imazapyr		Dichlorprop		Pic+2,4D(1:4)	Dic+2,4D(1:3)	Glyphosate	Tebuthiuron			Metsulfuron	Picloram				Fosamine		Pic+Tri(1:1)	Pic+dic(1:1)	Pic+clop(1:1)
Y		N		N		Y		Y	Y	N	N			Y	Y				Y		Y	Y	Y
N		Y		Y		N		N	N	Y	Y			N	N				N		N	N	N
Y		Y		Y		N		Y	Y	N	Y			Y	Y				N		Y	Y	Y
I/CS	ST	BS	ST	BS	ST	BS	FS	FS	FS	FS	BS	FS	ST	FS	BS	FS	I/CS	ST	BS	FS	FS	FS	FS
		R	R	R	R		R									I							
	S																	S					
		R	R	R	R								I										
		I	I	S	S	S	S	I	I	I			S	R	S	S				S			
													S										
													S			S		S					
								S								S		S			S		
							S-I						S			S		S			S	S	S
													S										
													S										
																		S					
S		I	I	S	S	I-R	I-R	I	R	I			I	R	S	S-I		S-I		I			
			I	S	S	S	I	I	I	S			I	S	S	S				I			
													I										
				S	S					I				S			S						
	S						R						S					S-I					
													S										
						I-R	I-R			S						S							
				I	I																		
						S-I	S-I			S	S-I				S	S-I				I			
			R	S	S	R	R	R	I	S			I	R	S	S				I			
										I													
							I									S							

TABLE 2 (Continued)*

Herbicide[‡,§]	2,4-D[†]			Triclopyr			Clopyralid		Bromacil			Tri+2,4D(1:2)	Dicamba	
	Y			Y			Y		N			Y	Y	
	N			N			N		Y			N	N	
	N			Y			Y		Y			Y	Y	
Species[#]	BS	FS	I/CS	BS	FS	I/CS	BS	FS	BS	FS	ST	FS	BS	FS
Virginia creeper (*Parthenocissus quinquefolia*)		S			S		R	R	R		R	S		S
Wait-a-minute-bush (*Mimosa biuncifera*)[a]		R								S	S			S
Walnut (*Juglans* spp.)	S	S							S	S	S			S
Waterlocust (*Gleditsia quatica*)			S											
Waxmyrtle, southern (*Myrica cerifera*)		I-R			I				I		I			R
Weaversbroom (*Spartium junceum*)	S	S												
Whitebrush (*Aloysia gratissima*)[b]		I								I	S		S-I	I
Willow (*Salix* spp.)[c]	S	S	S	I	I					I	I	I		I
black (*S. nigra*)[d]	S	S							S	S	S			S-I
coyote (*S. exigua*)	S	S							S		S			
peachleaf (*S. amygdaloides*)	S	S							S		S			
sandbar (*S. interior*)[e]	S	S							S		S			
whiplash (*S. caudata*)	S	S							S		S			
Witchhazel, common (*Hamamelis virginiana*)									S		S			S
Wolfberry Berlandier (*Lycium berlanderi*)		R			R									R
Wormwood, common (*Artemisia absinthium*)		S-I												
Yaupon (*Ilex vomitoria*)	S-I	R	S						S-I	S-I	S-I		S-I	I-R
Yaupon, desert (*Schaefferia cuneifolia*)														
Yellow-poplar (*Liriodendron tulipifera*)[f]			S-I						S		S			
Yerbasanta (*Eriodictyon* spp.)														
California (*E. californicum*)[g]		S-I												
Yucca (*Yucca* spp.)				S										I-R

* S, susceptible; S-I, susceptible to intermediate; I, intermediate; I-R, intermediate to resistant; R, resistant.

† Selective; Nonselective; Soil Effective.

‡ Application Methods: BS, basal spray; FS, foliar spray; I/CS, injection/cut surface treatment; ST, soil treatment; IPT, individual plant treatment.

§ Always check and double check precautions in this publication and on container label before using any herbicide.

[#] Amitrole T is labeled for poison ivy, poison oak, honeysuckle, kudzu, salmonberry, blackberry (western dewberry), big leaf maple, wild cherry, ash, locust, and sumac.

[a] Apply to individual plants when soil moisture is adequate.

[b] I to MCPA by FS.

[c] S to glyphosate by cut stump treatment.

[d] Retreatment often necessary.

[e] Retreatment often necessary.

[f] IPT with tebuthiuron.

[g] More sensitive to 2,4-D within 2 years after burning, or new plants 2 years old.

Dicamba		Hexazinone		Imazapyr		Dichlorprop		Pic+2,4D(1:4)	Dic+2,4D(1:3)	Glyphosate	Tebuthiuron			Metsulfuron	Picloram				Fosamine		Pic+Tri(1:1)	Pic+dic(1:1)	Pic+clop(1:1)
Y		N		N		Y		Y	Y	N	N			Y	Y				Y		Y	Y	Y
N		Y		Y		N		N	N	Y	Y			N	N				N		N	N	N
Y		Y		Y		N		Y	Y	N	Y			Y	Y				N		Y	Y	Y
I/CS	ST	BS	ST	BS	ST	BS	FS	FS	FS	FS	BS	FS	ST	FS	BS	FS	I/CS	ST	BS	FS	FS	FS	FS
				S	S				S	S			I	I									
							R									S		S					
							S									S							
				I	I		I-R			S							S-I						
	R	S	S							I			S	S	S	S		S					
		I	I	S	S	S	S	I	S	I		I	R							I			
								S									S						
							S																
										S								S					
													S			R							
				I	I	S-I	R						S		S	S-I		S					
													S										
		S	S	S	S					I			S				S	S					
			S												I-R								

use treated hay until it is harvested the next growing season. If triclopyr is applied at 2 qt/A or less other livestock can use hay harvested more than 7 days after treatment. If 2 to 4 qt/A are applied, do not harvest hay for 14 days after treatment. If triclopyr exceeds 4 qt/A, the hay is harvested the next growing season. Withdraw cattle from treated forage 3 days before slaughter.

Avoid drift of triclopyr on nontarget areas. A useful commercial mixture of 2,4-D with triclopyr (2:1) is Crossbow. Gardon 3A is the amine form of triclopyr and is commercially available at 3 lb/gal.

F. Glyphosate (Roundup and Touchdown)

Glyphosate is applied as a foliar spray to kill certain woody plants. Glyphosate is a water-soluble compound and is subject to washoff by rainfall. Glyphosate needs a 6-hr rain-free period for maximum results. Glyphosate is very effective using the injection or frill application methods on brush and trees. The Accord form may require the addition of surfactant to be the most effective. Glyphosate has limitations on some turf pasture and rangeland situations since it may injure or kill desirable forbs and grasses. It is considered a nonselective treatment unless directed sprays or special wipers are used for precise placement of the herbicide. Treated areas are not to be grazed within 14 days after treatment.

G. Tebuthiuron (Spike 20p)

Tebuthiuron can be used any time of year, but applications prior to rainfall during the growing season may be best. Great care should be taken to make certain tebuthiuron is not applied near the root zone of desirable trees or they may be severely damaged or killed. Tebuthiuron is most effective applied as pellets versus the wettable powder for individual plants, multistemmed clumps, or small stands of woody vegetation. Tebuthiuron pellets can also be broadcast by air or ground application. Tebuthiuron has a long soil residue, so allow 2 years after application before reseeding. Grazing is allowed in areas treated with 20 Ib/A (4 ai lb/A) or less of the product. Do not cut hay for livestock feed for 1 year after treatment.

H. Hexazinone (Vepar L)

Hexazinone is active through the foliage and soil. Soil application can be made with a spot gun and controls a large number of woody species. It is not very effective on persimmon or sassafras. It should not be used for honey mesquite control on marshy or poorly drained sites or on heavy clay soils. It is also effective as an injection treatment through the bark of undesirable trees such as black cherry, oak, and sweetgum. Hexazinone is highly water-soluble, so do not apply where water sources may become contaminated. Hexazinone is highly useful in forestry application for weed and hardwood control and release of conifers. Do not cut treated forage for hay or graze domestic animals on treated areas for 60 days after application. If the rate exceeds 3 gal/A do not cut treated vegetation for forage or hay or graze domestic animals for 1 year.

I. Bromacil (Hyvar X-L)

Bromacil is a highly useful herbicide, particularly on industrial sites, rights-of-way, and other noncrop areas, because at rates used to control woody plants it also kills many forbs and grasses. It controls a large number of woody species (Table 2). Its mode of action is similar to hexazinone and tebuthiuron. Woody plants treated with bromacil, hexazinone, and tebuthiuron may partiallly refoliate after the initial treatment several times for several months before the shrubs or trees are actually killed. The land manager must be cognizant of this slow action so that the treatment is not considered a failure or so the manager does not impose another treatment on the first one. Bromacil is a relatively inexpensive material of low toxicity with a fairly long soil residue. It can be applied broadcast to control such species as oak, pine, sweet gum, and willow. Basal treatments can be made to cottonwood, hackberry, maple, oak, poplar, red bud, sweet gum, wild cherry, and willow.

J. Fosamine (Krenite S)

Fosamine is a brush-control agent applied to the foliage, and treated susceptible plants do not produce foliage or grow the following spring. Rainfall will reduce its effectiveness if it occurs the same day following treatment. Fosamine is used on noncropland, including highway rights-of-way, storage areas, and utility and pipeline rights-of-way. Directed sprays of fosoamine can be used to treat only portions of susceptible woody plants for trimming effect, such as on utility rights-of-way. Fosamine is applied during full leaf expansion in the spring until full fall coloration on deciduous species to be controlled. Coniferous species may be treated at any time during the growing season.

K. Metsulfuron (Ally and Escort)

Metsulfuron is used as a foliar spray in the spring after full-leaf out. Metsulfuron does not control a broad spectrum of species. A nonionic surfactant added at 0.25% (v/v) is recommended. It controls or suppresses blackberry, dewberry, multiflora rose, ash, cherry, elm, willow, and buckbrush. It can be mixed with 2,4-D, dicamba, or 2,4-D and dicamba, as well as other herbicides, for controlling certain weeds. There are no grazing restrictions indicated on the Ally label.

L. Imazapyr (Arsenal and Chopper)

Imazapyr controls a large number of herbaceous and woody plants. Its main application is in site preparation for conifers and for conifer release. It can be used as a broadcast spray, or in understory broadcast applications, tree injection, frill or girdle treatments, thin-line basal and stem applications, cut-stump treatments, and low-volume basal bark sprays. Liquid Arsenal contains 4 ae lb/gal, and Chopper contains 2 ae lb/gal. Imazapyr resists leaching in soils and has a fairly long soil residue, making it an effective herbicide. Sprays of imazapyr may be removed from plant leaves by rainfall the first 24 hr of application, so the timing of application is important for the best results. Imazapyr is also useful in industrial settings in which weed and brush control is needed on utility installations, highways, parking lots, railroad tracks, and many other noncrop sites. Imazapyr is an imidazolinone herbicide of low mammalian toxicity because it affects acetohydroxyacid synthase activity in plants, disrupting the production of isoleucine, valine, and leucine amino acids.

M. Paraquat (Gramoxone Extra)

Paraquat is a nonselective, foliar-applied herbicide often used to control existing vegetation at planting in no-till. It has use for weed control in many crops, conservation reserve and federal set-aside programs, noncropland, and preharvest desiccation. It can be used alone or in combination with picloram for pricklypear control as an individual plant treatment. It is used to suppress weeds and existing sod to permit pasture and rangeland reseeding. Protective gear and clothing is required when sprayed. Paraquat is a restricted-use herbicide because of its toxicity.

N. Amitrole (Amitrole T and Amizol)

Amitrole T can be used to control poison ivy, poison oak, honeysuckle, kudzu, western salmonberry, dewberry, blackberry, big leaf maple, wild cherry, ash, locust, and sumac. Treatments can be made as sprays to the fully developed foliage in the spring until the plants begin to go dormant in the fall. The plants must be thoroughly wet to the ground line for best results. Amitrole is a restricted-use herbicide because of the potential ongogenic effects observed in laboratory animals.

VI. REVIEW OF APPLICATION METHODS

Woody plants are controlled with herbicides in different ways. Herbicides can be applied by

Spraying onto foliage and stems
Wiping onto foliage and stems
Spraying basal bark or stumps
Injection into the sapwood of trees by mechanical devices or through frills or notches cut into the tree
Soil application

A. Methods of Application

See Chapter 5 for details.

1. *Broadcast sprays.* Broadcast sprays applied to foliage and stems may be low-volume (9.4 to 280 L/ha) of total spray per acre. Ultralow vol-

ume is usually 4.7 L/ha or less. Low-volume sprays include both ground and aerial application. High-volume sprays (280 to 4675 L/ha) involve application of the herbicide for brush control along roads, rights-of-way, fence rows, and other areas, or to individual plants in forests, pastures, or rangeland. All foliage and stems are thoroughly wetted. Hand-carried or power equipment is used. High-volume sprays may be required for dense, hard-to-kill species.

2. *Individual plant treatment.* Treating individual plants can be done with foliar sprays, basal sprays, or soil treatment. Foliar sprays are applied when the plant has fully developed foliage and is growing rapidly. Individual plants are thoroughly wetted with good coverage of stem tissue. Basal sprays with herbicides consist of low-volume, streamline basal and conventional basal treatment. Low-volume basal uses a solution of 25% herbicide and 75% diesel oil. The spray solution is applied to the lower 30 to 46 cm of stem with a fan or hollow-cone nozzle to wet the stem but not to runoff, thus less total volume of spray is used per plant than in the conventional basal treatment, which applies spray solution to allow runoff with puddling at the base of the plant. Handheld or power equipment can be used.

   Stream basal application utilizes a mixture of 25% herbicide, 10% penetrant, and 65% diesel fuel oil or 25% herbicide and 75% diesel fuel oil. The penetrants are Cide-Kick II and Quick Step II. The mixture is sprayed in a band 7 to 10 cm wide with a straight stream nozzle to one or two sides of the stem near the ground level. The material must go completely around the stem. The best results are with one-stemmed plants less than 10 cm in diameter and with smooth bark. Thin-line basal bark treatment can be achieved with applications of undiluted herbicide (triclopyr) in a thin stream to all sides of the stem 15 cm above the plant base. Consult the herbicide label for specific application methods and mixing instructions.

   The ''Brush Buster'' technique can be used on honey mesquite and involves mixing clopyralid or triclopyr at 5% by volume of spray solution (water). A surfactant (0.25% v/v) and dye (0.25–0.50% v/v) are used to mark the sprayed plants. Diesel fuel oil at 5% v/v with an emulsifier can be substituted for the surfactant. Foliage is sprayed to wet but not to runoff. Garden, backpack, cattle sprayers, or sprayers mounted on small tractors or all-terrain vehicles (ATV) vehicles are used. The application of herbicides using a small orifice nozzle such as the Conejet 5500-X1 by Spraying Systems Co. is recommended. Pricklypear cactus and redberry and blueberry cedar can also be treated with this system using picloram. Cedar can also be controlled with spot sprays of hexazinone to the soil surface in late winter to midspring.

3. *Cut-surface and injection treatments.* Trees larger than about 13 cm in diameter often have bark too thick for basal sprays to penetrate. The herbicide can be applied to the sapwood through frills or notches cut into the bark with an ax, or it can be injected mechanically. Frills encircle the tree and act as cups to hold the herbicide. Undiluted amine forms of triclopyr, picloram, or 2,4-D can be used. Glyphosate undiluted or half-strength is also used. In the winter, resistant species may need closer injection spacing to kill the tree. If the trees are felled, the freshly cut surface of the stump can be treated with herbicide to prevent regrowth. The cut stump should be treated immediately after felling. Cover the entire stump on trees 10 cm or less in diameter. On larger trees only the outer 5 to 8 cm of stem needs herbicide coverage. If treatment is delayed, recut the stump for best results. Triclopyr and triclopyr + 2,4-D (1:2; (Crossbow) are good stump treatments mixed with diesel fuel oil. Water-soluble formulations are effective if mixed with diesel fuel oil and an emulsifier.
4. *Soil treatments.* Certain soil-active herbicides in pellet or spray form can be applied to the soil surface in grids or bands or may be broadcast. Rainfall is needed to move them into the root zone of target plants. Tebuthiuron and hexazinone are good examples of soil-applied materials that can be applied to a spot or individual tree basis with an exact delivery handgun applicator (spot gun). Spot or grid application minimizes the injury to forage plants.
5. *Wipers.* Herbicides such as glyphosate can be applied with hand-help wipers for both the best placement and less injury to any associated desirable plants. Specially designed tractor-mounted carpeted rollers can apply herbicides to weeds and brush if not over 2 meters tall. Rope-wick applicators are confined to herbaceous weeds since woody plants can cause excessive wear and breakage on such applicators.

VII. PREPARATION FOR HERBICIDE APPLICATION

Details on herbicide formulations, herbicide additives, spray equipment, and sprayer calibration were given in Chapter 5. This information is also provided on the herbicide label or from the university extension service and state experiment stations in your area.

Pesticide containers can be disposed of by triple rinsing and puncturing them before disposing of them in an approved burial site or sanitary landfill. Contact your agricultural chemical dealer or local university extension center for information. Follow local regulations.

VIII. EXAMPLE FOR HONEY MESQUITE CONTROL

Several options are available for controlling honey mesquite. A portion of publication B-1466 (McGinty et al., 1998) is given in Table 3 for the control of honey mesquite. Diesel fuel oil or kerosene can be used for individual plant treatment (IPT) by applying it around the base of the trunk any time the soil is dry and pulled away from the trunk. Other IPTs involve applications of hexazinone (Velpar L) to the soil surface or triclopyr (Remedy) at various mixtures with diesel fuel oil and/or additives for basal application. Mixtures of triclopyr + picloram, dicamba, or clopyralid are also used for basal trunk treatments. Broadcast sprays include 2,4-D, 2,4-D + picloram (Grazon P+D), 2,4-D + dicamba (Weedmaster), or tank mixes of 2,4-D + picloram, triclopyr + picloram, triclopyr + dicamba, or triclopyr + clopyralid. Tank mixtures of picloram + dicamba, picloram + clopyralid, or clopyralid applied alone are also used. The broadcast mixtures can also be applied as IPTs including clopyralid or picloram applied alone.

Table 2 reports current research and result demonstration findings on the response of brush and trees to the most promising herbicides available. It is not intended as a list of recommended herbicides nor is it complete. The herbicides listed include commonly used commercial herbicides. Research on chemical control of brush and trees is conducted by scientists in private industry and in State and Federal agencies. As research continues, new herbicides and methods of application are being developed. Consult your local County Agent or Extension Specialist, State Agricultural Experiment Station, or the U.S. Department of Agriculture for the latest recommended and approved herbicides for woody plant control.

IX. ENVIRONMENTAL PROTECTION

Some States have restrictions on the use of certain pesticides. Check your state and local regulations. Also, because registrations of pesticides are under constant review by the U.S. Environmental Protection Agency (EPA), consult your County Agricultural Agent or State Extension Specialist to be sure the intended use is still registered.

Mention of any trademark or proprietary product in this book does not constitute a warranty or guarantee of the product by the author or Texas A&M University and does not imply approval of that product to the exclusion of any other product that may also be suitable.

X. SUSCEPTIBILITY OF PLANTS TO HERBICIDE TREATMENTS

Brush and tree species vary in their relation to herbicides and the method of herbicide application.

TABLE 3 Herbicides for Controlling Honey Mesquite on Rangeland (English Units)

Brush controlled	Herbicide	Herbicide quantity (ai rate in parentheses): Broadcast rate per acre	Herbicide quantity (ai rate in parentheses): Individual plant treatment	Spray volume (per acre for broadcast, as described for individual plant)	Time to apply	Remarks
Mesquite, huisache, twisted acacia	Diesel fuel oil, kerosene		H[a]	Apply to base of trunk from 12 to 18 in. above soil surface. Apply until solution puddles on soil surface.	Any time soil is dry and pulled away from the trunk.	Apply sufficient oil to penetrate to plant bud zone. Diesel fuel oil does not evaporate as fast as kerosene.
Mesquite, huisache	Velpar L		H; 4.8 mL per 3 ft of canopy diameter or 1 in. of stem diameter at breast height		Late winter to midspring.	Apply undiluted Velpar L to soil surface within 3 ft of stem base. Use an exact delivery handgun applicator to apply the 4- to 8-mL dose per application shot. If plant size requires more than a single 4-to-8-mL application, apply subsequent applications equally spaced around the plant. Do not use on marshy or poorly drained sites or on soils classified as clays.
Mesquite	Remedy		VII; 2% in diesel fuel oil	Apply to base of trunk from 12 to 18 in. above soil surface down to soil surface. Apply until solution puddles on soil surface.	Any time soil is dry and pulled away from trunk.	
Mesquite, basal stem diameter $1\frac{1}{2}$ in. or less	Remedy		VH; 15% in diesel fuel oil	Apply to lower 12 to 18 in. of trunk to wet the trunk; do not spray to point of runoff. Apply completely around the trunk.	Any time. Optimum time is growing season when plants have mature leaves.	This is commonly called the low-volume basal applications method. Use a fan or hollow cone nozzle. Use only on plants with smooth bark and a trunk diameter less than 4 in.
Mesquite, basal stem diameter greater than $1\frac{1}{2}$ in.	Remedy		VH; 25% in diesel fuel oil			

Mesquite, basal stem diameter 1½ in. or less	Remedy		VH; 15% in diesel fuel oil 10% d,1 limonene (a penetrant) may be added to the mixture (see remarks)	Apply to the trunk in a 3- to 4-in.-wide band near ground level or at line dividing smooth bark from corky bark. Apply completely around the trunk.	Any time. Optimum time is during growing season when plants have mature leaves.	This is commonly called the streamline basal application method. Use a straight stream nozzle. Use only on plants with smooth bark and a trunk diameter less than 4 in. Addition of a penetrant to the mixture aids with coverage around the trunk. Trade names for d,1 limonene are Quick Step II, Cide-Kick II, and AD 100. Other penetrants may be effective but have not been tested on rangeland in Texas.
Mesquite (seedlings and saplings)	Remedy		VH; 5% in diesel fuel oil	Apply to lower 12 to 18 in. of trunk to point of runoff, but not to point of puddling.	May through August.	This is commonly called the low-volume basal application method. Use a hollow cone nozzle.
Mesquite (cut stumps)	Remedy		VH; 15% in diesel fuel oil	Spray the sides of the stump and the outer portion of the cut surface, including the cambium, immediately after cutting to thoroughly wet the stem and root collar area, but not to the point of runoff.	Any season of the year, except when snow or water prevent spraying to the ground line.	This is commonly called the cut-stump application. Apply with a backpack or knapsack sprayer using low pressures and a solid cone or flat fan nozzle.
Mesquite	2,4-D amine (including Hi-Dep) or low-volatile ester	L; 2 to 4 qt (2 to 4 lb)	M; 2% (4 lb/gal product)	2 to 4 gal oil-in-water emulsion as aerial spray (1 pt to 1 gal diesel fuel oil and water to make 2 to 4 gal/acre; 1 to 5 oil-to-water ratio is considered optimum); 20 to 25 gal oil-in-water emulsion (1/2 to 1 gal diesel fuel oil and water to make 20 to 25 gal/acre) or 20 to 25 gal water/acre plus surfactant (1 to 2 qt surfactant per 100 gal water) as ground broadcast. Thoroughly wet foliage; individual plant treatment.	Late spring to midsummer with mature leaves (dark green). Optimum period of application begins when soil temperature at a soil depth of 12 in. reaches 75°F and continues for 45 days thereafter.	Treatments will control many weeds. When using oil-in-water emulsion, use emulsifier. Using a treatment with a low control rating may result in a multistem growth that may be more difficult to control in the future.
	Grazon P + D	L; 1 to 1½ qt (0.6 to 0.9 lb)				
	Weedmaster	L; 1 to 1½ qt (1 to 1½ lb)				
	Tank mix Tordon 22K with 2,4-D amine or low volatile ester	L; ½ to ¾ pt (1/8 to 3/16 lb) Tordon 22K + 1 to 1½ qt (1 to 1½ lb) 2,4-D, 4 lb/gal product				
	Tank mix Banvel with 2,4-D amine or low-volatile ester	L; ½ to ¾ pt (1/4 to 3/8 lb) Banvel + ¾ to 11/8 qt (¾ to 11/8 lb) 2,4-D, 4 lb/gal product				

TABLE 3 Continued

Brush controlled	Herbicide	Herbicide quantity (ai rate in parentheses): Broadcast rate per acre	Herbicide quantity (ai rate in parentheses): Individual plant treatment	Spray volume (per acre for broadcast, as described for individual plant)	Time to apply	Remarks
	Remedy	L; 1 pt to 1 qt (1/2 to 1 lb)	M; 1%	2 to 4 gal oil-in-water emulsion as aerial spray (1 pt to 1 gal diesel fuel oil and water to make 2 to 4 gals/acre; 1-to-5 oil-to-water ratio is considered optimum); 20 to 25 gals oil-in-water emulsion (1/2 to 1 gal diesel fuel oil and water to make 20 to 25 gal/acre) or 20 to 25 gal water/acre plus surfactant per 100 gal water) as ground broadcast. Thoroughly wet foliage for individual plant treatment.	Late spring to midsummer with mature leaves (dark green). Optimum period of application begins when soil temperature at a soil depth of 12 in. reaches 75°F and continues for 45 days thereafter; when Reclaim is used alone or in a tank mix the period should continue for 60 days.	Use 1 qt/acre of Remedy or Banvel where 1 pt/acre has not provided successful control under good conditions. Use 1 pt/acre Tordon 22K plus 1/2 pt/acre Remedy, 1/2 pt/acre Banvel plus 1/2 pt/acre Remedy, 1 pt/acre Tordon 22K plus 1/2 pt/acre Banvel, 1 pt/acre Tordon 22K plus 1/3 qt/acre Reclaim and 1/3 qt/acre Reclaim only in west Texas. Banvel and Banvel mixtures have been more effective in west Texas than in other parts of the state. Mixtures with Tordon 22K or Banvel will control many weeds. Use mixtures that include 1/4 pt/acre Remedy and 1/3 pt/acre Reclaim and the fall application of Remedy only in Montague, Wise, Parker, Hood, Somervell, Bosque, Coryell, Lampasas, Burnet, Blanco, Kendall, Bandera, Real, Edwards and Val Verde Counties and those counties north and west of the named counties. Mixtures that include 1/2 pt Remedy and 2/3 pt Reclaim will give better control than mixtures with 1/4 pt Remedy and 1/3 pt Reclaim. When using oil-in-water emulsion, use emulsifier added to oil for proper emulsion. Using a treatment with a low control rating may result in a multistem growth that may be more difficult to control in the future.
	Banvel	L; 1 pt to 1 qt (1/2 to 1 lb)	M; 1%			
	Reclaim	M to H; 1/3 qt to 2/3 qt (1/4 to 1/2 lb)	VH; 1%			
	Tank mix Remedy with Tordon 22K	M; 1/2 to 1 pt (1/4 to 1/2 lb) Remedy; 1 to 2 pt (1/4 to 1/2 lb) Tordon 22K	H; 1/2% Remedy; 1/2% Tordon 22K			
	Tank mix Remedy with Banvel	L; 1/2 to 1 pt (1/4 to 1/2 lb) Remedy; 1/2 to 1 pt (1/4 to 1/2 lb) Banvel	M; 1/2% Remedy; 1/2% Banvel			
	Tank mix Remedy with Reclaim (see remarks)	M to H; 1/4 to 1/2 pt (1/8 to 1/4 lb) Remedy; 1/3 to 2/3 pt (1/8 to 1/4 lb) Reclaim	VH; 1/2% Remedy; 1/2% Reclaim			
	Tank mix Tordon 22K with Banvel	M; 1 to 2 pt (1/4 to 1/2 lb (Tordon 22K; 1/2 to 1 pt (1/4 to 1/2 1/2 lb) Banvel	H; 1/2% Tordon 22K; 1/2% Banvel			
	Tank mix Tordon 22K with Reclaim	H; 1 to 2 pt (1/4 to 1/2 lb) Tordon 22K; 1/3 to 2/3 qt (1/4 to 1/2 lb) Reclaim	VH; 1/2% Tordon 22K; 1/2% Reclaim			
	Tank mix Remedy, Reclaim, and Tordon 22K	M to H; 1/4 to 1/2 pt (1/8 to 1/4 lb) Remedy; 1/3 to 2/3 pt (1/8 to 1/4 lb) Reclaim; 2 pt (1/2 lb) Tordon 22K				

	Reclaim (see remarks)	H; 2/3 qt (1/2 lb)	VH; 1%		Aug. 1 to Sept. 30 with a soil temperature of 75°F or more at a soil depth of 12 in. Do not apply after a frost has occurred.	
	Tordon 22K		VH; 1 gal (2 lb)[b]	Apply with a carpeted roller.	Late spring through August with mature leaves (dark green). Best control during the period that begins when soil temperature at a soil depth of 12 in. reaches 75°F and continues for 45 days thereafter; when Reclaim is used alone or in a tank mix the period should continue for 60 days after soil temperature reaches 75°F.	Mesquite should be less than 6 ft tall and should pass under carpeted roller without breaking the main stem.
	Reclaim		VH; 2/3 gal (2 lb)[b]			
	Tank mix Tordon 22K with Reclaim		VH; 2 qt (1 lb) Tordon 22K; 1 1/3 qt (1 lb) Reclaim[b]			
Mixed brush (South Texas—will include several of the following: blackbrush, catclaw acacia, granjeno or spiny hackberry, huisache, mesquite, pricklypear, retama, skunkbush, tasajillo, twisted acacia)	Tank mix Tordon 22K with Remedy	M; 2 pt (1/2 lb) Tordon 22K; 1 pt (1/2 lb) Remedy	H; 1/2% Tordon 22K; 1/2% Remedy	4 gal oil-in-water emulsion as aerial spray (1 qt to 1 gal diesel fuel oil and water to make 4 gals/acre; 1-to-5 oil-to-water ratio is considered optimum); 20 to 25 gal oil-in-water emulsion (1/2 to 1 gal diesel fuel oil and water to make 20 to 25 gal/acre) or 20 to 25 gal water/acre plus surfactant (1 to 2 qt surfactant per 100 gal water) as ground broadcast. Thoroughly wet foliage for individual plant treatment.	Late spring to midsummer with mature leaves (dark green). Optimum period of application begins when soil temperature at a soil depth of 12 in. reaches 75°F and continues for 45 days thereafter; with the Reclaim tank mix the period should continue for 60 days after soil temperature reaches 75°F. If mesquite has 10% canopy cover or less, application may be made in spring or fall.	The mixture of 1 qt Tordon 22K plus 2/3 qt Reclaim will usually provide better results than the 1 qt Tordon 22K plus 1/3 qt Reclaim mixture. Mixtures will control most weeds. When using oil-in-water emulsion, use emulsifier added to oil for proper emulsion.
	Tank mix Tordon 22K with Reclaim	M; 1 qt (1/2 lb) Tordon 22K; 1/3 to 2/3 qt (1/4 to 1/2 lb) Reclaim	H; 1/2% Tordon 22K; 1/2% Reclaim			
	Tank mix Tordon 22K with Banvel	M; 2 pt (1/2 lb) Tordon 22K; 1 pt (1/2 lb) Banvel	H; 1/2% Tordon 22K; 1/2% Banvel			

[a] Treatment control ratings: VH—very high; H—high; M—moderate; L—low.
[b] Mix with 3 to 6 oz/surfactant and water to make 8 gal of mixture.
Source: McGinty et al. 1998.

The control ratings of the herbicides in the susceptibility table, Table 2, are as follows:

Susceptible (S)

Brush and tree species are considered SUSCEPTIBLE (S) to a herbicide if one application of the herbicide kills more than 70% of a stand.

Susceptible to Intermediate (S-I)

The species is considered SUSCEPTIBLE TO INTERMEDIATE (S-I) in its response to a herbicide if two applications of the herbicide are needed to kill at least 70 percent of a stand.

Intermediate (I)

Species considered INTERMEDIATE (I) in their response to a herbicide generally are top-killed by one or two treatments, but several more treatments are usually required to kill plants.

Intermediate to Resistant (I-R)

The species is INTERMEDIATE TO RESISTANT (I-R) in its response to a herbicide if the tops and sprouts of a species can be killed, but the roots continue to sprout—even after repeated application of the herbicide.

Resistant (R)

Species virtually unaffected by a herbicide.

XI. SUSCEPTIBILITY TABLE

The Susceptibility table (Table 2) lists the reactions of a number of common brush species to herbicides and methods of applications. The table can be used to determine which combination of material and method of application is best suited for each woody species. The phenoxy herbicides are usually the ester formulations in the table, unless otherwise indicated.

All references cited at the end of this chapter were used to construct Table 2.

REFERENCES

American Cyanamid. Chopper and Arsenal herbicide specimen labels. Parsippany NJ: American Cyanamid Co., 1998.

Baldwin FL, Boyd JW. Recommended chemicals for weed and brush control. MP-44. Little Rock, AR: Coop Ext. Serv., University of Arkansas, 1998, pp. 125–134.

BASF Corp. Banvel herbicide specimen label. Research Triangle Park, NC: BASF Corp., 1997.

Bovey RW. Response of Selected Woody Plants in the United States to Herbicides. Agric. Handbook no. 493. Washington, DC: USDA, ARS, 1977.

Bovey RW. Weed control problems, approaches and opportunities in rangeland. Rev Weed Sci 3:57–41, 1987.

Boyd J. Pasture brush control. FSA 2081. Little Rock, AR: Coop. Ext Serv., University of Arkansas, 1995.

Dewey SA, Whitson TD, Sheley R. Weed Management Handbook. Coop Ext. Serv., Bozeman, MT: Montana State University, Logan, UT: Utah State University and Laramie, WY: University of Wyoming, 1998.

Dow AgroSciences. Vegetation management specimen labels. Indianapolis, IN: Dow AgroSciences, Dow Elanco, 1996.

Duncan KW, McDainel KC. Chemical weed and brush control guide for New Mexico range lands. Las Cruces, NM: Coop. Ext. Serv., New Mexico State University, 1991.

DuPont. Vepar, Krenite S, Ally, and Hynar X specimen labels, Wilmington, DE: DuPont, Agricultural products division, 1997.

Johnson WG, Kendig JA. Weed and brush control guide for forages, pastures and noncropland. MP 581. Columbia, MO: Univ. Ext., University of Missouri, 1997.

Martin AR, Roeth FW, Wilson RG, Wicks GA, Klein RN, Lyon DJ. A 1998 Guide for herbicide use in Nebraska. EC 98-130 D. Lincoln, NE: Nebraska Coop. Ext. Serv., University of Nebraska, 1998, pp. 65–74.

Martin JR, Green JD. Weed management in grass pastures, hayfields and fencerows. AGR-172. Lexington, KY: Coop. Ext. Serv., University of Kentucky, 1998, pp. 6–10.

McGinty A, Cadenhead JF, Hamilton W, Hanselka WC, Ueckert DN, Whisenant SG. Chemical weed and brush control suggestions for rangeland. B-1466, College Station, TX: Texas Agric. Ext. Serv. Texas A&M University, 1998.

Monsanto. Roundup Pro and Accord herbicide specimen labels. St. Louis, MO: Monsanto Co., 1996.

Motooka P, Nagai G, Ching L, Kawakami G, Powley J. Summaries of herbicide trials for pasture, range and non-cropland weed control, 1997. Manoa, HI: Coop. Ext. Serv., University of Hawaii, 1988, 11–25.

PbI/Gordon Corp. Weed and brush control product specimen labels. Kansas City, MO: PbI/Gordon Corp., 1996.

Pennsylvania State University. The field crop weed control guide—1998. University Park, PA: College of Agric. Sci. Coop. Ext. Serv., Pennsylvania State University, 1998, pp. 83–84.

Regehr DL, Petersen DE, Ohlenbusch PD, Fick WH, Stahlman PW, Kuhlman DK. Chemical weed control for field crops, pastures, rangeland and noncropland, 1998. Rept. of prog. 797. Manhattan, KS: Coop. Ext. Serv. and Agric. Exp. Stn., Kansas State University, 1998, pp. 56–62.

Rhône Poulenc Ag. Co. Amitrole T herbicide specimen label. Research Triangle Park, NC: Rhône Poulenc Ag. Co., 1993.

Stritzke JF. Suggested weed control in pasture and range. In: 1998 OSU Extension Agent's Handbook of Insect, Plant Disease and Weed Control. Oklahoma Coop. Ext. Serv., Oklahoma State University, E-832. Stillwater, OK: 1998, pp. 205–212.

Tschirley FH. Problems in woody plant control evaluation in the tropics. Weeds 15:233–237, 1967.

Wrage LJ, Johnson PO. Weed control in grass pasture and range. Coop. Ext. Serv., FS. Brookings, SD: South Dakota State University, p. 525. 1997.

Zollinger RK. 1998 North Dakota weed control guide. Cir. W. 253. Fargo, ND: NDSU Ext. Serv., North Dakota State University, 1998, pp. 44–47.

13

Economics of Woody Plant Management

I. INTRODUCTION

The primary reason for woody plant control is to improve the land resource to reach a goal or goals. Land resource improvement usually involves improving the type of vegetation growing on the area for the purpose intended, such as improved and increased forage for livestock, conifer release, improved wildlife habitat, aesthetics, control of vegetation on roadways and utility lines for improved visibility and aesthetics, and fire prevention and aesthetics around buildings and industrial sites. Protecting roadways, railroads, electrical lines, and pipelines is necessary to protect the enormous investment in both these facilities and the services rendered by them.

As demands for forest products increase it has become apparent that artificial regeneration and intensive management is necessary to meet production requirements by controlling brush and low-value hardwoods. Adequate control of woody plants on grazing lands may be required for economic livestock production, but returns in livestock products may not always pay back the cost of treatment. In grazing lands allowing vegetation to revert back to climax or desirable vegetation may take too long or may never occur, as demonstrated by exclosures on many rangeland sites across the United States. In this case, interference by man's management practices is necessary to restore the intended goals. Take care of the biggest problems first, and if resources are limited treat only a portion of the problem area annually. Treatment of brush in strips or partially treated areas contains costs and improves wildlife habitat by providing protection in the untreated areas and food in the untreated and treated strips. Allowing wildlife harvest can add significantly to the income of the land resource. Woody plant control should also consider the integration of herbicides with other practices to contain costs and improve brush control, such as by fire or mechanical means. Brush control improves wildlife and livestock management by improving visibility and discourages predator activity.

It is important to realize that woody plant control is a continuous process, and treatment of a problem area is usually not a one-time event. Persistent woody plants may eventually reestablish a few years after brush control from rootstocks or seed. The land manager must be cognizant of these changes and individually or collectively treat missed or new plants before they become a problem. Recognizing potential problems and dealing with them early on (prevention) will save the land manager a great deal of money and time down the road.

Herbicides are used to make rapid ecological shifts for the land manager's purpose that may not occur by natural evolution. Another advantage of herbicides that may not always be considered in economic returns is the control of such factors as herbaceous, poisonous, injurious, and allergenic-producing weeds that may limit both wildlife and livestock production and cause human suffering. The presence of these weeds also limits the growth of endangered plants or the regeneration of desirable forbs and grasses. Good land management and weed and brush control promote soil and water conservation and improve income and land values, and everyone benefits. Too often the land is exploited for quick profits out of greed or ignorance of good stewardship. Good long-term management with an understanding of ecological principles balanced against sound economic expenditures and returns will be the most productive. Herbicides provide user-friendly tools to help accomplish weed- and brush-control goals and can be used as economics dictates by suppressing the most significant problems first followed by retreatment and maintenance where needed.

II. MESQUITE AND ASSOCIATED WOODY PLANTS

Valentine and Norris (1960) reported that the cost of mesquite control in New Mexico with ground spray on lightly to moderately infested rangeland, including labor, equipment, and material, ranged from $0.54 to $1.65 per acre and averaged $1.15 per acre. The costs were roughly proportionate to the mesquite infestation, which ranged from 10 to 154 plants per acre and averaged 82 plants per acre. Kill ranged from 38 to 90% and averaged 54% using 2,4,5-T, depending upon the site.

Hoffman (1967) indicated that the cost of aerial spraying for mesquite control in Texas with 2,4,5-T in diesel fuel oil and water diluent was $2.75 per acre and resulted in the following:

Increased calf weights of 40 lb per animal
Increased stocking rate of 30%
Improved forage grasses
$1 per acre saved in working livestock

Kuntz and Hoffman (1983) estimated the income losses on mesquite-oak rangelands from the loss of 2,4,5-T, silvex, and 2,4,5-T +picloram to be $341 million over a 16-year cumulative period, with a $454 million reduction in producers' incomes if these herbicides and dicamba were not used.

The loss of 2,4-D from use on pastureland, rangelands, alfalfa production, and noxious weeds was collectively and conservatively estimated at about $1 billion (Bovey, 1996).

Since 2,4,5-T and silvex have been lost, more expensive substitute herbicides are required for mesquite and associated brush control. The cost of herbicides is given in Table 1 in Chapter 12. 2,4-D is one of the most inexpensive materials, however; other herbicides mixed with 2,4-D are usually required for best results on woody plants.

Whitson and Scifres (1980) indicated that aerial application of 2,4,5-T consistently produced the highest annual rates of return in Texas, regardless of vegetation region. The unweighted average annual return was 16%. When 2,4,5-T was eliminated as a potential control measure, dicamba produced the highest annual rate of return—more than 11%.

The simple average of the highest average annual rate of return from each resource region for nonherbicide treatment was 5.7% (Whitson and Scifres, 1980). The average length of time required to recover all investment capital for treatment and additional livestock with 2,4,5-T was 8.5 years, or about half that for the next best herbicide treatment (16 years).

Vantassell and Conner (1986) developed a 15-year dynamic linear programming model to determine the economic feasibility of incorporating various alternative production methods and range sites in the Rolling Plains of Texas. Aerial spraying followed by prescribed burns was an economically effective method of controlling mesquite on productive clay loam soils, but was questionable on loamy prairie range sites. Control methods for sand shinnery oak and juniper did not appear economically viable. Ethridge et al. (1987), however, indicated that the discounted net return for the use of tebuthiuron on sand shinnery oak was generally positive with high and moderate calf prices and low and moderate discount rates. The optimum tebuthiuron treatment rate varies with calf prices, discount rate, and treatment costs.

Ethridge et al. (1984) developed an evaluation model for evaluating the economic feasibility of 2,4,5-T for honey mesquite control in the Texas Rolling Plains. The model estimated the net present value of added grass production from

2,4,5-T treatments over the life of the treatment. The gross value of treatment with 2,4,5-T was estimated using different combinations of the livestock price, top kill, canopy cover, and discount rate. The gross value of mesquite control varied from $22/ha to over $73/ha. These returns compare to current treatment costs of $22 to $25/ha.

Torrell and McDaniel (1986) used the equations of the Ethridge et al. (1984) model to define key physical, biological, and economic relationships for mesquite control. Although a positive grass response from mesquite control was anticipated for 5 years, the optimum economic retreatment schedule is only 4 years. An increase in beef prices shortens the optimal retreatment schedule, while an increase in treatment cost lengthens the optimal retreatment schedule. Implementing brush control during a favorable year for a high rate of kill is an important economic consideration.

Root plowing at Sonora resulted in a 2% annual real rate of return on the investment, while grubbing or aerial spraying were not economically viable options for increasing net income at either Sonora or Barnhart, Texas, from 1970 to 1980 (Whitson et al., 1984). Gross annual sales were increased 27% by root plowing, while increases from other methods were less than 2% compared to the control at both locations.

McBryde et al. (1984) indicated that profitable brush-management alternatives exist for all (four) brush types described in eastern south Texas (except mixed brush on shallow sites). The projected treatment cost has the largest influence on profitability, followed by prescribed burning.

Scifres et al. (1985) published an excellent bulletin on the integrated brush management system (IBMS) for south Texas. The bulletin contains eight chapters explaining how the IBMS evolved, how to use it in livestock and wildlife management, and how to evaluate its effectiveness. Using brush control to the best economic advantage is central to effective brush management; standard brush-control treatment may not pay back the cost of investment (Whitson and Scifres, 1980). A application of low-cost follow-up treatments may be necessary to extend the effectiveness of some brush-management schemes long enough to realize a profit (Scifres, 1985). Several possible alternative treatments following the initial treatment may be selected on the basis of the lowest-cost treatment that fits land-management strategies.

The first step in IBMS development is to identify management objectives for the land (Hamilton, 1985). Management objectives can be changed as interests or needs change. A comprehensive resource inventory of the range site is done to determine if brush control is needed, where it should be applied, and the species to be targeted. It must also become the basis for planning concomitant grazing management, wildlife management, and other system elements, as well as future maintenance needs and timing for the best vegetation composition.

Scifres and Hamilton (1985) developed the IBMS master-planning flowchart that includes all elements to consider in selecting brush-management alternatives. Working flowcharts must be updated periodically to include the latest technology in concert with livestock grazing and wildlife management goals.

Conner (1985) indicated that the first step in assessing benefits and costs is to estimate both the resources (labor, equipment, time, chemicals, etc.) required to implement each alternative practice or program and the changes in annual productivity (pounds of beef or wood, number of deer, etc.) expected to result from the practice.

The planning period should include the number of years after the initiation of a practice for which an increase in productivity over the current level can be expected (Conner, 1985). In practice this is usually 8 to 20 years.

Although specific predictions of production responses to brush management usually are not available, many range scientists, range conservationists, and ranchers have considerable experience in observing production responses. Whitson and Scifres (1980) developed a system based on a generalized response curve of Workman et al. (1965) for recovering, by interview data, to build comparative response curves. The response curve (Fig. 1) represents the change over time in the carrying capacity of a specific site or management unit resulting from a specific treatment or set of treatments.

As indicated by Conner (1985) the following estimates are needed to construct a response curve:

1. Pretreatment production of the site or management unit, indicated by point P_0 on the hypothetical curve (Fig. 1). If an entire management unit is to be evaluated, separate curves need to be developed by site, and the final analyses should be weighted by the proportion of each site in the management unit. This information comes from the resource inventory used to assess the production potential when developing an IBMS.
2. Expected treatment life (TL), defined as the length of time in years required for the production level to return to P_0. The point at which the treatment effect is exhausted is indicated as TE_0. In cases in which the initial treatment effect is prolonged or enhanced by follow-up treatment, the TL may be prolonged indefinitely. In such cases, the treatment effects (e.g., increased annual production levels) are projected through the last year of the planning period.
3. The maximum level of production (P_{max}) that will be achieved by the treatment for each major range site. The time that maximum production will be sustained is associated with the P_{max} value, TP_{max}, on the hypothetical response curve.

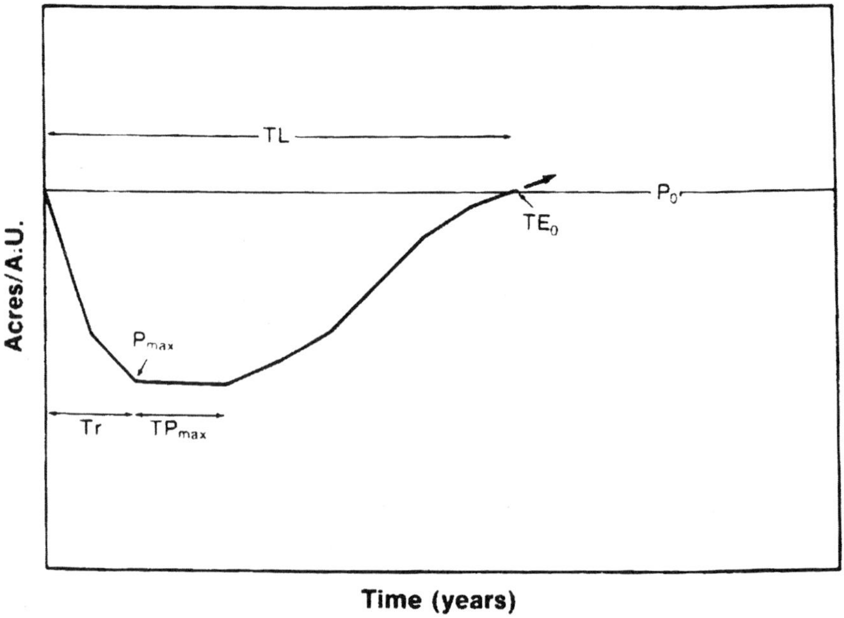

FIGURE 1 Components of a hypothetical response curve for an economic evaluation of brush-management alternatives. (From Conner, 1985.)

4. The time required to reach P_{max} after application of a given treatment at P_0. This is noted as Tr on the hypothetical response curve.

Several critical assumptions and considerations underlie the development of response curves (Conner, 1985). As additional data are accrued, these assumptions can be replaced by quantitative information.

1. The level of production without treatment is usually assumed to remain constant throughout the planning profile. This assumption simplifies the calculation of the changes in the annual production levels resulting from the treatments. If information is available from which to project annual production levels without the treatment, however, then these projections may be used instead of the constant P_0.
2. The current response curve procedure assumes average rainfall.
3. Brush management does not interact with other resource management practices. A radical change in other management practices (e.g., grazing management) may cause significant variation in the performance of brush-management treatments, therefore it is assumed that other

management practices are held constant throughout the planning period.

Changes in carrying capacity cannot be converted directly into changes in the number of pounds of beef produced. For example, in a cow–calf operation, the effects of brush treatment on the herd's annual calving percentage and the average calf weaning weight must also be estimated. Estimates of these changes can be obtained from producers with experience. A detailed example of how these changes in annual productivity are estimated can be found in McBryde et al. (1984).

The numerical response of game animals to brush treatments can be estimated with an adequate degree of accuracy. The tie between ecology and economics, however, is not as straightforward for wildlife as it is for livestock.

Economic evaluation involves partial budgeting to estimate the change in cash flow resulting from the application of the treatment for each year over the life of the treatment (Conner, 1985). These annual net cash flows (cash flow with treatment minus cash flow with no treatment) can then be discounted and summed for comparison with other improvement practices.

Most costs associated with brush management are for products and services (such as chemical and aerial applications) for which market prices are readily available. Furthermore, the market prices are subject to little variation over a wide range of conditions and locations.

Accurate estimates of benefits, however, are generally much more difficult to obtain. Tangible benefits usually result from increases in salable products and (or) reductions in the cost of producing salable products. Salable products are generally limited to livestock, livestock products, and hunting leases. Market (sale) prices of livestock products for each year over the life of the project may be estimated using average market prices after adjusting for the effects of inflation and cyclical price variation.

Estimates of the impact of brush-management alternatives on the annual costs associated with livestock production must also be obtained. Cost reductions resulting from brush-management practices may include the reduced levels of supplemental feeding and reduced labor needed for handling livestock.

Hunting lease costs are associated with deer management (inventories, feeding, etc.), the development and maintenance of hunt facilities (cabins, blinds, etc.), services rendered, insurance, and legal matters—all of which fluctuate with the intensity of game management.

Once the benefits and costs occurring each year over the life of each alternative are determined, they are discounted to account for the time value of money (i.e., the interest that could be earned if the money were invested in a savings account) and summed to indicate the relative economic feasibility of each alternative.

Conner (1985) showed two examples on a hypothetical 1000-acre pasture on south Texas rangeland with a 50% canopy cover of mixed brush. Two alternative were given: 1) aerial application of herbicide followed by periodic prescribed burning to suppress regrowth, and 2) roller chopping followed by periodic prescribed burning. A 12-year planning period was selected. The total cows, total cows added, percentage of calf crop and average weaning weights, average production per cow, increased production per cow, increased sales weight from year 0, sales weight from added cows, and total gross sales increase are constructed in a table for the 12-year period for range-improvement practice. The factors determining cattle investment costs include normal culling, buying cows and bulls to adjust the stocking rates based on carrying capacity. The initial brush-management costs must also be considered, as well as the cost per acre for the prescribed burning, which may be required three or four times during the 12-year period.

Determining the added annual variable costs from the range-improvement practice for added livestock and so forth may be needed. Other changes in revenue, such as hunting revenue or grazing deferment costs for forage for prescribed burning, change the cash flow (Conner, 1985).

The changes in revenue and costs for each year and the resulting net change in annual cash flow resulting from each program are recorded (Conner, 1985). These changes in annual cash flow are then discounted at the appropriate rate to account for the opportunity cost of money over the planning period. The present value factor for each year is determined by the formula $1/(1 + r)^n$, where r is the discount rate and n is the year. The investment in brush management should earn a return at least equal to the selected rate (10%, e.g.). This rate may be the cost in interest to borrow money on the interest one could earn if the money were invested elsewhere. If the accumulated net present value at the end of 12 years is zero, then the investment in brush management is equal to the discount rate specified; if negative, then the estimated return rate is less than the specified discount rate (Conner, 1985).

The cost of brush management changes with time, but this information can be used to decide which practice to select based on the expected rate of return.

Using the model indicated by Conner (1985), Scifres and Hamilton (1989) set up hypothetical case studies to determine treatment costs for mixed-brush management in south Texas. Six scenarios were involved: 1) aerial spraying in strips–60% of the pasture, 2) aerial spraying and retreatment, 3) aerial spraying followed by prescribed burning, 4) aerial spraying in strips—80% of the pasture, 5) aerial spraying 60% of the pasture in strips plus cool-season burns at years 7, 11, and 14, and 6) aerial spraying 60% of the pasture in strips plus burns in years 5, 9, and 13 on a sandy loam range site. For scenario 1, the economic performances of the treatment based on the internal rate of return (IRR) resulted in less than 0 at any cost of the herbicide and application above about $65/ha. The target IRR (discount rate) of 8% was not reached at a cost of over $44/ha for the treatment. In scenario 2, the IRR was positive at a cost of $80/ha, but the desired 8% return on investment was not reached at a treatment cost of $47/ha. In scenario 3, the treatment did not achieve the 8% return on the money invested. The actual IRR was 4%, at a cost of $64.25/ha for the treatment. In scenario 4, calf prices and all other assumptions were the same as for scenario 3. At the same treatment cost of $64.25/ha for the treatment, the IRR increased from 4 to 4.9% for the 80% treated area. In scenario 5, the IRR would be less than 0 for the cost of treatment at $43/ha, and the IRR of scenario 6 was less than 0 for herbicide and application costs greater than about $64/ha.

From the hypothetical scenarios Scifres and Hamilton (1989) concluded that the sandy loam site is best matched to the technical alternative being considered—aerial spraying picloram plus triclopyr at 1.1 kg/ha. The IRR generated presents the best opportunity for successful economic performance of the treatment alternative. The productivity of this site is great enough to create herbaceous fine fuel for effective burns. The study indicated the importance of matching available herbicide technology with the woody plant problems. The analysis also indicated that maintenance (prescribed burning) of the initial benefits of herbicide treatment is necessary for successful economic performance over the planning horizons compared to the treatment life of the herbicide alone.

Conner et al. (1990) designed ECON as a principal component of a Resource System Planning Model (RSPM) (Stuth et al., 1990) for rangeland resource management. It was designed to assist resource managers in estimating returns on investments in range improvement and/or grazing-management practices. ECON (Conner et al., 1990) is a computerized program to delineate annual costs and revenues of specific practices or a set of practices over a 1-to-20-year planning period.

More recently, the selection of the appropriate brush-management technologies should follow the examples given and can be made with the use of the EXSEL brush-management decision-aid software (Hanselka et al., 1996 a,b).

EXSEL (the Expert System for Brush and Weed Control Technology Selection) is now available on the Web free of charge (Rector, 1999). Access to this user-friendly technology or treatment selection process can be found at the TAMU Ranching Systems Group home page at http://cnrit.tamu.edu/rsg/ or at the Web site for the Center for Grazinglands and Ranch Management at http://cnrit.tamu.edu/cgrm. EXSEL can provide suggested alternatives for treating a brush or weed problem in Texas. The recommendations contained in the EXSEL program are intended for use in Texas only. The application and effectiveness of the recommended treatments outside Texas were not considered in the development of the program, and no usefulness outside Texas is intended or implied. The use of this program on the Web is funded by Dow AgroSciences.

The TAMU Ranching Systems home page shows a list of products that can be queried. To use this brush and weed program, select the EXSEL expert system for the selection of weed- and brush-control methods. An EXSEL 1.08 Web information page will appear and the user selects "Examine a Situation" for finding the methods of control or management for a particular weed or brush species. Additional selections at this site include user instructions about parameters, a controlled burn checklist, and an herbicide-mixing chart. At the conclusion of specifying a particular brush or weed situation, a brush- and weed-management report is generated, which can be printed for a client or personal file.

EXSEL is designed to consider a pest-management problem from a particular range site and the characteristics of a plant's growth in order to suggest the best recommendation for a certain brush or weed technology over another one. After describing a scenario and identifying any restrictions relative to method, the program provides several alternatives for a specific situation. The user may then decide among these alternatives by choosing the most desirable option. It may be the one for which materials and/or equipment are easily accessible or the one that suits the user's personal preference. At any point during the EXSEL program, the user may retrieve a report, which will indicate the most effective mechanical, chemical, or prescribed burn treatment for the identified site and target plant species.

Remember that your input is combined with the opinion of experts with an abundance of experience in brush and weed control (McGinty et al., 1998). It is therefore important to provide accurate information so that the valid suggestions for the management and control of the problem species will be included in the report. EXSEL is designed to offer the best recommendations. All possible treatments will not necessarily be included in the report.

III. SAGEBRUSH

In Oklahoma, Williams (1969) sprayed sand sagebrush and indicated that it is one of the most profitable range-improvement practices available. The cost and return per ha is dependent upon the site. It was suggested that poor sites with large, steep, sandy dunes be left untreated because of wind erosion.

Hyatt (1966) sprayed big sagebrush-infested areas in Wyoming and showed an increase in 10 lb/lamb and 16 lb/calf on an annual basis; the carrying capacity and watershed yields on the range were improved.

Controlling big sagebrush on a bunchgrass range in Oregon that was grazed during or after seed maturity of the principal grasses increased yearling grazing days 1.9 times and produced 2.3 times more beef per acre (Sneva, 1972). The IRR from beef and estimated costs was more than 50%. After 17 years the brush recovered almost to pretreatment conditions.

The economic feasibility of four Wyoming big sagebrush control methods (burning, spraying with 2,4-D, plowing and seeding, and rotocutting) were analyzed (Watts and Wambolt, 1989). Perennial grass response to big sagebrush treatment was measured to estimate a treatment-response function for each control method, which provided the biological informational base to develop the cost and benefits. The annualized present values of the net additional returns for single 2,4-D and burning treatments were $1.10/ha and $1.16/ha, respectively. When big sagebrush was retreated at optimal intervals, 2,4-D and burning generated annualized net additional returns of $2.88/ha and $2.57/ha, respectively. Roto-cutting was marginally feasible, and plowing was not feasible.

An approach was developed for estimating the economic optimum rate of initial overstory kill for increasing seasonal forage availability (Tanaka and Workman, 1988). The model was formulated using: 1) a biological production function relating understory production to the initial kill percentage, 2) a derived demand function for seasonal forage value, and 3) a cost of the overstory kill function for each control method. The general model may be applied to any ranching situation in which the understory forage production is constrained by undesirable overstory vegetation. The model was illustrated using the big sagebrush-crested wheatgrass vegetation type on a Utah cow–calf–yearling operation with prescribed burning, 2,4-D spraying, and tebuthiuron application as control methods. For the ranch analyzed, a big sagebrush kill rate between 92 and 100% is optimal, depending on the derived demand and cost-of-kill functions used. Kill rates that differ from the optimum cause significant opportunity costs to be incurred.

The optimum combination of three range improvements was determined for private lands on Utah ranches (Evans and Workman, 1994). The determination of which alternatives to implement must consider the total ranch operation. Linear programming (LP) makes it possible to simultaneously determine the profit-maximizing combinations of range improvements and how these improvements will affect the total ranch operation. The study examined (revegetation, burning, and chemical brush control) big sagebrush and pinyon-juniper (*Pinus* spp.-*juniperus* spp.) on upland loam and upland shallow loam range sites. The net present valuc analysis and an LP model were used to identify the most efficient alternative, the limiting constraints, and the optimum levels and combinations of alternatives. The optimal solution ran 238 brood cows, compared to 196 for the typical Utah ranch. Burning big sagebrush or pinyon-juniper infestations on crested wheatgrass (*Agropyron desertorum* Fisch. ex Link) foothill ranges was the most profitable range improvement. The annual net cash incomes after burning sagebrush or pinyon-juniper on the upland loam site were $37,873 and $37,770, respectively, compared to $31,278 on the typical Utah cow–calf operation. The optimal solution will change as the input and product prices change. The model was designed for application to specific ranches rather than to make general recommendations for the typical Utah ranch.

Fringed sagebrush is a native, drought-resistant, increaser species in Saskatchewan pastures and rangeland (Peat and Bowes, 1994). Conventional control by cultivating and reseeding is neither highly effective nor sustainable (Peat and Bowes, 1994). The isooctyl ester of 2,4-D showed sufficient control at 1.5 kg ai/ha. Two applications of 2,4-D yielded the greatest accumulation of grass over the length of the experiments. The economic threshold of fringed sagebrush is the density at which the yield is reduced by approximately 290 kg/ha; above this density it is economically viable to use 2,4-D at 1.5 kg/ha to control fringed sagebrush. Grazing practices can greatly aid in the management of fringed sagebrush by allowing the forage to outcompete the fringed sagebrush.

Much research concerning sagebrush control methods and forage response after control has been conducted due to the importance of sagebrush–grass-dominated rangelands for livestock and wildlife in the western United States (Bastian et al., 1995). Very little research has addressed the economic feasibility of sagebrush control at various levels of abundance. Bastian et al. (1995) estimated the economic threshold abundance of sagebrush based on the forage response data from a sagebrush control experiment in Carbon Country, Wyoming. The forage response data are based on the difference in herbage between treated and untreated experimental units from sites ranging in initial sagebrush canopy cover from 4 to 40%. Breakeven returns per AUM were estimated for each sagebrush canopy cover level, assuming 2,4-D or burning (for 28 to 40% canopy cover) as a control method, with the lives of the controls at 15, 20, and 25 years. These breakeven returns were compared to a net lease rate of $6.13/AUM. The results indicate the economic threshold abundance of sagebrush is 12%, assuming 2,4-D as the control method and a control longevity of 25 years, but the feasible sagebrush abundance increases as the longevity of the control decreases. If the longevity of the control only lasts 20 years, the sagebrush abundance must be at least 20% before treating sagebrush becomes economically feasible. If the longevity of the control is only 15 years, the sagebrush abundance must be at least 24% canopy cover before the treatment is economically viable. Given that the estimated costs of burning are almost half that of spraying with 2,4-D, all the scenarios that had enough biomass to sustain a burn (28 to 40%) indicated that sagebrush control by fire was economically viable.

IV. JUNIPER TREES

A cost evaluation was conducted of four alternatives for improvements on maturing western juniper (*Juniperus occidentalis*) woodlands (Young et al., 1982). The alternatives were: 1) the use of picloram to kill the trees with no further treatment, with a total cost of $78/ha ($31/acre); 2) picloram with sufficient limbing and/or removal of trees to allow passage of a rangeland drill for seeding at a cost of $448/ha ($179/acre); 3) mechanical clearing and burning of the trees at a cost of $595/ha ($237/acre); and 4) wood harvesting and slash disposal at a cost of $2,080/ha ($832/acre). The picloram and limb, mechanical, and wood-harvesting treatments provide mechanically seedable sites, but of considerably different quality in terms of ease of seeding and chances of seedling establishment. The mechanical treatment requires a large capital investment, while the wood-harvesting treatment requires a large amount of labor. Based on equivalent energy values, the wood-harvesting operation would produce a profit for the landowner who could afford to invest the labor. For specific woodland, a combination of treatments would be the most cost-effective.

V. BROOM SNAKEWEED

Even under relatively low beef prices, as was experienced in 1985, the economics of controlling a heavy infestation of broom snakeweed was estimated to be positive (McDaniel and Torell, 1987). As (expected) beef prices increase the economics of snakeweed control improves. If record-high beef prices occurred over a 5-year period, livestock benefits from controlling the snakeweed would exceed costs by 36% and the IRR would be a very competitive rate of nearly 18%.

Snakeweed control should be approached from both a biological and an economic perspective. The ultimate success of this range-improvement practice is measured by the land manager in terms of economic net return to the ranching enterprise and improved range conditions. Increasing the value of these factors determines the ultimate success of the practice.

VI. ASPEN POPLAR

The economic merit of controlling brush regrowth using combinations of 2,4-D and dicamba with different patterns of repeat applications was assessed for pastures in east-central Saskatchewan (Zentner and Bowes, 1991). Two experiments were conducted simultaneously in a community pasture located on a Waitville loam. The area, which was originally dominated by aspen poplar, had been cleared of trees and shrubs by bulldozing before the start of the tests. In the first experiment, started in 1981, 2,4-D ester plus dicamba and 2,4-D amine plus dicamba were foliar-applied to the brush regrowth. Single applications were compared with repeat sprayings in the first and second years (or both) following the initial herbicide application. A subset of these treatments was used in a second experiment started in 1983 to determine whether or not the productivity effects were influenced by growing conditions in the initial year of spraying. Herbicides effectively controlled the brush regrowth, which increased the yields of useable herbage (grasses plus forbs) by an average of 32 to 39%. The forage benefits did not decline with time to the end of the study in 1989. The herbicide treatments were generally more profitable when applied to younger stands of brush regrowth, with the 2,4-D ester formulations being more profitable than the 2,4-D amine formulations. The economic returns were the highest, although not always positive, for the treatment receiving a one-time application of 2,4-D ester at 2.2 kg/ha plus dicamba at 1.5 kg/ha for forage values between $25 and $75/ton (dry weight) and discount rates between 0 and 10%. In order to justify this treatment cost, the minimum value that must be obtained when utilizing the additional forage was $20/ton (dry weight) for the most profitable control treatment, assuming a 5% discount rate and the persistence of the treatment effects for up to 18 years. For the other treatments, breakeven forage values ranged between $37 and $70/ton under similar assumptions. Higher discount rates or a shorter effective treatment life increased the breakeven forage values. Repeat herbicide applications could not be justified under any of the economic scenarios.

VII. LANTANA

Six herbicide treatments are registered in South Africa for the control of *Lantana camara* 1. (Lantana) (Erasmus and Clayton, 1992). Three of the treatments were selected for a costing analysis of the initial control of lantana, based on their suitability for specific situations in which lantana commonly occurs as a weed. For the investigation, a work study was performed to separate the cost of preparative operations (slashing) and herbicide application. The volume of the registered concentration of herbicide that was applied was recorded to permit costing of the chemical. Approximately 27 mandays/ha were required to slash medium-to-dense stands of lantana and 3.5 mandays/ha for sparse infestations. For chemical treatment of the stumps, imazapyr (Chopper) in water was cheaper than picloram/triclopyr (Tordon Super) in diesel, owing to the respective costs of the carriers. The application of glyphosate (Roundup) to lantana coppice was also more expensive than imazapyr application to the stumps. The efficacy of all three herbicides was high. As reinfestation usually occurs in the cleared areas, follow-up treatment of lantana is essential.

VIII. FORESTRY

Carter and Holt (1978) showed that the north, south, Rocky Mountain, and Pacific Coast forest regions of the United States could markedly benefit from weed control. Seventy to 100% increases in merchantable wood production are easily realized. The types of weed control undertaken are type conversion, site preparation, release from competition, and timber stand improvement. Carter and Holt (1978) selected the Douglas Fir region and the southern pine region to compare the economic benefits of chemical, mechanical, and manual weed control. The net present value at an 8% discount rate after taxes was $198, $107, and $78 per acre for chemical, mechanical, and manual weed control in the Douglas Fir region, respectively, and $144, $92, and $83 per acre in the southern pine region, respectively. The benefit: cost ratio present value at an 8% discount rate after taxes was 3, 1.6, and 1.4 for the Douglas Fir region, and 3.5, 1.8, and 1.7 for the southern pine region for chemical, mechanical, and manual weed control, respectively. The most cost-effective chemical in the 1960s and the 1970s was 2,4,5-T, although other chemicals were available. Based on the net present values presented by Carter and Holt (1978) for the Douglas Fir and southern pine regions the difference between vegetation management with 2,4,5-T and the next best method was $50 to $90 per acre increased cost. Based on an average of $60 per acre they estimated that an additional $60 million annually would be required to accomplish the present level of forest rehabilitation without 2,4,5-T.

A silvicultural stand model for radiata pine in New Zealand compared net saw log breakeven stumpages of three postplanting grass weed-control treatments: 1) no spraying, 2) spot spraying, and 3) aerial spraying at 2 kg/ha hexazinone over one rotation (Glass, 1986). For each spray treatment an untended regime and a tended regime were evaluated. For both regimes, spot spraying was the most economical and aerial spraying the least economical postplanting grass-control treatment.

Anderson et al. (1986) found that an aerial application of hexazinone at label rates was the most cost-effective method of preparing cutover sites with a moderate to heavy residual stocking of hardwoods in a southern loblolly pine plantation. On sites on which little hardwood competition remained after harvesting, mechanical methods should be employed (normally shearing, piling, and burning) followed by disking. The measure of economic performance used was the IRR.

Herbicide applications were used to develop a natural regeneration reforestation plan for a 20-year-old loblolly pine plantation (Clason, 1992). The plan included intermediate harvesting, preharvest herbicide site preparation, shelterwood reproduction harvesting, and chemical precommercial thinning. Neither the timing nor the herbicide formulation of the preharvest herbicide applications affected plantation growth and yield. The mean merchantable volume and sawtimber growth from age 20 to 33 averaged 1,110 ft^3 and 3,180 board ft/acre. Naturally regenerated 3-year-old pine seedlings were chemically thinned to reduce seedling stocking rates and to adjust seedling distribution patterns. By age 33, the net value of the naturally regenerated stand exceeded an artificially regenerated stand by $607.98/acre. The present net-value differential implies that herbicides can be effectively incorporated into a natural regeneration reforestation plan.

Glover et al. (1994) found that an analysis of pine-release studies showed very high financial returns, particularly when there were large amounts of hardwood present in the pine stand at treatment. The analysis suggests that on production sites the use of silvicultural herbicides was definitely economically positive.

The factors addressed included the discount rate, site quality, competing vegetation, herbicide efficacy, pine survival, rotation age treatment cost, and wood product prices.

Miller et al. (1994) studied the growth of loblolly pine at 13 locations southwide as influenced by four extreme competition control treatments: NC–no control after chop-burn; HC–herbaceous weed control; WC–complete woody plant control; and TC–total competition control. The growth and yield were projected to age 25 for all sites and treatments using the NCSU Managed Pine Plantation Simulator (version 3.2). All treatments were profitable except for NC and HC at two locations. The profitability of TC would greatly depend on treatment costs.

Separate studies were conducted to evaluate the response of planted slash (*Pinus elliotii* Engelm.) and loblolly pines to midrotation competition control (Shiver and Warnell, 1994). The total control of competing vegetation resulted in gains of 0.25 and 0.35 cords per acre per year after 14 years and 8 years for slash pine and loblolly pine, respectively. The present value of these gains supports the hypothesis that midrotation control is economical at some stumpage prices and discount rates. The gains presented are average gains over all locations. Some slash pine soil classes never responded. For loblolly pine, top-slope positions did not respond as well as midslope or bottom-slope positions.

IX. STATUS OF HERBICIDES—FOREST USE

McCormack (1994) evaluated herbicide use in forest vegetation management and concluded that there is a distinct regional variation in the forestry uses of herbicides. Different land ownership patterns affect policies and practices, and crop tree species characteristics differ within and among regions. The recent decreases in the number of land areas treated have been attributed to budget reductions, changes in operating conditions, and pressures from the general public. Factors affecting future reductions in the amounts of herbicides used will include new chemistry, improved delivery technologies, the incorporation of vegetation management into all aspects of young stand silviculture, and the employment of alternative methods.

X. RIGHTS-OF-WAY HERBICIDE USE

There are over 3 million miles of paved and unpaved roadways, 170,000 miles of operating railroads, 6.1 million miles of electrical lines (87% distribution and 13% transmission), and 1.3 million miles of pipelines in the continental United States. The maintenance of vegetation along these rights-of-way requires treatment with herbicides on some 3.9 million acres per year (Newton, 1996). Because of the enormous investment in facilities, the protective role of these weed-management practices is of major economic importance.

Phenoxy herbicides are used in all major areas of road, railroad, and electrical rights-of-way management and in intensive forest production in the United States. Small amounts are used for pipeline maintenance. Approximately 1 million lb of 2,4-D are used annually in roadside vegetation maintenance. About 100,000 lb are used annually on railroad rights-of-way, about the same on electrical utility rights-of-way, and about 60,000 lb are used for pipeland maintenance. Some 22,000 lb of diclorprop are used for electrical rights-of-way, and diclorprop and MCPA are used along highways. Triclopyr, picloram, and dicamba use would increase in the absence of phenoxy herbicides. If the phenoxy herbicides were banned, the estimated increase in annual costs for vegetation management on rights-of-way from substituting alternative herbicides for 2,4-D would be $19,063,800, and $300,000 for substituting for diclorprop, thus the annual net societal loss from a ban of all phenoxy herbicides in rights-of-way was estimated to be $19.36 million for 1992.

XI. SUMMARY

The treatment of woody plant problems on rangeland and forests may be necessary for the economic production of livestock, wildlife, or forest products if weed problems become too severe. Economic production, however, may be confined to the more productive sites with minimal, infrequent, or no expenditure on the nonproductive sites. The land manager must decide on the goals he is to attain and the cost/benefit ratio that will involve improvements for aesthetics, fire suppression, or reasons other than economic return. Information and computer decision-making systems are available on the Web for some weed problems and from some state universities, industry, or the U.S. Department of Agriculture.

REFERENCES

Anderson WC, Hickman CA, Guldin RW. Cost effectiveness of hexazinone versus mechanical site preparation. Proc South Weed Sci Soc 39:220–227, 1986.

Bastian CT, Jacobs JJ, Smith MA. How much sagebrush is too much: An economic threshold analysis. J Range Mgt 48:73–80, 1995.

Bovey RW. Use of 2,4-D and other phenoxy herbicides on pastureland, rangeland, alfalfa forage, and noxious weeds in the United States, In: Biologic and Economic Assessment of Benefits from Use of Phenoxy Hebicides in the United States. NAPIAP report no. 1-PA 96. Washington, D.C.: USDA, 1996, pp. 76–86.

Carter MC, Holt HA. Alternative methods of vegetation management for timber production, Symposium on the Use of Hebicides in Forestry, Arlington, VA, 1978, pp. 125–150.

Clason TR. Herbicide usage in the natural regeneration of loblolly pine. Proc South Weed Sci Soc 45:189–193, 1992.

Conner JR. Technology selection based on economic criteria. In: Integrated Brush Management Systems for South Texas: Development and Implementation B-1493. College Station, TX: Texas Agri. Exp. Stn., Texas A&M University, 1985, pp. 47–54.

Conner JR, Hamilton Stuth JW, Riegel DA. ECON: An investment analysis procedure for range improvement practice. MP-1717. College Station, TX: Texas Agric. Exp. Stn., Texas A&M University, 1990.

Erasmus DJ, Clayton JNG. Towards costing chemical control of *Lantana camara* L.S. Afr J Plant Soil 9:206–210, 1992.

Ethridge DE, Pettit RD, Neal TJ, Jones VE. Economic returns from treating sand shinnery oak with tebuthiuron in West Texas. J Range Mgt 40:346–348, 1987.

Ethridge DE, Dahl BE, Sosebee RE. Economic evaluation of chemical mesquite control using 2,4,5-T. J Range Mgt 37:152–156, 1984.

Evans SG, Workman JP. Optimization of range improvements on sagebrush and pinyon-juniper sites. J Rang Mgt 47:159–164, 1994.

Glass BP. The cost/benefit of using hexazinone for selective grass control in radiata pine in Canterbury. New Zeal J For 30:115–120, 1986.

Glover GR, Quicke HE, Lauer DK. Economics of early vegetation control in southern pine forests. Proc South Weed Sci Soc 47:83–84, 1994.

Hamilton WT. Intiating IBMS. In: Integrated brush management systems for South Texas: development and implementation B-1493. College Station, TX: Texas Agric. Exp. Stn., Texas A&M University, 1985, pp. 9–14.

Hanselka CW, Hamilton WT, Rector BS. Integrated brush management systems for Texas. L-5164. College Station, TX: Texas Agri. Ext. Serv., Texas A&M University, 1996a.

Hanselka CW, Hamilton WT, Conner JR. Integrated brush management systems (IBMS): strategies and economics. B-6041. College Station, TX: Texas Agric. Ext. Serv., Texas A&M University, 1996b.

Hoffman GO. Mesquite control pays in Texas. Down to Earth 22:8–11, 1967.

Hyatt SW. Sagebrush control–Costs, results and benefits to the ranchers, J Range Mgt 19:42–43, 1966.

Kuntz TB, Hoffman GO. The biologic and economic assessment of 2,4,5-T use in forage production on pasture and rangelands in the United States (revised). In: Supplement to the Biologic and Economic Assessment of 2,4,5-T: A Report of the 2,4,5-T Assessment Team to the Rebutable Presumption Against Registration of 2,4,5-T. Washington, D.C.: USDA tech. bull. no. 1671. 1983, pp. 63–125.

McBryde GL, Conner JR, Scifres CJ. Economic analysis of selected brush management practices for Eastern South Texas. B-1468. College Station, TX: Texas Agric. Exp. Stn., Texas A&M University, 1984.

McCormack ML Jr. Reduction in herbicide use for forest vegetation management. Weed Tech 8:344–349, 1994.

McDaniel KC, Torell LA. Ecology and management of brown snakeweed. In: Capinera JL, ed. Integrated Pest Management on Rangeland. Boulder, CO: Westview Press, 1987, pp. 100–115.

McGinty A, Cadenhead JE, Hamilton WT, Hanselka WC, Ueckert DN, Whisenant SG. Chemical weed and brush control: Suggestions for rangeland. B-1466. College Station, TX: Texas Agric. Exp. Serv., Texas A&M University, 1998.

Miller JH, Zutter BR, Zedaker SM, Edwards MB, Newbold RA. Projected growth and economic returns from intensive vegetation control in loblolly pine plantations. Proc South Weed Sci Soc 47:84–85, 1994.

Newton M. Phenoxy herbicides in rights-of-way and forestry in the United States. In: Biologic and Economic Assessment of Benefits from Use of Phenoxy Herbicides in the United States. USDA-NAPIAP report no. 1-PA-96. 1996, pp. 165–178.

Peat HC, Bowes GG. Management of fringed sageland (*Artemisia frigida*) in Saskatchewan. Weed Tech 8:553–558, 1994.

Rector BS. Personal communication. College Station, TX: Texas Agri. Ext. Serv., Texas A&M University, 1999.

Scifres CJ, Hamilton WT, Conner JR, Inglis JM, Rasmussen GA, Smith RP, Stuth JW, Welch TG. Integrated brush management systems for South Texas: development and implemetation. B-1493. College Station, TX: Texas Agric. Exp. Stn., Texas A&M University, 1985.

Scifres CJ. IPMS: Ecological basis and evolution of concepts. In: Integrated Brush Management Systems for South Texas: Development and Implementation. B-1493. College Station, TX: Texas Agric. Exp. Stn., Texas A&M University, 1985, pp. 5–8.

Scifres CJ, Hamilton WT. Selecting IBMS components. In: Integrated brush management systems for South Texas: development and implementation. B-1493. College Station, TX: Texas Agric. Exp. Stn., Texas A&M University, 1985, pp. 47–54.

Scifres CJ, Hamilton WT. Factors affecting economic performance of herbicides. In: Management of South Texas Mixed Brush with Herbicides. B-1623. College Station, TX: Texas Agric. Exp. Stn., Texas A&M University, 1989, pp. 49–59.

Shiver BD, Warnell DB. Response of economics of mid-rotation competition control in southern pine plantations. Proc South Weed Sci Soc 47:85–92, 1994.

Sneva FA. Grazing return following sagebrush control in Eastern Oregon. J Range 25: 175–178, 1972.

Stuth JW, Conner JR, Hamilton WT, Riegel DA, Lyons BG, Myrick BR, Couch MJ. RSPM: A resource systems planning model for integrated resource management. J Biogeo 17:531–540, 1990.

Tanaka JA, Workman JP. Economic optimum big sagebrush control for increasing crested wheatgrass production. J Range Mgt 41:172–178, 1988.

Torrell LA, McDaniel KC. Optional timing of investment to control honey mesquite. J Range Mg 39:378–382, 1986.

Valentine KA, Norris JJ. Mesquite control with 2,4,5-T by ground spray application. Bull. 451. Agric. Exp. Stn., Las Cruces: New Mexico State University, 1960.

Vantassell LW, Conner JR. An economic analysis of brush control practices and grazing systems in the Rolling Plains of Texas, MP-1619. College Station, TX: Texas Agric. Exp. Stn., Texas A&M University, 1986.

Watts ML, Wambolt CL. Economic evaluation of Wyoming big sagebrush (*Artemisis tridentata*) control methods. Weed Tech 3:640–645, 1989.

Whitson RE, Scifres CJ. Economic comparisons of alternatives for improving honey mesquite infested rangelands. B-1307. College Station, TX: Texas Agric. Exp. Stn., Texas A&M University, 1980.

Whitson RE, Merrill LB, Wiedemann HT, Taylor CA Jr. Economic feasibility of brush control in the Edwards Plateau. MP-1554. College Station, TX: Texas Agric Exp Stn, Texas A&M University, 1984.

Williams HG. Costs and return in controlling sand sage: OSU Ext. Facts. No. 124. Woodward, OK: Oklahoma State University, 1969.

Workman DR, Tefertiller KR, Leinweber CL. Profitability of aerial spraying to control mesquite. Misc. pub. 425. College Station, TX: Texas Agric. Exp. Stn., Texas A&M University, 1965.

Young JA, Evans RA, Budy JD, Torell A. Cost of controlling maturing western juniper trees. J Range Mgt 35:437–442, 1982.

Zentner RP, Bowes GG. Economics of chemical brush control in pastures of east-central Saskatchewan, Can J Plant Sci 71:1133–1141, 1991.

14

Nonchemical Methods of Woody Plant Control

I. INTRODUCTION

No treatise on woody plant management would be complete without a discussion of nonchemical methods, since some of these methods are the only means to treat some woody plants or the only way to deal with some situations.

A. Mechanical Methods

1. Handheld Equipment

Hand methods of weed and brush control probably have been practiced since antiquity (Bovey et al., 1984). Although effective, they are slow, costly, and laborious. Hand methods include grubbing, cutting, girdling, and burning.

Grubbing consists of using a grubbing hoe, shovel, or similar tool to expose the root system partially or totally. The operation is difficult, time consuming, and effective if done properly. Axes or saws are effective on woody species that are killed when the top is removed, such as eastern red cedar and Ashe juniper. If species resprout the cut surfaces can be treated with herbicide.

Girdling, cutting a ring through the bark and cambium layer to prevent movement of nutrients to the roots, is practical in scattered stands. Large non-root-sprouting trees of 15 cm diameter or greater are effectively controlled during the summer. For improved control, herbicides can be applied to the cut ring. Portable chain or power saws and girdlers reduce labor, time, and cost in brush removal but have limited use in dense stands or large areas.

2. Large Equipment

a. Bulldozing Bulldozing, often called *dozing*, is a widely used method of brush control (Vallentine, 1989). Bulldozing is most commonly practiced in open stands of large trees and brush or on rocky soils where the use of other mechanical control methods is limited. It is not desirable in dense brush that sprouts from the roots after top removal. In moderately dense brush, chaining or raking can be done to smooth the land and prepare a seedbed. If raked, brush piles can be left in place for wildlife cover. Follow-up weed and brush control will be necessary within 2 to 5 years after bulldozing to kill resprouts and new seedlings.

The equipment and fuel required for bulldozing are costly. The soil is disturbed, and forage stands and wildlife habitat may be partially or heavily damaged. The soil may be subjected to wind and water erosion, depending on how well the job was done. Under some circumstances, leaving alternate undisturbed strips may be justified, particularly on steep or erodible lands or where wildlife is important (Scifres, 1980). Much clearing is done with straight dozer blades. Since it is inefficient and often ineffective, many modifications and attachments are available. Ideally, bulldozing removes brush and large trees by pushing or pulling the plants out of the soil. Special attachments to the straight blade include teeth, shearing blades, or U-shaped ''stingers'' that cut plants below the soil surface and lift out the roots.

Straight blade modification includes the rakelike ''stacker'' (Rechenthin et al., 1964) with teeth 13 to 36 cm apart mounted on a straight blade or in place of the solid blade. One advantage it has over the straight blade is that the brush rake displaces less soil and reduces the accumulation of soil in the brush piled for burning. Should small brush slip between the teeth of the blades an additional blade or root grubber can be welded across the bottom of the teeth to uproot the brush. The stacker is highly maneuverable and prepares a good seedbed. Some brush piles should be left for wildlife, and crushed and broken brush should be left to reduce wind and water erosion. The stacker has been useful for removing thick stands of pricklypear cactus in south Texas.

Large brush rakes and root rakes also are used in south Texas to clean up debris after other mechanical operations such as chaining or root plowing honey mesquite (Scifres, 1980). Some rakes are towed.

b. Grubbing The grubber consists of a U-shaped hydraulically controlled blade with an angle adjustment to sever roots 15 to 30 cm below the ground. Wiedemann and Cross (1981) developed a low-energy grubber to control redberry juniper, Ashe juniper, and other small weed trees on rangeland. Mounted on a 65-horsepower crawler tractor, the unit operates at low cost and selectively kills brush with minimal soil disturbance. Foam-filled tires have also allowed the use of wheeled farm tractors and wheeled loaders for grubbing small honey mesquite trees in areas in which thorns normally would puncture conventional tires (Wiedemann and Cross, 1982).

c. ***Chaining*** Chaining, a relatively inexpensive means of mechanically clearing brush (Rechenthin et al., 1964), does not remove all brush, and its major effect is temporary. Naval anchor chains are pulled behind two large crawler tractors across brush to be felled or pulled out. The chains may be 30 to 90 meters long and weigh 90 to 135 kg/meter. Cables, usually 4.3 to 5 cm in diameter, are not as effective as chains.

Many adaptations have been made to chains, such as welding on short lengths of angle iron, rails, or disks to make them more efficient (McKenzie et al., 1984). Such modifications included the Ely chain for big sagebrush control, swivels to allow chains to roll, ball-and-chain arrangements for steep terrain, the Dixie–Sager chain for better uprooting of brush and scarification of soil, and the disk chain to reduce draft. Chaining is most effective on large single-stemmed brush and trees growing on sandy, shallow, or moist soils, where many of the plants can be pulled out. Two-way chaining (chaining in opposite directions) increases effectiveness. In some situations the brush is bent over or sheared off at the ground level before being uprooted, and root sprouting occurs. In dense stands, the chain may also pull over large masses of brush, failing to uproot it.

Chaining can be used to advantage as a pretreatment to raking, root plowing, or goat grazing by knocking down dense brush to make it more accessible. Scifres et al. (1976) found that double chaining followed by raking and stacking mixed brush was more effective than one-way or double chaining. Chaining brush infested with pricklypear, cholla, and tasajillo may scatter the cactus, making the treatment undesirable if it is not followed by raking. Chaining results in poor seedbed preparation, and residual brush disrupts seeding.

Wiedemann and Cross (1984) developed a disk chain with a triangular pulling configuration that reduced draft requirements by 36% and increased the operating width by 23% compared to the two-tractor diagonal pulling method. The disk chain, pulled by a single tractor, appears to be cost-effective for preparing seedbeds on rough, log-littered, and root-plowed rangeland.

d. ***Railing*** Railing is a practice developed for the control of pricklypear and tasajillo in south Texas (Rechenthin et al., 1964). Three to nine railroad rails or heavy angle irons are welded or chained together and dragged in tandem in two or three sets behind a tractor. The rails are dragged over the area in one direction, then a second time in the opposite direction, crushing the cactus. Grazing cattle sometimes are used to clean up newly railed areas. Railing is the most effective in hot, dry weather. Repeated railing about 1 month after the original treatment removes new sprouts or missed plants.

e. ***Chopping*** Roller chopping is a quick method of knocking down and crushing brush. Like chaining, it is a temporary measure and kills few resprouting plants. Resprouting will occur soon after treatment unless the trees are nonsprouters. The equipment consists of a steel drum or cylinder around which cutter blades are attached. Weight can be added either by filling the roller drum with water or by adding steel or concrete ballast to the frame. Choppers vary in size, depending on the size of the brush. A weight of about 700 to 900 kg (1500 to 2000 lb) per linear foot of cutter blade is suitable for small and medium-size brush such as catclaw and blackbrush (Rechenthin et al., 1964). Larger choppers are needed for trees with stem diameters of 13 cm or more.

Chopping is useful only where repeated treatment is planned or where other methods will supplement brush control. Chopping followed with soil-applied herbicides did not improve control over herbicides applied alone (Meyer and Bovey, 1980). Chopping, which removes the top growth of brush and makes handling the livestock easier, may be useful where goats graze by allowing them to reach the leaves. Chopping is useful for seedbed preparation on log-littered land or where dense brush can be knocked down quickly and burned when dry. When chaining is not practical for tree felling, chopping is usually the most cost-effective method unless the trees are large.

f. ***Mowing and Shredding*** Mowing and shredding are temporary control methods for herbaceous weeds and small brush on pastures and rangeland. Mower types vary, but most consist of sharp rotary blades. Heavy-duty shredders can be used on large brush and small trees (8 to 10 cm diameter). Commercially available self-propelled cutters will cut mesquite stumps 35 to 43 cm in diameter (Hendon, 1971). Brush should be mowed in the spring and early summer. Mowing after July 1 may damage native grasses (Rechenthin et al., 1964). Bermuda grass can be mowed any time during the growing season. In humid areas, the principal benefit is to remove old growth, promote forage regrowth and quality, and reduce selective livestock grazing. Weeds and brush become established in the ungrazed areas. Normally, two to three mowings per year limit the spread of herbaceous or woody perennials in pastures, but do not eliminate them.

g. ***Root Plowing*** Root plowing consists of pulling a horizontal blade through the soil behind a large tractor at depths up to 45 cm to cut roots (Rechenthin et al., 1964). Root plowing was developed originally for land clearing. The blades must be operated 30 to 40 cm deep in the soil to kill large honey mesquite. Root plow blades vary from 1.8 to 4.3 cm wide, and fins welded to the top of the blade help lift the roots out of the soil. Fisher et al. (1972) indicated that finless root plow blades or thin blades gave effective control of honey mesquite and lotebush at a lower cost than conventional blades while allowing for better survival of perennial grasses.

Root plowing is useful in dense brush containing species that are resistant to herbicides or other mechanical control methods. Root plowing kills a high percentage of brush and can be done at any time. It is most effective during hot, dry periods but may be detrimental to the forage species at this time. Root plowing in central and northwest Texas is done predominantly during the winter when native forage species are dormant, allowing maximum survival.

If seeding is needed, it should be done during the optimum time for the area. After root plowing, the seedbed can be improved with tillage, such as chaining, roller chopping, or disking. Seeding is done by either aircraft or ground drills. Root plows equipped with grass seeders have had limited success. Root plowing is an effective brush-control treatment, but it is expensive, may destroy the grass stand, and may subject the soil to wind and water erosion. Some areas of brush should be left for wildlife habitat.

h. Disking Large disk plows or tandem disks will destroy stands of smaller brush as well as the grass stand. Disking is limited to tillable soils. Disking prepares a good seedbed, although compaction by a cultipacker, roller, or other implement may be desired to maximize soil–seed contact. Land-clearing disks are an adaptation of standard agricultural implements but weigh from 70 to 450 kg per blade. They are heavy-duty implements and are effective in uprooting, chopping, and mulching brush, even in dense stands. Commercial seeding attachments are available, but seeding usually is conducted as a separate operation.

B. Biological Control

With the high cost of herbicides and mechanical means for weed and brush control, biological control is attractive. In nature, biological control is already at work since plant species have natural enemies that feed on them, especially plants in their native habitat. Scientists recognize this natural phenomenon and attempt to restrain the enemies of desirable plants and augment their activity against undesirable plants. Schroeder (1983) stated that since a large number of economically important weeds were introduced from other parts of the world, biological control is concerned primarily with importing exotic organisms to control naturalized weeds. Rangelands offer unique possibilities for biological control because vast areas often are infested with one or more undesirable weed and brush species, many of foreign origin. Huffaker (1964) indicates that the primary objective in the biological control of weeds is not to eradicate but to use the plant's natural enemies to lower its density to a noneconomic level.

The procedure in developing biological control consists of determining the suitability of the weed problem for biological control, surveying the weed for natural enemies in both naturalized and native habitats, studying the biological characteristics and host relations of the natural enemies to determine how they may be used to solve the problem, implementing the natural enemies in weed control, and evaluating the effectiveness of the natural enemies (Harris, 1971). The most common method of implementing biological control has been through the introduction of new natural enemies. Other methods of implementation manipulate or augment already existing natural enemies of weeds (Rosenthal et al., 1991).

DeLoach (1991) indicated that the greatest advantage of augmentation is that it is active only where applied, so no damage occurs to plants in untreated areas. Augmentation effects can be ended by discontinuing it. The method is similar to herbicidal control in that the entire affected area must be treated at repeated intervals. Augmentation is unlikely to be economical in areas of low economic return, such as rangelands.

Biological weed control dates back to the 1860s, but it was little used until the 1920s. Some success with lantana and spectacular success with pricklypear and St.-John's-wort stimulated worldwide biological weed control research. Most biological control agents have been insects, but mites, snails, nematodes, fish, mammals, birds, plant pathogens, and even other plants also can be valuable (Rosenthal et al., 1991).

A lengthy and comprehensive procedure is followed to make certain that the insect or organism, especially if it is of foreign origin, is effective on the target weed and will not injure desirable plants or other organisms before release from quarantine into the field (DeLoach, 1991).

1. Control of Woody Plants with Insects

The best example of effective control by insects involves pricklypear control in Australia by the Argentine moth borer. Pricklypear was used as an ornamental and in hedges around homesteads in the late eighteenth century. By 1925, a total of 24 million ha was dominated by pricklypear. Mechanical and chemical control were too expensive. Although several insects were introduced, none was as successful as the Argentine moth borer. The larvae feed on the cactus by tunneling through the pads and root tissue. The damaged tissue is subjected to secondary attack by bacteria, fungi, and adverse weather (Schroeder, 1983; Rosenthal et al., 1984; Vallentine, 1989). Goeden et al. (1968) reported the successful reduction of pricklypear species (*Opuntia littoralis* and *O. oricola*) and their hybrids on Santa Cruz Island off the coast of Santa Barbara, California, using a cochineal insect (*Dactylopius* spp.). The insect was introduced from Mexico via Hawaii. Attempts to control pricklypear in Texas by biological means have been discouraged because some ranchers use pricklypear for livestock feed during periods of severe drought.

Insects have been used to control lantana, an escaped ornamental (Schroeder, 1983). Lantana, a perennial shrub native to Central America, has become a pest of rangeland and coconut plantations in Hawaii, Fiji, India, and Australia.

The lace bug was the most important control agent in Australia (Harley, 1973); however, a complex of organisms may be required to control lantana (Schroeder, 1983).

Gorse, a native of Western Europe and North America, has become a pest in New Zealand, Australia, and the United States (Hawaii, California, and Oregon). It is used extensively as an ornamental hedge but escaped to become a serious pasture weed. The gorse weevil (*Apion ulicis*) prevents seeding in Europe and New Zealand, but does not affect established plants. Several other insects have been tested, but none has proved as useful as *A. ulicis*. Although is has been released in Hawaii and on the Pacific coast of the United States, gorse control has not been outstanding (Holloway, 1964).

DeLoach (1990) indicated that the top weed candidates for biological control in the southwestern United States include snakeweed and broomweed, *Baccharis* spp., mesquite, salt cedar, and creosotebush. Promising foreign insects attack these species, and control strategies are being considered. Conflicts between beneficial and harmful potential should be resolved (DeLoach, 1990). Other possible biocontrol candidates include whitebrush, tarbrush, pricklypear, cholla, tasajillo, huisache, retama, and Macartney rose.

2. Control of Herbaceous Weeds with Insects

Another example of successful biological weed control is that of St.-John's-wort or klamath weed by a leaf-eating beetle, *Chrysolina quadrigemina*, and by the root-boring beetle (*Agrilus hyperici*) (Holloway, 1964). St.-John's-wort is an aggressive perennial forb accidentally introduced into the United States from Europe. This poisonous plant occurs mainly in California and the Pacific Northwest. Biological control of St.-John's-wort with the *Chrysolina* beetle has generally been excellent (Holloway, 1964; Rosenthal et al., 1991; Vallentine, 1989).

On western U.S. rangelands, puncture vine, tansy ragwort, musk thistle, rush skeletonweed, and toadflax have been controlled biologically (Julien, 1987; Quimby et al., 1991). Control is sometimes incomplete, and research continues. The U.S. Department of Agriculture-Agriculture Research Service (USDA-ARS) has nine laboratories concerned with developing biological control of at least 16 genera and more than 30 species of rangeland weeds (Quimby et al., 1991). They include snakeweed, creosotebush, tarbush, star thistles, toadflax, gorse, dryer's woad, Scotch thistle, poison hemlock, knapweeds, leafy spurge, salt cedar, St.-John's-wort, larkspur, whitetop, field bindweed, cocklebur, and bitter sneezeweed.

Rees and Spencer (1991) indicated that there are several Eurasian insects that attack leafy spurge on U.S. rangelands, and an additional eight insects are in various stages of clearance. With biocontrol, the goal is to reduce leafy spurge in North American plant communities to noneconomic levels, as are found in Eurasia.

Rees (1991) indicated that the seed heads of both musk and Canada thistles are attacked by a seed head weevil (*Rhinocyllus conicus*), while the larvae of a second weevil (*Trichosirocalus horridus*) feed on the growing point. The larvae of *Cheilosa corydon* burrow into musk thistle stems and roots. The larvae of the weevil *Ceutorhynchus litura* reduce the overwintering survival of Canada thistle by burrowing in the stems and roots. The larvae of the gall-producing fly *Urophora cardui* also burrow into Canada thistle stems and induce large galls. Galls act as a nutrient sink to stress the plant and in some cases to prevent flowering.

Rosenthal et al. (1991) are developing biological control measures for yellow starthistle and several knapweeds (diffuse, spotted, squarrose, and Russian) using insects, but indicate that a wide variety of natural enemies will be necessary to reduce weed populations to tolerable densities. Researchers are also investigating mites, nematodes, and pathogens. The integration of biological control with other range-management practices will be necessary to suppress the knapweeds.

3. Biocontrol with Plant Pathogens

Wilson (1969) described the use of plant pathogens in weed control. The successful control of pricklypear cactus in Australia is attributed to secondary attack by associated micro-organisms following the attack of *Cactoblastus cactorum. Gloeosporium lunatum* E&E and bacterial soft-rot organisms are considered the primary invaders following *C. cactorum* attacks. Fungi that cause disease in pricklypear also have been described, although their importance has not been fully assessed in relation to the decline of various cactus species.

Tainter (1972) evaluated oak wilt fungus as a selective silvicide in northern Arkansas. Post oak, white oak, blackjack oak, southern red oak, and black oak were inoculated with an Arkansas isolate of the oak wilt fungus in April 1971. The results indicated that oak wilt fungus effectively killed inoculated red oaks and prevented sprouting from the cut stump. The oak wilt fungus is reported to be nonhazardous to tree species other than those in the red oak group and may be useful as a selective silvicide since it does not spread by root grafts to untreated trees.

Van Arsdale and Halliwell (1970) reported that live oak decline is a complex of diseases that affect other oaks as well as live oak. A fungus (*Cephalosporium* spp.) has been identified as the major component of the disease complex. It causes a vascular wilt that is apparently root-graft-transmitted. Since live oak often is considered a desirable shade tree and produces acorns for wildlife, it is doubtful that live oak decline can be used as a means of biocontrol for oaks.

Templeton (1982) reported that rush skeletonweed has been successfully controlled in Australia by a rust fungus (*Puccinia chondrillina*) from the Mediterranean area, where the weed originated. *P. chondrillina* was released on the rush skeletonweed population in the western United States in 1978.

Quimby et al. (1991) indicated that northern joint vetch in Arkansas rice crops and strangler vine in Florida citrus orchards have been controlled with native pathogens. The major targets for biological control using plant pathogens are leafy spurge, musk thistle, the knapweeds, and the starthistles. A leaf-spotting fungus (*Alternaria* sp.) collected in North Dakota and Nebraska is being evaluated on leafy spurge. *Puccina* sp., rust disease, is being investigated on musk thistle, starthistles, and knapweeds (Quimby et al., 1991).

4. Selective Grazing

Selective grazing is a wide-scale practice to control undesirable weeds and brush with cattle, sheep, goats, horses, poultry, and some wildlife species. In some cases, the weeds may be so unpalatable, poisonous, or vigorous that grazing management is not practical.

In central Texas, California, Arizona, and other areas, goats use browse as a large part of their diet. Goats consume woody plants, such as live oak, blackjack oak, sumac, greenbriar, hawthorn, American beautyberry, yaupon, sweetgum, and retama. Brush and trees can be cut, chained, roller chopped, shredded, or bulldozed down when they are too tall for the animals to reach for foliage.

Goats are an effective means of brush management by grazing one goat per ha on a year-long basis (Vallentine, 1989). For short-term grazing, a 30-day period in which 12 to 20 animals per ha is required, depending upon both the kind of brush and its density (Rechenthin et al., 1964). The short-term approach, with high stocking rates, controls most brush with less damage to other forage plants. Three years are usually required for this system to reduce the brush to a manageable level. A lighter stocking rate can then be used to maintain the brush populations.

Warren et al. (1984) reported that Spanish goats grazing on mixed brush in the South Texas Plains used shrubs as food during the autumn, winter, and summer. The goats browsed blackbrush acacia, *Condalia* spp., guajillo, guayacan, and Berlandier wolfberry, suggesting their potential for use in conjunction with other brush-control practices to increase the efficiency of forage use on mixed-brush sites.

Goats were superior to cattle and sheep for brush control in New England (Wood, 1987). Goats destroyed small trees and saplings by debarking them. As the brush was destroyed, the grass increased. In Canterbury, New Zealand, gorse can be a year-round maintenance feed for goats (Radcliffe, 1986). Angora goats were used to control gorse in Australia in combination with sheep (Harradine and Jones, 1985). Goat grazing is not a simple operation since special goat-proof fences, management, and protection from adverse weather and predators are required. In some areas green foliage is not available year-round so animals must be sold or moved. Predators, such as bobcats, coyotes, and occasionally dogs, may injure or kill the goats, resulting in severe economic loss.

In Montana, 40 to 50% of a sheep's diet consists of leafy spurge in heavily infested pastures (Landgraf et al., 1984). Ewe weight gain in leafy spurge pastures was not different from weight gain in pastures free of leafy spurge. Sheep consumed leafy spurge without any apparent harmful effects, therefore sheep can be considered an effective biological control agent of leafy spurge.

Fay (1991) reported that some research showed that sheep refused to eat leafy spurge and suffered deleterious effects when they did eat it. At the present time, sheep are being used to control leafy spurge in Montana because of high herbicide costs and the lack of success of other biological control methods. Problems occurred because there are not enough sheep to control the weed adequately to prevent seed production without overgrazing infested areas. Sheep loss to predators and the containment of sheep on infested areas were also problems (Fay, 1991).

5. Fertilizer and Plant Competition to Control Weeds and Brush

Herbaceous weeds sometimes can be effectively eliminated or reduced by encouraging the growth of competing desirable vegetation. For example, broomsedge can be controlled on unproductive pastures in Virginia with annual nitrogen fertilizer applications for 5 years. Neither burning nor mowing controls the weed, which may make up 70% of the vegetation (Semple, 1970). Annual nitrogen fertilization after the first year maintained superior stands of five warm-season prairie grasses and reduced weed invasion at 12 locations in Nebraska (Warnes and Newell, 1969).

Johnston and Peake (1960) seeded crested wheatgrass on a 12-ha field infested with leafy spurge. While excellent stands of crested wheatgrass were established, they did not control the leafy spurge.

Controlling woody plants by fertilization has not been successful because fertilization may encourage rapid growth of the dominant brush. In some situations plants are used as browse and fertilization may increase production and palatability,but treatment costs may not be economical on rangelands (Bovey et al., 1984).

A vigorous, dense stand of grass and broadleaf species may discourage weed and brush encroachment, especially as seedlings, into rangelands. Once woody plants become established and dominant, however, they usually cannot be controlled or eliminated by other competitive plants.

C. Control with Fire

Fire is an ancient method of suppressing weedy vegetation. In the past, it occurred naturally in many areas. Whether man-made or natural, however, fire has been a strong influence on the vegetation type of an area. There is still great diversity of opinion and inconclusive data about prescribed burning as a valid weed-control method. It has been used with great benefit in some grassland situations but can have disastrous effects if uncontrolled. Burning is restricted by local or state air-pollution laws in some locations.

Scifres (1980), Vallentine (1989), and Kozlowski and Ahlgren (1974) have reviewed range improvement by burning. Wright and Bailey (1982) have evaluated fire for rangelands and forests in the United States and southern Canada. Fire has been used to control sagebrush in the western United States (Pechanec et al., 1965), convert chaparral to grasslands in California (Bently, 1967), control pinyon-juniper communities in the Southwest (Vallentine, 1989), and control mixed brush in south Texas (Box and White, 1969).

Fire can be used on rocky or rough terrain at moderate cost with fast coverage, as opposed to costly alternative methods (Scifres, 1980; Vallentine, 1989). It can also be used to maintain or extend the treatment life of initial brush-control methods (Scifres, 1980). Prescribed burning may increase palatability and forage production, kill or suppress undesirable plants, improve grazing use and distribution, and release plant nutrients for wildlife and livestock and plant growth. Much information has been accumulated on fire effects on desirable and undesirable vegetation, and the reader is referred to the literature for specific vegetation types and plant responses. Britton and Wright (1983) indicated that historically fire was a major factor that kept shrubs suppressed in grasslands. For fire to be effective, however, adequate vegetative fuel must be produced or other methods of weed and brush control must be considered. Wright and Bailey (1982) indicated that with an adequate fuel load, burroweed, broom snakeweed, creosotebush, young mesquite trees, and cacti can be suppressed, and tobosagrass, big sacaton, alkali sacaton, and mixed grass ranges can be managed. Black grama grasslands appear too delicate to manage with fire (Wright and Bailey, 1982). In Arizona, Cox (1988) found that burning or mowing big sacaton in any season removed the green foliage available to livestock and reduced the amount of green forage that may accumulate for at least two summer growing seasons.

Big sagebrush (a nonsprouting species) was killed in Wyoming when a uniform fire spread occurred (Smith et al., 1985). Green rabbitbrush and horsebrush resprouted. Greasewood was sometimes killed as a result of a high fuel load. Aspen, snowberry, and serviceberry resprouted, while bitterbrush did not (Smith et al., 1985). Wambolt and Payne (1986) found that sagebrush canopy was most effectively reduced by burning and that burning provided the most production from dominant forage species when compared to other methods.

Saw palmetto, a low-growing shrub dominating Florida rangeland, is only temporarily controlled by fire (Tanner et al., 1986). Burning saw palmetto reduced the carbohydrate content and increased the mortality if subsequent chemical or mechanical treatment was applied.

Converting chaparral to grasslands is done to improve fire control, to increase water yield, and to develop new forage for livestock and deer (Bentley, 1967). The control method depends upon the chaparral type, terrain, resistance of brush species to herbicides, and need for revegetation. Herbicides are preferred where practical for controlling the chaparral. In California, Bentley (1967) indicated that burning standing chaparral seldom removed the brush satisfactorily. Chaparral should be prepared to permit safe burning during weather unfavorable for ignition, buildup, and spread of fire to natural fuels. Chaparral will burn readily if crushed and compacted before broadcast burning. A bulldozer blade or roller chopper will compact the brush. Spraying chaparral with herbicides for fuel preparation is recommended only in light- or medium-density brush.

In pinyon-juniper communities, a uniform burn has been an effective, economical method of controlling nonsprouting junipers (Vallentine, 1989). A uniform burn depends upon the presence of enough juniper and understory plants to carry the fire. Barney and Frischknecht (1974) studied the vegetation changes following a fire of the pinyon-juniper type in Utah. Woodlands were well developed 85 to 90 years after the fire. Weedy annuals reached peak levels 3 to 4 years after fire. Intermediate successional stages followed a general pattern of perennial grasses, perennial grasses-shrubs (sagebrush), and perennial grasses-shrubs-trees.

Postsettlement invasion of trees and shrubs on a tallgrass prairie in the Kansas Flint Hills was assessed by using aerial photography, General Land Office survey data, and field observations (Bragg and Hulbert, 1976). On unburned sites, aerial photographs showed that combined tree and shrub cover increased 34% from 1937 to 1969; section-line data showed that tree cover alone increased 24% from 1856 to 1969. The data from two sites suggested that herbicide spraying only slowed the invasion rate. The percentage of woody plant cover varied with soil site, but increased most rapidly on lowland soils. The authors concluded that on the Flint Hills tallgrass prairie, regular burning (every 1 to 5 years) has restricted woody plant invasion to natural presettlement amounts, although the rate of woody plant invasion varies with soil type and topography.

On Oklahoma rangeland, burning had a negative effect on woody plant control (Elwell et al., 1974). Burning alone contributed little to the control of larger trees (post oak and blackjack oak). Fire increased the root sprouting of blackjack oak, sumac, and dogwood. Where 2,4,5-T was used to control woody plants, herbaceous vegetation increased and provided more fuel for fire. An increase in stands of decreaser (highly palatable to livestock) grass species occurred only in herbicide-treated plots. Burning either alone or combined with 2,4,5-T resulted in slight increases in grasses. Elwell et al. (1970) indicated that burning controlled small eastern red cedar trees but had little effect on post oak and blackjack oak. Engle and Stritzke (1991) developed a self-contained backpack propane burner to ignite larger eastern red cedar trees not controlled by the initial prescribed burning. Reburning larger trees on 32 ha-pastures ranged from 21 min to 9.5 hr at a cost comparable to that of chemical or mechanical methods.

Wright et al. (1976) found honey mesquite difficult to kill with fire on the High Plains and Rolling Plains of Texas. On upland sites in the Rolling Plains, 27% of the mesquite trees was killed by a single fire, but repeated fires at 5- and 10-year intervals killed 50% of the older trees. Seedlings 1 year of age were easily killed with fire, but plants 3 years of age were tolerant of fires.

In the Ashe juniper community, a minimum of 1200 kg/ha of fine fuel was needed to carry a fire to kill juniper seedlings and burn piles of bulldozed juniper (Wink and Wright, 1973). Grasses recovered quickly and soil erosion was minimal when burning was conducted during a wet winter and spring. During a dry winter and spring, however, burning increased drought stress, reduced herbaceous plant yield, and exposed the soil to wind and water erosion.

In south Texas, Scifres (1975) found that the best control of Macartney rose occurred when individual plants were treated with 2,4,5-T + picloram (1 : 1) and followed by a prescribed burn 18 months later. In British Columbia, Wikeem and Strang (1983) indicated that prescribed burning is being used increasingly for rangeland management, but fire ecological research is needed for intelligent use. Whisenant (1989) showed that frequent fires (<5 years apart) in cheatgrass ranges in Idaho retarded the normal vegetation replacement sequences, leading to vegetation that resembled that of less frequently burned areas. Reducing the frequency and size of fires should be the primary management objective.

White and Hanselka (1991) indicated that prescribed burning is a viable improvement practice for most Texas rangelands. When integrated with other practices, fire can be used to maintain a desired vegetation composition. Landers (1991) indicated that a detailed plan is required for a successful burn, including adequate fuel, appropriate fire control, interpersonal communications, a means for emergency calls, a proper burning procedure, and a follow-up plan.

D. Combinations of Control Methods

The selection of combinations of methods for weed and brush control is based on the maximum effectiveness with the lowest possible cost using good conservation practices. Using two or more control methods is referred to as an integrated weed- or brush-management system. Rechenthin et al. (1964) reported using goats for successful controlling a large number of brush species in Texas. If the brush is too large, it can be chopped, chained, or bulldozed down within reach of the goats.

Davis et al. (1974) reported that Gambel oak, an important component of several million ha in the foothill rangelands of Arizona, Colorado, New Mexico, and Utah, can be controlled with a mechanical treatment followed by goat grazing. The brush is first treated by roller chopping, bulldozing, or undercutting to allow the animals full access to all the foliage. Repeated annual defoliation by goats during late June and late August gave the best brush control.

Bentley et al. (1967) provided a list of brush-removal combinations for chamise-chaparral and mixed chaparral in California, along with the costs associated with each operation. For chaparral, crushing and burning is a common practice, followed by herbicide spraying 1 or 2 years after top removal.

Scifres (1975) proposed that undisturbed dense stands of Macartney rose can be controlled in south Texas by mechanical treatment followed by herbicide treatment 2 years later. Prescribed burning or herbicide treatment can be used for maintenance.

Scifres et al. (1983) concluded that herbicide–fire combinations improved rangeland supporting dense stands of running mesquite or whitebrush. The herbicide reduces the brush cover and releases herbaceous vegetation either for livestock or for fuel for burning. The burn expedites forage production, improves the botanical composition of herbage, and suppresses the brush. A burn at 3- to 5-year intervals was suggested to maintain desirable species. Wink and Wright (1973) used prescribed burning to reduce woody debris, suppress missed and newly established Ashe juniper plants, and enhance forage production after tree dozing. Ueckert et al. (1988) found that controlling pricklypear cactus after fires in lighter, fine fuel vegetation such as sideoats grama and buffalo grass was often not satisfactory. Fire followed by aerial sprays of picloram at 0.14 kg/ha reduced pricklypear cover by an average of 98% and represents an economically significant level of control.

Mayeux and Hamilton (1983) found that burning prior to the application of tebuthiuron synergistically enhanced the control of common goldenweed compared to each treatment applied alone. Buffelgrass yield was increased as much as threefold with the burn–herbicide combination.

Johnston and Peake (1960) investigated the effect of selective grazing by sheep and crested wheatgrass competition on leafy spurge control. While excellent crested wheatgrass stands were established, the competition did not control leafy spurge. Twelve years after seeding, ewes were placed on the pasture from May to September annually for 5 years. At the conclusion of the study, the basal leaf area of the leafy spurge was reduced by 98%, while the basal areas of the crested wheatgrass increased by 20%.

Chaining is an inexpensive method to knock down brush after herbicide treatment to hasten decomposition or to facilitate burning (Bovey et al., 1984).

Burning may be desirable to remove understory woody plants partially killed by foliar herbicide application. Individual plant treatments, such as grubbing, cutting, or herbicide application, may be used for maintenance of the plants missed by mechanical or other control methods.

Treatment combinations are selected on the basis of availability, weed species, cost, and effectiveness. The integrated approach to weed and brush control is recognized as an important area of research and field use.

E. When to Manage Weeds and Brush

The density and type of weed infestation warranting control are important, and require serious thought and calculation (Bovey et al., 1984). The decision to initiate weed and brush control depends upon the rainfall, terrain, vegetative composition, livestock production system, control techniques, and cost.

At some level, weeds and brush can become severe enough to limit forage production economics, and land managers must choose the most economical strategy of weed and brush management to continue livestock and wildlife production. Once efforts are made to control heavy infestations, continued efforts are necessary to control resprouts and scattered plants. Good management involves recognizing potential weed problems and controlling them promptly before they become expensive.

Weeds and brush left in strips maintain wildlife habitat and prevent wind and water erosion. The proper use and timings of herbicides, fire, or mechanical methods can control unwanted plants and enhance forage production. The improper use of chemicals or mechanical treatments such as bulldozing or root plowing also can destroy desirable vegetation. Where poor stands of low-quality forage occur, revegetation may be warranted. Land managers must select weed-control treatments that adequately return their investment.

F. Maintenance Control

Once an effective weed-and-brush-control method has been used, a maintenance program is required to prevent reinfestation (Bovey et al., 1984; Klingman et al., 1983). This commonly involves mowing or applying 2,4-D or other herbicides during the spring or summer to control herbaceous annual weeds. Woody plants and perennial herbaceous weeds sometimes can be managed by mowing or using herbicides, but often, because of their persistent nature and rapid recovery, individual plant treatment may be necessary. Depending on the species, sprays or pellets of the appropriate herbicides are used. Hand or power grubbers also are used, as well as suppression by fire and animals.

G. Forage Production and Revegetation

Some depleted ranges and pastures can be restored by improved management of the existing natural vegetation. Improved management, particularly better control of grazing, may be necessary to restore forage production. When undesirable vegetation limits forage production, removing weeds restores the grazing potential. Where residual desirable native vegetation is lacking, revegetation may be necessary.

II. FUTURE WEED AND BRUSH MANAGEMENT

The need to manage and manipulate weeds and brush on rangelands will not diminish. Range weeds will continue to dominate and will not be replaced with desirable vegetation without human interference. One only has to evaluate the immense number of ha infested with honey mesquite, junipers, chaparral, prickly-pear cactus, and many other plants to recognize their persistent nature, spread, and dominance. Future research may elucidate reasons for range weed aggressiveness and will be invaluable in management decisions, but may not lead to their containment. In other situations, weeds and brush may have some value in wildlife habitat and commercial ventures. Weeds and brush with no redeeming qualities are harmful to the environment and must be contained, however. One control method may be useful in one situation, but may have to be part of an integrated approach in other situations to be effective, economical, and environmentally sound.

Insects and plant pathogens are gaining attention for weed control on U.S. rangelands because of the vast areas infested. The USDA-ARS and several state experiment stations have programs and foreign cooperators in search of organisms that mainly attack introduced weeds.

Allelopathy may be defined as the biochemical interaction between plants (Anderson, 1996). Plants may compete with one another by releasing growth-inhibiting chemicals into the environment from their roots or leachate from dead, decaying vegetation, or plants may be affected by secondary compounds from microorganisms. Rice (1984) showed evidence of allelopathy in the natural environment, and numerous chemical groups have been isolated from terrestrial and aquatic plants (Putnam, 1988). Examples of allelopathy for use in weed control are limited because presently used synthetic herbicides are more effective and less costly.

Weed and brush management may be enhanced by intensive, short-duration grazing systems in which deer and cattle use woody plants more efficiently. Leaving brush in strips or blocks for wildlife cover is advocated where certain stocking ratios of livestock and deer are desired. Forthcoming research will be concentrated on developing the best stocking ratios of livestock and wildlife species. Exotic game animals also may have a place in weed- and brush-management systems. Goats will continue to be important in brush management.

Modifications to chains make them more effective for brush control. Such chains include the Ely chain for big sagebrush control, swivels to allow chains to roll, ball-and-chain arrangements for steep terrain, and the Dixie–Sager chain for better uprooting of brush and scarification of soil.

Wiedemann and Cross (1984) developed a disk chain with a triangular pulling configuration that reduced draft requirements by 36% and increased the operating width by 23% when compared to the two-tractor diagonal pulling method. The disk chain, pulled by a single tractor, appears to be a cost-effective method to prepare seedbeds on rough, log-littered, and root-plowed rangeland for grass establishment and the control of small brush.

The low-energy grubber is one of the most promising mechanical concepts for effective economical control of sparse to moderate stands of brush (Wiedemann and Cross, 1982).

Fire (prescribed burning) will continue to be important, especially in combination with other practices. Proper burning is effective for weed control and rejuvenation of forage species; however, skill and knowledge both of planning a burn and of fire behavior is essential. Future research will emphasize combinations of treatments for economical and effective weed control, as discussed in Section I.D.

III. SUMMARY

Weeds and brush infest most grazing land in the United States and are among the most serious barriers to economic livestock production. Weeds and brush reduce the quantity and quality of desirable forages, harbor insect vectors and predators, make handling livestock difficult, reduce land values, influence recreation opportunities, and increase production costs. Many unpalatable, mechanically injurious, and poisonous plants reduce grazing.

Mechanical control methods for woody plants include handheld saws, axes, and girdling and grubbing equipment. Hand methods are costly, laborious, and useful only for small areas. Practices requiring large equipment include bulldozing, chaining, railing, chopping, mowing, grubbing, root plowing, and disking. The selection of the method is determined by the available equipment, species and density, type of land, and terrain. The low-energy grubber is especially effective in removing small, scattered trees.

Biological control is possible with selective grazing by cattle, sheep, goats, poultry, and certain wildlife species. Insects and plant pathogens have the potential to control certain species.

Fire has been a natural event on many rangelands in the past and can be used alone or with other methods on some grassland situations for weed and brush control and the maintenance of desirable species, provided there is adequate fuel for burning.

Deciding when to control weeds or brush sometimes is difficult because many factors must be considered in addition to cost. Recognizing a potentially important weed and controlling it before it becomes a serious pest is the best approach.

Integrated weed- and brush-management systems will become more important in the future. Safer and more effective new herbicides, herbicide formulations, and herbicide mixtures will be developed. Improvements in mechanical equipment will continue. Biological control methods, including the use of grazing animals, insects, pathogens, allelopathy, and other agents, will gain greater attention. Fire will continue to be an important management tool in grazing lands and forests, especially when combined with other practices. Future weed- and brush-control methods will be more effective, less damaging to the environment, and more cost-effective.

REFERENCES

Anderson WP. Weed Science Principles. 3d ed. New York: West, 1996, pp. 18–21.

Barney MA, Frischknecht NC. Vegetation changes following fire in the pinyon-juniper type of west central Utah. J Range Mgt 27:91–96, 1974.

Bentley JR. Conversion of chaparral areas to grassland: Techniques used in California. Forest Service Agricultural handbook no. 328. USDA, 1967, pp. 1–35.

Bovey RW, Wiese AF, Evans RA, Morton HL, Alley HP. Control of weeds and woody plants on rangelands. AB-BU-2344. USDA and University of Minnesota, Minneapolis, 1984.

Box TW, White RS. Fall and winter burning of south Texas brush ranges. J Range Mgt 22:373–376, 1969.

Bragg TB, Hulbert LC. Woody plant invasion of unburned Kansas bluestem prairie. J Range Mgt 29:19–24, 1976.

Britton CM, Wright HA. Brush management with fire. In: McDaniel KC, ed. Proceedings of the Brush Management Symposium. Albuquerque, NM: Society of Range Management, 1983, pp. 61–68.

Cox JR. Seasonal burning and mowing impacts on *Sporobolus wrightii* grasslands. J Range Mgt 41:12–15, 1988.

Davis GG, Bartel TE, Cook CW. Control of Gambel oak sprouts by goats. J Range Mgt 28:216–218, 1975.

DeLoach JC. Terrestrial weeds. In: Habeck DH, Bennet FD, Frank JH, eds. Classical Biological Control in the Southern United States. South Coop. Series bull. no. 355. Gainesville, FL: IFAS Editorial, University of Florida, 1990, pp. 157–164.

DeLoach JC. Past success and current prospects in biological control of weeds in the United States and Canada. Nat Areas J 11:129–142, 1991.

Elwell HM, McMurphy WE, Santelmann PW. Burning and 2,4,5-T on post and blackjack oak rangeland in Oklahoma. bull. B-675. Stillwater, OK: Oklahoma Agric. Exp. Stn., 1970.

Elwell HM, Santelmann PW, Stritzke JF, Greer H. Brush control research in Oklahoma, bull. B-712. Stillwater, OK: Oklahoma Agric. Exp. Stn., 1974.

Engle DM, Stritzke JF. Igniting crowns of partially scorched juniper: 1983–1991. Cir. E-905. Range Research Highlights. Stillwater, OK: Oklahoma, Coop. Ext. Serv., 1991, pp. 15–16.

Fay PK. Controlling leafy spurge with grazing animals. In: James LF, Evans JO, Ralphs MH, Child RD, eds. Noxious Range Weeds. Boulder, CO: Westview Press, 1991, pp. 193–199.

Fisher CE, Wiedemann HT, Walter JP, Meadors CH, Brock JH, Cross BT. Brush Control Research on Rangeland. MP-1043. College Station, TX: Tex. Agric. Exp. Stn., 1972.

Goeden RD, Fleschner CA, Richer DW. Insects controlling pricklypear cactus. Calif Agr 22:8–9, 1968.

Harley KLS. Biological control of Lantana in Australia, Proceedings of the 3rd International Symposium on the Biological Control of Weeds, Montpelier, Canada, 1973, pp. 23–29.

Harradine AR, Jones AZ. Control of gorse regrowth by Angora goats in the Tasmanian Midlands. Austr J Exp Agric 25:550–556, 1985.

Harris P. Current approaches to biological control of weeds. Tech. Commun. Commonwealth Institute of Biological Control no. 4, 1971, pp. 67–76.

Hendon EP. Kershaw brush cutter. In: Noxious Brush and Weed Control Research Highlights—1971. Spec. Rep. 51. Lubbock, TX: Texas Tech University, 1971, p. 28.

Holloway JK. Projects in biological control of weeds. In: DeBach P, ed. Biological Control of Insects, Pests and Weeds. New York: Reinhold, 1964, pp. 650–670.

Huffaker CB. Fundamentals of biological weed control. In: DeBack P, ed. Biological Control of Insects, Pests and Weeds. New York: Reinhold, 1964, pp. 631–649.

Johnston A, Peake RW. Effect of selective grazing by sheep on the control of leafy spurge (*Euphorbia esula* L.). J Range Mgt 13:192–195, 1960.

Julien MH. Biological Control of Weeds: A World Catalogue of Agents and Their Target Weeds. 2nd ed. Farnham Royal, Slough, UK: Commonwealth Agric. Bureau, 1987, p. 144.

Klingman DL, Bovey RW, Knake EL, Lange AH, Meade JA, Skrock WA, Stewart RE, Wyse DL. Systematic Herbicides for Weed Control—Phenoxy Herbicides, Dicamba, Picloram, Amitrole and Glyphosate. AD-BU-2881. USDA and University of Minnesota, Minneapolis, 1983.

Kozlowski TT, Ahlgren CE, eds. Fire and Ecosystems. New York: Academic, 1974.

Landers RQ Jr. Planning a Prescribed Burn. L-2461. College Station, TX: Tex. Agric. Ext. Serv., 1991.

Landgraf BK, Fay PK, Havstad KM. Utilization of leafy spurge (*Euphorbia esula*) by sheep. Weed Sci 32:348–352, 1984.

Mayeux HS Jr, Hamilton WT. Response of common goldenweed (*Iscoma coronopifolia*) and buffelgrass (*Cenchrus ciliaris*) to fire and soil-applied herbicides. Weed Sci 31:355–360, 1983.

McKenzie D, Jensen FR, Johnsen TN Jr, Young JA. Chains for mechanical brush control. Rangelands 6:122–127, 1984.

Meyer RE, Bovey RW. Control of live oak (*Quercus virginiana*) and understory vegetation with soil-applied herbicides. Weed Sci 28:51–58, 1980.

Pechanec JF, Plummer AP, Robertson JH, Hull AC. Sagebrush Control on Rangelands. Washington, D.C.: USDA handbook no. 277. 1965.

Putnam AR. Allelochemicals from plants as herbicides. Weed Tech 2:510–518, 1988.

Quimby PC Jr, Bruckart WL, DeLoach CJ, Knutson L, Ralphs MH. Biological control of rangeland weeds. In: James LF, Evans JO, Ralphs MH, Child RD, eds. Noxious Range Weeds. Boulder CO: Westview Press, 1991, pp. 84–102.

Radcliffe JE. Gorse—A resource for goats? New Zeal J Exp Agr 14:399–410, 1986.

Rechenthin CA, Bell HM, Pederson RJ. Grassland restoration. Part II. Brush Control. Temple, TX: USDA, Soil Conservation Service, 1964.

Rees NE. Biological control of thistles. In: James LF, Evans JO, Ralphs MH, Child RD, eds. Noxious Range Weeds. Boulder, CO: Westview Press, 1991, pp. 264–273.

Rees NE, Spencer NR. Biological control of leafy spurge. In: James LF, Evans JO, Ralphs MH, Child RD, eds. Noxious Range Weeds. Boulder, CO: Westview Press, 1991, pp. 182–192.

Rice EL. Allelopathy. 2nd ed. New York: Academic, 1984.

Rosenthal SS, Campobasso G, Fornasari L, Sobhian R, Turner CE. Biological control of *Centaurea* spp. In: James LF, Evans JO, Ralphs MH, Child RD, eds. Noxious Range Weeds. Boulder, CO: Westview Press, 1991, pp. 292–302.

Schroeder D. Biological control of weeds. In: Fletcher WW, ed. Recent Advances in Weed Research. Slough, England: Commonwealth Agric. Bureaus, 1983, pp. 41–78.

Scifres CJ. Systems for improving Macartney rose infested Coastal Prairie Rangeland. MP-1225. College Station, TX: Tex. Agric. Exp. Stn., 1975.

Scifres CJ. Mechanical methods of brush management. In: Brush Management: Principles and Practices for Texas and the Southeast. College Station, TX: Texas A&M University Press, 1980, pp. 263–275.

Scifres CJ, Mutz JL, Durham GP. Range improvements following chaining of south Texas mixed brush. J Range Mgt 29:418–421, 1976.

Scifres CJ, Mutz JL, Rasmussen BA, Smith RP. Integrated brush management systems (IBMS): Concepts and potential technologies for running mesquite and whitebrush. B-145. Tex. Agric. Exp. Stn., 1983.

Semple AT. Control of undesirable vegetation, other pests and diseases: Eradication and control of weeds and brush. In: Grassland Improvement. Cleveland: Chemical Rubber Co. Press, 1970, pp. 204–244.

Smith MA, Dodd JL, Rodgers JD. Prescribed Burning on Wyoming Rangeland. B-810. Laramie, WY: Agric. Ext. Serv., University of Wyoming, 1985, pp. 2–10.

Tainter FH. Oak-Wilt: A Selective Silvicide in Arkansas? Fayetteville, AR: Arkansas Farm Res., University of Arkansas, 1972, pp. 8–9.

Tanner GW, Kalmbacher RS, Prevatt JW. Saw-Palmetto Control in Florida. Cir. 668. Gainesville, FL: Agric. Ext. Serv., University of Florida, 1986, pp. 4–5.

Templeton GE. State of weed control with plant pathogens. In: Charudattan R, Walker HL, eds. Biological Control of Weeds with Plant Pathogens. New York: Wiley, 1982, p. 38.

Ueckert DN, Petersen JL, Potter RL, Whipple JD, Wagner MW. Managing pricklypear with herbicides and fire. In: Sheep & Goat, Wool & Mohair, 1988. Prog. rept. 4570. College Station, TX: Texas Agric. Exp. Stn., 1988, pp. 10–15.

Vallentine JF. Range improvement by burning. In: Range Development and Improvements. 3rd ed. Provo, UT: Brigham Young University Press, 1989, pp. 168–214.

Van Arsdale EP, Halliwell RW. Progress in research in live oak decline. Plant Dis Rep 54:669–672, 1970.

Wambolt CL, Payne GF. An 18-year comparison of control methods for Wyoming big sagebrush in Southwestern Montana. J Range Mgt 39:314–319, 1986.

Warnes DD, Newell LC. Establishment and yield responses of warm season grass strains to fertilization. J Range Mgt 22:235–240, 1969.

Warren LE, Ueckert DN, Shelton M, Chamrad AD. Spanish goat diet on mixed-brush rangeland in the South Texas Plains. J Range Mgt 37:340–342, 1984.

Whisenant SG. Changing fire frequencies on Idaho's Snake River plains: Ecological and management implications, Proceedings of Symposium on Cheatgrass Invasion, Shrub Die-Off, and Other Aspects of Shrub Biology and Management, Las Vegas, April 5–7, 1989. USDA For. Serv., Gen tech. rept. INT-276. Ogden, UT: Intermountain Res. Stn., 1989, pp. 4–10.

White LD, Hanselka CW. Prescribed range burning in Texas. B-1310. College Station, TX: Texas Agric. Ext. Serv., 1991.

Wiedemann HT, Cross BT. Low-energy grubbing for control of junipers. J Range Mgt 34:235–237, 1981.

Wiedemann HT, Cross BT. Performance of front-mounted grubbers on rubber-tired equipment. In: Brush Management and Range Improvement Research, 1980–1981. CRP 3968-4014. PR-3982. College Station, TX: Tex. Agric. Exp. Stn., 1982, pp. 49–53.

Wiedemann HT, Cross BT. Influence of pulling configuration on draft of disk-chains. Trans ASAE 28:79–82, 1984.

Wikeem BM, Strang RM. Prescribed burning on B.C. rangelands: The state of the art. J Range Mgt 36:3–8, 1983.

Wilson CL. Use of plant pathogens in weed control, Ann Rev Phytopath 7:411–434, 1969.

Wink RL, Wright HA. Effects of fire on an Ashe juniper community. J Range Mgt 26: 326–329, 1973.

Wood GM. Animals for biological brush control. Agron J 79:319–321, 1987.

Wright HA, Bailey AW. Fire Ecology: United States and Southern Canada. New York: Wiley, 1982.

Wright HA, Bunting SC, Neunschwander LF. Effect of fire on honey mesquite. J Range Mgt 29:467–471, 1976.

15

Growing Woody Plants for Experimental Purposes

I. INTRODUCTION

Growing woody weed species under laboratory, growth chamber, greenhouse, and field nursery conditions is very useful in learning more about plant growth requirements, along with plant behavior, physiology, phenology, and response to herbicides and mechanical or physical treatment. Each species requires specific environmental conditions for survival and reproduction. The environment also influences the success or failure of herbicidal activity or other brush-control measures. In this chapter we will examine the requirements for growing woody plants on the rangelands and pastures of the Southwest, both indoors and in the field. The same principles can be applied at any location.

II. PLANT PROPAGATION FROM SEED

A. Honey Mesquite

Honey mesquite (*Prosopis glandulosa* Torr.) is propagated from seed, either by direct seeding or by transplanting greenhouse-grown plants in the field nursery. Seedlings found from natural wild stands can also be extracted from the soil and transported to the greenhouse or field nursery; however, seedling plants from natural stands are usually difficult to find. In Texas, seed pods are collected from the tree or ground in September and are sometimes fumigated to kill seed weevils. The fruit or pod is an indehiscent pod containing several seeds. Seed ripens to a shiny brown luster with an average oval shape of 5 mm wide by 7 mm long and 2 mm thick in the center. The pod, a linear, flat legume, is about 10 to 20 cm long and 1 cm wide at the widest point. The outer exocarp is hard, underlined by spongy mesocarp. Each seed is surrounded by a bony endocarp, which makes seed removal very difficult (Meyer et al., 1971). Seed removal is done with a scalpel or threshed and separated from the endocarp by a modified pearling machine and procedure described by Flynt and Morton (1969). Once the seed and pericarp tissue are threshed, wind cleaning is necessary to separate the seed from the pericarp. A small gravity separator removes the endocarp tissue, which is the same size as the seed. A 5.1-by-6.5-cm no. 30 grit sandpaper attached to the nonperforated portion of a screen that forms the periphery of the threshing chamber mechanically scarifies the seed. Mesquite has a hard seed coat, and it is necessary to scarify the seed to obtain the maximum germination. Seed germination can also be significantly increased by soaking the seeds in concentrated H_2SO_4 or placing them in boiling water and allowing them to cool as they soak for 24 hr (Flynt and Morton, 1969).

Once the seed is separated, cleaned, and scarified, it can be stored at 5°C for more than 3 years, while maintaining a germination percentage of more than 80%.

Honey mesquite can be seeded directly into soils in the greenhouse or field. In the field, seeds are placed about 2.5 cm deep and at 7.6-cm intervals in 3-meter-wide rows. Planting is done in April or May, when the soil moisture is adequate and the soil temperature is above 15°C. The seeds germinate rapidly, and seedlings become established with a long tap root. Seedlings may be thinned to 0.5-meter intervals in about 2 weeks and cultivated. Honey mesquite may grow to 1 meter tall the first season after planting. The root system of a 14-month-old plant was found to penetrate the soil to a depth of 1.5 meters, with lateral roots of 1.2 meters. Jackrabbits may feed heavily on new seedlings and destroy an entire planting, if allowed to do so.

Honey mesquite can be transplanted from the greenhouse. Two or more seeds are planted in 5-by-5-cm peat moss pots filled with a sandy loam soil or in Jiffy peat moss pellets. Plants are subsequently thinned to one per pot. After at least 6 weeks in a warm greenhouse (25 to 35°C), the plants are 10 to 20 cm tall and are ready to transplant. Also, older plants remaining from other studies can be removed from pots and transplanted from the greenhouse into the field nursery. April and May are the best months for transplanting, but with adequate soil moisture or irrigation, they can be transplanted at any time during the summer.

Honey mesquite grown in the greenhouse for herbicide efficacy or physiological studies can be planted into almost any type of soil from moderate to high fertility. We commonly used a rooting medium of pine bark, clay, and sand in a 1:1:1 ratio. Peat moss can be substituted for the pine bark while maintaining the other components. After 4 to 6 weeks of growth in the greenhouse, seedling honey mesquite responds very similarly to herbicides, whether field-grown or in natural stands. The more effective herbicides can be determined from small quantities of herbicides applied by hand or with sprays from a spray chamber as described by Bouse et al. (1970). Plants 4 to 6 weeks of age acquire the ability to resprout and regenerate from buds from stem tissue similar to field-grown mesquite. Pot size will vary according to the size of the plants desired and the length of time the plants are used. We commonly use plastic pots (12.7 cm in diameter by 12.7 deep), in which plants may attain a height of 0.5 to 0.8 meters over a period of 1 to 2 years. If the plants are kept for long periods of time, soil fertility is maintained with a standard nutrient preparation, and insects are controlled.

In the greenhouse, the growth of mesquite and usually other brush is slowed by the lower light intensity and air temperatures and shorter day length during the winter, as compared to the summer. The response to herbicides and other treatments may therefore be poor during these periods, and treatment should be postponed until a more favorable environment is present. Using supplemental lights and raising the air temperature of the greenhouse or using a growth chamber will improve the growing conditions.

B. Huisache

Huisache [*Acacia farnesiana* (L.) Willd.] is a legume similar in appearance to honey mesquite. The seeds are removed from the pod with the modified pearling machine, as was described for honey mesquite. Huisache seeds are ovoid, about 3 mm wide and 5 mm long. The seed is olive colored with a few dark brown or black seeds. The seed coat is extremely smooth and thick, and has an oily texture. Seed in central Texas is mature in late July and collected from the tree or beneath the tree. Germination is extremely poor without scarification. Seed can be mechanically scarified with sandpaper, but equal results can be accomplished with much less effort by soaking a large number of seeds in concentrated H_2SO_4 for 0.5 to 1 hr. The seed is then rinsed in tap water and allowed to dry. The seeds germinate 70% or more for 2 years or more when stored at 5°C. Like honey mesquite, huisache can be seeded directly in the field or transplanted from the greenhouse in April or May. Huisache has a more rapid growth rate than mesquite in the field, and may grow 2 meters tall during the first growing season and 4 to 5 meters tall after 2 years. Huisache is less subject to damage by pests (weeds, insects, rabbits, and fungi) in the greenhouse or field than honey mesquite. In the greenhouse, huisache is grown in a manner similar to honey mesquite. If huisache or honey mesquite becomes too large in the greenhouse or field to be useful, it can be pruned to the desired size and allowed to regrow. In the field nursery, huisache and mesquite can usually be treated with herbicides or used for physiology studies the second season after planting seeds or transplanting seedlings.

C. Live Oak

Live oak (*Quercus Virginiana* Mill.) is established from acorns planted directly in the greenhouse or in the field. Field plants may be established from transplanted seedlings grown in the greenhouse from acorns and from bare-rooted seedlings purchased from a commercial nursery.

The acorns are ellipsoid, obovoid, brownish, black, and shiny, and are about 1 to 1.5 cm long. The acorns are harvested in late fall and early winter by gathering them from beneath the tree. They are treated with a combination of fungicide and insecticide to minimize disease and wireworms. They are planted immediately before March or stored at 5°C. The acorns can be stored for 1 year, but lose most of their germination capability by the second year.

In the greenhouse, germination is slow. For the best results, the seedlings must be partially shaded until they are about 15 cm tall, otherwise the stems die back repeatedly. The plants can be grown under a greenhouse bench or covered with two layers of cheesecloth if exposed to direct sunlight.

Plants grow 10 to 15 cm tall in about 10 weeks in a warm greenhouse. If they are to be later transplanted in the field, acorns planted in sandy soil in the greenhouse are easier to extract. Plants from the greenhouse or commercial nurseries can be planted in February or March. Direct seeding in the field in rows, however, is less expensive, and a good stand can be obtained by seeding one to six acorns per meter of row. As in the greenhouse, live oak grows slowly in the field. Plants propagated from acorns or transplants grow 0.5 to 1 meter tall in the first season and 1 to 1.5 meters in the second. They are usually large enough for treatment if properly cared for the second year after planting.

D. Whitebrush

Whitebrush [*Aloysia gratissima* (Gillies & Hock) Troncoso] seeds/(achenes) are harvested by hand stripping them from the stem tips of plants in the greenhouse or field when mature. In the field, two crops of seeds are produced annually (May and October). The seeds can be stored for at least 4 years at 5°C. Seed can be germinated by placing it on sandy loam soil in a greenhouse flat and covering the flat with 0.5 cm of sand. After watering, the flats are placed in a cooler at 2°C for 3 days. The flats are then returned to the greenhouse, where a high percentage of seeds germinate. Few seeds germinate without the cold treatment. Seedlings grow best at 25 to 35°C. After about 2 months, the 2.5-to-5-cm-tall seedlings are transplanted into pots. After an additional 5 weeks' growth they are 15 to 30 cm tall and can be used for herbicide or physiological studies or transplanted into the field (Meyer, 1999).

Whitebrush can be transplanted anytime from April until September, when the soil is warm and moist. The plants are spaced at 0.6-meter intervals in rows spaced 3 meters apart. Irrigation may be necessary if transplanted in dry periods during the summer. Whitebrush will grow on all types of soil and will attain a height of 1 to 1.5 meters tall in one growing season. Whitebrush can be treated with herbicide or other control treatments during the second growing season.

E. Winged Elm

Winged elm (*Ulmus alata* Michx.) is propagated by seed (a samara), which is collected in February or early March before the leaves emerge. A tarpaulin is spread beneath the tree. The seeds are shaken or stripped from the branches. Seeds should be germinated soon after collection, but can be stored at 5°C up to 6 months. After 6 months in storage, germination decreases progressively, and after 1 year the seeds are not useable. Seed can be germinated in sandy loam soil in flats. After 6 weeks, the seedlings are about 5 cm tall and can be transplanted into individual pots. In 2 months the plants should be 15 to 30 cm tall and can be transplanted into the field in May. Winged elm can usually be treated with herbicide the year following transplanting in the field. Winged elm grows best in clay loam soils.

F. Texas Persimmon

Texas persimmon (*Diospyros texana* Scheele) produces a black fruit (depressed globose berry) 2 cm in diameter that contains three to eight triangular seeds (Meyer, 1974). The seeds, about 8 to 10 mm long and 1 to 2 mm thick, are dark red and glossy. The seed contains an embryo and food reserves stored as a hard, lustrous endosperm. Seeds germinate much more readily when washed free from the fruit pulp than when left in the fruit. The percentage of germination was reduced when seed was recovered from the digestive tracts of cattle, goats, sheep, or deer. Mechanical scarification, H_2O_4 treatment, or hot water did not improve germination. Seedling growth was slow in the greenhouse and subject to fungi attack. Seedlings grow the most rapidly at 27°C, but almost no growth occurred at 18°C. Plants of 45 to 60 cm tall have been obtained the first year of growth and in sufficient quantity for herbicide treatments, even though they grow slowly and erratically. No field plantings have been successful using seed. Transplanting has not been attempted into the field nursery.

G. Macartney Rose

Macartney rose (*Rose bracteata* Wendl.) has a low germination percentage. Seed (achenes), about 2 by 3 mm in size, are enclosed in a pulpy, hypanthium in hips 1.5 to 2.5 cm long. Seed germination is significantly improved by feeding the hips to cattle and recovering the seed from the manure (McCully, 1950). The improvement of germination by electrical treatments was unsuccessful (Nelson et al., 1978).

Dormancy is attributed to a hard seed coat. We have successfully propagated Macartney rose in the field nursery and greenhouse by transplanting plants from the field.

H. Yaupon

Yaupon (*Ilex vomitoria* Ait.) fruit consists of a subglobose shiny red drupe with a persistent sigma. Usually four seeds (nutlets)—about 1.5 mm long, obtuse, and ribbed—occur in the fruit. Yaupon seed is extremely slow in germination due to the hard endocarp surrounding the seed coat and an immature embryo (Fleming, 1970), and usually requires 1.5 to 3 years for completion under natural conditions. Attempts to speed up germination in the laboratory have largely been unsuccessful. Seedlings can be purchased from commercial nurseries or propagated from vegetative cuttings. Under our conditions (heavy clay soils—central Texas), however, field nursery transplants of yaupon seedlings have not been very successful.

I. Lotebush [*Condalia obtusifolia* (Hook.) Weberb.] and Agarito (*Berberis trifoliolata* Moric)

The seeds of these species occur in small berries collected from natural stands in south Texas. Germination and growth under greenhouse conditions has been erratic and very slow. Propagation and growth under field nursery conditions has not been attempted (Meyer, 1999).

III. PROPAGATION OF TRANSPLANTS

As indicated earlier, honey mesquite, huisache, live oak, whitebrush, winged elm, yaupon, loblolly pine (*Pinus taeda* L.), and other woody plant seedlings can be transplanted from the greenhouse in peat moss pots or from bare-rooted plants to the field. Other brush species, however difficult to grow from seed, may be propagated from root and crown tissue extracted from the field and transported to the woody plant nursery.

A. Macartney Rose

Macartney rose has been successfully vegetatively propagated from root segments. The plants are mowed, then plowed to lift the roots out of the soil. The roots are divided into sections about 8 cm long and planted every 0.5 meters in rows 3 meters apart.

The best results are obtained by transplanting in December to February, when the air temperature is cool. About 80 to 90% of the cuttings survive and produce vigorous plants. Macartney rose grows rapidly and can be sprayed in the fall of the first growing season. The plants tend to spread, and mowing or clean cultivation is required to restrict them to the rows.

B. Greenbriar

Greenbriar is propagated by transplanting sections of the rhizomes. The best results are obtained in January to March. The area is mowed and the rhizomes are brought to the soil surface by plowing. The rhizomes are chopped into 7-to-10-cm segments for planting. The segments can be stored in wet peat moss at 5°C for at least 2 months, but transplanting as soon as possible is recommended. The rhizomes are planted 2.5 cm deep at 0.3-meter intervals in rows 9 meters apart. Shoots are produced sooner if planted with a green stem segment protruding from the soil surface than if buried completely.

Greenbriar grows slowly and erratically. A month is required to produce shoots. If growth ensues, the plants may be treated with herbicides the year following planting.

C. Whitebrush

Whitebrush has been propagated in the field nursery by planting crowns with both stem and root tissue. Crowns are dug with a bulldozer and trimmed with an axe. The best survival occurred when the plants were dug and transplanted immediately in the fall.

Other plants, such as live oak or yaupon, can also be propagated vegetatively from root cuttings, which may be a useful method in some situations to grow plants of uniform age, size, and genetic homogeneity.

The species indicated in this chapter are major weed problems on Southwest rangelands and pastures. As outlined, the principles of growing woody plants in the laboratory, greenhouse, and field for experimental purposes can be applied to any area or location. The information given serves to illustrate the diversity and specific environmental requirements for each species to initiate and complete its life cycle.

IV. THE FIELD NURSERY

A field nursery is very useful in the preliminary evaluation of new herbicides, herbicide application equipment, and other methods of brush control. It is very helpful in obtaining data on the propagation of herbaceous and woody plants, growth habits, phenology, and physiology. These plants may be desirable or undesirable components of the grasslands of the Southwest. It is also useful in providing a location in which to study the fate (residues) of herbicides in soil and vegetation.

In developing herbicides for use in natural and wild stands of brush, the nursery has several advantages over field sites in preliminary investigations. Fewer plants and space are needed per treatment since the plants may be of similar age, size, genetic background, and physiological state. A similar environment also eliminates much variability in the response of brush and weeds to a given herbicide. The nursery may also have the added benefit of irrigation to sustain the plants through periods of drought. Experiments in the field are sometimes lost due to drought. The investigator usually has better control of the land in a nursery compared to field locations, subject to changes in ownership or in the cooperation of the landowner.

The nursery has some distinct disadvantages compared to the field since it is expensive to properly maintain, requiring much labor, specialized equipment, and time. Seedling plants are constantly subject to damage by livestock, rodents, insects, diseases, and weed competition. Such areas may require fencing and pesticide treatment. Some plants are difficult and expensive to propagate and maintain. Such plantings may require irrigation and frequent cultivation for weed control. Brush in natural stands requires little (or no) upkeep.

The woody plant nursery is usually restricted to one location, and the data obtained may not apply to the same species growing in a different environment.

V. EXPERIENCES WITH A FIELD NURSERY

Our nursery, located near Bryan, Texas, at the Texas A&M University Riverside Campus, consists of 28 ha between the runways of a former air base. The area was divided into 10 small fields varying in size from 2 to 6 ha. The soil is primarily a heavy clay loam (Udic pellustert). The area varies in slope from 0 to 3%. Heavy rainfall causes some flooding in low areas, but most of the land can be tilled within a few days after rainfall.

Conventional farm equipment was used in land preparation and maintenance. The land is normally plowed in the fall of the year preceding planting. The land is then bedded (listered) in rows 1 meter apart to provide an adequate seedbed. The following spring the land is disked and rebedded. About half the bed is leveled, and seeds or plants are planted on the level surface of the bed in a row. The distance between the planted rows is usually 3 meters.

Land with plants less than 1 year old is cultivated between rows with a two-row cultivator. Special shields are used between the innermost cultivating implements and the row to prevent burying the small plants. Cultivation is practiced when needed. Herbicides can be sprayed in a 25-cm band on some woody plants as soon as they emerge to control herbaceous weeds. On woody plants 1 year old or older weed control is accomplished by disking or mowing between the rows since the woody plants are too tall to cultivate over.

Irrigation requires significant labor, and has been accomplished by flood irrigation and the use of sprinkler irrigation on land too sloping for flood irrigation. Commercial fertilizer is sometimes used to enhance growth and seeding survival of the woody species.

A. Planting Methods

As indicated earlier, the seeds of honey mesquite and huisache can be seeded with a modified Planet Junior. The planter consisted of a 7.5-liter hopper attached to the frame of a tractor. A chain drive attached from the axle of the tractor drives an agitator in the bottom of the hopper to ensure a uniform flow of the seed through the plate opening in the bottom of the hopper. Planting plates (providing different-size orifices) to regulate the dispersal of the seed from the hopper were modified for honey mesquite and huisache seed. One row is planted at a time. Seeds are placed 10 to 15 cm apart and 2.5 cm deep.

Greenhouse transplants (honey mesquite, huisache, whitebrush, winged elm, live oak, loblolly pine, yaupon, etc.), woody plant root, stem sections, and acorns are planted with a tractor-mounted device. The device opens a furrow into which seedling plants or acorns are placed by hand and closes the furrow in one operation or pass of the tractor. Seedlings are planted about 0.6 meters apart in long rows spaced 3 meters apart. Bare-rooted seedlings and those in peat moss pots are planted with the same equipment.

Large root segments and crowns of woody plants such as whitebrush have been planted by opening a furrow in the row with a plow and planting the crowns by hand. The furrows can be closed with a disk.

B. Weed Control

Herbaceous weeds vigorously compete with emerging and young woody seedlings. Adequate weed control is the major problem in the maintenance of the nursery. Honey mesquite and huisache planted from seed are usually large enough to be cultivated within 2 weeks after seeding. Shields placed inside the cultivator shoes prevent burying the brush seedlings. Subsequent cultivations control the weeds between the rows but not those in the row. The weeds in the rows can be controlled with 2.2 to 3.4 kg/ha of simazine applied broadcast or in a band over the row after the honey mesquite and huisache are 5 to 15 cm tall.

Macartney rose roots, greenbriar rhizomes, and live oak acorns can be sprayed immediately after planting with 2.2 to 3.4 kg/ha of simazine. Weed control is maintained for 3 to 6 months, and the woody plants are cultivated about 6 to 10 months until the plants are about 1 meter tall. Transplants of seedling yaupon, winged elm, loblolly pine, or ash have not been sprayed with simazine. Whitebrush was damaged by simazine + paraquat sprayed directly on the soil.

Hand hoeing is an effective weed-control method, but is prohibitive due to the cost on a large scale.

Johnsongrass is one of the most serious weed problems in the nursery. Glyphosate has been used with some success. Repeated cultivation during dry periods has been partially successful in johnsongrass control in some areas.

C. Age and Size of Plants for Treatment

The primary objective in growing plants in the nursery was to obtain plants of proper vegetative size and physiological stage to give responses to herbicides similar to those in natural stands. Plants of 1 to 1.5 meters tall can be sprayed with a tractor-mounted sprayer, described by Flynt et al. (1970). In both the nursery and field, a second tractor-mounted sprayer was developed for larger trees, up to 4 meters in height (Flynt et al., 1970).

Huisache grows large enough to be sprayed or treated in the fall of the same year that it is planted. Honey mesquite, whitebrush, Macartney rose, winged elm, loblolly pine, yaupon, and ash can sometimes be treated the year after planting. Live oak and greenbriar grow more slowly than the other species and are usually treated when they are 2 years old. With the exception of loblolly pine, the woody plants can be trimmed with a rotary mower if they become too tall for treatment. Trees 1.5 to 5 meters in height can be treated with granular or pelleted herbicides or basal or subsurface treatments, as well as special spray equipment.

D. Equipment for Herbicide Application

Hand-carried and tractor-mounted equipment is used in the nursery. The hand-carried sprayer is used on small plots and areas where the tractor cannot be driven. The hand-carried sprayer consists of a 16-liter stainless steel tank that can withstand compressed air pressure up to 690 kPa. The tank is attached to a small hose outlet with a hand "shut-off" valve attached to a single nozzle or a 2- to 3-nozzle boom. The spraying pressure is usually 200 kPa at the nozzle, with a delivery rate of 11 to 224 L/ha.

Most spraying done with the tractor-mounted equipment on 1-to-1.5-meter-tall plants consists of a spraying platform attached to the back of the tractor. A large shield to prevent spray drift from wind, compressed air tanks, and a three-nozzle boom were the main components of the sprayer (Flynt et al., 1970). The equipment was designed in tandem so that two rows could be treated simultaneously or with the same pass of the tractor in order to save time. The tractor-mounted equipment is much faster and easier to operate than hand-carried sprayers. The shields enable treatments in winds up to 16 km/hr without significant drift. Since the rows are spaced only 3 meters apart, any drift to the next row must be prevented. Sprays were usually applied at 224 L/ha.

A tractor-drawn subsurface applicator was developed to apply soil-active herbicides (Bovey et al., 1976). The applicator was constructed with a large coulter (0.8 meter in diameter) to penetrate the soil to a depth of 0 to 20 cm and to cut through woody vegetation and roots. An injector-knife immediately behind the coulter wheel supported a spray nozzle to apply herbicide into the bottom of the slice made by the coulter. The injectors applied the herbicide in continuous narrow bands. The spacing interval of the bands depends upon the type and size of the brush being treated.

A tractor-mounted applicator was also constructed to apply granular herbicides in continuous bands at various spacings. The metering mechanism consisted of a rotating disc suspended directly over an opening in the bottom of a hopper (Flynt et al., 1976). The device was used in large brush on a large crawler tractor, as well as on nursery-grown plants.

E. Herbicide Evaluation

All woody plants were usually planted in rows spaced 3 cm apart in the nursery. The plot size was usually 1.5 meter wide by 6 meter long or longer, with a minimum of 10 plants per plot. Usually three or four replications (plots) are used per treatment in a randomized block design. Plants in a 1-meter border at either end of the plot were not rated.

Visual ratings were made by estimating the percentage of defoliation or desiccation of leaves within the same growing season after treatment. After 1 year or longer, control ratings are based on visual estimates of canopy reduction and/or the percentage of plants killed. Honey mesquite, huisache, live oak, Macartney rose, winged elm, greenbriar, loblolly pine, and yaupon have been rated for the percentage of stems killed. Whitebrush has been rated for both stem kill and the percentage of dead plants. More recently, 10 plants per plot have been rated at random in both nursery and natural stands for the percentage of canopy reduction. Plants showing no living foliage or sprouts were considered dead.

VI. SUMMARY

An irrigated woody plant nursery was established for the purpose of evaluating new herbicides, herbicide mixtures, formulations, carriers, adjuvants, and spray volumes for their defoliating and brush-control properties. Information was also obtained on the propagation, growth habits, phenology, and physiology of several problem woody species in the Southwest.

The cultivated plants were used to evaluate promising herbicide treatments that might be effective in controlling natural stands of brush and to supplement the data obtained from field experiments. The woody plant nursery provides several advantages over field sites in the preliminary evaluation of herbicides. First, fewer plants and less space are required per treatment, since the plants are of similar age, size, genetic background, and physiological state. Also, using a similar environment and soil type provides more uniform responses of the plants to herbicides. Second, one or more species from different locations and climatic areas can be grown and evaluated under the same environment in the nursery. Third, more observations and treatment applications for herbicide, ecological, and growth evaluation studies can be facilitated, since the nursery was close to office and greenhouse facilities in comparison to remote field sites. Fourth, irrigation can be used in the nursery on woody plants as an environmental variable with herbicides or other treatments and is sometimes needed during drought. Finally, the nursery is a source of soil and plant materials for greenhouse and laboratory studies and provides an area for herbicide residue research involving soils, plants, and water.

There are, however, several problems associated with operating a nursery. First, it is expensive to maintain, requiring much labor and specialized machinery. The nursery requires constant care because young plants are subject to livestock, rodent, and insect damage and may require fencing and pesticide treatment. Some woody species are difficult to propagate under cultivated conditions, requiring considerable time and experimentation to establish them. Second, weed control in young stands of brush is a constant problem. Brush in natural stands requires little or no upkeep. Third, a woody plant nursery is usually restricted to one location (environment) and may be a somewhat artificial situation; consequently, woody species may not always respond to a given treatment in nurseries as they do in natural stands. The nursery is not the final step in developing recommendations for brush-management treatments, but provides a useful link between greenhouse evaluation and large-scale field studies.

REFERENCES

Bouse LF, Francis HL, Bovey RW. Design of an herbicide sprayer for the laboratory. Washington, D.C.: USDA-ARS-165, 1970.

Bovey RW, Flynt TO, Meyer RE, Baur JR, Riley TE. Subsurface herbicide application for brush control. J Range Mgt 29:338–341, 1976.

Fleming GM. Germination of yaupon (*Ilex vomitoria* Ait.) seed. MS thesis, Texas A&M University, College Station, TX, 1970.

Flynt TO, Morton HL. A device for threshing mesquite seed. Weed Sci 17:302–303, 1969.

Flynt TO, Bovey RW, Baur JR, Meyer RE. Tractor-mounted sprayers for applying herbicides to brush. Weed Sci 18:497–499, 1970.

Flynt TO, Bovey RW, Meyer RE, Riley TE, Baur JR. Granular herbicide applicator for brush control. J Range Mgt 29:435–437, 1976.

McCully WG. Recovery and viability of Macartney rose (*Rosa brateata* Wendl.) seeds passed through the bovine digestive tract. MS thesis, Texas A&M University, College Station, TX, 1950.

Meyer RE. Morphology and anatomy of Texas persimmon (*Diosphyros tesana* Scheele). B-1147. College Station, TX: Tex. Agric. Exp. Sta. 1974.

Meyer RE. Personal communication. USDA-ARS (retired). College Station, TX, 1999.

Meyer RE, Morton HL, Haas RH, Robinson ED. Morphology and anatomy of honey mesquite. USDA-ARS tech. bull. no. 1423, 1971.

Nelson SO, Bovey RW, Stetson LE. Germination responses of some woody plant seeds to electrical treatment. Weed Sci 26:286–291, 1978.

16

Summary and Recommendations

I. INTRODUCTION

Woody plants and mixtures of woody plants or woody plants with herbaceous weeds present special problems on grazing lands, forested areas, rights-of-way, and other noncrop areas. Woody plants are sometimes very difficult to manage because of their large size, resistance to treatment, perennial nature, vigor, and sometimes rapid growth.

Their rapid encroachment on grazing lands is partially due to selective grazing or improper grazing, reduction of fire, seed transport by grazing animals, climate fluctuations, denudation for various reasons, and subsequent abandonment. Increased atmospheric CO_2 favoring the C3 broadleaf herbaceous and woody species over warm-season perennial grasses (C4 plants) may also contribute to the problem.

In many documented cases historical records and photographs indicated that shrublands, woodlands, and forested areas in North America have expanded and replaced what were once grasslands and savannas at the time of European settlement in the last 50 to 100 years.

Regardless of the cause, when woody plants become too numerous and dense, they may inhibit other desirable vegetation; reduce wildlife and livestock productivity and diversity, recreation, aesthetics, land values, soil water and runoff; harbor insect vectors and animal predators; and significantly increase the cost of maintenance.

It is estimated in the United States that at least one-third of range and pasture land is infested with woody plants, and that a major part of forested land is thus infested. Woody plants can be significant problems on rights-of-way, industrial sites, and other noncrop areas, with high costs for maintenance and control. Woody plants are a worldwide problem.

Woody plant control has changed during the last 50 years to a more holistic approach, from attempted eradication to brush management for multiple benefits, including wildlife habitat management, watershed enhancement, aesthetics, and improved livestock-carrying capacity or production of wood products.

Although only about 50 different woody species may constitute major problems in grazing, forested, and noncrop lands, about 380 problem plants are listed and discussed in Chapter 12.

II. HISTORY AND METHODS OF WOODY PLANT MANAGEMENT

Fire has been a tool for woody plant control and has influenced woody and herbaceous plant cover for many centuries. In North America fires started by natural means or by native people have been frequent (depending upon location) and sometimes of great benefit to the indigenous vegetation. Upon the arrival of Europeans fire suppression was soon practiced on grazing and forested land, but by the turn of the century catastrophic fires caused great concern. But experience and historical records showed that fire has a place in vegetation management. In the 1960s the National Park Service started using fire in its management programs. More recently, prescribed burning has been used to reduce excessive weed and brush growth in California chaparral and many other rangeland sites in the western United States. Southern fire scientists showed the benefits of controlled burning in pine plantations, creating a prescribed burning policy in the early 1940s. Similar practices soon followed in western U.S. forests.

Biological control of woody and herbaceous plants has been practiced since antiquity, since most plants have natural enemies of some kind. This natural feeding or predation of insects, pathogens, or grazing animals may suppress some plants and become potential control agents. Problem plants such as lantana, pricklypear cactus, gorse, Koster's curse, Crofton weed, banana poka, Scotch broom, blackberry, salt cedar, and Russian olive are controlled or partially controlled by specific insects. The search continues for insects and plant pathogens for woody plant as well as herbaceous weed control. On vast areas biological control may be the only feasible means to suppress noxious weeds.

Selective grazing by livestock and wildlife has also been practiced for centuries to control unwanted woody vegetation. Goats are one of the most effective species for woody plant destruction. They are also valuable for their meat, milk, and mohair production. Spanish goats are more aggressive grazers than Angora goats, but both are very useful for this purpose.

Mechanical methods of woody plant control have also been practiced for many centuries. Hand methods include grubbing, cutting, girdling, and burning.

Bulldozing, grubbing, chaining, railing, chopping, mowing and shredding, root plowing, and disking are done with power equipment. Hand methods are slow, laborious, and costly, but usually effective. With power equipment, bulldozing, disking, grubbing, and root plowing are the most effective, but are also the most costly. Chaining, roller chopping, and mowing are less effective but also less expensive.

There are several herbicides effective for woody plant control. They are covered in the next few sections and will not be discussed here. They are sometimes very effective alone or can be a very important component of integrated brush management systems (IBMS). In some situations the use of two or more control methods are more effective and economical to use than single treatments, and can be referred to as IBMS. An example of IBMS is California chaparral control, in which heavy browsing by livestock and big game can be followed by prescribed burning. Another method is brush crushing with a bulldozer before burning, followed by herbicides to kill sprouting brush a year or two after the fire. Many combinations of treatments may be available, as discussed in Chapters 2 and 14.

III. HERBICIDES USED

The herbicides discussed in this volume are amitrole, bromacil, clopyralid, 2,4-D, dicamba, dichlorprop, diquat, fosamine, glyphosate, hexazinone, imazapyr, metsulfuron, paraquat, picloram, tebuthiuron, and triclopyr.

Most of these materials have excellent activity on woody plants for their intended use. The user should understand their properties and effects to use them safely and effectively. Certain herbicide combinations, such as 2,4-D + triclopyr, 2,4-D + picloram, or dicamba or triclopyr + clopyralid, are also useful in reducing costs, enhancing their activity, or broadening the spectrum of weeds controlled.

Many of these herbicides provide good to excellent woody plant control. They were developed for other uses (e.g., perennial weed control), but have proven useful in brush control. The cost of development and reregistration has been high ($50 million plus at current costs), and any loss of use could cause economic hardship to applicators and growers. It is a serious deficit that no new herbicides for brush control have been developed since the late 1970s.

Herbicide 2,4,5-T (inexpensive) was available for weed and brush control until negative publicity from its use in Vietnam resulted in EPA suspension of some uses in 1974 and all uses in 1985. Increasingly complex laws, regulations, and costs to develop new herbicide chemistry for forestry, grazing, and noncropland have forced limited use of many of these compounds. Herbicides, however, will continue to be needed for the benefit of food, fiber, and animal feed production for many more years.

IV. METHODS OF APPLICATION

Woody plants can be controlled by 1) applying herbicides to foliage, 2) wiping herbicides onto stems and foliage, 3) spraying basal bark or stumps, 4) injecting herbicides into sapwood by mechanical devices or frills or notches cut into the tree, and 5) using soil application. A recent popular and low-cost technique called ''Brush Busters'' for individual plant treatment involves the use of special nozzles and techniques to treat honey mesquite, pricklypear cactus, redberry, and blueberry juniper (Chap. 5). For large areas with dense brush, however, aerial application must be considered. Many commercial preparations are available as adjuvants to control spray drift. They are added in small amounts to the spray solution in ecologically sensitive areas. They reduce drift by reducing fine droplets when sprayed. They are commonly polyacrylamides (polymers).

Surfactive agents and other adjuvants can be added in small amounts ($<5\%$) to spray solutions to improve herbicide activity by modifying spray solution characteristics by reducing interfacial surface tension on leaves, enhancing both spray retention and/or leaf and stem penetration.

Whether handheld, ground, or aerial, spray equipment must be in good repair and calibrated to deliver the proper rates per ha to comply with label requirements. Various types of spray equipment are discussed in Chapter 5, as well as techniques to clean application equipment.

V. THE SIGNIFICANCE OF HERBICIDE RESIDUES

Once applied, herbicides undergo several means of biodegradation in the environment. Herbicides vary in their rate of dissipation (Chap. 6). For example, 2,4-D, dicamba, dichlorprop, and fosamine are rapidly degraded in plants, soil, and water sources, whereas materials such as bromacil, hexazinone, imazapyr, picloram, and tebuthiuron are more persistent in the environment. The persistent nature of these materials make them effective for woody plant control, however, since woody plants may resprout from dormant buds on the stem several months after the initial application and recover. Soil-applied materials such as tebuthiuron may cause damage to desirable trees if applied to overlapping root systems. Picloram (a restricted-use herbicide) will cause damage to desirable broadleaf crops if allowed in irrigation or runoff water. Reading the herbicide label should alert the user of any necessary precautions. Injury to desirable forbs on range and forested areas may occur during the first season from foliar- or soil-applied herbicides, but forb diversity and number usually recover by the second growing season.

VI. TOXICOLOGY AND SAFETY

Most herbicides used in woody plant management are of moderate to low toxicity to warm-blooded animals. The exceptions are the bipyridiniums, diquat, and paraquat. These compounds are highly toxic, with an LD_{50} of 230 and 150 mg/kg body weight, respectively, in rats. Numerous studies have shown (Chap. 7) that no adverse effects occur to man, laboratory animals, or wildlife if the label rates of herbicides and the proper protective gear for humans are used.

Actual human exposure tests have been done measuring the amount of herbicide excreted in the urine of pilots, mechanics, batchmen, hand applicators, supervisors, and observers applying phenoxy and other herbicides. In all cases a safety factor of several hundred to several thousand times exists below the no observable effect level (NOEL) for all workers, regardless of job, protective clothing, or method of application.

As indicated earlier the pesticide label is the single most important safety tool for herbicide users. The Federal Environment Pesticide Control Act (FEPCA) requires that herbicide use be consistent with the herbicide label. Information found on the label is given in Chapter 7. If environmental hazards exist, precautionary statements apply, or if information on reentry of a field is needed, it is given on the pesticide label. The complete instructions, the product or trade name, and the chemical name including the active ingredient plus application rates are indicated.

The Federal Insecticide Fungicide and Rodenticide Act (FIFRA) of 1947 regulates the distribution, sale, and use of pesticides in the United States. It has been amended several times to provide the safe use of pesticides and is administered by EPA. Many other acts and laws regulate or affect pesticide and herbicide use and manufacturing, as indicated in Chapter 7. Will future regulation of pesticides encourage the development of new, improved products or will overregulation curtail pesticide use and the search for new chemistry? Future success of food production may depend upon innovative pest manipulation, including chemicals.

An example of pesticide loss by political rule is that of 2,4,5-T, an inexpensive, effective brush control agent banned by EPA in 1985. It was implicated for use with 2,4-D in jungle defoliation in Vietnam in the 1960s as Agent Orange in Operation Ranch Hand. It contained minute amounts of 2,3,7,8-tetrachlorodibenzo-*p*-dioxin (TCDD) and caused extensive political controversy. It was implicated in producing health problems to the people of Vietnam, military personnel, herbicide applicators, and people involved in its manufacture. None of the Ranch Hand veterans was diagnosed with chloracne or health problems related to Agent Orange. Some Vietnam War veterans claim serious health problems that they believed were related to Agent Orange. After the Vietnam War it was discovered that polychlorinated benzodioxins and benzofurans are generated in incinerators, cigarette smoke, car and truck mufflers, and charcoal grilled steaks and are widespread in the environment. This finding should change attitudes toward 2,4,5-T as a problem chemical in the environment, since it has not been proven to be a problem after 40 years of widespread use and scientific investigation.

An expert EPA panel concluded that at most there was a weak link between cancer (non-Hodgkin's lymphoma) and 2,4-D. The available studies indicate that 2,4-D is rapidly excreted from the body and is not metabolized to reactive intermediates. It does not produce genotoxic or neurological effects in animals, and it is not a carcinogen.

VII. HERBICIDE EFFECTS ON PLANTS

A. Physiological Effects

A knowledge of the factors affecting the absorption and translocation of an herbicide in woody plant control is essential for optimum results. Lethal amounts of foliar- or soil-applied herbicide must be absorbed by leaf or root tissue for translocation within the plant to the site of action. Clopyralid, 2,4-D, dicamba, dichlorprop, picloram, and triclopyr are auxinlike, growth regulator herbicides with similar modes of action. Amitrole inhibits the accumulation of chlorophyll and carotenoids in the light, but the specific site of action has not been determined. Bromacil, hexazinone, and tebuthiuron inhibit photosynthesis, whereas imazapyr and metsulfuron inhibit acetolactate synthase (ALS), a key enzyme in the synthesis of isoleucine, leucine, and valine. Glyphosate inhibits 5-enolpyruvylshikimate-3-phosphate (EPSP) synthase, which produces EPSP from shikimate-3-phosphate and phosphoenolpyruvate in the shikimic acid pathway. Diquat and paraquat are contact herbicides that form free radicals in plants by reducing the ion and subsequent autooxidation to yield the original ion, at which time hydrogen peroxide and other phytotoxic by-products are formed. Research suggests that the hydroxyl radical is responsible for paraquat-induced peroxidation and related phytotoxic symptoms. The mode of action of fosamine is not well understood, but in honey mesquite fosamine strongly inhibits mitosis.

B. Woody Plant Effects

Specific herbicides are required to control specific woody plant problems. This is because certain woody species are biochemically susceptible to certain herbicides and one or a few materials are not sufficient for the brush-management arsenal. Chapter 12 lists more than 370 woody species and their responses to specific herbicides, methods of application, and chemicals, along with the proper timing of application for successful treatment. Proper identification of the weed problem and fitting the proper chemical treatment to the problem is paramount for success. Follow-up treatments are usually necessary to maintain original control. Reregistration of presently used products by EPA is essential for woody plant management because no new chemistry is evident at this time. Combinations of certain chemicals are highly desirable for reducing costs, enhancing the effects on problem weeds, or expanding the spectrum of weeds controlled in many situations. Herbicides are also essential components of IBMS. The development of successful brush-management systems requires the cooperation of industry, state and federal scientists, and EPA and other agencies.

C. Effect on Crop Plants

An added benefit of woody plant control is the control of herbaceous weeds, including poisonous plants associated with the woody plants. Because of this, herbaceous crops and desirable vegetation may be extremely sensitive to the herbicides used. Desirable plants adjacent to weed- and brush-treated areas may be damaged by herbicide vapors or spray drift. For this reason, low-volatile esters or amine forms of the growth-regulator-type herbicides are used. Spray drift or accidental application of herbicides can also cause injury to broadleaf agronomic and horticultural crops. Rates as low as 0.01 kg/ha of 2,4-D at the seedling stage of cotton can delay maturity and reduce yield. Most other herbicides are not as injurious as 2,4-D on cotton. Soybeans tolerated 2,4-D amine in early growth stages, but soybeans became more sensitive as they matured. Soybean yields were reduced, however, at all stages of growth by 0.28 kg/ha of 2,4-D. Picloram and dicamba were more injurious than 2,4-D to soybeans at prebloom. At flowering, dicamba at 0.035 kg/ha and picloram at 0.009 kg/ha reduced soybean yield by about 50%. Tomato and root crops were more sensitive to 2,4-D than other crops tested. Other crops and herbicides are discussed in Chapter 10.

VIII. ECOLOGICAL EFFECTS OF WOODY PLANT MANAGEMENT

Numerous studies show that grass and forb production can be significantly increased by adequate woody plant control in many different climatic regions with differing woody plant problems. Forb production may be reduced the first growing season, but usually recovers by the second growing season. Increased forage production sometimes returns to pretreatment levels in a few years if the brush recovers and follow-up treatments are not administered. Herbicide treatment may not be justified in light or scattered brush stands or on poor sites. Spraying strips or patches of brush and leaving alternate untreated brush may significantly increase livestock forage as well as maintain wildlife habitat and food plants while increasing small animal numbers and plant diversity.

In forest land herbicides are used to create and maintain wildlife habitats, create mixed and uneven aged stands of plants, restore damaged landscapes, control exotic, noxious, and poisonous plants, maintain recreational areas, and manage rights-of-way for multiple uses. To avoid economic loss in forests, controlling shrubs and hardwood species is sometimes necessary. The beneficial effects of weed control and conifer release is usually long-term (10 years or more). In rangeland, forest, and noncroplands herbicides and other control methods may temporarily reduce wildlife numbers, not because of toxicity but because of reduced food plants. The effects of herbicide use on microorganisms, insects, nematodes, aquatic organisms, birds, small mammals, and large mammals are discussed in Chapter 11.

The economics of woody plant control with herbicides and nonchemical woody plant control, along with growing woody plants for experimental purposes are discussed in Chapters 13, 14, and 15, respectively.

IX. POLITICAL AND PUBLIC SUPPORT NEEDS

Many native and exotic weeds have become tremendous problems on grazing and forested lands. Both legislators and the public need to be aware of the loss of productivity and the problems that these weed and brush species cause. Some common herbaceous and woody species on rangeland include leafy spurge, perennial snakeweed, yellow starthistle, medusa head, dyers woad, spotted knapweed, dalmation toadflax, yellow toadflax, black henbane, tansy mustard, tansy ragwort, juniper, sagebrush, rabbitbrush, oakbrush, mesquite, and salt cedar, just to name a few that dominate millions of ha in the western United States. In forested areas there is a variety of herbaceous weeds as well as hardwood weedy species that inhibit conifer production and multiple uses of the land. Miller (1995) lists seven exotic trees and shrubs, four vines, and four grasses that may occur in southern forests. If not controlled or contained these plants may become huge problems. Randall (1991) indicated that the Nature Conservancy, a nonprofit conservation organization, operates the largest system of privately owned preserves in the world. The presence and spread of nonnative plants and animals is a problem at many locations and in some cases weed problems are the single greatest threat to the species or communities that the preserves were designed to protect. For these reasons research and education programs need to be strengthened.

X. FUTURE TRENDS AND PRIORITIES

Herbicides have limitations on rangelands because of cost. On selected sites, however, herbicides will continue to be very useful. On certain sites they are the only means possible. Herbicides have unjustly received bad press (Bovey and Young, 1980), limiting their use and increasing the cost of their application. Harris and Reid (1994) indicated that herbicides are useful in California to economically control exotic plants that destroy the native habitats on which many endangered species depend.

Even though herbicide use on rangelands or forests may be very infrequent (every 20 years) compared to cropland, future use will be further reduced but delivery systems will be more efficient, as in the individual plant treatments

(McGinty and Ueckert, 1995) and the herbicide carpeted roller (Mayeux and Crane, 1984). Aircraft can now apply small amounts of pelleted herbicides (Bouse et al., 1980), and new developments in aerial sprays improve spray deposit and reduce drift. Electronic sensors, monitors, and control systems on ground sprayers can spray only the target species with precision.

New generation herbicides such as the sulfonylurea and related compounds control weeds at extremely low rates (a fraction of an ounce per acre). Other new developments will follow in finding new and improved chemistry, formulations, and delivery systems. Herbicide research in agronomic cropping systems has been well supported by private, state, and federal funding. Herbicide research for rangelands has been well supported in the past, but now funding is essentially nonexistent. To be successful we must retain the older herbicides and continually search for new chemistry.

More research is needed on 1) drift control from ground and aircraft sprays, 2) evaluation of weather conditions before and after herbicide treatment, 3) proper herbicide timing, 4) improved placement of herbicides and application equipment, 5) better systems to predict herbicide success on woody plant and weed control, and 6) systems to reduce application cost.

Mechanical methods of weed control are the most common on rangelands, including handgrubbing, mowing, or bulldozing. These practices have been developed by man's necessity to control weeds and brush. Although there is adequate handheld and power equipment for weed and brush control (Vallentine, 1983), there are always ways to improve the current technology. Mechanical methods will continue to be an important means to control brush and weeds on rangelands alone and in combination with other methods, including herbicides.

Prescribed burning for weed and brush control will continue to be important in some regions of the United States, especially in connection with other practices. Historically fire was a major natural event in controlling weeds and brush and renewing forests and grasslands. Prescribed burning has become more attractive because of the high costs of other practices. It can definitely be beneficial in remote areas with enough fuel to carry the fire and to prevent catastrophic fires in range and forestry sites where weed and brush growth may become excessive.

Biological control, especially with insects, is receiving a lot of attention in the USDA and in some state programs. Weed and brush management may also be enhanced by intensive, short-duration grazing systems in which deer, goats, and cattle can use woody and herbaceous plants more efficiently. Enough vegetative cover needs to be left to protect wildlife, but must also be compatible with livestock use. Forthcoming research should develop the best stocking ratios of livestock and wildlife. More creative use should be made of exotic game and goats.

Some success has been gained with insects and plant pathogens for controlling weeds, but most programs have had limited success. Control with insects and plant pathogens is highly expensive (until acceptable control is achieved), extremely slow, uncertain, and risks the escape of organisms that may feed on desirable vegetation. Even though these biocontrol organisms are an environmental risk, they may be the only means to suppress vast acreages of range weeds, and further work should be encouraged.

One area needing renewed and increased effort is the revegetation of pastures and rangeland. Research work in the past has been sporadic and poorly supported. This involves investigating native and introduced plant species for the establishment and maintenance of stands under grazing use. Weed control is usually necessary for seedling establishment and can be accomplished by herbicides, mowing, or cultivation. The use of forage crops with allelopathic properties to suppress weeds should be further studied. Ecological restoration of rangelands is high-priority research and should be given considerable attention and support.

Further research on integrated weed and brush management systems should be given high priority, looking at all possible combinations of mechanical, chemical, fire, biological, and plant competition/allelopathy/revegetation to fit the site, cost, and purposes of the land manager. Combinations of treatments are selected on the basis of availability, weed species, cost, effectiveness, and other factors. Computer decision-making programs such as EXSEL (Hamilton et al., 1993) will be continually improved as data input and technology dictate.

Finally, to know the life cycles, ecology, and physical and physiological characteristics of each weed and brush species we are dealing with is essential to understanding their nature of encroachment and persistence on rangelands, forest lands, and noncrop areas.

REFERENCES

Bouse LF, Carlton JB, Brusse JC. Dry materials metering system for aircraft. Trans ASAE 25:316–320, 1980.

Bovey RW, Young AL. The Science of 2,4,5-T and Associated Phenoxy Herbicides. New York: Wiley, 1980.

Hamilton WT, Welch TG, Myrick BR, Lyons BG, Stuth JW, Conner JR. EXSEL: Expert system for brush and weed control technology selection. In: Heatwole CD, ed. Application of Advanced Information Technologies: Effective Management of Natural Resources. Proceedings of the ASAE Conference Spokane, WA, 1993, pp. 391–398.

Harris V, Reid TS. Use of herbicides for endangered species protection, Proceedings of the California Weed Conference, San Jose, CA, 1994, pp. 24–25.

Mayeux HS Jr, Crane RA. Application of herbicides on rangelands with a carpeted roller: Control of goldenweeds (*Isocoma* spp.) and false broomweed (*Ericamere austrotexana*). Weed Sci 32:845–849, 1984.

McGinty A, Ueckert D. Brush Busters: How to beat mesquite. L-5144. College Station, TX: Texas Agricultural Experiment Station & Extension Service., Texas A& M University, 1995.

Miller JH. Exotic plants in southern forests: Their nature and control. Proc South Weed Sci Soc 48:120–126, 1995.

Randall JM. Exotic weeds in North America and Hawaiian natural areas: The Nature Conservancy. In: McKnight BN, ed. Plan of Attack in Biological Pollution: The Control and Impact of Invasive Exotic Species Symp. Indianapolis, 1991, pp. 159–172.

Vallentine JF. Mechanical brush control methods. In: McDaniel KC, ed. Brush Manage Symposium. Lubbock, TX: Texas Tech University Press, 1983, pp. 53–59.

Index